Application of Infrared Thermography in Sports Science

Jose Ignacio Priego-Quesada
Editor

Application of Infrared Thermography in Sports Science

Second Edition

Springer

Editor
Jose Ignacio Priego-Quesada
Department of Physical Education and Sports
University of Valencia
Valencia, Spain

ISBN 978-3-031-93310-3 ISBN 978-3-031-93311-0 (eBook)
https://doi.org/10.1007/978-3-031-93311-0

This Springer imprint is published by the registered company Springer Nature Switzerland AG
The registered company address is: Gewerbestrasse 11, 6330 Cham, Switzerland

As an editor, this book is dedicated to all the authors of the various chapters, whose contributions have made its completion possible.
On a more personal note, I dedicate this book to Rosa for her unwavering support and love throughout the editing and writing process (and throughout my entire professional career), and to Marina and Alba, who fill my life with excitement, laughter, and joy.

Foreword by McCafferty

Infrared thermography is an increasingly accessible technique, and therefore this second edition of *Application of Infrared Thermography in Sports Science* is extremely timely given major advances in technology and greater use of thermal imaging by researchers and practitioners. Indeed, since the first edition in 2017 there have been more than 1600 publications in sports science involving infrared thermography/thermal imaging (Web of Science). In this edition, the authors have substantially revised the content in all chapters and included four new chapters to reflect the diversity of applications and new developments in technology and analysis techniques.

Although the basics of thermal imaging can be mastered within a relatively short period of time, understanding how skin surface temperature is determined both by the physical principles of radiation and underlying physiology is challenging. This book clearly outlines the fundamentals of infrared thermography so that, firstly, valid surface temperature measurements can be made, and secondly, how they must be correctly interpreted to be able to draw conclusions about what skin surface temperature reveals about the body. Therefore, one of the most important new chapters in this edition is "Foundations of Human Thermoregulation and Skin Temperature" (Chap. 3) that outlines how temperature of the skin surface is determined by changes in blood flow for the maintenance of homeothermy. Sports and exercise medicine have standardised methods and protocols for measurement of human skin temperature, providing detailed methodologies for human assessment in a number of contexts (Chap. 4). With the growing use of infrared thermography in sports medicine, Chap. 5 has been fully updated to highlight recent applications for both injury prevention and injury risk management. Furthermore, there has been considerable interest in the use of infrared thermography to detect muscle damage, delayed onset muscle soreness and fatigue (Chap. 6), and how sport disciplines may alter asymmetry in skin temperature (Chap. 8) that warrant these new chapters. A new chapter (Chap. 10) examines how the use of dynamic thermography (through cooling or warming prior to evaluation) can reveal further information about a human's physiological state. These new chapters provide extremely valuable information for the sports science community supporting athletes. Many of the applications of thermography in human sports are also widely used in veterinary medicine and therefore it is extremely useful to see how thermography is being successfully applied to detect injury, performance and welfare in equestrian sport (Chap. 15). This chapter not only highlights the benefits of infrared thermography in equestrian science but may stimulate new research areas in human sports science and medicine. The final revised chapter "Issues and Future Developments of Infrared Thermography in Sports Science" (Chap. 16) demonstrates how sports science is leading the way in adopting the latest technology including low-cost thermal devices, 3D thermography, hyperspectral imaging and implementation of machine learning but also highlights the importance of applying principles of quality assurance to ensure reliability of measurements.

I would recommend this book not only for those currently using or planning to use infrared thermography in sports science but to other researchers using thermal imaging in different areas of human and animal science. Infrared thermography has always been an interdisciplinary technique and there is much to learn from other fields.

Editor-in-Chief, Journal of Thermal Biology Dominic J. McCafferty
University of Glasgow,
Glasgow, Scotland

Foreword by Carpes

Over recent years, there has been a remarkable increase in the use of infrared thermography in sports sciences. While I consider myself and my colleagues to be active contributors to this growing field, it is important to acknowledge that we joined a movement—one that we were far from being the first to initiate. This growing interest has been driven by pioneering scientists who played a crucial role in introducing this imaging technique to the field of sports sciences. Many of these foundational studies are highlighted throughout the different chapters of this second edition of *Application of Infrared Thermography in Sports Science*, alongside more recent publications that have contributed to advancing not only the methods for applying infrared thermography but also new theories and applications explored in these studies. The second edition retains the original aim of the book: to serve as a key reference for those conducting research on the application of infrared thermography in sports science. However, we look forward to this second edition advancing further and reaching those who are applying thermography in the field. This updated edition includes four new chapters that delve into highly specialized applications of the technique. In addition to these new chapters, several existing chapters have undergone substantial revisions and updates. On average, with the new content and the revisions made, approximately 50% of this edition presents fresh insights. As a result, this edition is structured into 16 chapters, each starting with an abstract and an introduction to the topic being discussed. Each chapter also includes conclusions and perspectives, providing readers with a clear understanding of the next steps for advancing the field. This new edition solidifies the book's leading position in the scientific literature on infrared thermography, making it an essential resource for the libraries of researchers and practitioners alike—whether they are already applying infrared thermography in their work or are looking to integrate it into their routine. Just as the first edition sparked curiosity and inspiration, the second edition will continue to engage and challenge scientists in the field.

Felipe P. Carpes

Applied Neuromechanics Research Group,
Laboratory of Neuromechanics
Federal University of Pampa
Uruguaiana, RS, Brazil
President of the International Society of Biomechanics
Uruguaiana, RS, Brazil

Preface

This second edition of the book *Application of Infrared Thermography in Sports Science* arises from the need to address the significant progress made in the application of infrared thermography in sports sciences since the first edition was published in 2017. It seems like such a short time ago, but looking back, I realize how much more we know now compared to a few years ago. We have gained deeper insights into metric reliability, key methodological aspects, and asymmetries before and after exercise in non-pathological conditions, among other topics. However, while we have made great strides in some applications, many uncertainties still remain. This rapid scientific evolution has made it necessary to update this book, both by revising content from the first edition and by adding new chapters.

One of the objectives of this book is to serve as a guide for newcomers using infrared cameras in sports science research centers, laboratories, or sports facilities. Like many others, when I first started using infrared thermography in a scientific context, my initial step was to review the scientific literature to ensure methodological rigor. At that time, many of the methodological aspects that are now part of our standard practices were not well documented. While I believe I managed to apply the methodology reasonably well, my hypotheses were completely incorrect due to a lack of understanding of how skin temperature responds to physical exercise. My main concern was ensuring methodological accuracy, but I believe that this is not always the case—many people start using infrared cameras without proper knowledge simply because they are easy to operate. In many instances, infrared thermography is introduced into laboratories, clinics, or sports clubs as a new, inexpensive, and intriguing tool, leading professionals to use it without adequate knowledge of its methodology, the physics of heat transfer and thermoregulation, or existing research in sports science. Consequently, many "first-time" studies contain clear errors in design, methodology, and interpretation. The perceived ease of using an infrared camera is likely one of the primary reasons for this issue.

I have been conducting research using infrared thermography since 2011, first as a pre-doctoral researcher and later as a full-time professor at the University of Valencia. During this time, my colleagues and I have developed various lines of research, including the application of infrared thermography in cycling, the combination of infrared thermography with biomechanics, its potential use for detecting muscle damage, and methodological investigations. Many of these studies are presented in the different chapters of this book. With each study, I have learned more about human thermoregulation during exercise and the possible applications of infrared thermography in sports science. Each study has helped me refine my approach and minimize errors. I hope this book will accelerate the learning curve for future researchers using infrared thermography in sports science. Therefore, the aims of this book are:

- To demonstrate the applicability of infrared thermography in sports science
- To update the reader on current knowledge in various fields of application
- To provide foundational knowledge for the proper use and interpretation of thermographic results

I hope this book serves as a primary resource for students new to infrared thermography in sports science, as well as an update for those who have been involved in the field for a long time. While the book is primarily targeted at sports researchers, it may also be valuable for sports technicians working in medical and sports centers. Ultimately, the goal is to equip readers with fundamental knowledge and the latest updates on the application of infrared thermography in sports science.

In recent years, new findings have emerged regarding skin temperature dynamics during exercise, the applicability of infrared thermography in sports science, and methodological advancements in the field. These findings, discussed throughout the book, provide fresh insights into how infrared thermography can be used in sports science, how thermoregulation influences sports performance, how injuries affect skin temperature, and how temperature can be used to evaluate garments and sports equipment, among other applications.

The book is organized into 16 chapters. The first chapter provides a historical introduction to the application of infrared thermography in sports science. Chapters 2, 3, and 4 focus on providing the basic background, methodological knowledge for proper infrared thermography use, and the foundations of human thermoregulation. The final chapter discusses current challenges and potential developments in infrared thermography within sports science, aiming to guide future research and development. Chapters 5, 6, 7, 8, 9, 10, 11, 12, 13, 14, and 15 each focus on a specific application of infrared thermography in sports science. Each chapter begins with an introduction to the topic, followed by a discussion, and concludes with key takeaways. Additionally, each chapter addresses methodological aspects relevant to its specific subject. Efforts have been made to keep the text as clear and concise as possible while ensuring it remains comprehensive.

I hope this book will be a valuable addition to the libraries of those who use infrared thermography in their sports research or professional practice. Furthermore, I hope it will help guide and enhance future studies, ultimately advancing scientific knowledge in this field.

Valencia, Spain
March, 2025

Jose Ignacio Priego-Quesada

Acknowledgments

The authors of this book are grateful to the different editorials for providing the permission rights to show some of the figures published in their articles.

Thanks are also given to Prof. Felipe Carpes and Prof. Dominic McCafferty for their very kind opening words in this book.

Some acknowledgments are provided by the authors for some of the chapters:

Chapter 6. Alvaro S. Machado contribution was funded by Conselho Nacional de Desenvolvimento Científico e Tecnológico (CNPq, process 169461/2023-8).

Chapter 8. The authors are grateful to Ana María Mifsut Climent for her contribution in the design of Figs. 8.4, 8.5, 8.6, 8.7, 8.8, and 8.9. Joaquín Martín Marzano-Felisatti's contribution was funded by a predoctoral grant from the Ministry of Universities of Spain (ref FPU20/01060). Alberto Hermo Argibay's contribution was funded by "Generalitat Valenciana, Conselleria de Innovacion, Universidades, Ciencia y Sociedad Digital" (grant PROMETEO CIPROM/2022/32).

Chapter 12. Research Activity (IMAMCA/2025) on which this chapter was based was supported by Instituto Valenciano de Competitividad Empresarial e Innovación (IVACE+i) and Valencian Regional Government (GVA) funding.

Chapter 15. Figures 15.10, 15.11, 15.12, 15.13, 15.14, 15.16, 15.17, and 15.18 have been published in Soroko and Davies-Morel (2016) and are reproduced here with the kind permission of CABI Publishers (UK). Highly deserved thanks also go to Andrzej Soroko for his graphic skills in preparing other thermograms for this publication.

Contents

Contributors

Inmaculada Aparicio-Aparicio Research Group in Sports Biomechanics (GIBD), Department of Physical Education and Sports, Universitat de València, Valencia, Spain

Alberto Hermo Argibay Foundation for the Promotion of Health and Biomedical Research in the Valencian Region (FISABIO), Department of Endocrinology and Nutrition, University Hospital Doctor Peset, Valencia, Spain

Javier Arnáiz Lastras Sports Department, Faculty of Physical Activity and Sport Sciences—INEF, Universidad Politécnica de Madrid, Madrid, Spain

PEMA THERMO GROUP S.L., ThermoHuman, Madrid, Spain

Aaron J. E. Bach School of Health Sciences and Social Work, Griffith University, Southport, QLD, Australia

Felipe P. Carpes Applied Neuromechanics Research Group, Laboratory of Neuromechanics, Federal University of Pampa, Uruguaiana, RS, Brazil

Clara Carrión-González Research Group in Sports Biomechanics (GIBD), Department of Physical Education and Sports, Universitat de València, Valencia, Spain

Nina Čebulj-Kadunc Institute of Preclinical Sciences, Veterinary Faculty, University of Ljubljana, Ljubljana, Slovenia

Willian da Silva Escuela de Kinesiologia, Facultad de Ciencias, Pontificia Universidad Católica de Valparaíso, Valparaíso, Chile

Mina C. G. Davies Morel Reader Emeritus, Department of Life Sciences, Aberystwyth University, Aberystwyth, UK

Rosa María Cibrián Ortiz de Anda Biophysics and Medical Physics Group, Department of Physiology, University of Valencia, Valencia, Spain

Alberto Encarnación-Martínez Research Group in Sports Biomechanics (GIBD), Department of Physical Education and Sports, University of Valencia, Valencia, Spain

Víctor Escamilla Galindo Sports Department, Faculty of Physical Activity and Sport Sciences—INEF, Universidad Politécnica de Madrid, Madrid, Spain

PEMA THERMO GROUP S.L., ThermoHuman, Madrid, Spain

Ismael Fernández-Cuevas Sports Department, Faculty of Physical Activity and Sport Sciences—INEF, Universidad Politécnica de Madrid, Madrid, Spain

PEMA THERMO GROUP S.L., ThermoHuman, Madrid, Spain

Damiano Formenti Department of Biotechnology and Life Sciences (DBSV), University of Insubria, Varese, Italy

Damien Fournet DECATHLON SportsLab, Thermal Sciences Laboratory, Villeneuve d'Ascq, France

Joaquim Gabriel LABIOMEP, INEGI, LAETA, University of Porto, Porto, Portugal

Faculdade de Engenharia, Universidade do Porto, Porto, Portugal

Juan Carlos González García Instituto de Biomecánica de Valencia (IBV), Valencia, Spain

Marina Gil-Calvo Faculty of Physical Activity and Sports Sciences, Universidad de León, León, Spain

Research Group in Sport Biomechanics (GIBD), Department of Physical Education and Sports, University of Valencia, Valencia, Spain

Pedro Gómez Carmona Sports Department, Faculty of Physical Activity and Sport Sciences—INEF, Universidad Politécnica de Madrid, Madrid, Spain

PEMA THERMO GROUP S.L., ThermoHuman, Madrid, Spain

Ricard Barberà Guillem Instituto de Biomecánica de Valencia (IBV), Valencia, Spain

George Havenith Environmental Ergonomics Research Centre, Loughborough University, Loughborough, UK

Kevin Howell Microvascular Diagnostics, Royal Free Hospital, London, UK

Irene Jimenez-Perez Research Group in Sport Biomechanics (GIBD), Department of Physical Education and Sports, University of Valencia, Valencia, Spain

Biophysics and Medical Physics Group, Department of Physiology, University of Valencia, Valencia, Spain

Marcos Roberto Kunzler Applied Neuromechanics Research Group, Laboratory of Neuromechanics, Federal University of Pampa, Uruguaiana, RS, Brazil

Consuelo Latorre-Sánchez Instituto de Biomecánica de Valencia (IBV), Valencia, Spain

Álvaro S. Machado Applied Neuromechanics Research Group, Laboratory of Neuromechanics, Federal University of Pampa, Uruguaiana, RS, Brazil

Joaquín Martín Marzano-Felisatti Research Group in Sports Biomechanics (GIBD), Department of Physical Education and Sports, University of Valencia, Valencia, Spain

Robert D. Meade Department of Epidemiology, T.H. Chan School of Public Health, Harvard University, Boston, MA, USA

Arcangelo Merla Department of Engineering and Geology (INGEO), University "G. d'Annunzio" of Chieti-Pescara, Pescara, Italy

Fergus K. O'Connor School of Health Sciences and Social Work, Griffith University, Southport, QLD, Australia

Rafael Mengual Ortolá Instituto de Biomecánica de Valencia (IBV), Valencia, Spain

Sonia Gimeno Peña Instituto de Biomecánica de Valencia (IBV), Valencia, Spain

Pedro Pérez-Soriano Research Group in Sport Biomechanics (GIBD), Department of Physical Education and Sports, University of Valencia, Valencia, Spain

David Perpetuini Department of Engineering and Geology (INGEO), University "G. d'Annunzio" of Chieti-Pescara, Pescara, Italy

Jose Ignacio Priego-Quesada Research Group in Sport Biomechanics (GIBD), Department of Physical Education and Sports, University of Valencia, Valencia, Spain

Biophysics and Medical Physics Group, Department of Physiology, University of Valencia, Valencia, Spain

Rosario Salvador-Palmer Biophysics and Medical Physics Group, Department of Physiology, University of Valencia, Valencia, Spain

Roberto Sanchis-Sanchis Research Group in Sports Biomechanics (GIBD), Department of Physical Education and Sports, University of Valencia, Valencia, Spain

Adérito Seixas LABIOMEP, INEGI, LAETA, University of Porto, Porto, Portugal

Faculdade de Ciências da Saúde, Universidade Fernando Pessoa, Porto, Portugal

Carlos Sendra-Pérez Research Group in Sports Biomechanics (GIBD), Department of Physical Education and Sports, Universitat de València, Valencia, Spain

Department of Education and Specific Didactics, Jaume I University, Castellon, Spain

Maria Soroko-Dubrovina Institute of Animal Breeding, Wrocław University of Environmental and Life Sciences, Wrocław, Poland

Ricardo Vardasca LABIOMEP, INEGI, LAETA, University of Porto, Porto, Portugal

ISLA Santarém – Instituto Politécnico, Santarém, Portugal

LABIOMEP, CIFI2D, Faculty of Sport, University of Porto, Porto, Portugal

João Vilas-Boas LABIOMEP, CIFI2D, Faculty of Sport, University of Porto, Porto, Portugal

Chapter 1
Introduction: Historical Perspective of Infrared Thermography and Its Application in Sport Science

Jose Ignacio Priego-Quesada, Rosa María Cibrián Ortiz de Anda, Pedro Pérez-Soriano, and Rosario Salvador-Palmer

Abstract Within the various chapters of this book, the methodology and different applications of infrared thermography in sport science will be discussed. But what are the origins of infrared thermography? What has its development been like in sport science? The present introductory chapter of the book aims to show the historical developments of infrared thermography and, fundamentally, its application in sport science.

1.1 Introduction

Infrared thermography features and its applications are constantly evolving. In recent decades, the technique has seen significant development. Furthermore, approximately 30 years ago, the use of infrared thermography in sport science was uncommon. However, in recent years, many sport laboratories have acquired an infrared camera and have begun to undertake thermographic assessments. Knowledge of the evolution of thermography during the twentieth century in

J. I. Priego-Quesada (✉)
Research Group in Sport Biomechanics (GIBD), Department of Physical Education and Sports, University of Valencia, Valencia, Spain

Biophysics and Medical Physics Group, Department of Physiology, University of Valencia, Valencia, Spain
e-mail: j.ignacio.priego@uv.es

Rosa María Cibrián Ortiz de Anda · R. Salvador-Palmer
Biophysics and Medical Physics Group, Department of Physiology, University of Valencia, Valencia, Spain
e-mail: rosa.m.cibrian@uv.es; rosario.salvador@uv.es

P. Pérez-Soriano
Research Group in Sport Biomechanics (GIBD), Department of Physical Education and Sports, University of Valencia, Valencia, Spain
e-mail: pedro.perez-soriano@uv.es

J. I. Priego-Quesada (ed.), *Application of Infrared Thermography in Sports Science*,
https://doi.org/10.1007/978-3-031-93311-0_1

general, and in sports science in particular, can provide a perspective that helps in the present and future use of thermography.

Infrared thermography has a large number of applications. As will be shown in the different chapters of this book, infrared thermography in sports science has applications in various fields: sport medicine, analysis of thermoregulation, clothing and sport equipment assessment, and so on. Outside the field of sports, thermography is used in a wide range of sectors. Infrared thermography is used in the military, policing and security work, enabling night vision and surveillance. In electrical installations, nuclear power plants, and other factories, it is used, for example, for the monitoring of electrical components. In medicine, it is used for the diagnosis of injuries and diseases. In firefighting, infrared cameras are used to locate people/animals in heavy smoke and for detecting hot spots in forest fires. Infrared imaging is used in building inspections to detect issues related to insulation and water leaks. There are also infrared satellites that measure the ocean temperatures continuously. Although this chapter will be focused as much as possible on sport and human assessment, we believe that it is necessary to use the chapter to show other different applications. Sometimes these applications may appear to be far from the field of sport science. However, they are necessary to know because they may contribute to ideas for future developments and research in sport science.

This chapter aims to provide a general overview of both the past and present of infrared thermography in sport science. However, in order to achieve this objective, it is first necessary to review the origins of infrared thermography and its applications in other fields. One of these fields, very close to sport science, is medicine, and the beginning of the use of infrared thermography in medicine will be reviewed.

1.2 Origins: Discovery of Infrared Radiation, the First Thermography, and the Development of Infrared Sensors

Infrared thermography is an image technique used to record infrared radiation, which permits estimation of a body's surface temperature. Any object at a temperature above absolute zero (0 K or −273.15 °C) emits energy—electromagnetic radiation—depending on its temperature [1, 2]; for the ~37 °C of the human body, the emitted energy is in the infrared region of the electromagnetic spectrum (thermal radiation) [3, 4]. Aspects related to the physics principles of infrared thermography will be discussed in Chap. 2.

Infrared radiation was discovered in 1800 by Sir Frederick William Herschel (born in 1738 in Hannover, Germany; died in 1822 in Slough, England), an English–German astronomer (Fig. 1.1). It is well known by all scientists working with thermography that William Herschel was the astronomer who discovered infrared radiation. However, there are some interesting facts that may be unfamiliar to some. Herschel was a professional musician and the director of an orchestra; he could be considered to be a professional musician and an amateur astronomer or scientist.

Fig. 1.1 William Herschel and the experiment in which he discovered infrared radiation

Some of the discoveries he made in his dedication to astronomy was the observation of the planet Uranus in 1781, as well as the discovery in 1783 that the sun was not static as it had always previously been believed, but rather that the sun moves by dragging with it the planets of the solar system.

After learning a little more about William Herschel, we will discuss his discovery of infrared radiation. Herschel was interested in learning how much heat passed through the different colors by watching the sun filters. He set this objective because he observed that the amount of heat transmitted depended on color. Herschel thought that the colors themselves could filter out different amounts of heat, so he devised an original experiment to test his hypothesis. He directed sunlight through a glass prism to create a visible spectrum (a rainbow, a division of the light in different colors). He placed three mercury thermometers in the spectrum obtained by a glass prism in order to measure the heat emitted by each color [5] (Fig. 1.1). Herschel used three thermometers with blackened bulbs (to better absorb heat) and, for each color of the spectrum, placed one bulb in a visible color while the other two were placed beyond the spectrum as control samples. He found that the heat increased from the violet to the red color of the spectrum. However, he also observed that next to the red spectrum, when there was no light, the temperature was higher [5]. This experiment from Herschel was very important because it marked the first time that someone had demonstrated that there were types of energy that are invisible to our eyes. Herschel called this radiation "Calorific Rays," a term that remained popular throughout the nineteenth century before being replaced by the modern term "infrared radiation."

The following text is an extraction of the conclusion of the article in which Herschel explained the results of his experiment [5]:

> To conclude, if we call light, those rays which illuminate objects, and radiant heat, those which heat bodies, it may be inquired, whether light be essentially different from radiant heat? In answer to which I would suggest, that we are not allowed, by the rules of philosophizing, to admit two different causes to explain certain effects, if they may be accounted for by one. A beam of radiant heat, emanating from the sun, consists of rays that are differently refrangible. The range of their extent, when dispersed by a prism, begins at violet-coloured light, where they are most, refracted, and have the least efficacy. We have traced these calorific rays throughout the whole extent of the prismatic spectrum; and found their power increasing while their refrangibility was lessened, as far as to the confines of red-coloured light. But their diminishing refrangibility, and increasing power, did not stop here; for we have pursued them a considerable way beyond the prismatic spectrum, into an invisible state, still exerting their increasing energy, with a decrease of refrangibility up to the maximum of their power; and have also traced them to that state where, though still less refracted, their energy, on account, we may suppose, of their now failing density, decreased pretty fast; after which, the invisible thermometrical spectrum, if I may so call it, soon vanished. If this be a true account of solar heat, for the support of which I appeal to my experiments, it remains only for us to admit, that such of the rays of the sun as have the refrangibility of those which are contained in the prismatic spectrum, by the construction of the organs of sight, are admitted, under the appearance of light and colours; and that the rest, being stopped in the coats and humours of the eye, act upon them, as they are known to do upon all the other parts of our body, by occasioning a sensation of heat.

After the death of William Herschel, his son, John Frederick William Herschel (born in 1792 in Slough, England; died in 1871 in Collingwood, England), repeated the experiments of his father and made an image using solar radiation in 1840 [6]. This image was achieved by focusing solar radiation onto a suspension of carbon particles in alcohol using a lens, a method known as evaporography [6]. He obtained an image that he called a "thermogram" (Fig. 1.2), a term that is still in use today

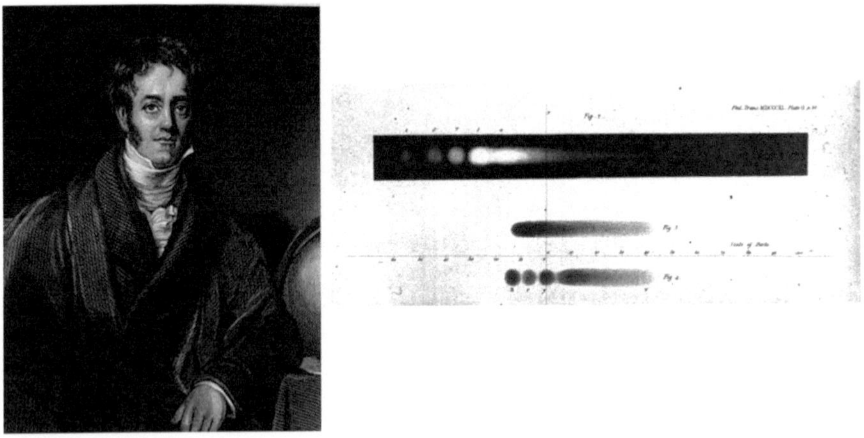

Fig. 1.2 John Herschel and the thermogram performed by him using the evaporography method. (Figure modified from Ring [6])

for the infrared thermography technique [6]. John Herschel was an important mathematician, astronomer, chemist, inventor, and photographer (Fig. 1.2).

After John Herschel, some of the most important advances were related to the development of infrared detectors. Of course, the determination of the physical laws that explained the relationship between temperature and infrared radiation had great importance, but these laws will be explained in Chap. 2 of this book.

A bolometer is a device that measures electromagnetic radiation by the increase in the resistance of an electrical conductor. Samuel Pierpont Langley (born in 1834 in Roxbury, USA; died in 1906 in Aiken, USA) was the inventor of the balometer, originally called Langley's bolometer, in 1880 [2, 7]. Langley's bolometer followed a continuous process of development over 20 years, increasing its sensitivity by 400 more times from the first prototype [7]. The last bolometer of Langley was able to detect the heat from a cow at a distance of ~400 m [7]. It is considered that the invention of the balometer was relevant and important in the development of the posterior infrared cameras [2].

After the development of the bolometer, one of the most important steps was considered to be the development of the infrared photon detectors during the twentieth century [7]. The first infrared photoconductor was developed in 1917 [7], but the origins of the modern infrared detector technology are considered to have been developed during World War II [6, 7]. These detectors were developed at the end of World War II for use by snipers to enable features such as rudimentary night vision [8]. These sensors involved the conversion of infrared radiation in visible light, due to the emission of electrons from near-infrared cathodes by the visible phosphors [8]. At the same time, another infrared detector was made from indium antimonide and was mounted at the base of a small Dewar vessel to allow cooling with liquid nitrogen [6, 8]. This device had the limitation of requiring a constant supply of liquid nitrogen, which was impractical for battlefield, but not an important inconvenience in other fields such as medicine [8].

1.3 Antecedents and Contemporaries: Infrared Thermography in Medicine and Other Applications

As was discussed in the previous section, modern infrared detectors were developed during World War II. The sensor was developed in this period for its military applications [2, 4, 9]. Its main use in the military field was, and remains, night vision. This utility is also highlighted when the objective is to detect people in dark environments and complex spaces. A clear example of this is border and maritime space supervision (Fig. 1.3).

After infrared thermography had been used in military applications, the technology was released for civilian uses [1, 9]. In 1934, human skin was described to be a good emitter of infrared radiation [10]. After that, infrared thermography started to be considered as a potential technique for skin temperature measurement and, then,

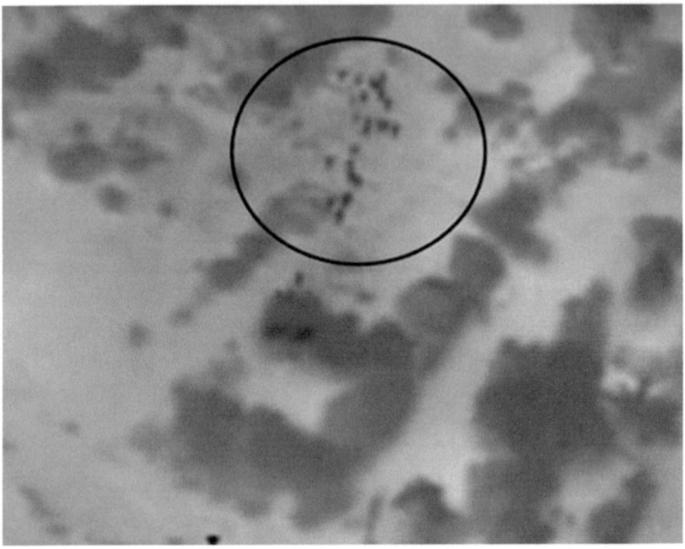

Fig. 1.3 Use of infrared camera in a helicopter as part of the border control of a country. Inside the *circle* it can see *small dots* that are people walking across the border during the night

as a possible way to enable diagnostic imaging in medical science [9]. In this regard, one of the applications of infrared thermography that is very related to sport science is medicine, which is discussed below.

Although Vardasca and Simoes [2] reported that the first human thermographic image was done by Czerny in 1928 in Germany [11], Ring [6] described that the greatest advances of the techniques were performed in the 1940s and 1950s; these advances allowed a better quality of thermal images, but were very crude in comparison with modern thermographies (Fig. 1.4). Due to the advances of the technique in those years, infrared thermography is considered to have been used in medicine since the 1960s [1, 9, 12, 13]. It is considered that the first medical infrared images were taken at The Middlesex Hospital in London and The Royal National Hospital for Rheumatic Diseases in Bath, UK, between 1959 and 1961, using a prototype system called "Pyroscan" [6, 8]. This system represented an improvement in the quality of the image and "printed" the image line-by-line on electro-sensitive paper, but it took a lot of time to record each image (between 2 and 5 min) [6, 8]. The first applications of infrared thermography in medicine in that decade were the diagnosis of peripheral vascular diseases [14], its use in obstetrics and gynecology [15], and the detection of ischemic neuropathies [16], among others. A current overview of all the applications that infrared thermography has in medicine is provided by the interesting reviews performed by Ring and Ammer [17] and Lahiri and colleagues [9], both of which were published in 2012. These reviews extracted the following applications of infrared thermography in medicine: assessment of the thermoregulation, breast cancer detection, diagnosis of diabetic neuropathy and vascular disorder, fever screening, dental diagnosis, dermatological applications,

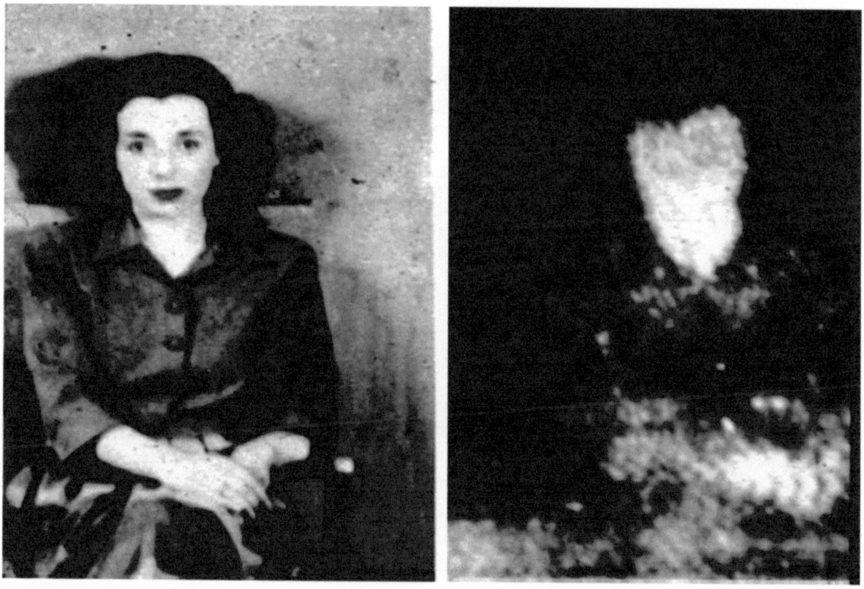

Fig. 1.4 Photography and thermography of one woman taken in 1949. (Figure obtained from Ring [6])

blood pressure monitoring, diagnosis of rheumatic diseases, diagnosis of inflammatory arthritis, assessment of fibromyalgia, diagnosis of dry eye syndrome and ocular diseases, diagnosis of liver diseases, complementary test in the treatment of kidney, gynecology applications, and assessment of the psychological state [9, 17].

Of the above applications, cancer detection was given a higher priority in the hope that it could be a valid screening technique [6]. In this sense, the earliest assessment of the temperature distribution of a breast cancer was reported by Ray Lawson in 1956 using the evaporograph system [18] (Fig. 1.5a). The author said in his publication: "In its present state of development, the Evaporograph has a high resolution for objects having sudden or abrupt temperature changes, however slight (such as blacked-out cities at night), but where the temperature shows a smooth fall-off, the oil image lacks sensitivity. This remarkable device is quite capable of much future development." Lawson, one year later, appreciated the potential of infrared thermography in the detection of breast cancer [20]. Different research groups have worked on this topic since the suggestion of Lawson in 1957 to the present day. Figure 1.5b shows a current breast thermography in order to illustrate how the technique has improved since Lawson's days. However, more research into infrared thermography is still required in order to improve its sensitivity and specificity in breast cancer screening and diagnosis [21]. Most of the recent research is focused on applying dynamic thermography (assessment after a stressor, for example, a cold stress test) [22] or the use of machine and deep learning [23].

In the beginning of infrared thermography's application in medicine, the technique was often discredited, mainly because the technology at the time was

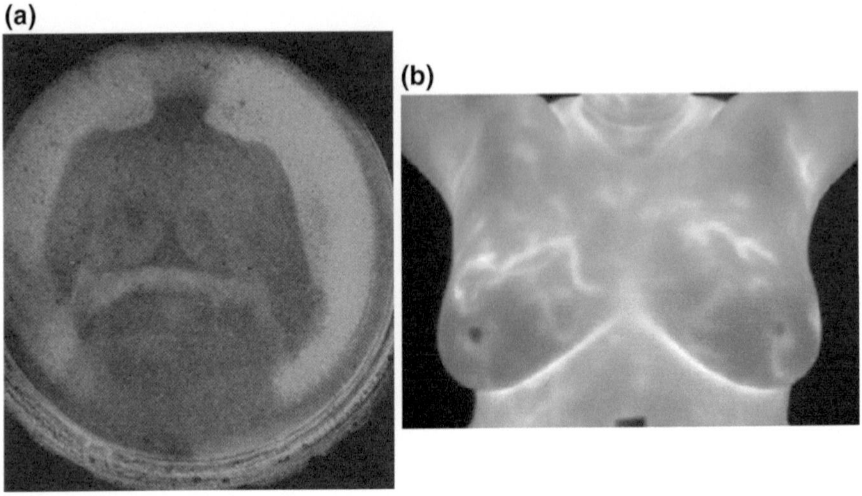

Fig. 1.5 Evolution of the thermal detection of breast cancer: (**a**) evaporograph image in 1956. (Figure obtained from Lawson [18]) and (**b**) current breast thermographies. (Figure obtained from Silva et al. [19])

inadequate [24]. However, the improvement of the quality of the camera and the analysis of the thermal data over the years have led to the increasing use and efficiency of thermal imaging as a tool for diagnostic imaging procedures [24]. In this sense, infrared thermography was recognized as a feasible diagnostic and analytical tool by the American Medical Council in 1987 [3, 24]. After that, the application of infrared thermography in the assessment of skin temperature in medicine has been acknowledged and promoted by different worldwide thermographic associations, such as the American Academy of Medical Infrared Imaging, the International Academy of Clinical Thermology, the International Thermographic Society, the European Association of Thermology, the Northern Norwegian Centre for Medical Thermography, and the German Society of Thermography and Regulation Medicine [3, 25]. The main objective of these groups was and remains to develop reliable standardized methods and appropriate protocols for infrared thermography for clinical application [3].

Infrared thermography is used in other applications related to medicine (e.g., veterinary science) but also in other fields such as engineering. Although these applications are further away from sport science, all of them are based on the assessment of heat transference. A superficial knowledge of its use can sometimes provide useful ideas for thermography assessment in sport science. For a review of a high number of applications of infrared thermography across different fields, the book of Vollmer and Möllmann [26] is recommended. In this chapter, some examples of representative and diverse applications are commented on below:

- Veterinary science. Infrared thermography has, in this area, a similar utility as in medicine. However, due to some characteristics, it has become very important in veterinary science. The difficulty of knowing the pain and the patient's condition without invasive techniques is higher in veterinary science than in medicine, and thus thermography is a low-cost, fast, efficient, and simple method that allows knowledge of the status of the animal's injury [27] (Fig. 1.6). Additionally, the use of invasive methods produces anxiogenic responses in the animal that can affect the results [27]. Furthermore, it is considered that, in the veterinary field, thermography assessment can be useful to estimate the physiological state of animals in situations of stress, fertility, welfare, metabolism, health, and disease detection [27].

- Predictive maintenance of factories. This application is widely used in electrical and nuclear facilities as well as in different factories. This kind of application is particularly important when the factories need to work continuously without stoppages occurring in their activity due to breakdowns. In this context, infrared thermography helps routine inspections, where it is possible to visualize abnormalities in the components [28]. This display of anomalies would be able to repair or replace components before they produce serious damage to the operation of the factory (Fig. 1.7). In addition, infrared thermography in this context allows the person to carry out inspection at a safe distance from hazardous components (e.g., those with high temperature or radioactivity) and to measure the temperature without contact. Another use in this field, especially in the petroleum

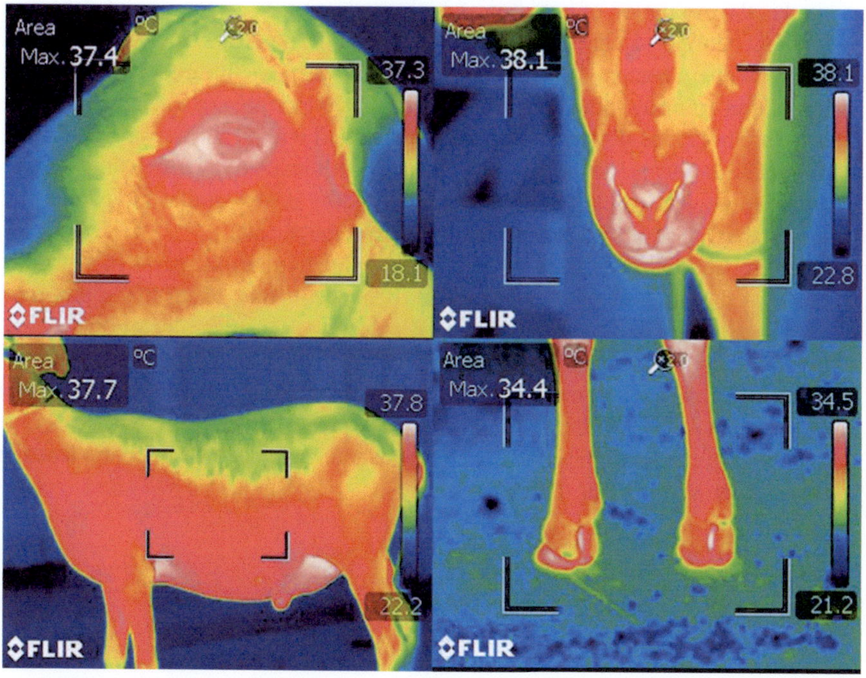

Fig. 1.6 Examples of thermographic images of sheep used to predict physiological stress. (Figure obtained from McManus et al. [27])

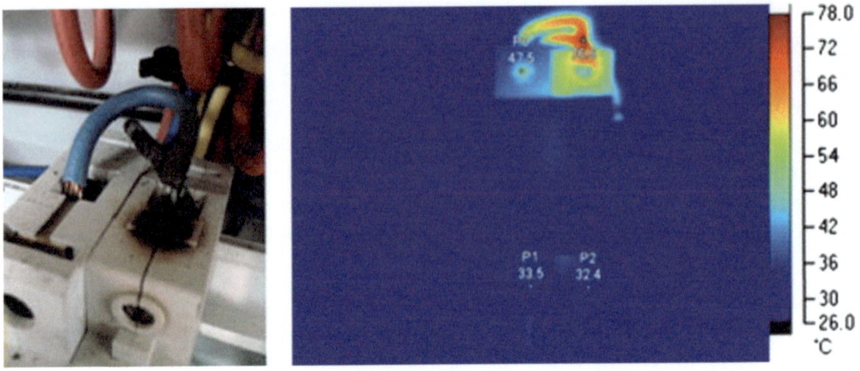

Fig. 1.7 Example of an oxidized connection with a high temperature, which can be a cause of breakdown. (Figure obtained from Jadin and Taib [29])

chemical industry, is to determine the level of liquids and solids in the deposits [26].

- Building Inspection. Thermography is considered to be an excellent tool for diagnosing the state of buildings [30–32] (Fig. 1.8). Evaluation of insulation of different materials and/or the location of water leaks, and thus humidity, are specific uses of infrared thermography in this field (Fig. 1.9). One of the techniques used is the evaluation of heat flow, which consists of the generation of a thermal gradient between the inside and outside of the building. An example of this is that in the winter, heaters heat the inside. Thus, through infrared thermography inspection from the outside of the building, it is possible to visualize whether there are hot areas. These hot areas are associated with heat transfer from inside to outside and therefore represent a problem with insulation in the walls or windows. If the whole outside surface of the building is at a similar temperature, it means that the inside is properly isolated from the outside.
- Gas detection industry. Infrared cameras are able to visualize gases in a specific range of the infrared spectra (Chap. 2). Because of this, using the appropriate camera, it is possible to visualize specific gases such as ethanol, carbon dioxide, propanol, or sulfur hexafluoride, among others. This provides interesting applications of infrared thermography in the industry (Fig. 1.10). Two of the main applications are (1) to check whether factories are not producing toxic gases or if the levels of the gases expelled are adequate and (2) to check if there are leaks in the structures to store or transport gas.

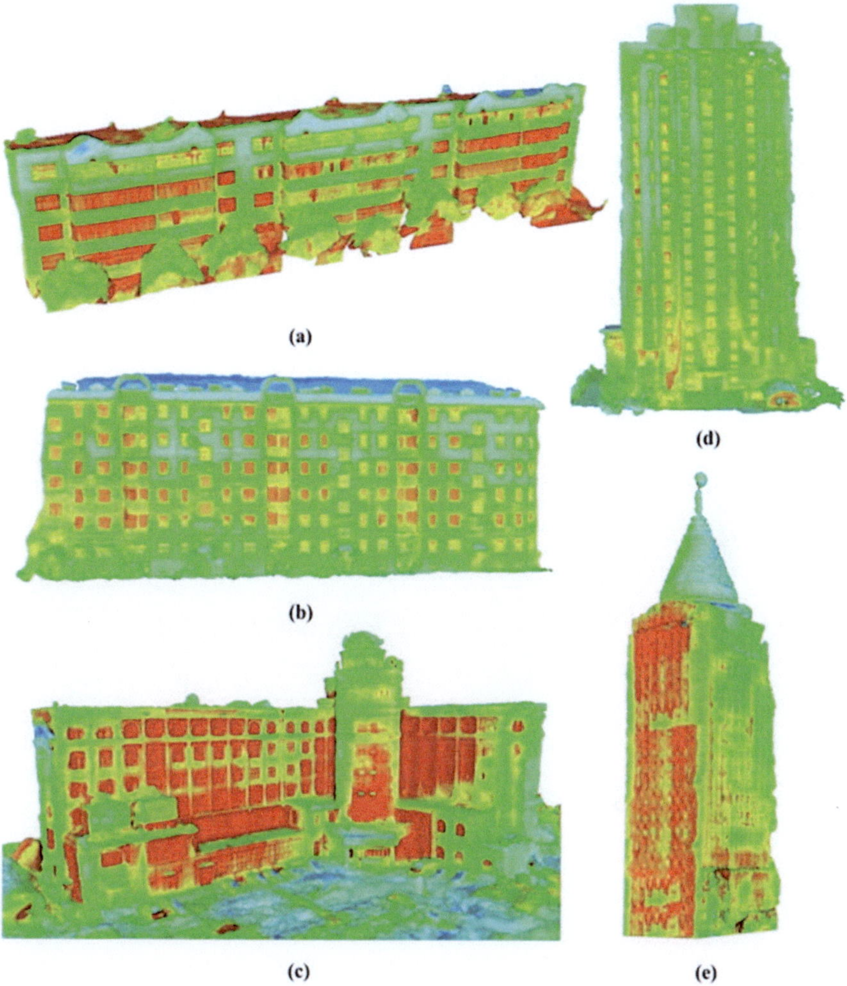

Fig. 1.8 3D thermography models performed by an infrared camera and unmanned aerial vehicle. (Figures obtained from Zheng et al. [32])

1.4 Infrared Thermography in Sport Science: Past and Present

In sport science, the studies using thermography started approximately 10 years after medicine. Furthermore, the first studies in sport science were related to its application in sport medicine.

The first study about thermography in sport science was in 1975, titled "Thermography in sport injuries and lesions of the locomotor system due to sport" by Keyl and Lenhart in the German journal *Fortschritte der Medizin* [35]. This

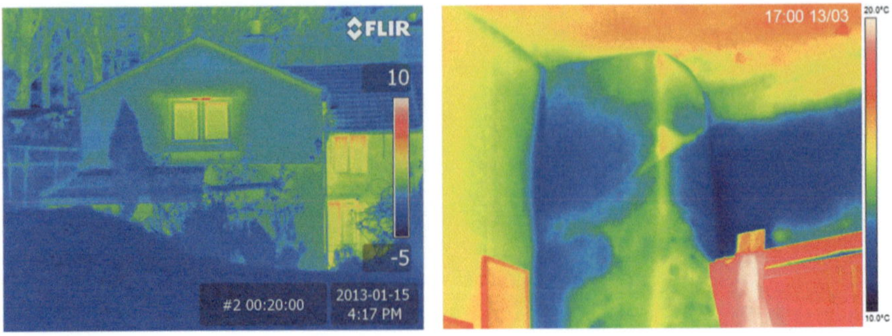

Fig. 1.9 Examples of thermographies in the building inspection. (Figures obtained from Fox et al. [33])

Fig. 1.10 Thermography of the gas emission in a paper factory. (Figure obtained from Sedlák et al. [34])

study was written in German. In this study, the authors examined 82 patients and 50 athletes with different injuries. The main result of the study was the observation of hyperthermia in the injured area.

The next study in the field of sport medicine was in February of 1977, and it was also performed by German authors and published in German language. The study was titled "Skin-thermography with fluid crystals in orthopedics and sport-medicine" and it was performed by Lelik, Solymossy, and Kézy and published in the journal *Zeitschrift für Orthopädie und Unfallchirurgie* [36]. Although we did not have the possibility to access the paper, it is possible to show the English abstract: "Thermography with fluid crystals is an effective diagnostic aid for the differentiation between bacterial and non-bacterial inflammation i.e. for the differential diagnosis between joint damage due to overloading and irritative arthrosis. Thermography enables us to assess severity and extent of the condition and to prove the success of treatment."

Just three months after the previous publication, the first study using infrared thermography in sport science in the English language by English authors was published in *The Journal of Physiology*. The study was titled "Skin temperature during running—a study using infra-red colour thermography" and the authors were Clark and Mullan [37]. This study was not oriented to the medical application as the previous studies were, and its objective was to assess the validity of skin temperature measurements under outdoor conditions. Two runners participated in the experiment. They ran for approximately 1 h and 15 min, one day in an outdoor track and another day in a climatic chamber. Measurements were performed before running and at 15-minute intervals during running, with an infrared thermography camera and thermocouples, in order to compare the data of both systems. In the climatic chamber, they also assessed the effect of the wind speed (no wind vs. 4.5 m s^{-1}). Figure 1.11 shows examples of thermographies obtained in this study. They obtained the following results [37]:

1. They analyzed the body temperature distribution before running. They described that the body temperature distribution depended on the following factors: body composition (i.e., body structures or regions with a higher proportion of fat presented lower temperatures), cutaneous blood flow (related to the warming of the hands), sweat rate evaporation, and muscle role (i.e., higher skin temperatures were recorded in the regions of the active muscles).

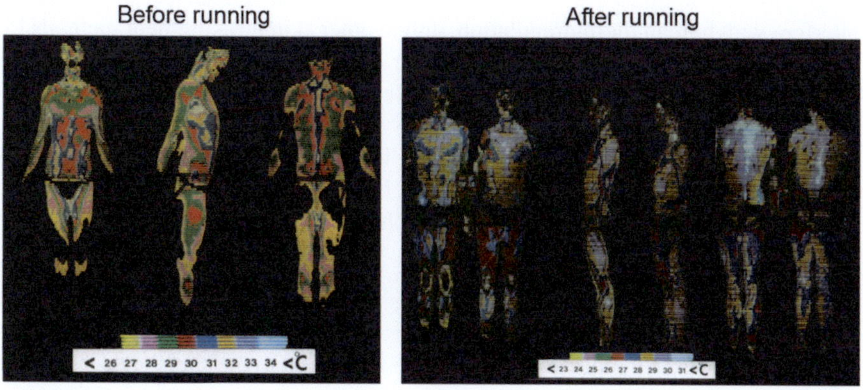

Fig. 1.11 Thermographies performed in the study of Clark et al. [37]

2. They observed a decrease in the skin temperature during running. Although this variation is different in each body region, an average value could be considered as 5 °C. This decrease was associated with the sweat evaporation.
3. They observed in the images that the emissivity of the skin was not influenced by the presence of sweat.
4. Similar temperatures before and during running were observed in the climatic chamber and in the outdoor track.
5. Infrared thermography could underestimate areas near the side of the body, due to the curvature of the body.
6. Thermography and thermocouples agreed, within 1.5 °C, on the mean skin temperature, but in specific regions could differ by 4 °C. The authors suggested that although thermography provides a more complete analysis, it is more expensive and consumes more time.

After observing the results and their discussion, it is interesting how this study from 1977 obtained results and discussed different topics that keep being topics of interest until the present, such as the body temperature distribution [38–40], the effect of the sweat on skin emissivity [41–43], and the differences between thermocouples and thermography [42, 44, 45], among others.

In order to know the trend of the number of publications performed in sport science using infrared thermography, we performed a search in the PubMed (Medline) database. The keywords used in the online search included "thermography," "thermovision" or "thermal imaging," and "exercise" or "sport." Boolean operators, "OR" and "AND," were used to combine within and between the search terms of the subject areas.

Figure 1.12 shows the result of this online search which evidenced an exponential increase in the number of publications. In this sense, the number of publications in the first 5 years of this decade (2021–2025: 122 articles) is approximately nine times more than in the previous decade (2001–2010: 14 articles published). The progressive increase in the number of studies using infrared thermography in sport science could be explained by some of the technique's advantages. Infrared thermography is a noninvasive method that can be used at a distance, and does not interfere with human thermoregulation, unlike other skin measurement methods [3, 42, 44]. Additionally, this important increase in the number of studies in the last 15 years could be explained by the great decrease in the price of infrared cameras compared to the 90s.

In addition to the number of publications, it is important to know in which fields or applications the studies were published, in order to understand which are the most important uses of infrared thermography in sport science. In this sense, we defined the following categories:

- Thermophysiology: thermography studies focused on the assessment of human thermoregulation. All the studies that assessed the effect of exercise on skin temperature, the skin temperature distribution in relation to the sport or exercise, the differences in skin temperature between different groups (e.g., trained vs.

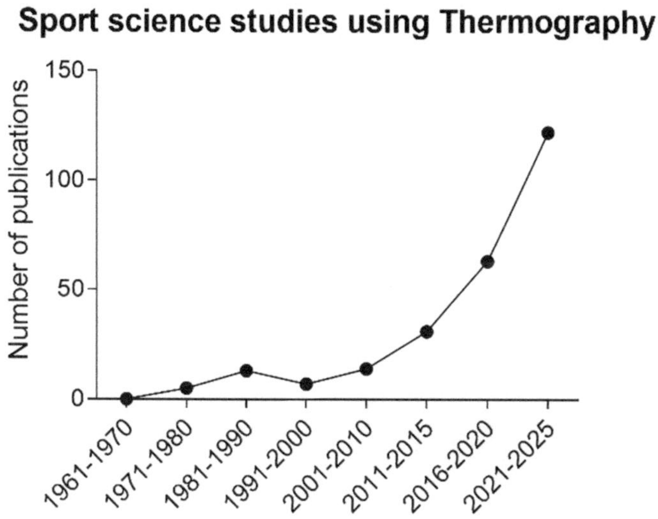

Fig. 1.12 Number of papers published in PubMed (Medline) using infrared thermography in sport science. (Note: This figure was created in February of 2025)

untrained; young vs. old; woman vs. men), the effect of sport performance (e.g., VO_{2max} on skin temperature), and so on, were included in this category.

- Sport medicine: thermography studies related to the assessment of the effect of injury or disease in humans. In addition, this category included all studies that presented a clinical/medical perspective.
- Animals and sport: thermography studies conducted with animals that are involved in an exercise or a sport. Papers with a clinical perspective involving animals were also included in this category. In this sense, this category would include the previous two categories, but with an animal focus.
- Clothing: thermography studies focused on the assessment of clothing in a specific sport or during exercise.
- Methodology/developments: studies that investigated the methodological aspects of infrared thermography and development of the technique. Studies that compared infrared thermography with other techniques (e.g., thermocouples), analysis of the data, determination of the regions of interest, guidelines, and so on, were included in this category.
- Reviews: reviews and overview papers about infrared thermography in sport science were included in this category.

Papers obtained were categorized into the defined applications or fields (Fig. 1.13). During the approximately 50 years that infrared thermography has been used in sports science, the application that has been most used is thermophysiology, representing 38% of the total number of papers. The second most commonly used application has been sport medicine, with 26% of the total papers, and the third application is animals and sports with 17% of the total papers. In other application

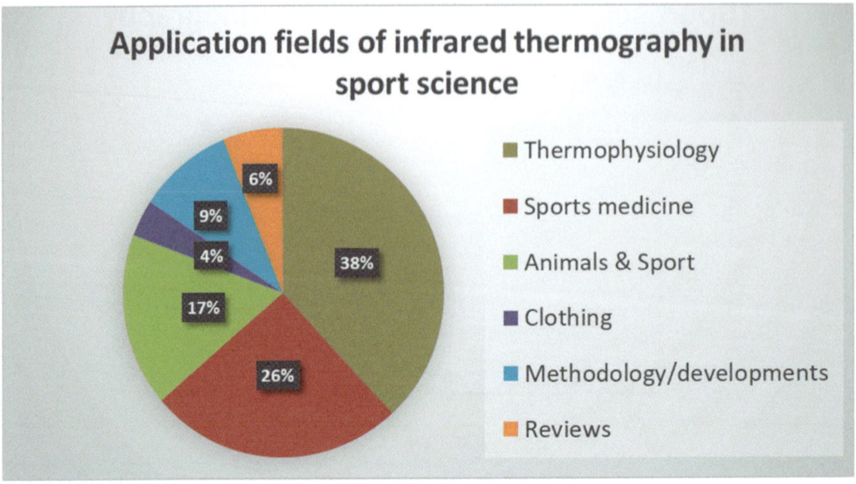

Fig. 1.13 Percentage of the papers published using infrared thermography in sport science application fields

Table 1.1 The five most relevant publications using infrared thermography in sport science and relevant publications identified in the first edition of this book

Rank	Title	Authors and year	No of citations
1	An overview of recent application of medical infrared thermography in sports medicine in Austria [3]	Hildebrandt et al. (2010)	290
2	Thermographic imaging in sports and exercise medicine: A Delphi study and consensus statement on the measurement of human skin temperature [46]	Moreira et al. (2017)	286
3	Thermal imaging of cutaneous temperature modifications in runners during graded exercise [47]	Merla et al. (2010)	173
4	Changes in eye temperature and stress assessment in horses during show jumping competitions [48]	Valera et al. (2012)	121
5	Thermal imaging of exercise-associated skin temperature changes in trained and untrained female subjects [49]	Formenti et al. (2013)	111
–	Dynamic thermography: Analysis of hand temperature during exercise [50]	Zontak et al. (1998)	106
–	Skin temperature during running—a study using infra-red colour thermography [37]	Clark et al. (1977)	76

Relevancy was defined by the number of citations (data from Scopus until February 2025)

fields, the percentage of published articles is much smaller, but no less important, as will be shown in the different chapters of this book. It is important to note that only papers that are focused on sport science were included in the methodology and review categories.

Finally, the five most relevant papers that used infrared thermography in sport science will be presented (Table 1.1). Relevancy was defined as the papers that have been cited the most times. We have also kept in the table some publications that were among the most cited in the first edition of this book in 2016. These publications are discussed below.

The paper of Hildebrandt et al. [3] is an overview of the technological advances of infrared thermography in sport medicine and its technical requirements. It presents the physical principles of the technique and a study of the intra-examiner reproducibility of knee temperature. It is very interesting that this chapter presents case studies to illustrate the clinical applicability and limitations of infrared thermography in sport medicine. The relevancy of this study could be explained because it is a complete publication about the application of thermography in sport medicine, which shows the principles, the methodology of the technique, and different images and examples that demonstrate the potential use of infrared thermography in this field (Fig. 1.14).

The article "Thermographic imaging in sports and exercise medicine: A Delphi study and consensus statement on the measurement of human skin temperature" by Moreira et al. [46] employed the Delphi method, in which a panel of 24 experts in the application of infrared thermography in sports science, physiology, physiotherapy, and medicine from 13 different countries participated to develop standardized guidelines. The result was the "Thermographic Imaging in Sports and Exercise Medicine" (TISEM) checklist, a set of 15 standardized items covering participant demographics, environmental and camera setup, and data recording and analysis. The significance of this study lies in its effort to standardize thermography protocols in sports and clinical settings. The checklist is intended to help researchers and practitioners reduce methodological biases, enhance data accuracy, and ensure

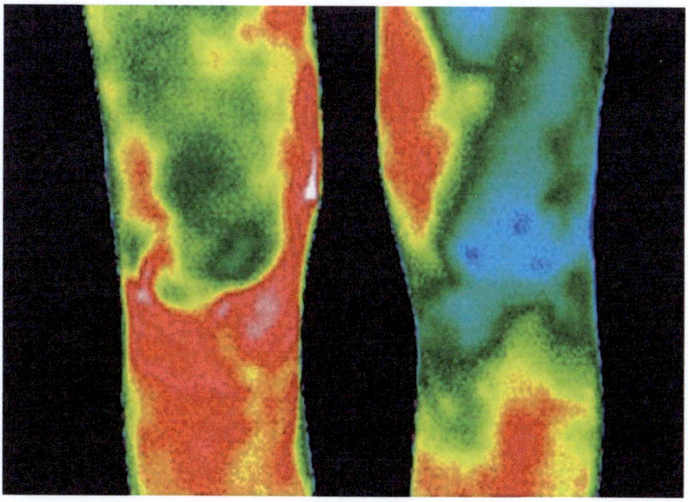

Fig. 1.14 Example of an injury (anterior cruciate ligament rupture in the right knee) presented in the article of Hildebrandt [3]

consistency in thermographic studies. The fact that in approximately 7 years this paper has achieved 286 citations shows the great applicability of the TISEM checklist.

The third most cited study was performed by Merla and colleagues in 2010, titled "Thermal Imaging of Cutaneous Temperature Modifications in Runners During Graded Exercise" [47]. In this study, they assessed the skin temperature dynamics of 15 participants during an incremental test until maximal heart rate. They observed a decrease in skin temperature during all the incremental tests and an increment until basal values in the recovery (Fig. 1.15). The main strength of this study was that it used the new generation of infrared thermography cameras, with a higher resolution and sensitivity. This study could probably be considered to be one of the first studies with this new generation of infrared cameras, which were more powerful and also cheaper. It may attract the attention of a lot of sport scientists in terms of the possibilities of the infrared camera in sport science.

The article "Changes in Eye Temperature and Stress Assessment in Horses During Show Jumping Competitions" by Valera et al. [48] measured eye temperature using infrared thermography and salivary cortisol levels in 16 Spanish Sport Horses over three days of competition. Measurements were taken three hours before, immediately after, and three hours after each competition event. The findings showed that eye temperature increased significantly immediately after competition and returned to baseline after three hours, whereas salivary cortisol levels did not show significant differences across phases. The researchers found a strong correlation between eye temperature and cortisol levels in specific time intervals, suggesting that infrared thermography can effectively detect acute stress responses in horses. The study results support the potential of thermography as a practical tool

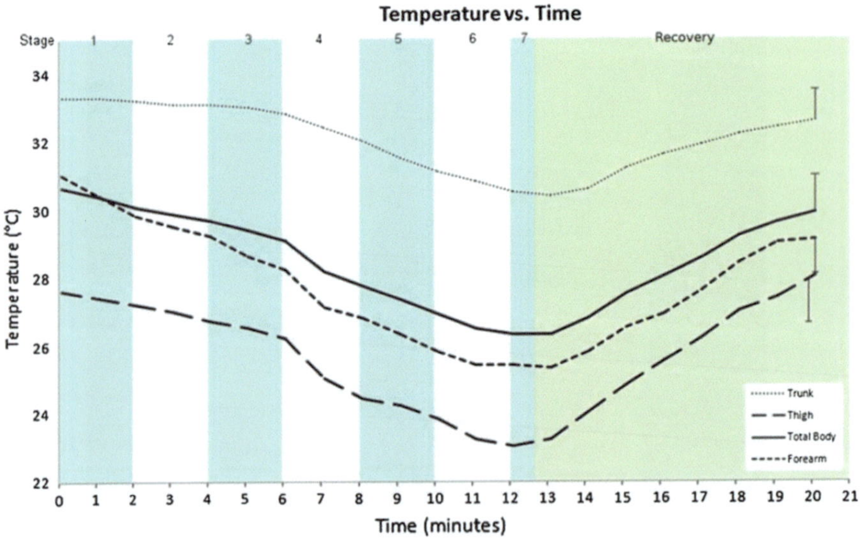

Fig. 1.15 Results obtained in the study of Merla et al. (Figure obtained from Merla et al. [47])

for monitoring horse welfare in competitive environments, allowing for better stress management and performance optimization.

The fifth most relevant article is "Thermal imaging of exercise-associated skin temperature changes in trained and untrained female subjects" [49] that examined the effects of physical training on thermoregulation by comparing the skin temperature responses of trained and untrained female participants during localized exercise (standing heel raise for 2 min). The results showed that trained athletes exhibited a significantly faster increase in skin temperature compared to untrained participants, indicating a more efficient thermoregulatory response. This study is significant as it highlights infrared thermography as a valuable tool for assessing thermoregulation and vascular adaptations to training.

Two papers are included, which were among the most referenced in the first edition of this book. Firstly, a relevant piece of research was the study performed by Zontak et al. [50], which had the objective of characterizing skin temperature response to exercise (bicycle ergometry) using infrared thermography. They decided to measure the skin temperature of the hand in 10 participants because this area has a high skin blood innervation, resulting in a high variation of the skin temperature due to exercise (Fig. 1.16a). Two experimental protocols were assessed, an incremental workload and a stable workload. In the incremental workload, they observed a decrease in skin temperature during all exercise; and in the stable workload, a decrease in the skin temperature was first observed with a posterior increase in the skin temperature in the middle of the test (Fig. 1.16b). The authors suggested that these results depend on the vasodilator/vasoconstrictor balance of the skin blood flow, which depends mainly on the blood requirements for the active muscles and the skin (heat loss requirements). This study is very interesting and relevant due to this discussion regarding the effect of the vasodilation and vasoconstriction on skin temperature.

Finally, the publication of Clark et al. [37], which was previously commented on, is also considered highly relevant. Although the main limitation of this study is that it was performed with only two participants, its strengths are (1) it was the first study written in English with infrared thermography in sport science; (2) the study deals with different aspects such as the effect of wind, the comparison of infrared thermography with thermocouples, or the comparison between performing exercise in a laboratory or outdoors; and (3) the study discusses many aspects, from the temperature distribution to the effect of the curvature of the body or sweat on thermal data. These seven studies provide a quick overview of the application of infrared thermography in sport science. More specifically, these studies are a good example of the three categories in sport science with the highest number of papers, as noted in Fig. 1.12: thermophysiology, sport medicine, and animals and sports.

However, a question may be called to mind after this review: What is the most recent application of infrared thermography in sport science? To investigate this, we decided to select and discuss four studies published in 2024 (Table 1.2).

The first study to comment on is the publication performed by our research group, titled "Effect of different volumes of exercise on skin temperature responses over the following 24 hours" [51]. This study aimed to determine whether

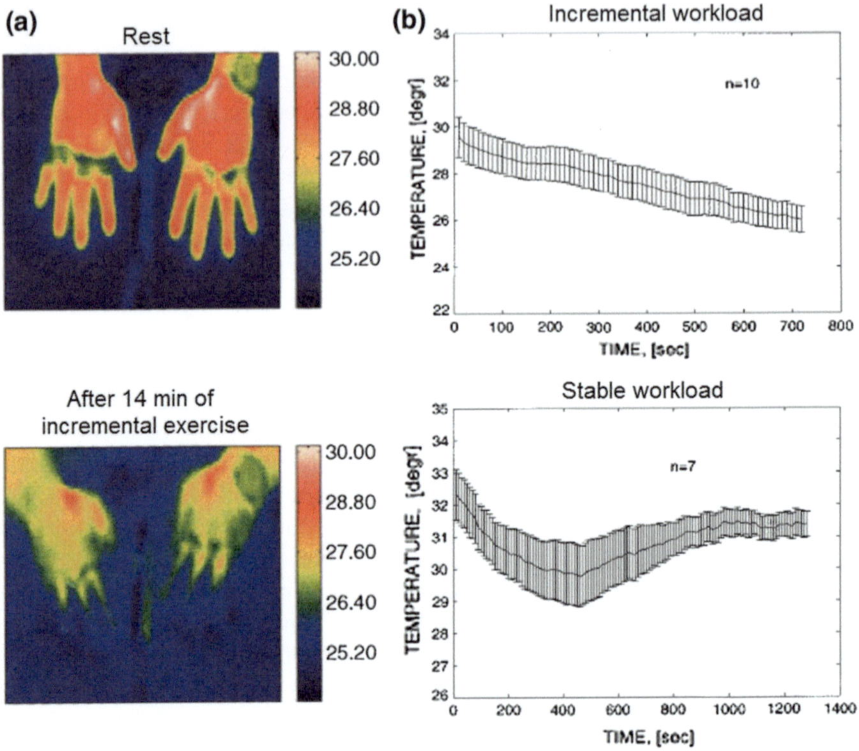

Fig. 1.16 (**a**) Example of hand thermographies from the study of Zontak et al. and (**b**) graphs of the skin temperature dynamics in their experiments. (Figure adapted from the figures of Zontak et al. [50])

Table 1.2 Four selected publications of 2024 using infrared thermography in sport cience

Title	Authors	Journal
Effect of different volumes of exercise on skin temperature responses over the following 24 h [51]	Da Silva et al.	Journal of thermal biology
Physical activity and temperature changes of Asian elephants (*Elephas maximus*) participating in eco-tourism activities and elephant polo [52]	Tilley et al.	PloS one
Firefighter uncompensable heat stress results in excessive upper body temperatures measured by infrared thermography: Implications for cooling strategies [53]	Coehoorn et al.	Applied ergonomics
Knee temperature remains abnormal in patients successfully treated with anterior cruciate ligament reconstruction: An infrared thermography analysis [54]	De Marziani et al.	Journal of experimental Orthopaedics

exercise-induced muscle damage and the timing of temperature assessments influence skin temperature readings post-exercise. The study involved 24 trained male military personnel, divided into three groups: Control Group (no exercise), Exercise

Group (performed a military combat simulation without inducing muscle damage), and Exercise and Muscle Damage Group (completed the same combat simulation plus 5 sets of 20 drop jumps to induce muscle damage). The key finding of this study was that the Exercise Group increased their skin temperature obtaining higher skin temperatures 24 h post-exercise, but the Exercise and Muscle Damage Group and Control group had similar responses with no differences between pre-exercise and 24 h post, which suggested that muscle damage may suppress skin temperature increases (Fig. 1.17).

The article "Physical activity and temperature changes of Asian elephants (*Elephas maximus*) participating in eco-tourism activities and elephant polo" by Tilley et al. [52] investigates the thermoregulatory responses of captive Asian elephants engaging in different physical activities (Fig. 1.18). The research was conducted on ten female elephants housed in an eco-tourism facility in Nepal, with six participating in an elephant polo tournament and four engaging in more familiar eco-tourism activities such as walking and grass cutting. Thermal images were captured before and after activities, and the results revealed no significant differences in overall skin temperature responses between polo-playing elephants and those engaged in eco-tourism activities. However, significant temperature variations were observed across different body regions, with the ear pinna showing the most pronounced changes, reinforcing its role as a key thermal window for heat dissipation. This study is particularly relevant because it demonstrates that elephants are capable

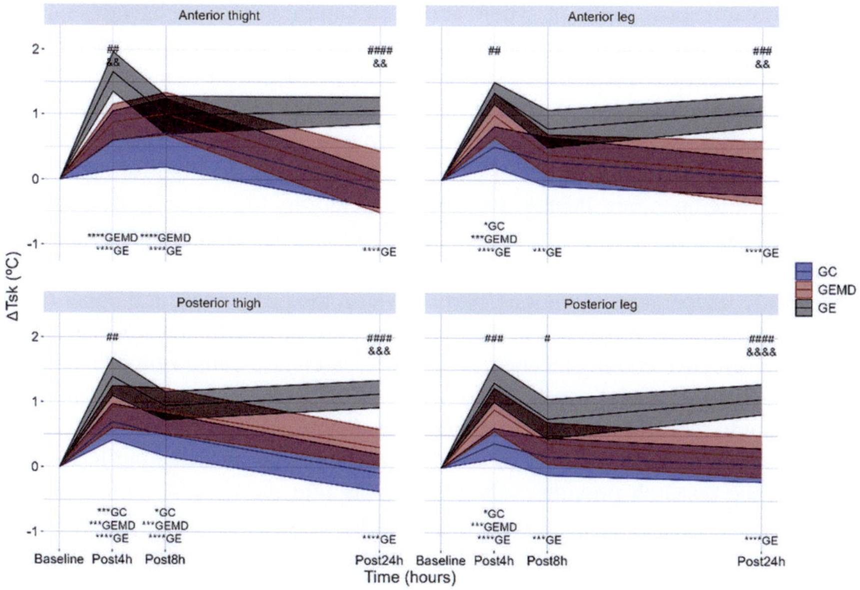

Fig. 1.17 Skin temperature variation at different time moments for three different groups assessed. GC control group, GEMD group exercise and muscle damage, GE group exercise. (Figure obtained from Da Silva et al. [51])

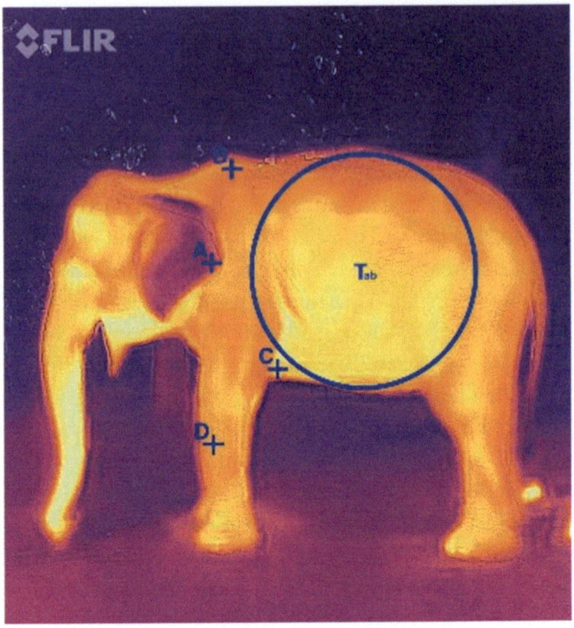

Fig. 1.18 Example of an elephant thermography. (Figure obtained from Tilley et al. [52])

of regulating their skin temperature effectively during physical exertion, regardless of activity type. However, it also raises concerns about potential thermoregulatory limitations in captive settings, where access to natural cooling behaviors, such as bathing and shade-seeking, might be restricted because tourism and human activities are prioritized.

The article "Firefighter uncompensable heat stress results in excessive upper body temperatures measured by infrared thermography: Implications for cooling strategies" explores the potential of infrared thermography as a tool for monitoring upper body temperature during firefighting operations [53]. The study analyzed upper body skin temperature responses before, during, and after simulated fire suppression activities. The results indicated that firefighters experienced significant upper body temperature elevations, particularly in the head, neck, and upper torso, due to the insulating properties of their protective clothing, which restricts evaporative and convective cooling mechanisms. The research also demonstrated that traditional cooling strategies, such as passive rest in a cool environment, may be insufficient to rapidly reduce skin temperature and restore thermal homeostasis. By demonstrating the effectiveness of infrared thermography in detecting heat stress patterns, the research provides valuable insights into the possible application of infrared thermography in the monitoring of firefighters.

Finally, the article "Knee temperature remains abnormal in patients successfully treated with anterior cruciate ligament reconstruction: an infrared thermography analysis" explores the long-term skin temperature responses in the knee following anterior cruciate ligament reconstruction, in baseline and after exercise conditions (single-leg squat). The study aimed to determine whether individuals who had undergone anterior cruciate ligament reconstruction exhibited persistent alterations in knee skin temperature, compared to their uninjured knee and to a control group with no history of knee injury. The results revealed that the operated knee consistently displayed temperature asymmetries when compared to the uninjured side before and during exercise. Specifically, the reconstructed knee tended to be either significantly warmer or cooler, depending on factors such as underlying inflammation, vascular alterations, or changes in metabolic activity within the joint. These findings suggest that although patients achieved functional recovery, their persistent thermal asymmetries may indicate subclinical inflammation or altered blood flow, which could influence long-term joint health and risk of osteoarthritis development.

1.5 Conclusions

Infrared thermography began with the discovery of infrared radiation in 1800 by Sir William Herschel, the development of infrared sensors between 1880 and 1917, and, subsequently, the development of infrared cameras in the 1940s and 1950s. After this, infrared thermography started to be used in human assessment and, more specifically, in medicine. It is possible to consider the 1970s as the beginning of infrared thermography in sport science; since then, the number of publications has been exponentially increasing, with the number of studies published between 2021 and 2025 double the amount published in the whole previous decade. Almost all of the research has been performed in three kinds of applications: thermophysiology, sport medicine, and animals and sports. However, infrared thermography presents other possible applications such as the sport clothing assessment/design, or the improvement of the technique and the analysis for sport science, among others. In conclusion, the application of infrared thermography in sport science presents an interesting past, an important present application, and a more promising future application.

References

1. Jones BF (1998) A reappraisal of the use of infrared thermal image analysis in medicine. IEEE Trans Med Imaging 17:1019–1027. https://doi.org/10.1109/42.746635
2. Vardasca R, Simoes R (2013) Current issues in medical thermography. In: Topics in medical image processing and computational vision. Springer, Berlin, pp 223–237
3. Hildebrandt C, Raschner C, Ammer K (2010) An overview of recent application of medical infrared thermography in sports medicine in Austria. Sensors 10:4700–4715
4. Ng E-K (2009) A review of thermography as promising non-invasive detection modality for breast tumor. Int J Therm Sci 48:849–859
5. Herschel W (1800) Experiments on the refrangibility of the invisible rays of the Sun. By William Herschel, LL. D.F.R.S. Philos Trans R Soc Lond 90:284–292

6. Ring EFJ (2006) The historical development of thermometry and thermal imaging in medicine. J Med Eng Technol 30:192–198. https://doi.org/10.1080/03091900600711332
7. Rogalski A (2002) Infrared detectors: an overview. Infrared Phys Technol 43:187–210
8. Ring FJ, Jones BF (2012) Historical development of thermometry and thermal imaging in medicine. In: Biomedical signals, imaging, and informatics. CRC Press. Taylor and Francis, Boca Ratón., EEUU, p 2–1, 2–6
9. Lahiri BB, Bagavathiappan S, Jayakumar T, Philip J (2012) Medical applications of infrared thermography: a review. Infrared Phys Technol 55:221–235
10. Hardy JD (1934) The radaition of heat from the human body. J Clin Invest 13:615–620
11. Czerny M (1929) Über Photographie im Ultraroten. Z Für Phys 53:1–12. https://doi.org/10.1007/BF01339378
12. Ring EFJ, Ammer K (2000) The technique of infrared imaging in medicine. Thermol Int 10:7–14
13. Ammer K (2008) The Glamorgan protocol for recording and evaluation of thermal images of the human body. Thermol Int 18:125–144
14. Skversky NJ, Herring AB, Baron RC (1964) Thermography in peripheral vascular diseases. Ann N Y Acad Sci 121:118–134. https://doi.org/10.1111/j.1749-6632.1964.tb13691.x
15. Birnbaum SJ, Kliot D (1964) Thermography—obstetrical applications. Ann N Y Acad Sci 121:209–222. https://doi.org/10.1111/j.1749-6632.1964.tb13697.x
16. Marinacci AA (1965) Thermography in the detection of non-neurological imitators of peripheral nerve complexes; a preliminary report of ischemic (anoxic) neuropathy. Bull Los Angel Neurol Soc 30:1–11
17. Ring EFJ, Ammer K (2012) Infrared thermal imaging in medicine. Physiol Meas 33:R33–R46. https://doi.org/10.1088/0967-3334/33/3/R33
18. Lawson R (1956) Implications of surface temperatures in the diagnosis of breast cancer. Can Med Assoc J 75:309–310
19. Silva LF, Santos AASMD, Bravo RS et al (2016) Hybrid analysis for indicating patients with breast cancer using temperature time series. Comput Methods Prog Biomed 130:142–153. https://doi.org/10.1016/j.cmpb.2016.03.002
20. Lawson R (1957) Thermography; a new tool in the investigation of breast lesions. Can Serv Med J 8:517
21. Vreugdenburg TD, Willis CD, Mundy L, Hiller JE (2013) A systematic review of elastography, electrical impedance scanning, and digital infrared thermography for breast cancer screening and diagnosis. Breast Cancer Res Treat 137:665–676
22. Jacob G, Jose I, S., S. (2023) Breast cancer detection: a comparative review on passive and active thermography. Infrared Phys Technol 134:104932. https://doi.org/10.1016/j.infrared.2023.104932
23. Torres-Galván JC, Guevara E, Kolosovas-Machuca ES, Oceguera-Villanueva A, Flores JL, González FJ (2022) Deep convolutional neural networks for classifying breast cancer using infrared thermography. Quant InfraRed Thermogr J 19(4):283–294. https://doi.org/10.1080/17686733.2021.1918514
24. Costello J, Stewart IB, Selfe J et al (2013) Use of thermal imaging in sports medicine research: a short report: short article. Int Sport Med J 14:94–98
25. Hildebrandt C, Zeilberger K, Ring EFJ, Raschner C (2012) The application of medical infrared thermography in sports medicine. Ultrasound 10:2
26. Vollmer M, Möllmann K-P (2011) Infrared thermal imaging: fundamentals, research and applications. Wiley, Weinheim
27. McManus C, Tanure CB, Peripolli V et al (2016) Infrared thermography in animal production: an overview. Comput Electron Agric 123:10–16. https://doi.org/10.1016/j.compag.2016.01.027
28. Venegas P, Ivorra E, Ortega M, de Ocáriz IS (2022) Towards the Automation of Infrared Thermography Inspections for Industrial Maintenance Applications. Sensors 22(2):2. https://doi.org/10.3390/s22020613

29. Jadin MS, Taib S (2012) Recent progress in diagnosing the reliability of electrical equipment by using infrared thermography. Infrared Phys Technol 55(4):236–245
30. Kim H, Lamichhane N, Kim C, Shrestha R (2023) Innovations in building diagnostics and condition monitoring: a comprehensive review of infrared thermography applications. Buildings 13(11):11. https://doi.org/10.3390/buildings13112829
31. Larbi Youcef MHA, Feuillet V, Ibos L, Candau Y (2022) In situ quantitative diagnosis of insulated building walls using passive infrared thermography. Quant InfraRed Thermogr J 19(1):41–69. https://doi.org/10.1080/17686733.2020.1805939
32. Zheng H, Gao G, Zhong X, Zhao L (2022) Monitoring and diagnostics of buildings' heat loss based on 3D IR model of multiple buildings. Energ Buildings 259:111889. https://doi.org/10.1016/j.enbuild.2022.111889
33. Fox M, Coley D, Goodhew S, De Wilde P (2015) Time-lapse thermography for building defect detection. Energ Buildings 92:95–106. https://doi.org/10.1016/j.enbuild.2015.01.021
34. Sedlák D, Straško B, Janiga M (2015) Thermal imaging in analysis of local climate in industrialized area—exploring new methodology. Oecologia Mont 24:74–77
35. Keyl W, Lenhart P (1975) Thermography in sport injuries and lesions of the locomotor system due to sport. Fortschr Med 93:124–126
36. Lelik F, Solymossy O, Kézy G (1977) Skin-thermography with fluid crystals in orthopedics and sport-medicine (author's transl). Z Für Orthop Ihre Grenzgeb 115:105–108
37. Clark RP, Mullan BJ, Pugh LG (1977) Skin temperature during running—a study using infrared colour thermography. J Physiol 267:53–62
38. Fournet D, Ross L, Voelcker T et al (2013) Body mapping of thermoregulatory and perceptual responses of males and females running in the cold. J Therm Biol 38:339–344. https://doi.org/10.1016/j.jtherbio.2013.04.005
39. Chudecka M, Lubkowska A (2015) Thermal maps of young women and men. Infrared Phys Technol 69:81–87. https://doi.org/10.1016/j.infrared.2015.01.012
40. Priego Quesada JI, Lucas-Cuevas AG, Salvador Palmer R et al (2016) Definition of the thermographic regions of interest in cycling by using a factor analysis. Infrared Phys Technol 75:180–186. https://doi.org/10.1016/j.infrared.2016.01.014
41. Ammer K (2009) Does neuromuscular thermography record nothing else but an infrared sympathetic skin response? Thermol Int 19:107–108
42. Priego Quesada JI, Martínez Guillamón N, Rosa M, de Anda CO et al (2015) Effect of perspiration on skin temperature measurements by infrared thermography and contact thermometry during aerobic cycling. Infrared Phys Technol 72:68–76. https://doi.org/10.1016/j.infrared.2015.07.008
43. Priego-Quesada JI, Machado AS, Gil-Calvo M, Jimenez-Perez I, Ortiz C, de Anda RM, Salvador Palmer R, Perez-Soriano P (2020) A methodology to assess the effect of sweat on infrared thermography data after running: preliminary study. Infrared Phys Technol 109:103382. https://doi.org/10.1016/j.infrared.2020.103382
44. de Andrade FA, dos Santos Amorim PR, Brito CJ et al (2014) Measuring skin temperature before, during and after exercise: a comparison of thermocouples and infrared thermography. Physiol Meas 35:189
45. James CA, Richardson AJ, Watt PW, Maxwell NS (2014) Reliability and validity of skin temperature measurement by telemetry thermistors and a thermal camera during exercise in the heat. J Therm Biol 45:141–149. https://doi.org/10.1016/j.jtherbio.2014.08.010
46. Moreira DG, Costello JT, Brito CJ, Adamczyk JG, Ammer K, Bach AJE, Costa CMA, Eglin C, Fernandes AA, Fernández-Cuevas I, Ferreira JJA, Formenti D, Fournet D, Havenith G, Howell K, Jung A, Kenny GP, Kolosovas-Machuca ES, Maley MJ et al (2017) Thermographic imaging in sports and exercise medicine: a Delphi study and consensus statement on the measurement of human skin temperature. J Therm Biol 69:155–162. https://doi.org/10.1016/j.jtherbio.2017.07.006

47. Merla A, Mattei PA, Di Donato L, Romani GL (2010) Thermal imaging of cutaneous temperature modifications in runners during graded exercise. Ann Biomed Eng 38:158–163. https://doi.org/10.1007/s10439-009-9809-8
48. Valera M, Bartolomé E, Sánchez MJ, Molina A, Cook N, Schaefer A (2012) Changes in eye temperature and stress assessment in horses during show jumping competitions. J Equine Vet Sci 32(12):827–830. https://doi.org/10.1016/j.jevs.2012.03.005
49. Formenti D, Ludwig N, Gargano M, Gondola M, Dellerma N, Caumo A, Alberti G (2013) Thermal imaging of exercise-associated skin temperature changes in trained and untrained female subjects. Ann Biomed Eng 41(4):863–871. https://doi.org/10.1007/s10439-012-0718-x
50. Zontak A, Sideman S, Verbitsky O, Beyar R (1998) Dynamic thermography: analysis of hand temperature during exercise. Ann Biomed Eng 26:988–993
51. Da Silva W, Godoy-López JR, Machado ÁS, Lemos AL, Sendra-Pérez C, Brejano MG, Carpes FP, Priego-Quesada JI (2024) Effect of different volumes of exercise on skin temperature responses over the following 24 hours. J Therm Biol 123:103923
52. Tilley HB, Murphy D, Wierucka K, Wong TC, Surreault-Châble A, Mumby HS (2024) Physical activity and temperature changes of Asian elephants (Elephas maximus) participating in ecotourism activities and elephant polo. PLoS One 19(5):e0300373. https://doi.org/10.1371/journal.pone.0300373
53. Coehoorn CJ, St. Martin P, Teran J, Cowart H, Waite L, Newman S (2024) Firefighter uncompensable heat stress results in excessive upper body temperatures measured by infrared thermography: implications for cooling strategies. Appl Ergon 120:104342. https://doi.org/10.1016/j.apergo.2024.104342
54. De Marziani L, Orazi S, Boffa A, Andriolo L, Di Martino A, Zaffagnini S, Filardo G (2024) Knee temperature remains abnormal in patients successfully treated with anterior cruciate ligament reconstruction: an infrared thermography analysis. J Exp Orthop 11(3):e70012. https://doi.org/10.1002/jeo2.70012

Chapter 2
Physics Principles of the Infrared Thermography and Heat Transfer

Jose Ignacio Priego-Quesada, Rosario Salvador-Palmer, and Rosa María Cibrián Ortiz de Anda

Abstract Although it is easy to capture a thermal image with infrared thermography, it is necessary to have the basic knowledge about how it works and the physical laws relating to it, as well as the radiative characteristics of the different bodies, and how heat is transferred between space and bodies. This knowledge is essential to establishing a logical hypothesis, using the camera rigorously, and interpreting the thermal data correctly. The aim of this chapter is to present the basic physical principles of infrared thermography and heat transfer.

2.1 Introduction

In recent years, infrared thermography has become more popular in sport science due to several applications, such as injury prevention and screening [1–3], performance assessment [4–6], and clothing assessment [7–9].

Many researchers have recently begun studies using infrared thermography in sport science for different reasons. First, it has many advantages: infrared thermography is a noninvasive method that can be used at a distance and does not interfere with human thermoregulation, unlike other skin measurement methods [1, 10, 11]. Second, it has great applicability, as will be shown in the different chapters of this

J. I. Priego-Quesada (✉)
Research Group in Sport Biomechanics (GIBD), Department of Physical Education and Sports, University of Valencia, Valencia, Spain

Biophysics and Medical Physics Group, Department of Physiology, University of Valencia, Valencia, Spain
e-mail: j.ignacio.priego@uv.es

R. Salvador-Palmer · Rosa María Cibrián Ortiz de Anda
Biophysics and Medical Physics Group, Department of Physiology, Universitat de València, Valencia, Spain
e-mail: rosario.salvador@uv.es; rosa.m.cibrian@uv.es

J. I. Priego-Quesada (ed.), *Application of Infrared Thermography in Sports Science*,
https://doi.org/10.1007/978-3-031-93311-0_2

book. And finally, another important reason is the decrease in price of infrared cameras in the last 20 years.

In the journal peer review process of different recent studies in sport science using infrared thermography performed by the authors of these chapters, we observed a high number of studies with important issues that make publication difficult. Most of these issues are related to methodological aspects or with establishing logical reasoning to support the study hypotheses and to interpret the results. In this context, it is necessary that new users of infrared thermography in sport science have the necessary physics background to understand the technique, in order to avoid methodological issues. In addition, it is important to always remember that infrared thermography is commonly used to calculate surface temperatures, and, in this sense, knowledge about heat transference and thermophysiology is necessary to establish logical hypotheses and to correctly interpret the thermal results.

This chapter will present the basic physical knowledge of infrared thermography and heat transference. Because this book is not oriented toward physical scientists, all aspects will be discussed in a simple way in order to provide scientists in sport science the minimum knowledge necessary for their study. In order to study the physical aspects in more detail, it is therefore recommended to use other sources. Likewise, in this chapter, thermoregulatory aspects will be displayed briefly related to heat transfer; Chapter 3 will explain human thermoregulation in more detail, and other chapters will apply to the different topics.

2.2 Heat Transfer

First of all, it is important to define the concept of heat and temperature. Heat is energy in transit that passes from a warm body to a cold body, and its unit is the joule (J). J/s or Watt is the common unit used to measure heat flow per unit time. A hot body decreases its internal energy, and a cold body increases due to heat transfer. On the other hand, the temperature is a measure of the internal energy of bodies. Temperature can be defined as the "measure" of the average kinetic energy of one body [12]. There are different scales for measuring temperature, and the most well known are the degree Celsius, Kelvin, and degree Fahrenheit. The degree Celsius is used by most countries in the world. The degree Fahrenheit is used in the Anglo-Saxon world, and it can be changed to the Celsius scale using the following expression:

$$^{\circ}\mathrm{C} = \left(^{\circ}\mathrm{F} - 32\right) * \frac{5}{9} \tag{2.1}$$

Equation 2.1 Fahrenheit to Celsius formula

However, in the international system of units, the temperature is expressed using the Kelvin scale (K). This scale is called absolute, because it is not possible, by imperative of the laws of thermodynamics, to have bodies with lower values than

Fig. 2.1 Heat flow: the heat is transferred from the hotter side to the cooler

0 K (−273.16 °C). The Kelvin scale is used in all aspects related to the physics of thermography.

The term heat, therefore, should be understood as heat transfer and only occurs when there is a temperature difference between two bodies. Therefore, the heat is always transferred following a gradient, from the hotter body to the cooler. This transference of energy is called heat flow or heat flux (Fig. 2.1). When two bodies are at the same temperature, they are in thermal equilibrium [13]. In addition, when one body has a constant temperature, it is considered that it is in a thermal equilibrium, and it could be a result of the environment being at the same temperature as the body, or because it has a constant supply of heat that keeps it at a constant temperature even though the environment is different (as in the case of the core temperature of the human body).

Heat flow is an important concept in the human body. As will be shown in the next chapters, the thermal gradient between the core and the skin and between the skin and the environment is important in the rate of heat transference. For example, the greater the difference between core temperature and skin temperature, the easier the transference from the core to the skin. On the contrary, if both temperatures are similar, the heat dissipation from the core will be more costly [14, 15]. Another example might be the temperature of the skin and the environment. In hot environments where the air temperature is similar to or warmer than the skin, the body increases temperature due to the heat transfer by radiation.

Furthermore, the transference of heat is commonly explained by three mechanisms: conduction, convection, and radiation. The three mechanisms are explained below, putting them in the context of how they occur in the human body.

Conduction: Heat transfer by the contact of two solid bodies at different temperatures. The amount of heat transferred by conduction is given by Fourier's law. This law states that the rate of heat conduction through a body per unit of cross-sectional area is proportional to the temperature gradient existing in the body, being the proportionality constant material conductivity.

In the human body, the conduction transference is between the different structures of the body, and also between the human body and clothing. Regarding heat transference through the human body tissues, it is a slow process; in the limbs, it is mainly dependent on the temperature gradient between the muscle and skin and the thermal conductivity of muscle [16]. For this reason, the transference of heat by conduction is facilitated when the temperature gradient between the muscle and the skin is increased, occurring mainly by sweat evaporation during exercise or during exercise in cold environments [14, 16]. Breathability and insulation properties of clothing will have an effect on this type of heat transfer. However, conduction heat loss is usually considered negligible unless the skin is in contact with highly conductive surfaces for a prolonged duration [17].

Convection: Heat transfer by the contact of a solid body with a liquid or gaseous element. The convective heat exchange between the body and the fluid is given by Newton's cooling law and depends on the temperature gradient and surface contact area of both, and the convective heat transfer coefficient (determined by the density, heat capacity, thermal conductivity, and velocity of the fluid) [17].

In the human body, this convection could be explained by the heat dissipation via the blood flow; when blood flows through the core, it is heated, and that blood in turn heats the skin as it passes through. Convection also explains the effect of the wind on the skin. Wind convection also facilitates sweat evaporation [17]. This heat transfer system is very important when the body is immersed in water. In this setting, it is considered that 100% of heat loss occurs via convection due to the contact between the skin and the water. In the sports world, it is of interest to consider clothing insulation because it can promote or reduce convective heat loss.

Radiation: Heat transferred by the body is by the emission of electromagnetic radiation. All of the objects with temperature above absolute zero (0 K or −273.16 °C) emit electromagnetic radiation according to the Stefan–Boltzmann law (see Sect. 2.4), which indicates that the energy radiated is proportional to the fourth power of its temperature. Furthermore, objects absorb electromagnetic energy emitted by the environment. Since the emission of energy implies a decrease in internal energy and thus the temperature of the object, if the object maintains a constant temperature (thermal equilibrium), it is because the energy emitted is compensated by the energy absorbed, but if a body is at a higher or lower temperature than the environment, the net radiative heat transfer is the difference between the absorbed and the emitted radiation.

Commonly, the human body emits more heat radiation than it absorbs, due to the environment being cooler. However, in close proximity to warm objects or hot environments (e.g., a fire), this radiation is absorbed by the body, resulting in the body heating [13]. During outdoor exercise, the most common source of radiation is the sun [17].

Radiation is heat transfer that detects thermography, while conduction and convection are important for proper thermography protocols. Furthermore, conduction and convection could affect skin temperature, and therefore human radiation.

To the three heat transfer processes detailed (convection, conduction, and radiation), it is necessary to add human body thermoregulation in terms of the sweat evaporation and respiration. Sweat evaporation is considered to be the main mechanism of heat dissipation during exercise. Evaporation results in a temperature decrease of the skin surface [1]. It is important to understand that it is the evaporation of sweat that is responsible for the temperature decrease and not the sweat production itself. Sweat rate is related to the ambient temperature, exercise intensity, and absolute heat production [17]. In addition, the amount of evaporation is inversely proportional to the relative humidity of the air [13]. In an environment with 100% relative humidity, humidity could be condensed in the body, resulting in a heat gain, at the same time as heat is lost by evaporation [13]. During respiration (specifically in expiration), heat is lost to the environment via convection and evaporation [17]. Heat loss by respiration is higher in cold and dry environments; however, there is a low contribution of respiration in whole-body heat loss [17].

All these discussed processes are shown graphically in Fig. 2.2. Following the first law of thermodynamics, the heat balance equation is commonly used in thermoregulation studies, where heat production is on one side of the equation and heat loss is on the other side. More specifically, this equation is defined as follows:

$$M - W = (K + C + R + ESK) + S \tag{2.2}$$

Equation 2.2 Heat balance equation, where M is the rate of metabolic heat production, W is the rate of mechanical work, K is the rate of conductive heat loss, C is the rate of convective heat loss from the skin, R is the rate of radiative heat loss from the skin, ESK is the rate of evaporative heat loss from the skin, and S is the rate of body heat storage

All the concepts from Fig. 2.2 that have been discussed are expressed in the heat balance equation. However, there is one concept that we have not yet discussed, and that is heat production. Heat production is the difference between metabolic heat production and the rate of mechanical work (M-W). Metabolic heat production was defined by Cramer and Jay [17] as "the rate of free energy released from the catabolism of carbohydrate, fat, and amino acids to resupply adenosine triphosphate (ATP) for cellular activities such as biosynthesis, transport, and muscular contractions."

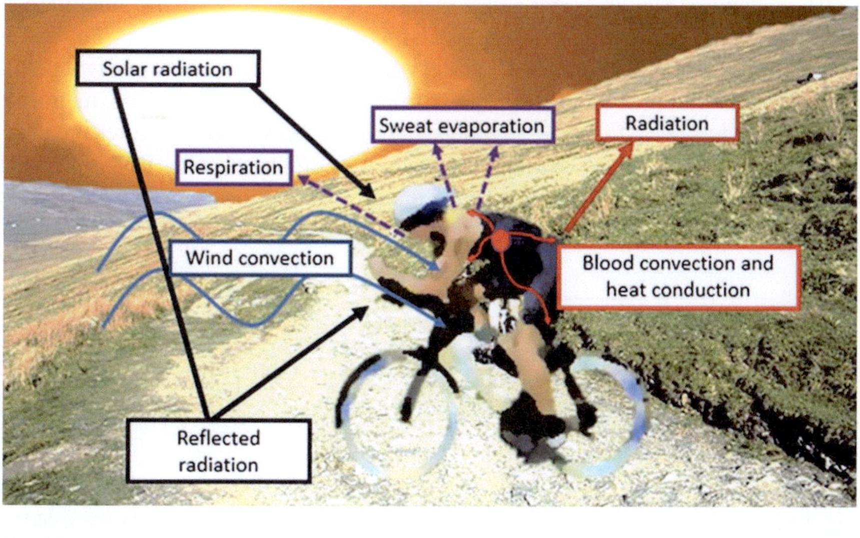

Fig. 2.2 Heat exchange between the body and the environment during exercise

Some of this energy is converted into external work (W) and the rest is converted into heat [17]. For this reason, heat production is the difference between all the energy produced (M) and the energy used for external work (W). The body is very inefficient in transforming the energy into mechanical work, and between 30% and 70% of the energy produced, depending on activities, results in thermal energy [16]. Heat production is usually calculated by the combination of ergometers, which measure the work, and indirect calorimetric measurements (gas exchange) that measure metabolic heat production [17]. Exercise results in an increment in heat production that the body needs to dissipate. We can exemplify how heat production increases with the intensity of the activity using different values of heat production. The heat production of sleeping, walking, cycling at 250 W, and running at 16 km/h are 1.0, 4.0, 13.3, and 20.0 W/kg, respectively.

As will be shown in the following section, the emitted radiation of the human body temperature is in the range of infrared radiation in the electromagnetic spectrum and is called thermal radiation.

2.3 Electromagnetic Spectrum and Infrared Radiation

Electromagnetic radiation is a way of energy propagation through a vacuum or a material medium without mass transport. An electromagnetic wave can be defined as the propagation of the vibration of an electric field (E) and magnetic field (B), which together form an angle of 90° (Fig. 2.3). The propagation direction is perpendicular to both the plane of vibration of the electric field and the vibration of the magnetic field. It is therefore a transverse wave whose propagation velocity in the medium is 2.99792458 * 10^8 m/s (this value is usually rounded to 3.10^8 m/s).

The quantities that characterize the electromagnetic waves are the amplitude of the electric field and the magnetic field and the frequency of the wave. The wavelength (λ) is the spatial period of the wave, which is the distance between two points with identical vibration state (from pulse to pulse), measured in meters, micrometers, nanometers, etc. (Fig. 2.3). Frequency (f) is a quantity that measures the number of repetitions of the wave per unit time, and its unit is s^{-1} (Hz). Both quantities are related through the fundamental equation of wave motion:

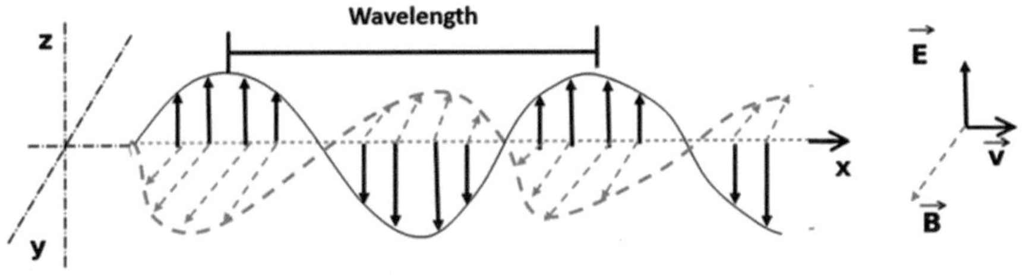

Fig. 2.3 Representation of an electromagnetic wave with the vibration of the electric field (E), magnetic field (B), and the direction of the propagation of the wave

$$\lambda = \frac{c}{f} \tag{2.3}$$

Equation 2.3 Relationship between the wavelength and the frequency, where λ is the wavelength, c is the electromagnetic wave velocity (speed of light = 3.10^8 m/s), and f is the frequency

It can be observed in (2.3) that when the frequency increases, the wavelength decreases.

The electromagnetic spectrum comes from the concept that electromagnetic radiation has different characteristics depending on its wavelength and frequency. The electromagnetic spectrum is a representation of the frequency distribution of all the electromagnetic waves (Fig. 2.4). Regarding a body, the electromagnetic spectrum is all the electromagnetic radiation that is emitted (emission spectrum) or absorbed (absorption spectrum). The electromagnetic spectrum extends from the radiation of lower energy (or longer wavelength), such as the radio waves, the microwaves, the infrared rays, or the visible light, to electromagnetic radiation with higher energy (and shorter wavelength), such as ultraviolet light (UV), X rays, and gamma rays. Most waves are commonly known to everyone. For example, X-rays are known for their medical applicability. UV is known due to the ozone hole, and it is widely known that a high incidence of this radiation increases the likelihood of skin cancer. The remaining types of radiation, such as the microwave or radio waves, and their applications are also known to everyone.

For the different types of radiation, different imaging techniques were developed in the medical field, such as the X-ray or the arthroscopy [1] (Fig. 2.5).

The radiation that will be the focus of this chapter is infrared radiation. This radiation is also known as thermal radiation, and its wavelength range is from

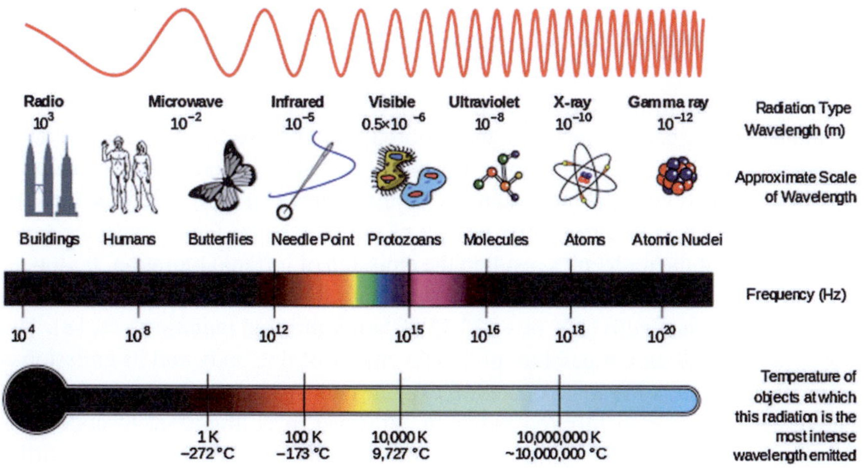

Fig. 2.4 Representation of the electromagnetic spectrum. (Figure modified from the free media repository of Wikimedia Commons)

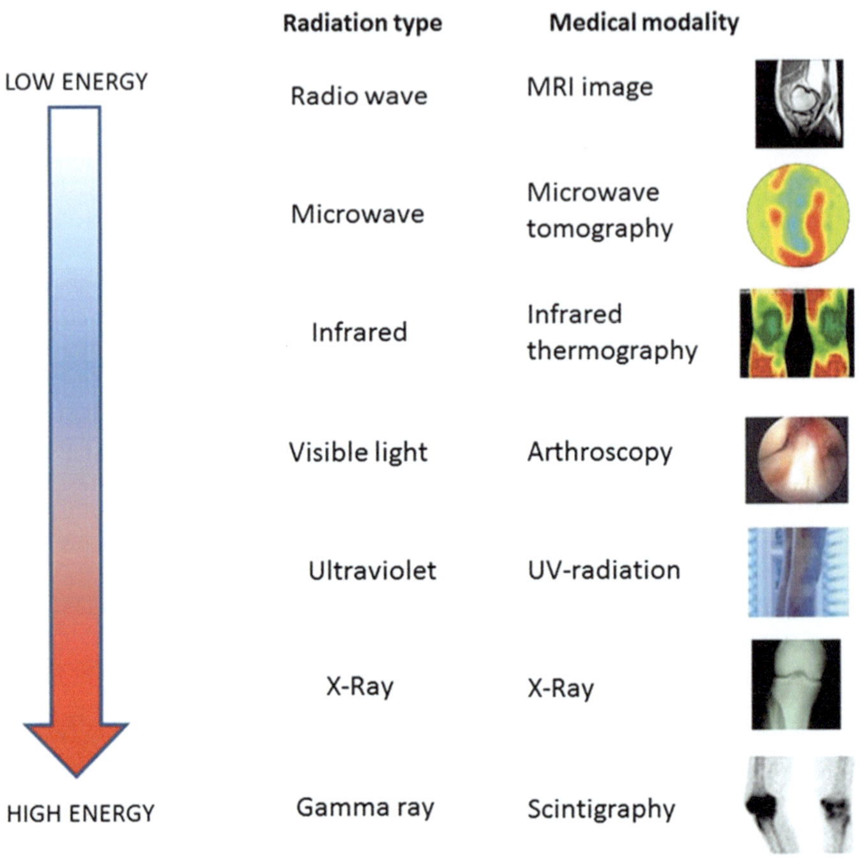

Fig. 2.5 Medical imaging techniques and their relation to the electromagnetic spectrum. (Figure modified from Hildebrandt et al. [1])

760 nm to 1 mm. Infrared radiation is known as thermal radiation because there is a relationship between temperature and infrared radiation. In Sect. 2.2, temperature was defined as the average kinetic energy of one body [12]. In other words, temperature is associated with the movement of the molecules in a body. More heat means more movement and more kinetic energy of the molecules. Changes in the speed and agitation of the molecules result in the emission of infrared radiation. Following on from this concept, it is easier to understand that all bodies that present a temperature above absolute zero (0 K or −273.15 °C) emit infrared radiation [13, 18]. This relationship between temperature or kinetic energy of one body and its emission of infrared radiation is the first concept to explain why it is possible to calculate temperatures from emitted infrared radiation. However, it is important to understand that bodies with very high temperature (more than 1000 K) are capable of emitting other radiations (visible and even UV), and bodies are heated during the absorption of their radiation. However, in bodies in which the temperature is close to the

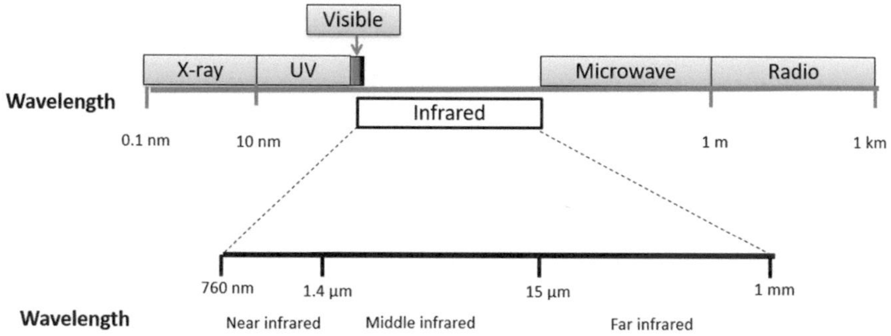

Fig. 2.6 Electromagnetic spectrum focusing on the infrared radiation spectrum

environmental temperature, the energy radiated is from the infrared wavelength of the electromagnetic spectrum [19].

On the other hand, infrared spectrum is usually classified in three ranges: near infrared, middle infrared, and far infrared (Fig. 2.6). The main reason for this division is that objects emit more radiation in one region than in others. For example, the human body emits most of its radiation in the infrared of long wavelength (8–15 μm) [1, 20]. However, different gases and other elements can be transparent in the far infrared but are visible in the near or middle infrared. Because of this, there are specific cameras to view specific wavelengths [21, 22]. Cameras with detectors for long-wavelength infrared are the most commonly used because they register most bodies, including humans. However, there are also infrared cameras for infrared of short and medium wavelengths, which are much more expensive but they are capable of visualizing specific gases (Fig. 2.7). There is also one specific window of the middle infrared (between 5 and 7.5 μm) that is rarely used because it is the region where the atmosphere absorbs infrared radiation, and the air is opaque for infrared cameras [19].

Regarding infrared radiation, infrared thermography was developed. The next section is focused on the laws governing the emission of infrared radiation by bodies.

2.4 Physical Laws of Infrared Radiation

The physical laws of infrared radiation and their relationship to temperature are based on the theory of the black body. The concept of the black body was introduced by Gustav Kirchhoff in 1860. A black body is a theoretical object that absorbs all incident electromagnetic radiation, and its absorption is equal to its emission. None of the incident radiation is reflected or passes through the black body. The black body also has a uniform surface and a uniform temperature. The black body is therefore an ideal, perfect emitter of infrared radiation. Three properties are determined for the idea of the black body [21]:

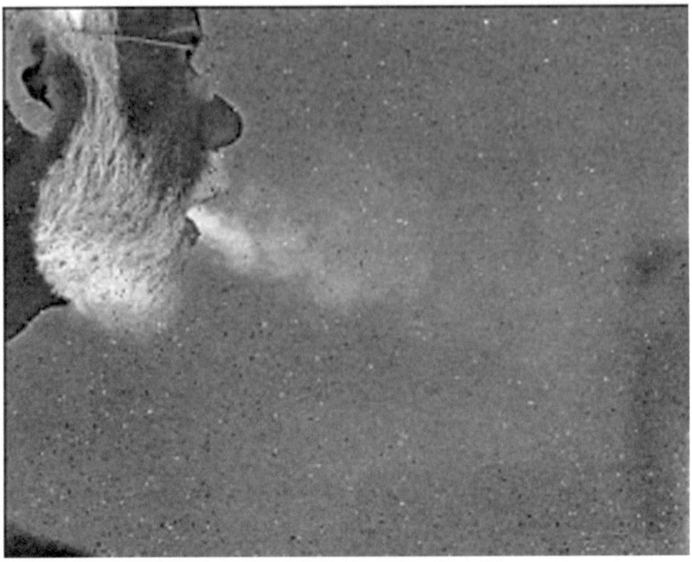

Fig. 2.7 Infrared camera with a detector for visualizing the middle infrared, and then the CO_2 exhaled in the respiration. (Figure extracted from Walther and Weimann [23])

1. A black body absorbs all the incident radiation, independently of the wavelength and the direction of the radiation.
2. For a specific temperature and wavelength, there is no surface that can emit more energy than a black body.
3. The emitted radiation of a black body depends on the wavelength; however, its radiance does not depend on the direction, it behaves like a Lambertian radiator (an object in which the surface has the same radiance from any angle).

Originally, black bodies were accomplished by a hollow sphere, covered internally with an insulating surface (in matte black) with a small aperture (Fig. 2.8a). This small aperture acts as a black body because any radiation that penetrates through it undergoes multiple reflections inside the cavity, each of which is partially absorbed, such that this radiation is completely extinguished before finding the exit. Currently, there are electronic instruments known as black body sources that are instruments with a stable and accurate temperature surface, with a high emissivity, that are configured with a specific temperature emission (Fig. 2.8b).

The quantity characterizing the emission of electromagnetic radiation by a body is called the total emissive power, E; that is, the energy emitted per unit of time and surface. Because this total emissive power is not equal in all the wavelengths (λ), there is also the spectral emissive power, E_λ, which is the energy emitted by the unit of surface, time, and range of wavelength.

The graphical representation of the experimental curves corresponding to the spectral emissive power of the black body for each wavelength at different temperatures is shown in Fig. 2.9. This figure shows how the body emits radiation at

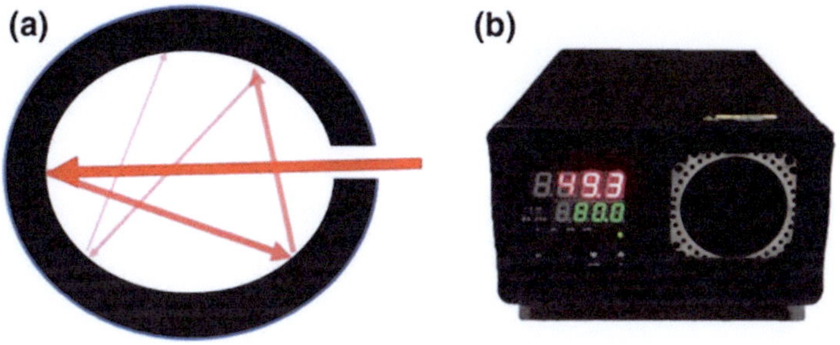

Fig. 2.8 (**a**) Origin of the concept of realization of a black body. (**b**) Current electronic black body source

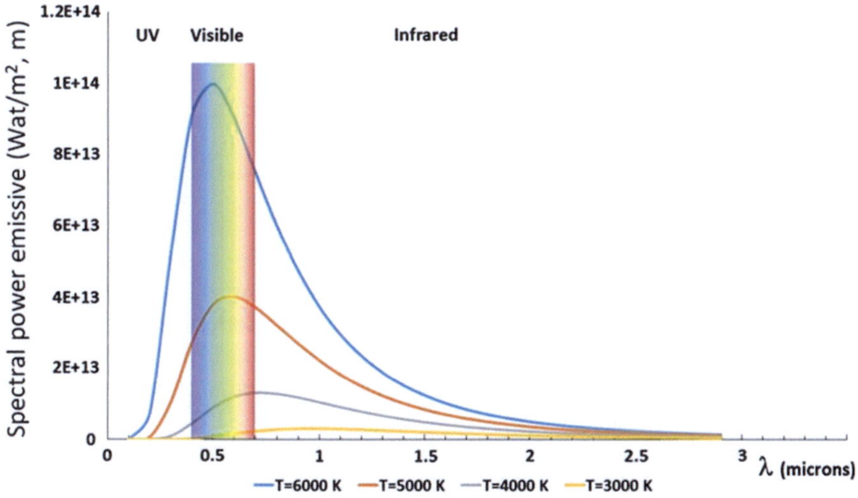

Fig. 2.9 Spectral emissive power of the black body for each wavelength at different temperatures

different frequencies according to its temperature. For example, the sun, because it is at a temperature between 5000 and 6000 K, emits its radiation in the ultraviolet, visible and infrared spectrum; however, its emission peak is in the range of visible radiation (526 nm, corresponding to the green–yellow colors of the visible light). Humans visualize the sunlight as white because this emission peak is at the center of the visible radiation [21]. As the temperature is lower in the body, its emission is higher in the infrared region. A black body with a temperature close to the room temperature does not emit in the visible range and emits all its radiations as infrared radiation [13] (Fig. 2.10). Thus, it is understood that for the 37 °C of human body (370 K), the maximum emitted energy is 9.3 μm, and therefore all its radiation is infrared radiation; more specifically, the 90% of the emitted infrared radiation is in the infrared of long wavelength [1, 20] (Fig. 2.10).

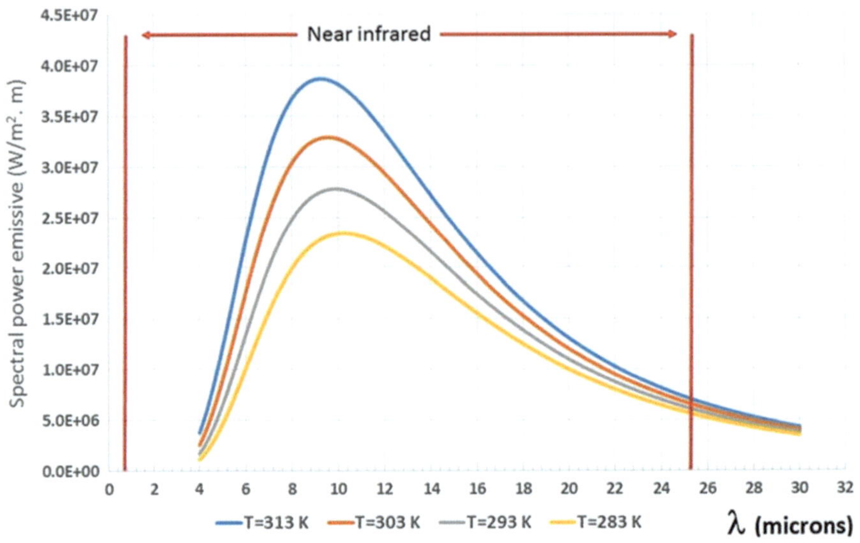

Fig. 2.10 Spectral emissive power of the black body at the following temperatures: 313 K (40 °C), 303 K (30°), 293 K (20°), and 283 K (10°). At these temperatures, all spectral emissive power are inside the infrared spectrum

Theoretical interpretation of these curves was not performed until 1900, representing a drastic change in the conception of physics, from classical physics to modern physics. It was also considered to be a pioneering result of modern physics and quantum theory [21]. However, the laws that govern thermography were experimentally established much earlier.

Wien's displacement law was established by Wilhelm Wien in 1893. This physical law established that the wavelength of the peak of the blackbody radiation curve decreases as the body temperature is increased. The equation of Wien's displacement law is

$$\lambda_{\max} = \frac{a}{T} \tag{2.4}$$

Equation 2.4 Wien's displacement law, where $\lambda_{\max}$ is the wavelength of emission peak in meters, a is Wien's displacement constant ($2.897 * 10^{-3}$ m K), and T is the absolute temperature in Kelvin

Representation of Wien's displacement law in the experimental curves corresponding to the spectral emissive power of the black body for each wavelength at different temperatures is shown in Fig. 2.11.

Stefan–Boltzmann's law was deduced by Josef Stefan in 1879 on the basis of experimental measurements made by John Tyndall and was derived from theoretical considerations by Ludwig Boltzmann in 1884. Stefan–Boltzmann's law expresses

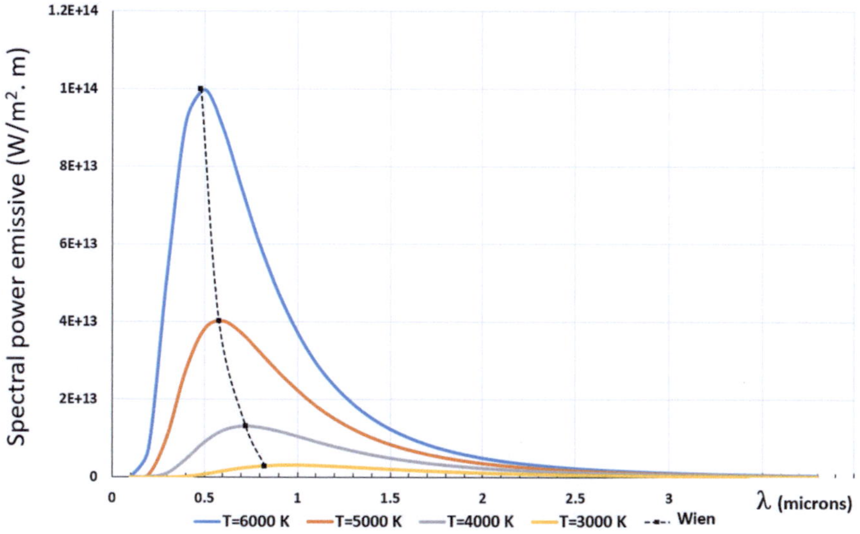

Fig. 2.11 Representation of Wien's displacement law at different temperatures of the black body

that the total emissive power or radiated energy from a black body is proportional to the fourth power of its absolute temperature:

$$E = \sigma * T^4 \tag{2.5}$$

Equation 2.5 Stefan–Boltzmann's law, where E is the total emissive power (W/m^2), σ is the Stefan–Boltzmann's constant ($5.67 * 10^{-8}$ W/m^{-2} K^{-4}), and T is the temperature in Kelvin

Small changes in temperature resulted in big changes in the emissive power. This is really easy to see in Stefan–Boltzmann's equation because the temperature is expressed as the fourth power. This relationship is very important because it explains how the calculation of the temperature from the emissive power is very sensitive and allows differentiation of areas at different temperatures, as they will have different emissive powers. It can be considered that the Stefan–Boltzmann's law is the fundamental law that governs infrared thermography.

For real surfaces, the Stefan–Boltzmann's equation is modified with the incorporation of emissivity (2.6). The concept of emissivity and also the consequence of this parameter in the equation will be explained in the next section. Most of the infrared cameras and radiation thermometers made the calculations using the modified Stefan–Boltzmann's equation [13]:

$$E_0 = \varepsilon * \sigma * T^4 \tag{2.6}$$

Equation 2.6 Modified Stefan–Boltzmann's Eq. (2.5), where the value of the emissivity (ε) is incorporated

Wien's displacement law and Stefan–Boltzmann law were obtained experimentally before 1900. However, these laws were corroborated by the law established by Max Planck in 1900:

$$E_{\lambda}(\lambda,T)=\frac{2hc^{2}}{\lambda^{5}}\frac{1}{e^{\frac{hc}{\lambda kT}}-1} \tag{2.7}$$

Equation 2.7 Planck's law, where $E_{\lambda}(\lambda, T)$ is the spectral radiance as a function of temperature and wavelength, λ is the wavelength of emission in meters, T is the temperature in Kelvin, h is the Planck constant ($6.6256 * 10^{-34}$ Js), c is the speed of light in the medium ($3 * 10^{8}$ ms^{-1}), and k is the Boltzmann's constant ($1.38054 * 10^{-23}$ W s K^{-1})

Integrating Planck's law for all frequencies leads to Stefan–Boltzmann's law [18]. Applying the function maximum condition to the Planck equation (first derivative equal to zero), Wien's displacement law is obtained.

After knowing the physical laws that explain the calculation of the temperature using infrared radiation, it is important to understand the radiative characteristics of one body in terms of its absorptivity, emissivity, reflectivity, and transmissivity.

2.5 Radiative Characteristics of the Bodies

Infrared radiation has a behavior very similar to visible radiation. For this reason, its propagation can be described in the same way as geometric optics [21]. In homogeneous materials, infrared radiation propagates in a straight line as if it were rays. Furthermore, the geometrical properties of reflection and refraction are also applied to infrared radiation, so that geometrical characteristics of the surfaces, such as uniform or roughened surfaces, have an effect on the emission and absorption of infrared radiation [21]. In the visualization of the human body, these considerations are important in understanding why the delimitation of the body in the image presents a lower estimation of the temperature. In this sense, the loss of infrared radiation in the outline of a body was identified as an important source of error (Fig. 2.12) [24–26].

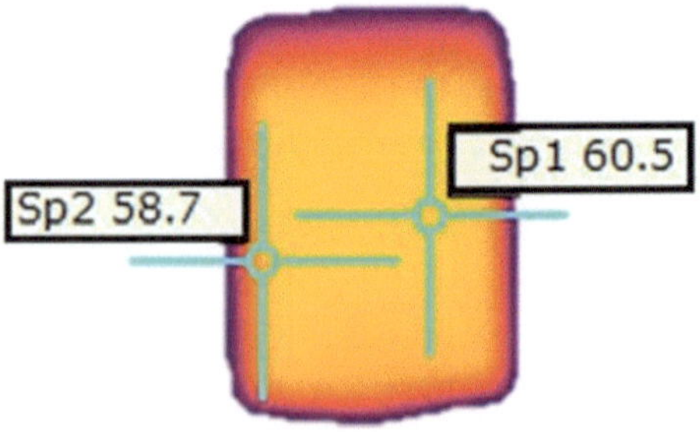

Fig. 2.12 Effect of curvature in the infrared emission and the temperature calculation of an object with a homogenous temperature surface

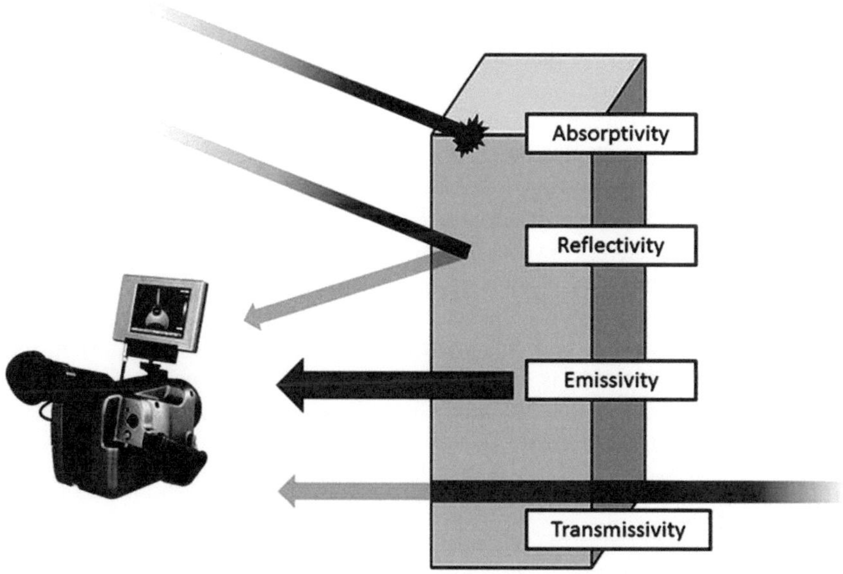

Fig. 2.13 Radiative characteristics of one body

Infrared radiation that arrives on a body can be absorbed, reflected, or pass through the body, depending on its physical characteristics (Fig. 2.13). These characteristics define the behavior of one body in relation to the infrared spectrum:

- Absorptivity (α): capacity of one body to absorb infrared radiation.
- Emissivity (ε): capacity of one body to emit its own infrared radiation. This parameter corresponds to the absorptivity ($\alpha = \varepsilon$).
- Reflectivity (ρ): capacity of one body to reflect the infrared radiation that comes from the environment.
- Transmissivity (τ): capacity of one body to let pass through it the radiation. If radiation is not transmitted through the body, the body is therefore called opaque.

The proportion of these components in a body is expressed by the following equations:

$$\alpha + \rho + \tau = 1 \tag{2.8}$$

Equation 2.8 Destinations of the incident radiation in a body: absorbed (α), reflected (ρ), or transmitted radiation (τ)

$$\varepsilon + \rho + \tau = 1 \tag{2.9}$$

Equation 2.9 Sources of the outgoing radiation of a body: emitted (ε), reflected (ρ), or transmitted radiation (τ)

As we discussed in the previous section, a black body is an ideal with values of emissivity and absorptivity equal to 1, and without reflectivity and transmissivity, and is therefore considered to be the perfect infrared emitter surface.

In the calculation of the temperature of a body, it is essential, in order to have an accurate calculation, that the body presents a zero transmissivity and an emissivity as close as possible to 1. Bodies with high values of reflectivity will result in temperature calculations with many errors because most of the outgoing radiation comes from the environment.

Behavior of different bodies can be different in the visible region compared with the infrared. And for this reason, it is essential to know the radiative characteristics of the bodies. One important example is water. In the visible region, water is more transparent and reflective. However, in the infrared, the characteristics of water are different. In the infrared, the water is opaque, with a higher emissivity than 0.9 [27]. If we do not know the radiative characteristics at the beginning of a thermography swimming study, we will probably want to measure while the swimmer is submerged in water, but this is impossible because of the opacity of the water (Fig. 2.14a). Another example is glass. Glass is transparent in visible light, but is opaque in infrared. Therefore, we cannot see anything that is behind glass (Fig. 2.14b). An example of this is that the infrared cameras used at airports as a

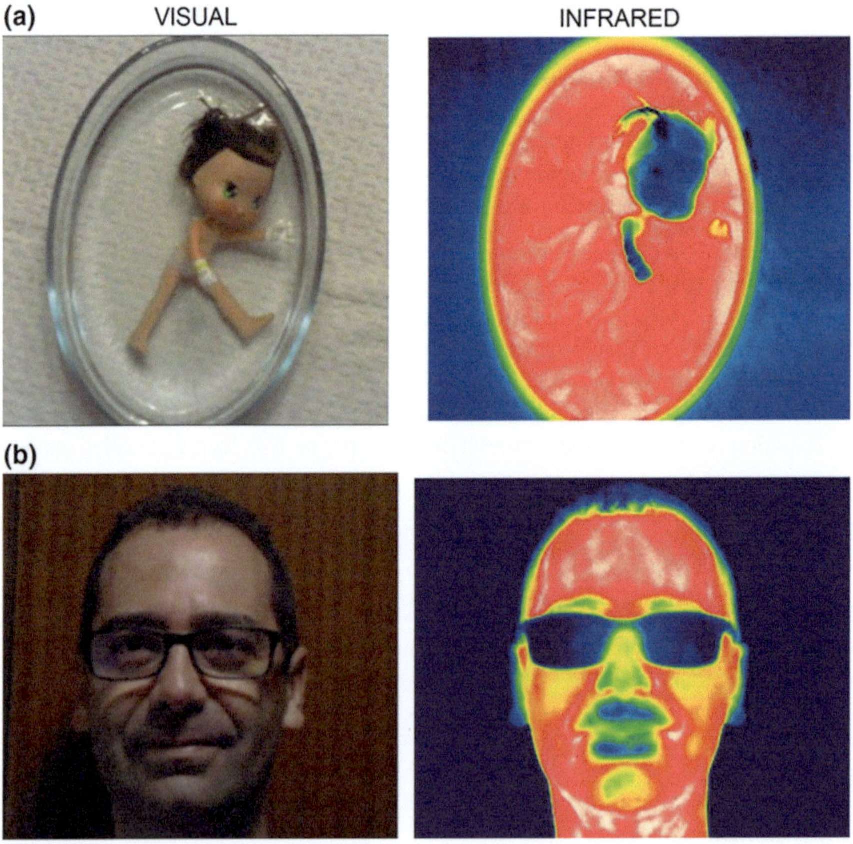

Fig. 2.14 Examples of the different behavior of some objects in the infrared and visible spectrum. Infrared opacity of water (**a**) and glasses (**b**)

fever screening method require glasses to be removed, in order to measure the temperature of the inner canthi of the eyes.

In this sense, knowing the emissivity of a body is of vital importance. Emissivity is a number ranging from 0 to 1, and it can be interpreted as the ratio of the actual amount of infrared energy emitted compared with the theoretically perfect amount that could be emitted (black body) [28]. This means that one object with an emissivity value of 0.7 emits only the 70% of the maximum theoretical amount of infrared energy, and the other 30% of outgoing radiation comes from the energy reflected by the body from other sources. It must therefore be taken into account that the energy that reaches the camera corresponds to both the energy emitted by the body and the energy reflected by it from the environment (Eq. 2.10) to properly determine the temperature of the analyzed body:

$$E = \varepsilon \sigma T^4 4 + (1 - \varepsilon) \sigma T_r^4 \tag{2.10}$$

Equation 2.10 Energy that reaches the thermographic camera corresponding to the energy emitted by the body due to its temperature T and the energy it reflects from the surrounding temperature T_r (reflected temperature)

The aim of this modification is to consider only the percentage of radiation from the body and to remove from the calculation the percentage of the reflected temperature. Table 2.1 presents the emissivity values of different materials.

In sport science, the most important "material" is usually the skin. Skin emissivity was determined with values between 0.97 and 0.99 with a standard deviation of 0.01 by different studies [29, 36, 37]. In scientific studies, an emissivity of 0.98 is commonly established. This means that the skin is a good emitter of infrared radiation, and it is possible to accurately calculate its temperature.

However, for the thermography assessment of sport equipment and garments, it is necessary to know the emissivity of their materials in order to measure their temperatures correctly. In this sense, two strategies could be considered. First, it is of

Table 2.1 Emissivity values of different materials

Material	Emissivity value
Skin	0.98 [29]
Enamel of human teeth	0.65 [30]
Cotton fabric	0.95 [31]
Water	~0.9 [27]
Aluminum (polished)	0.02 [32]
Black paper	0.9 [32]
Building materials: Plaster	0.57–0.84 [33]
Building materials: Porous stone	0.61–0.9 [33]
Building materials: Marble	0.68–0.94 [33]
Wood	0.85–0.91 [34, 35]
Cork	0.85 [34]
Granite	0.81 [34]

interest to review the scientific literature to determine whether previous studies have measured the emissivity of materials. The second strategy is to calculate the emissivity of a material or to check whether the emissivity measurement obtained is similar to that observed in the literature. The second strategy is always recommended because although it is possible to find the emissivity value of one material, this value can be different in a given equipment/garment due to a different composition or a different condition of the surface (e.g., polished, oxidized, or scaled). Although there are more accurate methods, a method that is easy to perform and therefore can be carried out in any laboratory will be explained. This method consists of the following steps:

1. Attach a material with a known emissivity to the object of interest. Tapes with a known emissivity (with emissivity values between 0.95 and 0.99) are commonly used.
2. Measure the reflected temperature (see Chap. 4 for the detailed procedure) and include it in the camera or software settings. Reflected temperature will affect the emissivity estimated, especially if there is a low value.
3. Heat the object of interest with the aim of having a higher contrast between the object and the environment. Ideally, this heat should be homogenous. For this reason, hot plates from laboratories are recommended.
4. Use the infrared camera to determine the temperature of the material with known emissivity.
5. Set the emissivity of the object of interest until obtaining the same temperature as the material with known emissivity. This is the estimation of the emissivity of the object of interest.
6. Repeat the experiment different times, at different temperatures, and with different samples of your object of interest, in order to obtain a robust average of the emissivity.

An example of this procedure was performed in the work of Jimenez-Perez et al. [38], who determined, after measuring 15 different footwear, that the overall emissivity of the running footwear sole was 0.97.

After all these physical concepts, it is possible to better understand how thermal imaging cameras can measure temperatures from the detection of infrared radiation. In addition, the importance of knowing the radiative characteristics of an object with the aim to appreciate whether it is possible to measure its temperature and the error during the measurement.

2.6 Conclusions

Infrared thermography researchers in sport science need to have knowledge about the physical principles of infrared thermography and heat transference. This knowledge is necessary in order to understand the operation of infrared camera, to avoid and to know methodological issues, to establish logical hypotheses, and to correctly interpret the thermal results. Different conclusions can be extracted from this chapter:

- All the bodies that present a temperature above absolute zero emit infrared radiation, and infrared cameras are capable of capturing this radiation. However, depending on the body, its emission is in a specific range of the infrared spectrum.

For some specific applications, such as the visualization of gases, special cameras are needed.

- Radiated energy from a black body is proportional to the fourth power of its absolute temperature. This relationship explains the high sensitivity of thermal cameras.
- It is necessary to know the radiative characteristics of a body in order to know if it is possible to calculate its temperature and to design a study without errors. In this sense, the estimation of the temperature will be more accurate in bodies with high emissivity values. Skin is a good emitter of infrared radiation (emissivity of 0.98), which is important in sport science studies, because it is commonly the "material of interest." However, for other materials, it is necessary to know their emissivity values in order to correctly calculate the temperature.
- It is necessary to understand heat transference mechanisms in thermography studies. More specifically, in human studies, it is necessary to understand the mechanisms involved in thermoregulation.

References

1. Hildebrandt C, Raschner C, Ammer K (2010) An overview of recent application of medical infrared thermography in sports medicine in Austria. Sensors 10:4700–4715
2. Escamilla-Galindo V-L, Felipe JL, Alonso-Callejo A, Van-der-Horst R, de la Torre-Combarros A, Minafra P, Fernández-Muñoz D, Fernández-Cuevas I (2024) Return-to-play criteria based on infrared thermography during anterior cruciate ligament rehabilitation in football players. Biol Sport 42(2):161–167. https://doi.org/10.5114/biolsport.2025.144295
3. Gómez-Carmona P, Fernández-Cuevas I, Sillero-Quintana M, Arnaiz-Lastras J, Navandar A (2020) Infrared thermography protocol on reducing the incidence of soccer injuries. J Sport Rehabil 1-6:1222. https://doi.org/10.1123/jsr.2019-0056
4. Chudecka M, Lubkowska A (2010) Temperature changes of selected body's surfaces of handball players in the course of training estimated by thermovision, and the study of the impact of physiological and morphological factors on the skin temperature. J Therm Biol 35:379–385
5. Duc S, Arfaoui A, Polidori G, Bertucci W (2015) Efficiency and thermography in cycling during a graded exercise test. J Exerc Sports Orthop 2:1–8
6. Aylwin P, Havenith G, Cardinale M, Lloyd A, Ihsan M, Taylor L, Adami PE, Alhammoud M, Alonso J-M, Bouscaren N, Buitrago S, Esh C, Gomez-Ezeiza J, Garrandes F, Labidi M, Lange G, Moussay S, Mtibaa K, Townsend N et al (2023) Thermoregulatory responses during road races in hot-humid conditions at the 2019 Athletics World Championships. J Appl Physiol 134(5):1300–1311. https://doi.org/10.1152/japplphysiol.00348.2022
7. Fournet D, Ross L, Voelcker T et al (2013) Body mapping of thermoregulatory and perceptual responses of males and females running in the cold. J Therm Biol 38:339–344. https://doi.org/10.1016/j.jtherbio.2013.04.005

8. Priego Quesada JI, Lucas-Cuevas AG, Gil-Calvo M et al (2015) Effects of graduated compression stockings on skin temperature after running. J Therm Biol 52:130–136. https://doi.org/10.1016/j.jtherbio.2015.06.005
9. Raccuglia M, Heyde C, Lloyd A, Hodder S, Havenith G (2019) The use of infrared thermal imaging to measure spatial and temporal sweat retention in clothing. Int J Biometeorol 63(7):885–894. https://doi.org/10.1007/s00484-019-01701-5
10. de Andrade FA, dos Santos Amorim PR, Brito CJ et al (2014) Measuring skin temperature before, during and after exercise: a comparison of thermocouples and infrared thermography. Physiol Meas 35:189
11. Priego Quesada JI, Martínez Guillamón N, Cibrián O, de Anda RM et al (2015) Effect of perspiration on skin temperature measurements by infrared thermography and contact thermometry during aerobic cycling. Infrared Phys Technol 72:68–76. https://doi.org/10.1016/j.infrared.2015.07.008
12. Parsons K (2002) Human thermal environments: the effects of hot, moderate, and cold environments on human health, comfort and performance, 2nd edn. CRC Press
13. Vardasca R, Simoes R (2013) Current issues in medical thermography. In: Topics in medical image processing and computational vision. Springer, Berlin, pp 223–237
14. Cuddy JS, Hailes WS, Ruby BC (2014) A reduced core to skin temperature gradient, not a critical core temperature, affects aerobic capacity in the heat. J Therm Biol 43:7–12. https://doi.org/10.1016/j.jtherbio.2014.04.002
15. Andrade MT, Nunes-Leite MMS, Bruzzi RS, Souza CH, Uendeles-Pinto JP, Prado LS, Soares DD, Gonçalves DAP, Coimbra CC, Wanner SP (2023) Predicting the body core temperature of recreational athletes at the end of a 10 km self-paced run under environmental heat stress. Exp Physiol 108(6):852–864. https://doi.org/10.1113/EP091017
16. González-Alonso J (2012) Human thermoregulation and the cardiovascular system. Exp Physiol 97:340–346. https://doi.org/10.1113/expphysiol.2011.058701
17. Cramer MN, Jay O (2016) Biophysical aspects of human thermoregulation during heat stress. Auton Neurosci Basic Clin 196:3–13. https://doi.org/10.1016/j.autneu.2016.03.001
18. Jones BF (1998) A reappraisal of the use of infrared thermal image analysis in medicine. IEEE Trans Med Imaging 17:1019–1027. https://doi.org/10.1109/42.746635
19. Astarita T, Carlomagno GM (2012) Infrared thermography for thermo-fluid-dynamics. Springer, Berlin
20. Ng E-K (2009) A review of thermography as promising non-invasive detection modality for breast tumor. Int J Therm Sci 48:849–859
21. Vollmer M, Möllmann K-P (2011) Infrared thermal imaging: fundamentals, research and applications. Wiley, Weinheim
22. Blunck DL (2022) Review: applications of infrared thermography for studying flows with participating media. Exp Thermal Fluid Sci 130:110502. https://doi.org/10.1016/j.expthermflusci.2021.110502
23. Walther M, Weimann G (2006) Infrared imaging with InAs/GaSb type-II superlattices. Phys Status Solidi A 203:3545–3549. https://doi.org/10.1002/pssa.200622385
24. Watmough DJ, Fowler PW, Oliver R (1970) The thermal scanning of a curved isothermal surface: implications for clinical thermography. Phys Med Biol 15:1. https://doi.org/10.1088/0031-9155/15/1/301
25. Salter DC (1976) The effect of obliquity in clinical thermograms. Phys Med Biol 21:980
26. Ammer K (2008) The Glamorgan Protocol for recording and evaluation of thermal images of the human body. Thermol Int 18:125–144
27. Mitchell HJ, Salvaggio C (2003) MWIR and LWIR spectral signatures of water and associated materials. In: Proceedings of SPIE image exploit. Target recognition, algorithms multispectral hyperspectral ultraspectral image, vol IX, pp 195–205
28. Bernard V, Staffa E, Mornstein V, Bourek A (2013) Infrared camera assessment of skin surface temperature—effect of emissivity. Phys Med 29:583–591. https://doi.org/10.1016/j.ejmp.2012.09.003

29. Steketee J (1973) Spectral emissivity of skin and pericardium. Phys Med Biol 18:686
30. Kells BE, Kennedy JG, Biagioni PA, Lamey PJ (2000) Computerized infrared thermographic imaging and pulpal blood flow: part 1. A protocol for thermal imaging of human teeth. Int Endod J 33:442–447
31. Carr WW, Sarma DS, Johnson MR et al (1997) Infrared absorption studies of fabrics. Text Res J 67:725–738
32. Marinetti S, Cesaratto PG (2012) Emissivity estimation for accurate quantitative thermography. NDT E Int 51:127–134. https://doi.org/10.1016/j.ndteint.2012.06.001
33. Avdelidis NP, Moropoulou A (2003) Emissivity considerations in building thermography. Energ Buildings 35:663–667. https://doi.org/10.1016/S0378-7788(02)00210-4
34. Barreira E, Almeida RMSF, Simões ML (2021) Emissivity of building materials for infrared measurements. Sensors 21(6):6. https://doi.org/10.3390/s21061961
35. Pitarma R, Crisóstomo J (2021) Determination of wood emissivity using active infrared thermography. In: Ksibi M et al (eds) Recent advances in environmental science from the Euro-Mediterranean and Surrounding Regions, EMCEI 2019. Environmental science and engineering, 2nd edn. Springer, Cham. https://doi.org/10.1007/978-3-030-51210-1_65
36. Togawa T (1989) Non-contact skin emissivity: measurement from reflectance using step change in ambient radiation temperature. Clin Phys Physiol Meas 10:39
37. Sanchez-Marin FJ, Calixto-Carrera S, Villaseñor-Mora C (2009) Novel approach to assess the emissivity of the human skin. J Biomed Opt 14.:24006–24006–6. https://doi.org/10.1117/1.3086612
38. Jimenez-Perez I, Gil-Calvo M, Salvador-Palmer R, Ortiz C, de Anda RM, Pérez-Soriano P, Priego-Quesada JI (2021) Footwear outsole temperature may be more related to plantar pressure during a prolonged run than foot temperature. Physiol Meas 42(7):074004. https://doi.org/10.1088/1361-6579/ac0fbe

Chapter 3
Foundations of Human Thermoregulation and Skin Temperature

Aaron J. E. Bach, Fergus K. O'Connor, and Robert D. Meade

Abstract The skin plays a crucial role as both sentinel and effector, enabling rapid vasomotor adjustments and powerful sweating responses, and maintenance of homeostasis during both heat and cold stress. The dynamic interplay between core and skin temperatures, along with short- and long-term adaptations, helps humans thrive in diverse environments. Understanding the role of the skin in thermoregulation informs clinical and performance strategies, particularly as climate change challenges human thermal adaptation.

3.1 Introduction

The goal of this chapter is to provide a brief overview of the mechanisms supporting human body temperature regulation, with an emphasis on skin temperature throughout. Body temperature regulation is coordinated by the body's thermoregulatory system, the primary goal of which is to maintain body temperature within the relatively narrow range required optimal enzyme function and cellular integrity and to prevent damage to tissues and organ systems [1, 2]. Basal internal (or "core") temperature is generally in the range of 36.5–37.5 °C, though actual values vary between and within individuals [2, 3]. Thermal stress, whether stemming from an increase in heat gain due to exercise (which increases internal heat generation) or exposure to a hot environment, or due to an elevation in heat loss from cold exposure or water immersion, can cause internal body temperature to stray from basal levels. In response, the thermoregulatory system initiates powerful autonomic and

A. J. E. Bach (✉) · F. K. O'Connor
School of Health Sciences and Social Work, Griffith University, Southport, QLD, Australia
e-mail: a.bach@griffith.edu.au

R. D. Meade
Department of Epidemiology, T.H. Chan School of Public Health, Harvard University, Boston, MA, USA

J. I. Priego-Quesada (ed.), *Application of Infrared Thermography in Sports Science*,
https://doi.org/10.1007/978-3-031-93311-0_3

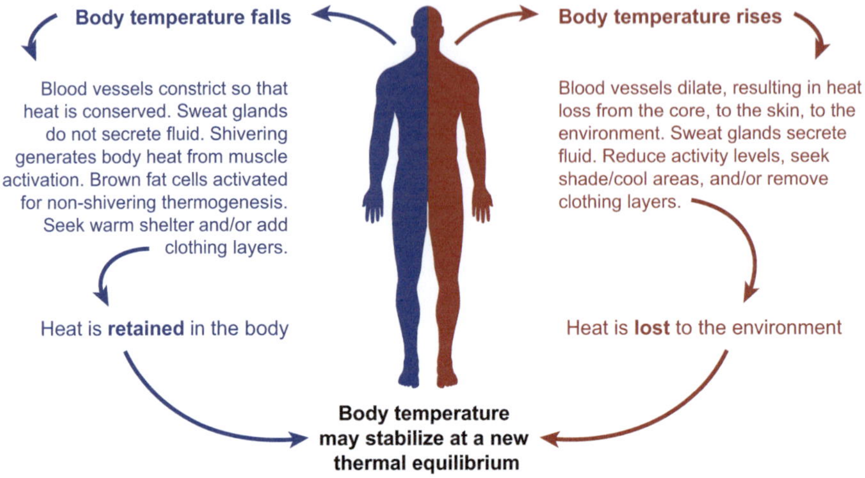

Fig. 3.1 Physiological and behavioural responses to heat imbalance

behavioural mechanisms to prevent further changes in body temperature to prevent organ damage and dysfunction [4, 5] (Fig. 3.1).

The thermoregulatory system is also highly malleable and adaptable. For example, regular heat exposure (e.g. repeated exercise in a hot environment) can induce physiological adaptations improving heat tolerance, including reductions in resting body temperature and earlier and more intense activation of heat loss responses [3, 6]. Dehydration and associated increases in extracellular osmolality cause reductions in sweat production during heat stress [7, 8] that can help to limit further water loss and thereby maintain effective circulating volume and blood pressure [9]. With illness, the thermoregulatory system can cause a controlled rise in core temperature (fever) to activate the immune system and inhibit the growth of pathogens [2]. All this to say that thermoregulation is a highly dynamic process, and its functioning may provide some insights into other physiological systems and states.

In line with the focus of this book, the skin plays an important role in body temperature regulation. The skin acts as the body's thermal interface with the environment and a key thermoregulatory organ, both for sensing temperature changes and controlling skin–environment heat exchange. Given its ease of access, it is also the site for measurements of temperature and provides important information on underlying thermal processes. However, it is important to note that as the interface, skin temperature can be highly variable and is influenced by many factors, including but not limited to morphology, internal temperature gradients (e.g. between the core and the skin) and pathophysiological processes (e.g. fever), skin–environment heat fluxes (both sensible and evaporative), and other environmental and contextual factors (e.g. clothing insulation). Understanding this complex interplay between skin and temperature regulation requires diving beneath the surface, quite literally, into the mechanisms that keep us running neither too hot nor too cold.

3.2 Mechanisms of Human Thermoregulation

The thermoregulatory system regulates body temperature through a complex negative feedback loop that acts to balance the rate of heat gain by the body's internal metabolic process with the rate of heat exchange between the skin and the environment. In this subsection, the passive physiological processes governing heat exchange along these pathways are introduced; later sections discuss how the thermoregulatory systems actively modify heat production and exchange to regulate body temperature.

Heat exchange between the body and the environment can be modelled using the human heat balance equation [1]:

$$S = (M - W) - (C_{\text{skin}} + R_{\text{skin}} + K_{\text{skin}}) - E_{\text{skin}} - (C_{\text{res}} + E_{\text{res}})\,\text{Watts or W·m}^{-2}$$

where S is the rate of body heat storage, $M - W$ is the rate of metabolic heat production (metabolic rate, M, minus external work, W), and C_{skin}, K_{skin}, R_{skin}, and E_{skin} are the rates of convective, conductive, radiative (together known as dry heat exchange), and evaporative heat loss from the skin. The rates of convective and evaporative heat loss from the respiratory tract, C_{skin} and E_{res}, are also shown. However, the role of respiratory heat exchange in thermoregulation is considerably smaller than that of the skin and will therefore not be discussed further. All components of the heat balance equation can be expressed in watts (W) or watts normalized to the body surface area (W·m^{-2}).

3.2.1 *Heat Production*

Humans are endothermic (or "warm-blooded") animals, maintaining a relatively constant core temperature despite perturbations in external conditions. Endothermy arises from a high metabolic rate [2]. Because the biochemical processes supporting cellular function are inefficient, most of the energy converted from carbohydrates, fats, and proteins to support cellular function is ultimately liberated as waste heat [1, 10, 11]. Resting metabolic rate in humans is ~60 W·m^{-2} [12], driven primarily by energy consumption in major organs, including the brain (~20%), liver (~20%), heart (~10%), and skeletal muscles (~20%) [13]. It is important to note here that the contribution of the brain, heart, and liver is due to their high mass-specific metabolic rates despite their small size; skeletal muscle, on the other hand, has a high overall mass but mass specific metabolic rate equivalent to only ~5% of those other organs [14]. During intense exercise, however, increases in muscular metabolism can result in rates of metabolic heat production as high as 1400 W·m^{-2} in highly trained individuals [15]. Metabolic heat production can also be increased during cold-defence, with metabolic rates as high as four times resting values observed during shivering [16].

3.2.2 Pathways of Skin–Environment Heat Exchange

3.2.2.1 Convection, Conduction, and Radiation (Dry Heat Exchange)

Convection, conduction, and radiation are collectively described as dry or sensible heat exchange [1, 10] (Fig. 3.2). Briefly, convection refers to heat transfer resulting from movement of a fluid over the skin (e.g. air or water), conduction is the heat transfer occurring between objects in direct contact (e.g. between the ground and soles of the feet), and radiation is the transfer of heat through electromagnetic radiation (e.g. between the sun and skin on a hot day). The rate of heat transfer through each of these processes occurs in proportion to the temperature gradient between the skin and the ambient environment [1, 10]. Skin temperature in thermoneutral conditions is ~32–33 °C; dry heat loss is therefore augmented by low cool objects placed in direct contact with the skin, or radiative heat sinks at sub-skin temperatures (e.g. the night sky on a clear night). As environmental temperatures increase, however, the skin–environment temperature gradient is narrowed, and the rate of dry heat gain is progressively attenuated, approaching 0 W when ambient and skin temperatures converge (at ~35–37 °C) [17]. In these situations, dry heat gain is experienced.

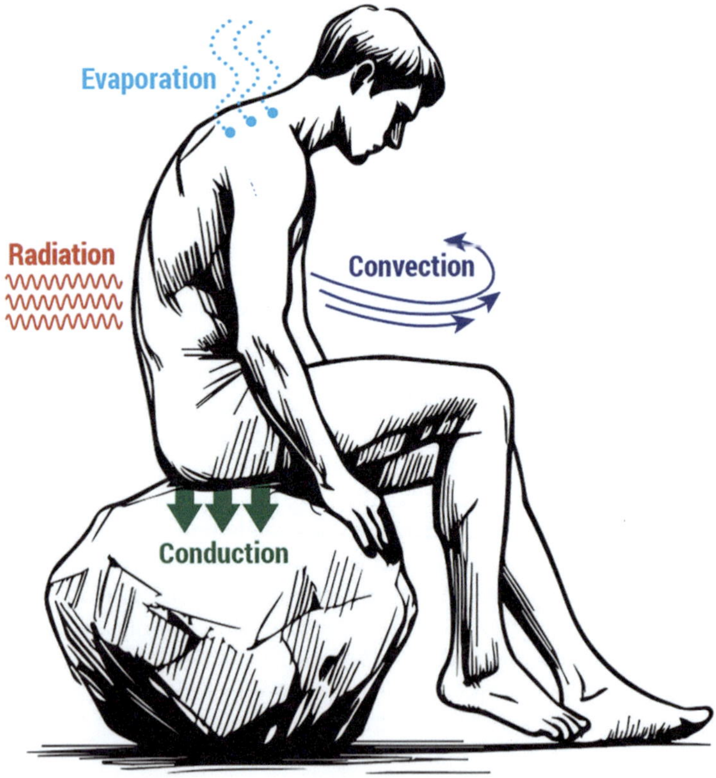

Fig. 3.2 Pathways for heat gain/loss in humans

It is important to note that the convention of grouping convection, conduction, and radiative heat exchange together stems from their dependence on the skin–environment temperature gradient. However, heat exchange along each of the three pathways is not necessarily similar and does not even have to flow in the same direction. For example, a person standing outdoors in light clothing on a clear spring day (~26–30 °C) may experience increased convective heat loss due to a light breeze. However, net dry heat loss might be kept at basal levels if the same person moves into direct sunlight. During exercise, these dry heat exchange mechanisms become particularly relevant as metabolic heat production increases substantially. For instance, the metabolic cost of work generates heat which can see core temperature elevate beyond 39 °C, creating steeper temperature gradients between the body and environment. However, the effectiveness of dry heat loss during exercise is largely dependent on environmental conditions; in hot environments where ambient temperatures approach skin temperature, these pathways may actually contribute to heat gain rather than heat loss, necessitating greater reliance on evaporative cooling mechanisms [17]. The influence of heat loss potential depending upon ambient conditions will be discussed further in subsequent sections.

3.2.2.2 Evaporative Heat Loss

Evaporative heat loss is driven by the evaporation of sweat (or other fluids) from the skin [18, 19]. Evaporative heat loss is also termed "insensible" heat loss since, unlike dry heat loss, it does not occur according to the skin–environment temperature gradient. Instead, evaporative heat loss is driven by the skin–environment water vapour pressure difference. This means that evaporative heat transfer can effectively contribute to heat loss in all but the hottest and humid environments (rarely experienced on earth). Given the high latent heat of sweat vapourization (~2426 J of heat removal per gram evaporated) and high maximal sweat rates (discussed below), sweating is the primary autonomic thermoeffector in humans [10]. However, increases in ambient moisture (humidity) will narrow the skin–environment water vapour pressure gradient, causing sweat to remain on the skin or drip off the body and provide no cooling benefit [20].

3.2.2.3 Temperature Sensation and Thermoregulatory Control

Disequilibrium between heat loss and heat gain, whether due to exposure to adverse thermal conditions (heat or cold), the performance of exercise, or some other mechanism, causes a change in body heat content and thereby tissue temperatures. It is the role of the afferent arm of the thermoregulatory system to sense such changes so that an appropriate response to establish heat balance can be mounted [21–23]. Temperature sensation is accomplished via an array of thermoreceptor neurons distributed throughout the body, including in the tissues of the central nervous system and skin [21, 24], and likely other areas such as the gastrointestinal track and skeletal muscles [25–27]. Interestingly, thermoreceptors found in the central nervous system are primarily warm sensitive, whereas the peripheral thermoreceptors (in the skin and deep body tissues) are mostly sensitive to cold [2]. The thermosensitivity of these neutrons is likely owed primarily to the presence of transient receptor potential (TRP) channels. Numerous TRP subtypes are involved in temperature sensation, including TRPM8 [28] and TRPM2 [29], and possibly TRPV1, TRPV3, and TRPV4 [30, 31], though evidence on the involvement of these latter sensors (or lack thereof) is inconclusive [2, 26]. Various other channels, including calcium-activated chloride channels and various sodium and potassium channels, may also play a role in temperature sensing [2]. Whatever the contributors, the wide variety of thermosensitive channels with overlapping response characteristics allows the thermoregulatory system to encode changes in temperature over a wide range.

Tissue temperature information coded by the various thermosensors is relayed to the central thermoregulatory centres within the central nervous system via afferent nerve tracks [2]. The preoptic anterior hypothalamus acts as the central regulatory centre, integrating feedback on the status of the body temperature and eliciting an appropriate response [32]. The functional architecture of the neural pathways involved in thermoregulation is beyond the scope of this chapter, but interested readers are referred to some of the many excellent resources on this topic [2, 33]. The important takeaway is that the afferent arm of the thermoregulatory system allows for constant monitoring of body temperatures which, when coupled with the efferent arm to be discussed in subsequent sections, allows for the optimization of heat loss responses at the skin in a bid to ensure the core temperature remains within the relatively narrow range required for homeostasis.

3.3 Skin's Role in Temperature Regulation

3.3.1 Structure and Function of the Skin

The skin's critical role in thermal balance is often overlooked, yet a stable core temperature depends heavily on the body's largest organ. Through a complex interplay of structural layers, specialized cells, and physiological feedback loops, the skin is the site for human heat exchange, insulation, and sensory input. This enables humans to withstand diverse climatic conditions and maintain core temperature within a narrow, optimal range [2, 3]. By effectively buffering the body against thermal extremes, the skin ensures that other organ systems can operate with precision and reliability.

3.3.2 Layers of the Skin and Their Thermal Properties

The skin is composed of three primary layers: 1) epidermis, 2) dermis, and 3) hypodermis, each contributing a distinct element to thermoregulatory control (Fig. 3.3). Collectively, these layers balance between protection, insulation, and physiological responsiveness.

I. *Epidermis:* As the outermost layer, the epidermis is often perceived as a protective barrier, yet it also contributes subtly to maintaining thermal stability. Its keratinized surface cells are continuously shed and replaced, and provide low thermal conductivity that helps insulate the body from abrupt environmental temperature changes [34, 35]. Although the epidermis is avascular, it contains small openings, or pores, through which sweat can emerge from underlying glands. These pores serve as critical release points, allowing evaporative cooling to occur at the skin's surface. Although less directly involved than the deeper layers, the epidermis is the interface with the environment for controlling heat flow; its integrity and hydration levels influence the efficiency of subsequent, more dynamic thermoregulatory responses.

II. *Dermis:* Below the epidermis lies the dermis, a complex and active layer that forms a sophisticated vascular system, which is critical to human thermoregulation. The thickness of this layer can vary across different body regions [36].

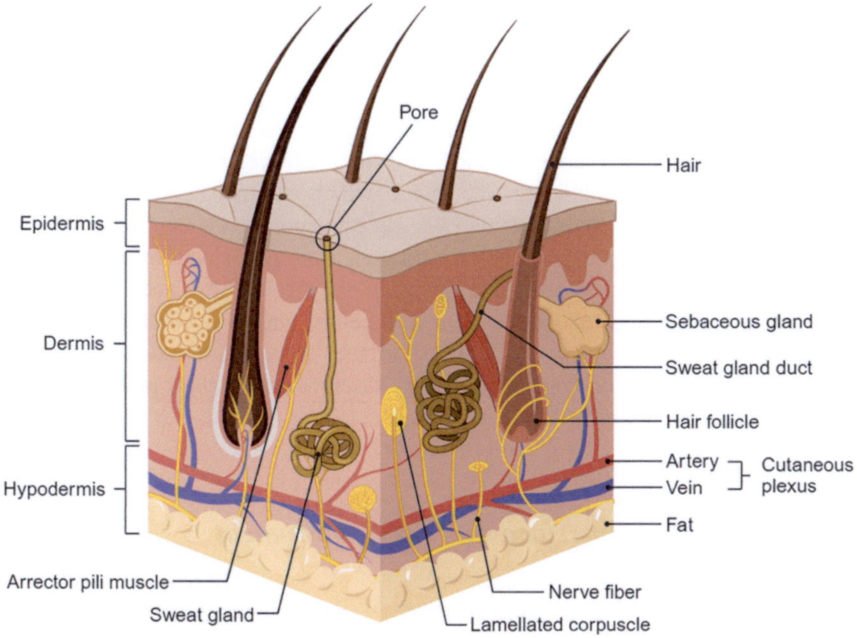

Fig. 3.3 Anatomy of the skin. (Created in BioRender. Bach, A. (2025) https://BioRender.com/f60n598)

Regardless, this layer houses a dense network of blood vessels that can swiftly adjust their diameter to direct blood flow towards or away from the surface, depending on environmental and/or metabolic demands. Alongside these vessels, specialized sweat glands (particularly eccrine glands), sebaceous glands, and hair follicles interweave through the dermis. Within this layer, several structures play key roles:

- *Blood vessels:* A dense vascular network finely tunes blood flow to the skin's surface. Under heat stress, vasodilation increases cutaneous perfusion to offload excess heat, while vasoconstriction conserves body warmth when exposed to cold stress.
- *Sweat glands:* Widely distributed eccrine glands secrete fluid onto the skin surface. As sweat evaporates, it substantially enhances latent heat loss, the primary defence against overheating [37].
- *Thermoreceptors:* Sensory nerve endings embedded in the dermis continuously gauge subtle changes in skin temperature. These receptors transmit signals centrally, prompting rapid reflex adjustments and ensuring the body's thermoregulatory responses remain well calibrated [38] while informing conscious and subconscious behavioural thermoregulation.

The dermis, therefore, is not just a structural support layer, it is an ever-active thermoregulatory hub, continuously integrating signals and adjusting blood flow and sweat production in real time to buffer the body against dynamic environmental demands.

III. *Hypodermis:* At the skin's deepest level, the hypodermis, primarily composed of adipose tissue, serves as a strong thermal insulator. By curtailing the flow of heat from the core to the surface, the hypodermis helps preserve energy, reduces stress on cardiovascular and metabolic systems, and ensures stable internal temperatures. The hypodermis does not just help retain internal heat in cold conditions; it also moderates the rate at which external heat influences core temperature in hot environments [39]. By preventing a sudden influx of environmental heat, the hypodermis contributes to a more stable internal thermal environment, allowing other thermoregulatory mechanisms, such as vasodilation and sweating, to operate more efficiently and in a controlled fashion. Beyond insulation, the hypodermis contributes to structural integrity, cushioning deeper tissues, and providing energy reserves essential for prolonged environmental or metabolic challenges.

3.3.3 Skin as an Interface with the Environment

The skin's extensive surface area and active physiology make it a primary regulator of the body's thermal equilibrium. Thermoreceptors sense even subtle localized and/or environmental shifts, triggering reflexive responses that recalibrate blood flow and sweat rates almost instantaneously [40–42]. By blending these physiological adjustments with its insulating properties, the human skin deftly moderates heat exchange with the surroundings, enabling people to adapt to a remarkable spectrum of environmental extremes [43].

3.3.3.1 Vasomotor Responses

Among the most sophisticated thermoregulatory tools at the skin's disposal are vasomotor responses, fine adjustments in blood vessel diameter that alter skin blood flow. Governed primarily by the autonomic nervous system, vasomotor control ensures that humans can navigate both extremes of temperature and subtle shifts in thermal conditions throughout the day.

3.3.3.2 Vasoconstriction Versus Vasodilation

The delicate balance between conserving body heat during cold exposure and promoting heat dissipation when overheated relies heavily on the skin's dynamic vascular control. This is achieved through coordinated vasoconstriction and vasodilation responses, governed predominantly by the sympathetic nervous system and finely tuned by the hypothalamus [44]. When the body confronts cold stress, sympathetic adrenergic nerves release norepinephrine, binding to α-adrenergic receptors on the smooth muscle within cutaneous blood vessels. This interaction prompts the vessels to narrow, reducing blood flow to the skin, and effectively limiting the amount of heat radiated away from the body's core [45, 46]. Even pronounced vasoconstriction in response to cold rarely threatens skin viability with ischaemia unless the cooling stimulus, and the circulatory reduction it provokes, approaches near maximal shunting for a prolonged period [45].

On the other hand, during heat stress, the thermoregulatory system shifts strategy. It first reduces active vasoconstriction, followed by activation of a separate, active vasodilator system that widens skin blood vessels. Here, sympathetic cholinergic nerves release acetylcholine and other co-transmitters, triggering cutaneous vessels to relax and dilate [44]. This expansion increases skin blood flow dramatically, channelling more heat towards the surface where it can be liberated into the environment. In fact, under maximal heat stress, skin blood flow can exceed $8\ L\cdot min^{-1}$, a remarkable circulation shift that underscores the high priority placed on preventing core overheating [47]. Both vasoconstriction and vasodilation represent not just opposing end points of a physiological spectrum, but components of a highly adaptable, context-dependent system, continually recalibrated by the hypothalamus in response to internal and external cues [48].

3.3.3.3 Impact on Heat Dissipation

In warm environments, the vasodilatory response emerges as a central player in promoting efficient heat loss. By increasing cutaneous blood flow, the skin's temperature increases, creating a more favourable gradient for heat to flow outward. Heat can then dissipate via radiation, convection, and when sweat production is triggered, evaporation. Local skin heating alone can induce pronounced vasodilation, demonstrating the capacity for regional control. Applying heat of approximately 42 °C to a localized area can independently drive substantial blood flow increases to that site, regardless of the overall core temperature [49]. In healthy individuals, this local responsiveness allows the body to tailor its strategy, directing vasodilation to where it is most needed, such as preserving heat and maintaining central blood volumes in the cold, and directing blood to peripheral sites for heat dissipation.

However, accommodating such large shifts in blood distribution requires tradeoffs. As more blood is sent to the skin, the cardiovascular system must adjust by reducing blood flow to other regions, such as the splanchnic and renal circulations, ensuring systemic stability [50]. This redistribution can be challenging. The heart must work harder to meet the skin's sudden demand for increased perfusion, while other organs temporarily cope with less blood supply. Although this short-term reprioritization preserves thermal homeostasis, it also comes with heightened cardiovascular strain, particularly if heat exposure is prolonged, the individual is dehydrated, and/or suffering from an underlying cardiac disease [51].

Vasodilation and sweating operate synergistically. Increased skin blood flow delivers more warmth to the surface, intensifying the potential for evaporative cooling as sweat glands release fluid that vapourizes and removes heat. By working in concert, these thermoeffectors ensure the body can cope with a wide range of environmental stresses and activity levels. Nevertheless, even this sophisticated system has its boundaries. In extreme conditions, persistent vasodilation may not suffice to avert hyperthermia. Furthermore, aging, chronic diseases, certain medications, and other factors can diminish vasodilatory capacity or sweating responses, making some populations more vulnerable to heat stress [52, 53].

3.3.3.4 Sweating and Evaporative Cooling

Humans possess a unique and highly efficient cooling mechanism through sweat evaporation, which is a key factor enabling our species to traverse extreme environments and maintain unmatched mammalian endurance capacities. Unlike many animals, humans have eccrine sweat glands distributed across their nearly hairless skin [36], allowing efficient whole-body cooling during sustained physical activity. This adaptation permits humans to shed excess heat at relatively high metabolic outputs, enabling prolonged exertion in hot environments [54]. Remarkably, over more extreme distances, humans can even outperform other species known for their endurance. For instance, while quadrupeds like horses are faster over short distances, humans can outrun them over ultra-endurance (>90 km) events, where efficient heat dissipation from sustained energy demands plays a crucial role [55, 56]. This ability likely evolved as an adaptation to persistence hunting on the savannas, where early humans relied on their capacity to outlast prey in the heat through superior thermal regulation [57]. While vasomotor responses facilitate heat transfer from the core to the skin's surface, it is ultimately the evaporation of sweat that provides the most powerful means of heat dissipation, accounting for almost all heat loss during exercise in the heat [58, 59].

3.3.3.5 Mechanism of Sweat Production

Eccrine sweat glands, numbering approximately two to four million [60], are critical for thermoregulation through evaporative cooling. These glands, distributed across glabrous and non-glabrous skin, comprise a coiled secretory portion and a duct extending to the skin's surface. The secretory portion consists of clear cells, dark cells, and myoepithelial cells, each playing distinct roles in sweat production and transport.

The production of sweat is initiated by the stimulation of clear cells through acetylcholine binding to muscarinic receptors. This triggers intracellular calcium release, leading to an osmotic gradient that facilitates water movement into the gland's lumen [61]. The initial secretion, isotonic with plasma, undergoes substantial modification in the duct, where sodium and chloride ions are actively reabsorbed. This process, regulated by Na+/K+ ATPase activity and influenced by aldosterone, results in a hypotonic final sweat with sodium concentrations significantly lower than plasma levels [62].

Sweat gland function is modulated by both central and peripheral factors. Central thermoregulatory control involves input from thermoreceptors and subsequent integration in the hypothalamus, while local influences include the effects of osmolality [63] and skin temperature [49]. Plasticity in glandular response is observed with repeated stimulation, as seen in heat acclimation, which enhances per-gland sweat production, hypertrophy, cholinergic sensitivity, and sodium reabsorption efficiency, key adaptations for maintaining electrolyte balance during prolonged sweating [61].

3.3.3.6 Efficiency in Different Conditions

The effectiveness of evaporative cooling through sweating varies considerably depending on both environmental and physiological factors. Under optimal conditions, low humidity, moderate air movement, and temperatures below skin temperature, evaporative cooling can theoretically remove up to 1700 watts of heat loss [54]. However, this theoretical maximum is rarely achieved in real-world settings due to various limiting factors.

The rate of evaporative heat loss is fundamentally driven by the water vapour pressure gradient between the skin surface and the surrounding air. At typical maximal skin temperatures of ~35 °C, the vapour pressure at the skin surface is approximately 5–6 kPa (Fig. 3.4). This creates an important and often misunderstood role of relative humidity as it pertains to sweat evaporation. For instance, even in conditions of 100% relative humidity, evaporative cooling remains possible as long as the air temperature (and thus its water vapour pressure) remains below skin temperature. At an air temperature of 25 °C, the saturated vapour pressure is approximately 3 kPa, creating a 2–3 kPa gradient that enables sweat evaporation despite the "100% relative humidity" that might instinctively suggest otherwise. However, this relationship reverses at air temperatures exceeding skin temperature, where even at lower relative humidity values, the higher vapour pressure of the ambient air can completely prevent evaporation and may even result in condensation on the skin surface, and thus results in accelerated heat gain or uncompensable heat stress.

Air movement significantly influences evaporative efficiency by disrupting the boundary layer of saturated air that forms near the skin surface. Higher air velocities not only enhance evaporative (latent) heat loss but also help maintain a wide vapour pressure gradient between the skin and the environment. However, when environmental temperatures exceed that of the skin, fans may instead promote convective (dry) heat gain [64, 65]. This relationship becomes particularly critical for older adults, who experience an age-related decline in sweating capacity, potentially

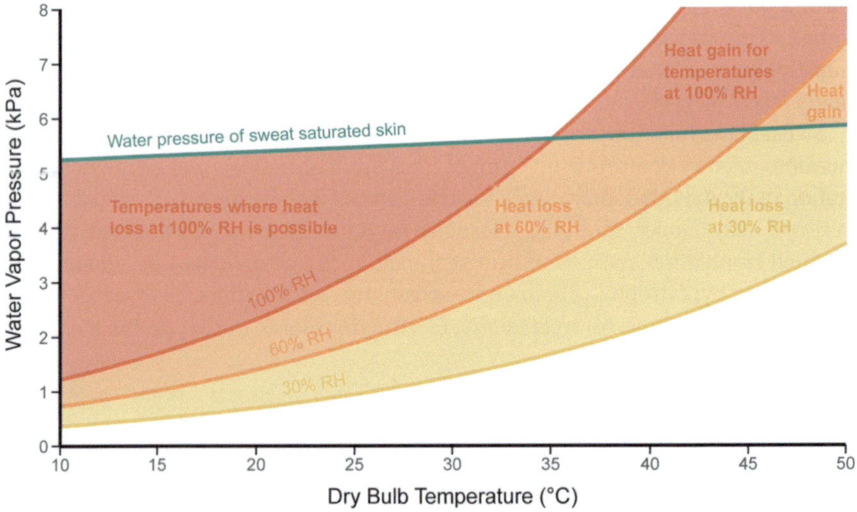

Fig. 3.4 Importance of water vapour pressure, heat loss/gain potential of sweat under different relative humidities. The area shaded below the green line represents a positive evaporative gradient, where the water vapour pressure of the sweat-saturated skin is higher than the environmental vapour pressure. This gradient allows effective heat loss through sweat evaporation. Conversely, the area shaded above the green line represents conditions where the skin vapour pressure is lower than the environmental vapour pressure, making evaporation impossible and leading to condensation on the skin. Fortunately, such conditions are rarely experienced by humans on Earth

compromising their ability to benefit from evaporative cooling [66]. Despite this, the ability of fans to improve net heat loss via evaporation dictates that any increases in heat gain from convection can be offset by increased evaporation of sweat in many populations under various combinations of temperature and humidity. While conjecture remains regarding the exact temperature threshold for which increased evaporation can no longer compensate for convective heat gain under various sub-populations, recent laboratory trials have shown fans do not exacerbate heat strain when temperatures are ≤38.0 °C, though improvements in heat dissipation are small [67, 68]. The relationship between air movement, air temperatures, and evaporative cooling becomes particularly relevant when considering clothing's impact, as garments can significantly impede air flow and create microclimates that reduce evaporative efficiency. Available surface area becomes particularly crucial in this context, an adult human typically has 1.8–2.0 m^2 of skin surface area, but effective evaporative surface area can be significantly reduced by clothing, body position, or skin wettedness patterns.

Individual factors also substantially influence sweating efficiency. Heat acclimatization can increase both maximum sweat rates and the timing of sweat onset while improving electrolyte conservation [69]. Physical fitness typically enhances sweating responses, with trained individuals often showing earlier sweat onset and higher sustainable sweat rates compared to their untrained counterparts [70]. Conversely, ageing impairs sweat gland function, with older adults typically showing reduced sweat rates and delayed onset of sweating [71].

The distribution of sweating across the body surface exhibits notable regional variations that affect overall cooling efficiency. Areas such as the chest and back typically show higher sweat rates than the extremities, though these patterns can shift with acclimatization and training [72]. Understanding these regional differences becomes particularly important when designing protective clothing or cooling interventions, as targeting high-sweat areas for enhanced evaporation can maximize cooling effectiveness.

Importantly, the relationship between sweat production and evaporative cooling is not linear. Beyond certain ambient thresholds, increased sweat production may not enhance cooling if environmental conditions limit evaporation. This "wasteful sweating" can accelerate dehydration without providing additional thermal benefit [20] highlighting the importance of understanding both the capabilities and limitations of the human sweating response when developing strategies to cope with heat stress.

3.4 Responses to Thermal Stress

As highlighted within the introduction to this chapter, the thermoregulatory system regulates core temperature through a complex negative feedback loop that acts to balance the rate of heat gain by the body's internal metabolic process with the rate of heat exchange between the skin and the environment. In this subsection, we will discuss the physiological and behavioural responses to acute thermal stress, in the form of heat and cold exposure and physiological adaptations from prolonged exposure to hot and cold environments.

3.4.1 Responses to Cold Exposure

3.4.1.1 Shivering and Non-shivering Thermogenesis

Shivering is an involuntary response to cold exposure that occurs when the body senses a significant and fairly rapid reduction in core temperature [16]. As a second-line physiological response, shivering begins after initial mechanisms, such as vasoconstriction and increased metabolism, have been activated. Once the hypothalamus detects a drop in core temperature, it triggers shivering [16]. This response involves rapid, periodic contractions of skeletal muscles, which generate metabolic heat in an effort to counteract heat loss and stabilize core temperature [73]. Shivering necessitates a significant energetic contribution and thus, more efficient mechanisms (vasoconstriction [discussed above], behavioural adaptations [discussed below]) are prioritized. To this point, while shivering is an effective thermoregulatory mechanism, it is energy intensive and can lead to extreme fatigue if sustained for prolonged periods of time.

Non-shivering thermogenesis is the physiological process of heat production in the absence of muscular activity (e.g. shivering, skeletal muscle contraction during exercise). Non-shivering thermogenesis primarily occurs in brown adipose tissue and plays a critical role in core temperature regulation during exposure to cold stress [74]. In this context, brown adipose tissue produces heat via thermogenin (an uncoupling protein present in mitochondria), whereby protons are permitted to flow back into the mitochondrial matrix without generating adenosine triphosphate (ATP), and instead release heat [75]. In comparison to shivering, non-shivering thermogenesis is more efficient and is thus relied upon for prolonged heat generation during sustained cold exposure [76]. Although significant inter-individual variation exists, brown adipose tissue activation, and thus non-shivering thermogenesis, generally declines with advancing age [77]. However, acclimation or acclimatization to cold environments has been shown to enhance non-shivering thermogenesis by upregulating brown adipose tissue activity, thereby improving non-shivering heat production and increasing cold tolerance in acclimated individuals [78]. Of interest, by measuring changes in skin temperature, infrared thermography is being increasingly utilized as a non-invasive tool to indirectly assess brown adipose tissue activity [79].

3.4.1.2 Behavioural Adaptations

Behavioural adaptations to cold exposure exist in the form of heat-conserving behaviours (e.g. postural adaptations), heat-generating behaviours (e.g. exercising), and environmental modifications (e.g. seeking shelter or heat sources) [80]. These behavioural responses are both conscious and subconscious. For example, while postural adjustments such as curling up, hunching over, and moving to the foetal position are subconscious actions which reduce the amount of body surface exposed to cold environments, adding additional layers of clothing, lighting a fire, or commencing physical activity are likely conscious actions which are planned or actioned based on individual awareness of cold exposure. Despite a clear common goal (core temperature regulation), complex interplay between conscious and subconscious actions in response to cold exposure exists, which are often dependent on the magnitude of cold exposure [81]. Indeed, while the initial responses to cold are typically subconscious, such as subtle adjustments in posture, these actions are often followed by conscious efforts to seek, retain, or generate warmth [82]. As a result, behavioural responses to cold are often impaired in individuals with sensory impairments [83] or when severe hypothermia ensues.

3.4.2 Responses to Heat Exposure

3.4.2.1 Enhanced Sweating

During exposure to heat stress, thermoreceptors in the skin detect environmental stimuli and relay afferent information to the hypothalamus, which then processes this input to initiate appropriate thermoregulatory responses [84]. Upon sensing an increase in core temperature, the sympathetic nervous system activates sweat glands (located in the dermis), which then secrete fluid onto the skin surface [85]. Depending on environmental conditions, the sweat subsequently evaporates. As sweat evaporates from the skin surface, latent heat is absorbed from the body, cooling the skin, a process termed evaporative heat loss [10]. The rate at which sweat evaporates from the skin surface is determined by the gradient between the skin and air vapour pressure, such that, in environments characterized by high humidity, the process of evaporation is impaired, and sweat drips from the skin surface, providing no cooling benefit. In hot-dry environments, the vapour pressure gradient between the skin and air is greater, and sweat readily evaporates, providing effective cooling [86]. Sweating, along with the subsequent evaporation of sweat from the skin surface, is critical for maintaining thermal equilibrium (albeit sometimes at elevated levels) whereby sweating is the primary method of heat dissipation during exposure to hot environments [10].

3.4.2.2 Behavioural Adaptations

In the same manner as behavioural adaptations to cold exposure, during periods of heat exposure, individuals will consciously and subconsciously modify their behaviour [87]. These adaptations, driven largely by perceptions of thermal discomfort, occur in the form of reducing metabolic heat production (i.e., reducing physical activity), consuming more fluids, postural adaptations (limiting skin contact with hot surfaces, spreading limbs to maximize body surface area to air flow), cool-seeking behaviours (e.g. seeking shaded outdoor environments, utilizing air conditioning indoors), and the removal of additional clothing layers [88].

While the primary aim of behavioural adaptations to heat exposure is to maintain a stable core temperature, both skin temperature and skin wettedness play critical roles in the perception of heat [89, 90], and subsequent behavioural responses to heat exposure. In this context, the skin represents a direct reflection of how the body exchanges heat with the environment. The relation between high skin temperature and thermal discomfort (primary stimulus for behavioural thermoregulation) is well understood [91], whereby in response to sensation of even small increases in temperature, afferent information is relayed from thermoreceptors in the skin surface to the hypothalamus, whereby thermoregulatory (both biological and behavioural) responses are initiated. In hot-humid environments where skin wettedness is high (as secreted sweat does not freely evaporate from the skin surface), despite not possessing hygroreceptors and thus an inability to directly detect sweat accumulation on the skin surface [92], humans will engage in cool-seeking behaviours independently of increases in core and skin temperatures [93].

3.4.3 Adaptation Mechanisms

3.4.3.1 Heat Acclimation

Heat acclimation is the process through which repeated exposure to hot environmental conditions, eliciting strong sweating responses and elevations in core and skin temperatures, enables the body to adapt to hot environments, inducing biological adjustments that reduce the deleterious effects of heat stress. Heat acclimation interventions have traditionally involved undertaking exercise in hot environments, although emerging evidence has highlighted the efficacy of passive acclimation interventions via immersion in hot water [94] and sauna bathing [95] in healthy older adults and older adults with coronary artery disease. The onset of the biological adaptations is fairly rapid, with heart rate adaptations occurring as quickly as 4–5 days following the commencement of the intervention, while remaining adaptions (reductions in core and skin temperatures, increased skin blood flow, sweating and evaporative heat loss, decreased sweat sodium concentration, plasma volume expansion, etc.) are generally complete following 10–14 days of repeated exposure [96] (Fig. 3.5). Together, these thermoregulatory adaptations increase thermal

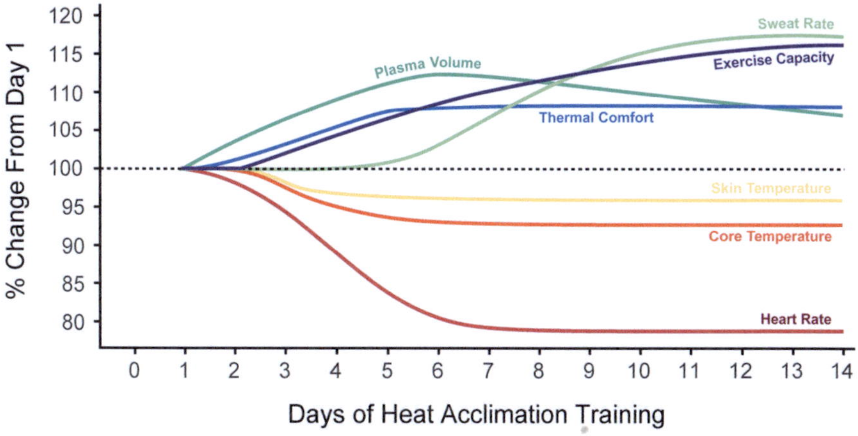

Fig. 3.5 Theoretical time course of heat acclimation adaptation. (Adapted from Périard et al. [96])

tolerance and decrease physiological strain for a given heat load, therefore decreasing the likelihood of heat-related injuries and illness, and heat stroke during exposure to hot environments.

3.4.3.2 Short-Term Versus Long-Term Adaptations

Both short- and long-term adaptations to hot environments improve an individual's thermal tolerance by enhancing the body's ability to cope with heat stress. However, these adaptations result in distinct physiological changes depending on the duration and nature of heat exposure.

Certain common changes are observed following both short- and long-term exposures to hot environments, including reductions in resting skin temperature, earlier onset of sweating, and cardiovascular adjustments [97]. These changes collectively improve the efficiency of thermoregulation, facilitating a reduction in physiological strain for a given level of heat stress (compared to non-acclimated/acclimatized individuals). Specifically, a lower resting skin temperature reduces the thermal gradient between the body and the external environment, allowing for more efficient heat exchange [98]. The earlier onset of sweating increases the body's capacity for evaporative cooling, while cardiovascular adjustments, such as improved plasma volume and cardiac output, enhance the delivery of blood to the skin for heat dissipation [99].

However, some adaptations differ significantly between short-term heat acclimation (typically achieved through weeks of controlled heat exposure) and long-term acclimatization (occurring over months to years of living in hot environments). Short-term heat acclimation protocols are often associated with a reduction in resting core temperature [96], which provides a larger buffer against heat gain during exposure to high temperatures. This reduction helps prevent rapid overheating and enhances endurance performance in the heat [96]. In contrast, long-term acclimatization (likely requiring more than 12 months of continuous residence in a hot environment) has been observed to result in higher resting core temperatures [100]. While this may seem counterintuitive, it represents an adaptive advantage. An elevated resting core temperature increases the core-to-skin temperature gradient during heat exposure. This enhanced gradient facilitates more efficient heat transfer from the body's core to the skin and, subsequently, to the surrounding environment [98]. As a result, long-term acclimatization improves the body's ability to dissipate heat effectively, maintaining stable body temperatures even under extreme heat stress [97].

This divergence in physiological adaptations underscores the complexity of the human body's response to thermal environments. Short-term adaptations are primarily focused on immediate improvements in heat tolerance, while long-term adaptations reflect a more sustained and optimized approach to managing chronic heat exposure. Both types of adaptations are critical for survival and performance in hot climates, albeit with different mechanisms and outcomes.

3.5 Conclusions

The skin plays a central role in human temperature regulation. Through its complex layers and specialized structures, the skin serves as both sentinel and effector in maintaining thermal homeostasis. The dynamic nature of thermoregulatory responses, from rapid vasomotor adjustments to the remarkable cooling power of our sweating response, enables humans to thrive across diverse thermal environments and circumstances. These adaptations are further enhanced through both short- and long-term adaptations, demonstrating the plasticity of our thermoregulatory system.

Understanding the skin's role in temperature regulation has significant implications beyond basic physiology. The principles outlined in this chapter inform approaches to managing thermal stress in both clinical and performance settings, while highlighting the importance of considering individual variations in thermoregulatory capacity. As climate change poses increasing challenges to human thermal adaptation, this knowledge becomes ever more critical for developing effective strategies to protect vulnerable populations and optimize human performance in thermally challenging environments. The skin is a sophisticated thermal interface with the environment, coupled with its integration into broader physiological systems, it continues to reveal crucial insights into human homeostatic mechanisms.

Bibliography

1. Gagge AP, Gonzalez RR (2010) Mechanisms of heat exchange: biophysics and physiology. Compr Physiol:45–84
2. Romanovsky AA (2018) The thermoregulation system and how it works. Handb Clin Neurol 156:3–43
3. Taylor NA, Tipton MJ, Kenny GP (2014) Considerations for the measurement of core, skin and mean body temperatures. J Therm Biol 46:72–101
4. Meade RD, Atkins WC, Bach AJE, Foster J, Hutchins KP, McKenna ZJ et al (2024) Human heat resilience in a warming climate: biophysical and physiological underpinnings of heat vulnerability and personal cooling strategies. One Earth 7(8):1343–1350
5. Tipton MJ (2016) Environmental extremes: origins, consequences and amelioration in humans. Exp Physiol 101(1):1–14
6. Rodrigues P, Orssatto LBR, Gagnon D, Dahhak A, Hecksteden A, Stewart IB et al (2024) Passive heat therapy: a promising preventive measure for people at risk of adverse health outcomes during heat extremes. J Appl Physiol 136(4):677–694
7. Meade RD, Notley SR, D'Souza AW, Dervis S, Boulay P, Sigal RJ et al (2019) Interactive effects of age and hydration state on human thermoregulatory function during exercise in hot-dry conditions. Acta Physiol 226(1):e13226
8. Sawka MN, Young AJ, Francesconi RP, Muza SR, Pandolf KB (1985) Thermoregulatory and blood responses during exercise at graded hypohydration levels. J Appl Physiol 59(5):1394–1401
9. Persson PB, Persson AB (2018) Water is life. Acta Physiol 224(2):e13173
10. Kenny GP, Jay O (2013) Thermometry, calorimetry, and mean body temperature during heat stress. In: Terjung R (ed) Comprehensive physiology [Internet], 1st edn. Wiley; [cited 2025 Jan 14], pp 1689–1719. Available from: https://onlinelibrary.wiley.com/doi/10.1002/cphy.c130011
11. Notley SR, Mitchell D, Taylor NAS (2023) A century of exercise physiology: concepts that ignited the study of human thermoregulation. Part 1: foundational principles and theories of regulation. Eur J Appl Physiol 123(11):2379–2459
12. Gagge AP, Burton AC, Bazett HC (1941) A practical system of units for the description of the heat exchange of man with his environment. Science 94(2445):428–430
13. Elia M (1992) Organ and tissue contribution to metabolic rate. In: Energy metab tissue determinants cell corrolaries, pp 61–77
14. Wang Z, Ying Z, Bosy-Westphal A, Zhang J, Schautz B, Later W et al (2010 Dec) Specific metabolic rates of major organs and tissues across adulthood: evaluation by mechanistic model of resting energy expenditure. Am J Clin Nutr 92(6):1369–1377
15. Saltin B, Astrand PO (1967) Maximal oxygen uptake in athletes. J Appl Physiol 23(3):353–358
16. Haman F, Blondin DP (2017) Shivering thermogenesis in humans: origin, contribution and metabolic requirement. Temperature 4(3):217–226
17. Notley SR, Mitchell D, Taylor NAS (2023) A century of exercise physiology: concepts that ignited the study of human thermoregulation. Part 2: physiological measurements. Eur J Appl Physiol 123(12):2587–2685
18. Shibasaki M, Crandall CG (2010) Mechanisms and controllers of eccrine sweating in humans. Front Biosci S2(2):685–696
19. Taylor NA, Machado-Moreira CA (2013) Regional variations in transepidermal water loss, eccrine sweat gland density, sweat secretion rates and electrolyte composition in resting and exercising humans. Extreme Physiol Med [Internet]. [cited 2025 Jan 14];2(1). Available from: https://extremephysiolmed.biomedcentral.com/articles/10.1186/2046-7648-2-4
20. Jay O, Cramer MN, Ravanelli NM, Hodder SG (2015) Should electric fans be used during a heat wave? Appl Ergon 46:137–143
21. Hellon RF (1972) Central thermoreceptors and thermoregulation. In: Neil E (ed) Enteroceptors [Internet]. Berlin, Heidelberg: Springer Berlin Heidelberg; [cited 2025 Jan 18],

pp. 161–86. (Handbook of sensory physiology; vol. 3/1). Available from: http://link.springer.com/10.1007/978-3-642-65252-3_5
22. Jessen C (1996) Interaction of body temperatures in control of thermoregulatory effector mechanisms. In: Terjung R (ed) Comprehensive physiology [Internet], 1st edn. Wiley; [cited 2025 Jan 18], pp. 127–38. Available from: https://onlinelibrary.wiley.com/doi/10.1002/cphy.cp040107
23. Taylor N, Werner J, Mekjavic IB (2008) Concepts in physiological regulation: a thermoregulatory perspective
24. Hensel H (1974) Thermoreceptors. Annu Rev Physiol 36(1):233–249
25. Morris NB, Bain AR, Cramer MN, Jay O (2014) Evidence that transient changes in sudomotor output with cold and warm fluid ingestion are independently modulated by abdominal, but not oral thermoreceptors. J Appl Physiol 116(8):1088–1095
26. Romanovsky AA, Almeida MC, Garami A, Steiner AA, Norman MH, Morrison SF et al (2009) The transient receptor potential Vanilloid-1 channel in thermoregulation: a thermosensor it is not. Pharmacol Rev 61(3):228–261
27. Todd G, Gordon CJ, Groeller H, Taylor NAS (2014) Does intramuscular thermal feedback modulate eccrine sweating in exercising humans? Acta Physiol 212(1):86–96
28. Almeida MC, Hew-Butler T, Soriano RN, Rao S, Wang W, Wang J et al (2012) Pharmacological blockade of the cold receptor TRPM8 attenuates autonomic and behavioral cold defenses and decreases deep body temperature. J Neurosci 32(6):2086–2099
29. Tan CH, McNaughton PA (2016) The TRPM2 ion channel is required for sensitivity to warmth. Nature 536(7617):460–463
30. Fromy B, Josset-Lamaugarny A, Aimond G, Pagnon-Minot A, Marics I, Tattersall GJ et al (2018) Disruption of TRPV3 impairs heat-evoked vasodilation and thermoregulation: a critical role of CGRP. J Invest Dermatol 138(3):688–696
31. Vizin RCL, Scarpellini CDS, Ishikawa DT, Correa GM, De Souza CO, Gargaglioni LH et al (2015) TRPV4 activates autonomic and behavioural warmth-defence responses in Wistar rats. Acta Physiol 214(2):275–289
32. Boulant JA (1981) Hypothalamic mechanisms in thermoregulation. Fed Proc 40(14):2843–2850
33. Morrison SF, Nakamura K (2019) Central mechanisms for thermoregulation. Annu Rev Physiol 81(1):285–308
34. Buettner K (1951) Effects of extreme heat and cold on human skin. II. Surface temperature, pain and heat conductivity in experiments with radiant heat. J Appl Physiol 3(12):703–713
35. Vendrik AJH, Vos JJ (1957) A method for the measurement of the thermal conductivity of human skin. J Appl Physiol 11(2):211–215
36. Rushmer RF, Buettner KJK, Short JM, Odland GF (1966) The skin: the most accessible tissue of the body serves as a potential focus for multidisciplinary research. Science 154(3747):343–348
37. Lind AR (1963) A physiological criterion for setting thermal environmental limits for everyday work. J Appl Physiol 18(1):51–56
38. Hensel H (1981) Thermoreception and temperature regulation. Academic Press, London. 321 p. (Monographs of the physiological society)
39. Stolwijk JAJ, Hardy JD (1965) Skin and subcutaneous temperature changes during exposure to intense thermal radiation. J Appl Physiol 20(5):1006–1013
40. Minson CT, Berry LT, Joyner MJ (2001) Nitric oxide and neurally mediated regulation of skin blood flow during local heating. J Appl Physiol 91(4):1619–1626
41. Nadel ER, Bullard RW, Stolwijk JA (1971) Importance of skin temperature in the regulation of sweating. J Appl Physiol 31(1):80–87
42. Wingo JE, Low DA, Keller DM, Brothers RM, Shibasaki M, Crandall CG (2010) Skin blood flow and local temperature independently modify sweat rate during passive heat stress in humans. J Appl Physiol 109(5):1301–1306

43. Pandolf KB, Sawka MN, Gonzalez RR (1988) Human performance physiology and environmental medicine at terrestrial extremes. No Title [Internet]; [cited 2025 Jan 14]. Available from: https://cir.nii.ac.jp/crid/1130282271611215872
44. Johnson JM, Minson CT, Kellogg DL (2014) Cutaneous vasodilator and vasoconstrictor mechanisms in temperature regulation. In: Terjung R (ed) Comprehensive physiology [Internet]. 1st edn. Wiley; [cited 2025 Jan 16], pp. 33–89. Available from: https://onlinelibrary.wiley.com/doi/10.1002/cphy.c130015
45. Edholm OG, Fox RH, Macpherson RK (1957) Vasomotor control of the cutaneous blood vessels in the human forearm. J Physiol 139(3):455–465
46. Kellogg DL, Johnson JM, Kosiba WA (1989) Selective abolition of adrenergic vasoconstrictor responses in skin by local iontophoresis of bretylium. Am J Physiol-Heart Circ Physiol 257(5):H1599–H1606
47. Rowell LB (1974) Human cardiovascular adjustments to exercise and thermal stress. Physiol Rev 54(1):75–159
48. Savage MV, Brengelmann GL (1996) Control of skin blood flow in the neutral zone of human body temperature regulation. J Appl Physiol 80(4):1249–1257
49. Pergola PE, Johnson JM, Kellogg DL, Kosiba WA (1996) Control of skin blood flow by whole body and local skin cooling in exercising humans. Am J Physiol-Heart Circ Physiol 270(1):H208–H215
50. Rowell LB, Brengelmann GL, Blackmon JR, Murray JA (1970) Redistribution of blood flow during sustained high skin temperature in resting man. J Appl Physiol 28(4):415–420
51. Crandall CG, González-Alonso J (2010) Cardiovascular function in the heat-stressed human. Acta Physiol 199(4):407–423
52. Bach AJE, Cunningham SJK, Morris NR, Xu Z, Rutherford S, Binnewies S et al (2024) Experimental research in environmentally induced hyperthermic older persons: a systematic quantitative literature review mapping the available evidence. Temperature 11(1):4–26
53. Meade RD, Akerman AP, Notley SR, McGinn R, Poirier P, Gosselin P et al (2020) Physiological factors characterizing heat-vulnerable older adults: a narrative review. Environ Int 144:105909
54. Gagnon D, Crandall CG (2018) Sweating as a heat loss thermoeffector. In: Handbook of clinical neurology [Internet]. Elsevier; [cited 2025 Jan 16], pp. 211–32. Available from: https://linkinghub.elsevier.com/retrieve/pii/B9780444639127000138
55. Bramble DM, Lieberman DE (2004) Endurance running and the evolution of Homo. Nature 432(7015):345–352
56. Minetti AE (2003) Efficiency of equine express postal systems. Nature 426(6968):785–786
57. Liebenberg L (2006) Persistence hunting by modern hunter-gatherers. Curr Anthropol 47(6):1017–1026
58. Nielsen M (1938) Die Regulation der Körpertemperatur bei Muskelarbeit. Skand Arch Für Physiol 79(2):193–230
59. Sawka MN, Wenger CB, Pandolf KB (1996) Thermoregulatory responses to acute exercise-heat stress and heat acclimation. In: Terjung R (ed) Comprehensive physiology [Internet], 1st edn. Wiley; [cited 2025 Jan 16], pp. 157–85. Available from: https://onlinelibrary.wiley.com/doi/10.1002/cphy.cp040109
60. Sato K, Kang WH, Saga K, Sato KT (1989) Biology of sweat glands and their disorders. I. Normal sweat gland function. J Am Acad Dermatol 20(4):537–563
61. Baker LB (2019) Physiology of sweat gland function: the roles of sweating and sweat composition in human health. Temperature 6(3):211–259
62. Sato K, Dobson RL (1970) The effect of intracutaneous d-aldosterone and hydrocortisone on human eccrine sweat gland function. J Invest Dermatol 54(6):450–462
63. Fortney SM, Wenger CB, Bove JR, Nadel ER (1984) Effect of hyperosmolality on control of blood flow and sweating. J Appl Physiol 57(6):1688–1695

64. Meade RD, Notley SR, Kirby NV, Kenny GP (2024) A critical review of the effectiveness of electric fans as a personal cooling intervention in hot weather and heatwaves. Lancet Planet Health 8(4):e256–e269
65. Ravanelli NM, Hodder SG, Havenith G, Jay O (2015) Heart rate and body temperature responses to extreme heat and humidity with and without electric fans. JAMA 313(7):724
66. Gagnon D, Romero SA, Cramer MN, Jay O, Crandall CG (2016) Cardiac and thermal strain of elderly adults exposed to extreme heat and humidity with and without electric fan use. JAMA 316(9):989
67. Chaseling GK, Vargas NT, Hospers L, Barry H, Harwood A, Graham C et al (2024) Simple strategies to reduce cardiac strain in older adults in extreme heat. N Engl J Med 391(18):1754–1756
68. O'Connor FK, Meade RD, Wagar KE, Harris-Mostert RC, Tetzlaff EJ, McCormick JJ et al (2024) Effect of electric fans on body core temperature in older adults exposed to extreme indoor heat. JAMA 332(20):1752–1754
69. Taylor NAS (2014) Human heat adaptation. In: Terjung R (ed) Comprehensive physiology [Internet], 1st edn. Wiley; [cited 2025 Jan 16], pp. 325–65. Available from: https://onlinelibrary.wiley.com/doi/10.1002/cphy.c130022
70. Lee JB, Kim TW, Min YK, Yang HM (2014) Long distance runners present upregulated sweating responses than sedentary counterparts. Lucia A, editor. PLoS ONE 9(4):e93976
71. Inoue Y, Nakao M, Araki T, Murakami H (1991) Regional differences in the sweating responses of older and younger men. J Appl Physiol 71(6):2453–2459
72. Smith CJ, Havenith G (2019) Upper body sweat mapping provides evidence of relative sweat redistribution towards the periphery following hot-dry heat acclimation. Temperature 6(1):50–65
73. Nakamura K, Morrison SF (2011) Central efferent pathways for cold-defensive and febrile shivering. J Physiol 589(14):3641–3658
74. Himms-Hagen J (1984) Nonshivering thermogenesis. Brain Res Bull 12(2):151–160
75. Fedorenko A, Lishko PV, Kirichok Y (2012) Mechanism of fatty-acid-dependent UCP1 uncoupling in brown fat mitochondria. Cell 151(2):400–413
76. Van Marken Lichtenbelt WD, Schrauwen P (2011) Implications of nonshivering thermogenesis for energy balance regulation in humans. Am J Physiol-Regul Integr Comp Physiol 301(2):R285–R296
77. Yoneshiro T, Aita S, Matsushita M, Okamatsu-Ogura Y, Kameya T, Kawai Y et al (2011) Age-related decrease in cold-activated brown adipose tissue and accumulation of body fat in healthy humans. Obesity 19(9):1755–1760
78. Gordon K, Blondin DP, Friesen BJ, Tingelstad HC, Kenny GP, Haman F (2019) Seven days of cold acclimation substantially reduces shivering intensity and increases nonshivering thermogenesis in adult humans. J Appl Physiol 126(6):1598–1606
79. Law J, Chalmers J, Morris DE, Robinson L, Budge H, Symonds ME (2018) The use of infrared thermography in the measurement and characterization of brown adipose tissue activation. Temperature 5(2):147–161
80. Castellani JW, Young AJ (2016) Human physiological responses to cold exposure: acute responses and acclimatization to prolonged exposure. Auton Neurosci 196:63–74
81. Falla M, Micarelli A, Hüfner K, Strapazzon G (2021) The effect of cold exposure on cognitive performance in healthy adults: a systematic review. Int J Environ Res Public Health 18(18):9725
82. Flouris AD, Cheung SS (2009) Human conscious response to thermal input is adjusted to changes in mean body temperature. Br J Sports Med 43(3):199–203
83. Christogianni A, Bibb R, Davis SL, Jay O, Barnett M, Evangelou N et al (2018) Temperature sensitivity in multiple sclerosis: an overview of its impact on sensory and cognitive symptoms. Temperature 5(3):208–223

84. Sawka MN, Leon LR, Montain SJ, Sonna LA (2011) Integrated physiological mechanisms of exercise performance, adaptation, and maladaptation to heat stress. Compr Physiol 1(4):1883–1928
85. Shibasaki M, Wilson TE, Crandall CG (2006) Neural control and mechanisms of eccrine sweating during heat stress and exercise. J Appl Physiol 100(5):1692–1701
86. Kenny GP, Tetzlaff EJ, Journeay WS, Henderson SB, O'Connor FK (2024) Indoor overheating: a review of vulnerabilities, causes, and strategies to prevent adverse human health outcomes during extreme heat events. Temperature 11(3):203–246
87. Vanos J, Guzman-Echavarria G, Baldwin JW, Bongers C, Ebi KL, Jay O (2023) A physiological approach for assessing human survivability and liveability to heat in a changing climate. Nat Commun 14(1):7653
88. Flouris AD, Schlader ZJ (2015) Human behavioral thermoregulation during exercise in the heat. Scand J Med Sci Sports 25(S1):52–64
89. O'Connor FK, McGarr GW, McCourt ER, Meade RD, Kenny GP (2024) Foot immersion with and without neck cooling reduces self-reported environmental symptoms in older adults exposed to simulated indoor overheating. Temperature 11(4):318–332
90. O'Connor FK, McGarr GW, Harris-Mostert RC, Boulay P, Sigal RJ, Meade RD et al (2025) Effects of pedestal-mounted electric fans on self-reported symptoms and mood-state in older adults exposed to indoor overheating during a simulated heatwave. Appl Physiol Nutr Metab. In Press
91. Frank SM, Raja SN, Bulcao CF, Goldstein DS (1999) Relative contribution of core and cutaneous temperatures to thermal comfort and autonomic responses in humans. J Appl Physiol 86(5):1588–1593
92. Filingeri D, Havenith G (2015) Human skin wetness perception: psychophysical and neurophysiological bases. Temperature 2(1):86–104
93. Vargas NT, Chapman CL, Ji W, Johnson BD, Gathercole R, Schlader ZJ (2020) Increased skin wetness independently augments cool-seeking behaviour during passive heat stress. J Physiol 598(13):2775–2790
94. Janetos KMT, O'Connor FK, Meade RD, Richards BJ, Kirby NV, McCormick JJ et al (2025) Short-term warm-water immersion for improving whole-body heat loss in older men. Med Sci Sports Exerc
95. Debray A, Gravel H, Garceau L, Bartlett AA, Chaseling GK, Barry H et al (2023) Finnish sauna bathing and vascular health of adults with coronary artery disease: a randomized controlled trial. J Appl Physiol 135(4):795–804
96. Périard JD, Racinais S, Sawka MN (2015) Adaptations and mechanisms of human heat acclimation: applications for competitive athletes and sports. Scand J Med Sci Sports 25(S1):20–38
97. Tochihara Y, Wakabayashi H, Lee JY, Wijayanto T, Hashiguchi N, Saat M (2022) How humans adapt to hot climates learned from the recent research on tropical indigenes. J Physiol Anthropol 41(1):27
98. Cuddy JS, Hailes WS, Ruby BC (2014) A reduced core to skin temperature gradient, not a critical core temperature, affects aerobic capacity in the heat. J Therm Biol 43:7–12
99. Rodrigues S, O'Connor F, Morris N, Chaseling G, Sabapathy S, Bach A (2025) Passive heat therapy for cardiovascular disease: current evidence and future directions. Appl Physiol Nutr Metab 50:1
100. Lee JY, Wakabayashi H, Wijayanto T, Hashiguchi N, Saat M, Tochihara Y (2011) Ethnic differences in thermoregulatory responses during resting, passive and active heating: application of Werner's adaptation model. Eur J Appl Physiol 111(12):2895–2905

Chapter 4
Methodological Aspects of Infrared Thermography in Human Assessment

Jose Ignacio Priego-Quesada, Marcos Roberto Kunzler, and Felipe P. Carpes

Abstract Infrared thermography presents some important advantages in the determination of skin temperature, as it is a safe, noninvasive, and noncontact technique of relatively easy use and with wide applications in the field of sports sciences. Like many other techniques, valid measurement in thermography requires following strict methodological steps from data acquisition to analysis and interpretation. In this chapter, we discuss the methodological recommendations that must be followed when acquiring thermic images, along with some practical examples and recommendations based on the current literature to ensure proper image acquisition and data processing.

4.1 Thermography: Advantages and Limitations

Over the last 10 years, infrared thermography has become a popular technique to determine the temperature of human skin during exercise [1, 2]. The reasons accounting for its popularization were the technological advances permitting assessment in a variety of sporting conditions and the more accessible price of the cameras. Previously used mostly with engineering applications, infrared thermography became known by sports scientists who were interested in using this technique for field assessment of physical activity and sport.

J. I. Priego-Quesada (✉)
Research Group in Sport Biomechanics (GIBD), Department of Physical Education and Sports, University of Valencia, Valencia, Spain

Biophysics and Medical Physics Group, Department of Physiology, University of Valencia, Valencia, Spain
e-mail: j.ignacio.priego@uv.es

M. R. Kunzler · F. P. Carpes
Applied Neuromechanics Research Group, Laboratory of Neuromechanics, Federal University of Pampa, Uruguaiana, RS, Brazil
e-mail: marcoskunzler@unipampa.edu.br; carpes@unipampa.edu.br

J. I. Priego-Quesada (ed.), *Application of Infrared Thermography in Sports Science*,
https://doi.org/10.1007/978-3-031-93311-0_4

Many recent studies are available presenting data from experiments that can illustrate the different applications of thermography in the assessment of exercise performance and sport injuries [3–7]. However, thermography in sports is still a topic of interest and there are many fundamental discussions concerning different methodological aspects (e.g., determination of the regions of interest and strategy for data analysis), protocols of data acquisition, and the interpretation of results concerning its physiological or even mechanical meaning in the context of sports performance. While the wide use of infrared thermography is notable, there are still matters of concern that need further research. The good news is that the implication of the skin temperature information on athletic training is endorsed by basic research. For example, high skin temperatures are known to reduce the temperature gradient between the skin and the core, limiting heat transfer and producing a negative effect on aerobic performance [8–10]. However, while the infrared thermography presents some methodological advantages, it also has some limitations.

Methodological strengths and weaknesses of the technique have to be considered when designing, developing, and analyzing data from experiments involving infrared thermography. De Andrade Fernandes et al. [11] described some of the advantages and limitations of using thermography in the assessment of skin temperature in humans. In Table 4.1, De Andrade Fernandes et al. [11] described advantages and limitations of the technique, which we complement here considering the current state of the literature. We also include some additional comments concerning the relevance of each advantage and limitation.

One important advantage of infrared thermography over other methods to determine skin temperature is that infrared thermography is considered to be a distance technique, which means that the measurement can be performed without interaction with the environment or subject. This advantage is essential in some applications, such as the assessment of electric or nuclear facilities, in which the evaluator can analyze the components without being exposed to hazards. In the context of human assessment, this characteristic is also beneficial in order to determine skin temperature without interference in the processes of heat exchanges via convection, conduction, radiation, and evaporation [11–13]. This is an inconvenient aspect of the sensors based on thermal contact, such as thermistors and thermocouples (see the circle in Fig. 4.1). It is known that the interaction between these sensors, the skin, and the environment can reduce the reliability of the measurement [14]. Additionally, the method used to attach these sensors to the skin (e.g., the clinical tape used) has been shown to affect the local heat transfer [15, 16], and hence the local thermal regulation and skin temperature might be altered [11–13, 17, 18].

The noninvasive and nonharmful characteristics of infrared thermography make it a very attractive technique in medical areas, since it allows human physiological study without the risk invasive radiation present in other imaging techniques (e.g., X-rays) [19, 20]. One attractive application of thermography is to detect musculoskeletal injuries [4, 21]. However, it is not considered a diagnostic technique due to its nonspecificity, although it has been used as a complementary tool in physical examination [3, 22, 23].

Table 4.1 Advantages and limitations of infrared thermography in human assessment (modified and adapted from De Andrade Fernandes et al. [11]) and its implication for using or not using thermography in assessment protocol

Advantages	Why is it an advantage?	Limitations	Why is it a limitation?
1. Noninvasive method 2. Distance method 3. It does not interfere with human thermoregulation 4. Image technique 5. Freedom of movement during exercise 6. Possibility to define the region of interest with small or large regions 7. High sensitivity, accuracy and reproducibility 8. Possibility to record video when using specific camera models 9. Availability of low-cost cameras 10. Possibility of using machine and deep learning for data analysis	1. There is no physical interaction with the subject. It is possible to measure each subject as many times as needed, since it is not harmful 2. It is possible to take the measurements without interference with the actual performance of the subject 3. There is no interaction that can affect skin temperature during measurement (e.g., isolation and reduction of sweat evaporation for the tape attachment of contact sensors) 4. Qualitative analysis of thermal images can provide good results for some applications 5. Being able to measure without additional contact with the subjects (or with wires), movement patterns have more chance of occurring naturally 6. As many different regions as desired can be measured and monitored. Is not a single point measure, as with thermocouples 7. It can provide very good and confident results, if the experiment is performed according to the methodological requirements 8. Measurement can be performed at different times or even during a performance that can result in significant variation of skin temperature 9. It makes technology more accessible to everyone, as well as being able to provide cameras for on-site measurement (e.g., at the person being assessed at home or sport facilities) 10. Machine learning for image analysis can be applied quickly and effectively to numerous applications and reduce user interference during data analysis	1. High cost of the high-definition cameras 2. Low resolution and confidence of the data of the low-price cameras 3. The need for specific training of the user in the control of the different factors that can affect the measurements 4. The need to stop exercise and take the images during a static position when the camera does not have the video option, or the exercise space could affect the measurement 5. Long time required to analyze the thermal images because the software provided by the manufacturers for image analyses are not specific to human assessment 6. Machine and deep learning analysis are not accessible for all users or have a high economic cost associated	1. The greater the definition needed, the higher is the cost of the hardware. It limits access to the technique 2. Cheaper cameras can be good for learning the technique, but will not provide results good enough to draw scientific conclusions. More studies are necessary to understand how to ensure the quality of the data obtained from low-cost cameras 3. Interuser variability is high and affects the measurement. Therefore, specific training is necessary, which may increase costs in implementing the technique 4. Measuring skin temperature during exercise can be more "realistic" and can present more reliable results than static measurements, but the technology is still limited for this 5. Data analysis is still limited when considering big databases 6. Most of the machine learning software developed are not available in a user-friendly environment or not economically available to everyone

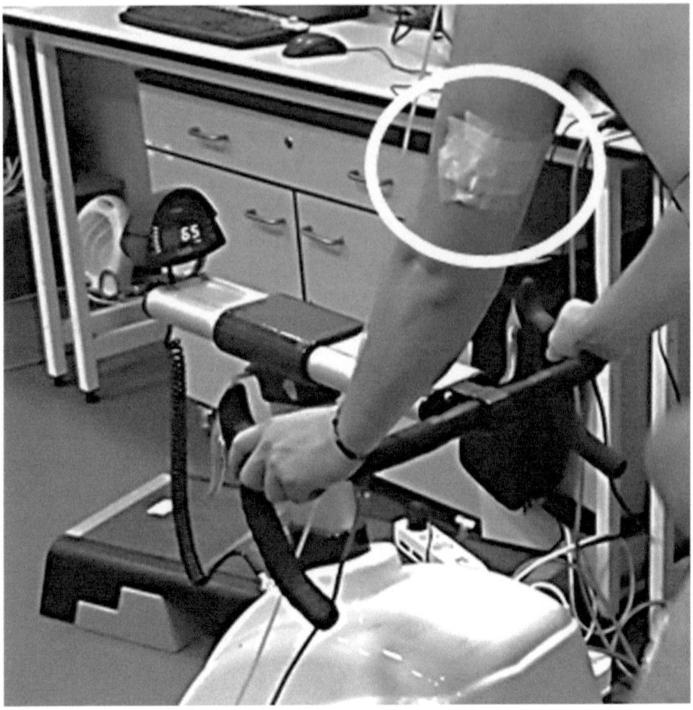

Fig. 4.1 Thermal contact sensor attached to the skin during cycling (inside the *circle*)

The thermography measurements also attract interest from researchers investigating pathologies and injuries. Some physiological disorders are associated with an alteration of skin temperature in the patient [19, 20, 24]. In the context of sports injury, there are some studies discussing the potential associations between local temperature and specific injuries [3, 4, 19, 25]. However, it remains clear that further investigation is needed to provide evidence if its application in the field reduces the number of injuries or days lost by injury, and other parameters about its use (e.g., percentage of false positives and percentage of false negatives). We consider that most of the inconsistencies in discussing the associations between thermography outcomes and injuries rely on the fact that infrared thermography allows the determination of surface temperatures, and inferences regarding tissue temperature may lead to flawed conclusions. While very useful for discussing heat dissipation, inferences on central or tissue temperature are not possible based solely on infrared thermography measurements of the body skin. Some studies used infrared cameras and argue that measured direct data from deep tissues, such as muscles, but this approach is limited due to all the processes of heat exchanges via convection, conduction, radiation, and evaporation, which make the temperatures from the internal tissues and the superficial skin temperature differ [26]. Furthermore, neuromuscular electrical activity, even measured by surface electromyography in the muscle, does not show a significant relationship with skin temperature in most of the lower limb muscles involved in the pedalling gesture [27]. The information about heat dissipation in the different body sites and the responses to exercise are important in the context of sportswear design [28, 29], preventing heat stress risk during competition [9, 30], or in the assessment of the capacity of heat dissipation due to the physical fitness level [7, 31, 32]. There are many situations in which information regarding skin temperature is useful, and its application in the field as well as research to further develop the technique should focus on those.

To draw consistent conclusions, accurate data are essential. To achieve this, a rigorous methodology must be followed [23, 33, 34]. Infrared thermography camera captures the infrared radiation, and the different levels of radiation define a thermal image. However, the proper determination of the temperature from this infrared radiation depends on technical and methodological factors that can influence the capture and the determination of temperatures (e.g., angle and distance of the camera and emissivity). Therefore, in addition to clearly understanding the physiological basis behind the image measurement, it is very important to strictly follow a protocol for the image acquisition. The main aspects that must be taken into account during image acquisition will be discussed in the next sections.

4.2 Methodology for Infrared Thermography Assessment in Humans

Different organizations proposed standard recommendations for the use of thermal imaging in studies with humans. Some examples are the technical guidelines published by the American Academy of Thermology in 1986 [35], the guidelines for neuromusculoskeletal thermography published by the same association in 2006 [36], and the Glamorgan Protocol published in 2008 by the European Association of Thermology [33]. Another reference for the area is the paper titled "The Thermographic Imaging in Sports and Exercise Medicine" (TISEM) checklist [37] in which the authors present a set of 15 items that includes the most important methodological aspects that must be taken into account and that are very useful during the design of a thermographic evaluation, as well as when disseminating the results. However, different methodological aspects remain unclear. To this end, recent scientific articles analyzed specific aspects of the methodology and thermography outcomes in human assessment [12, 38–44]. Taking together guidelines and publications, the main methodological aspects for the assessment of humans with thermography should attempt to:

1. Reduce the error of measurement of the infrared thermography camera (minimum error is commonly established by the manufacturers at ±2%).
2. Reduce the variability in the measure of the skin temperature between participants.
3. Increase the reproducibility of the measurements.

To achieve these three methodological concerns, both technical and methodological requirements must be considered and, more importantly, must be carefully administrated to improve the measurements and therefore lead to valid conclusions. The next sections will introduce each of these requirements.

4.2.1 Determination of the Region of Interest (ROI)

A proper determination of the ROI is one of the most important steps in the design of a thermography study. The determination of the ROI will depend on the purpose of the study and must be determined in the early phase of study design. The ROI definition will affect other aspects of the methods, such as the body position during the thermal room adaptation or the position of the camera. The variables of interest, such as descriptive statistics, will be extracted from each ROI obtained from the thermal image. Determination of the ROI is important because it can also help in improving data acquisition quality, especially when camera resolution is limited. ISO normative determines that to reduce error, an ROI should consider a minimum of 25 pixels [45, 46]. The camera should be placed as close as possible to the ROI in order to increase the number of pixels of measurement [47]. Additionally, the size of the ROI will directly influence the temperature measurement, as a smaller ROI will present a higher temperature than a larger ROI due to the greater influence of hot spots in smaller areas [12]. In this sense, using cameras placed at a larger distance and with lower resolution (and therefore fewer pixels for measurement) will negatively affect the ability to obtain reliable data on thermal asymmetries [41].

The Glamorgan Protocol was published in 2008 with the purpose of standardizing the ROI determination in thermographic studies [33]. This document presents the definition of 90 different ROIs. Although the Glamorgan Protocol provides important guidance in the determination of the ROI, there are a number of studies with very specific demands. Because of this, researchers have developed their own criteria to define ROIs, with different geometries and methodologies [34, 38, 44]. Differences in skin temperature between ROIs are due to the tissue composition, peripheral blood flow irrigation, muscular activity, and capacity of sweating [5, 27, 42, 44, 48], and these factors are important to take into account during the definition of the ROI.

The reproducibility in the determination of the ROIs for different images is another important factor in ensuring the accuracy of the results, especially when different subjects are compared. Different strategies were observed in the literature in order to increase the reproducibility of thermographic data. Some researchers use software with an automatic ROI selection feature [29, 49–51]. However, these software might not be accessible to all users, and for this reason, one of the most used strategies is the determination of ROI according to anatomical proportions [27, 52] or body segments [53–56]. In any case, it is important to ensure reproducibility by defining the criteria for the definition of the ROI, and performing training before the analysis [38]. Ideally, these methodological steps should also be reported in papers addressing infrared thermography measures.

Mathematical methods can also be used in the determination of ROIs. One example of this is the Tmax method proposed by Ludwig et al. [39]. In this method, the operator selects an ROI and the software will automatically select the five warmest pixels within the ROI, with a minimum distance of at least five pixels between each. The software selects an area of 5 x 5 pixels around each of the five pixels previously selected, determining the mean temperature across the 125 pixels included in the selection [39]. The authors suggested that the main advantage of this method is the use of the same number of pixels for the calculation, removing the effect of the different anatomical sizes between participants [39]. It is also possible to determine the ROIs using markers in anatomical references or outer delimiting areas of the body surface (Fig. 4.2) [19, 42, 57]. This strategy can be useful in ensuring the reproducibility of the delimitation of the ROI, and it facilitates further data analysis.

A current interest is the implementation of automatic analysis of thermographic data. Infrared thermography combined with machine learning techniques is increasingly being used to enhance the automation and accuracy of thermal image analysis, particularly in the detection of ROIs in sport science and biomedical applications. Deep neural networks allow the segmentation of ROIs by learning distinctive thermal patterns, which improves the reproducibility and comparability of

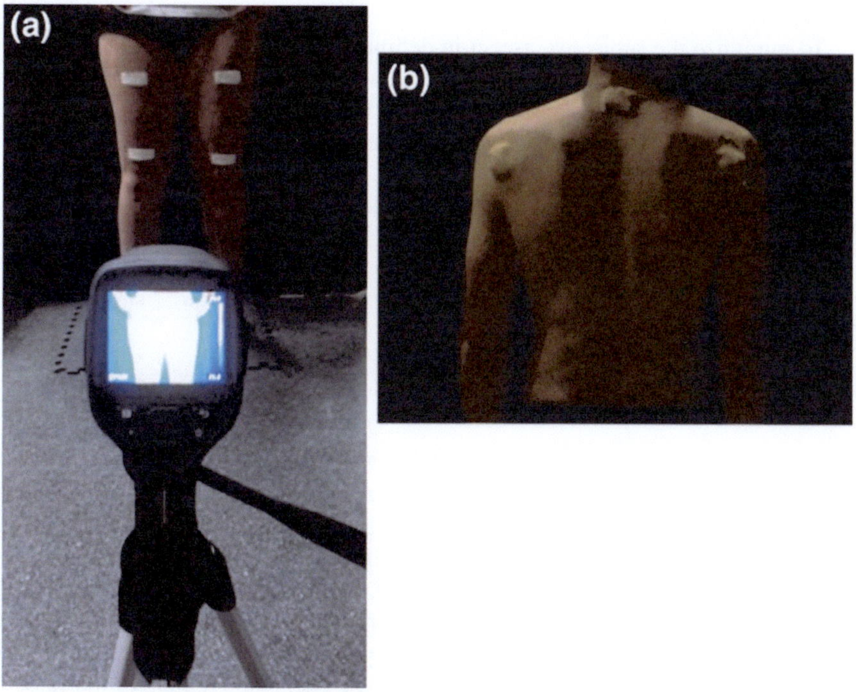

Fig. 4.2 Examples of the use of markers to delimit the ROIs in the quadriceps (**a**) and back (**b**)

thermographic assessments [58]. One of the key advantages of this approach is its ability to process large datasets, identifying subtle thermal variations that might be overlooked in manual analyses [58]. Moreover, automated ROI selection increases the objectivity of thermographic studies, mitigating intra- and interobserver variability [2]. Despite these benefits, machine learning based thermal analysis still faces limitations, including the need for extensive labeled datasets to train deep learning models and the potential for errors in cases of motion blur or low-resolution thermal images [58]. Future research should focus on integrating advanced tracking algorithms and hybrid machine learning models that incorporate physiological priors to improve ROI detection and thermal pattern recognition in real-time applications [58].

After this recapitulation of information, we can provide some overall recommendations for the manual determination of ROIs:

- To define, before starting the measurements, the criteria for the ROIs determination.
- To record the maximum number of pixels of the ROI in the thermal image, put the camera as close as possible to the region (always considering that the image must be focused). A higher number of pixels will provide more robust descriptive statistical data relating to the ROI.
- Whenever the objective of the study permits, use large ROIs instead of small ROIs. The large ROIs will reduce the possible effects of a single pixel of the ROI, such as hot spots.
- To minimize the subjective aspect in defining the ROIs and increase reliability, strategies such as using software with automatic ROI selection, marking points on the skin, or using anatomical references for delimitation can be helpful.
- To determine the ROIs, consider aspects like tissue composition, muscular activity, and capacity of sweating into account. We recommend avoiding measurements in the same ROI, regions in which the muscles are close to the skin (e.g., thigh or leg), and regions with a higher proportion of bone and connective tissue close to the skin (e.g., knee or Achilles).

4.2.2 Methodological Aspects of the Camera

The choice of camera is an important step toward the valid use of thermography. Cameras for thermographic assessment can be cooled or uncooled cameras (Fig. 4.3). Cooled cameras normally operate at cryogenic temperatures [59]. The older cameras were cooled by the addition of liquid nitrogen, argon gas, or sterling cooler, resulting in a limitation of the angle of operation of the camera [59–61]. In this case, a high camera angle could put the liquid nitrogen or other cooling system in contact with the electronics and cause damage. Other cooling systems (e.g., thermoelectric systems) were introduced to overcome the angle problem of the camera [59, 60]. The cooling system is usually associated with power and time taken to cool

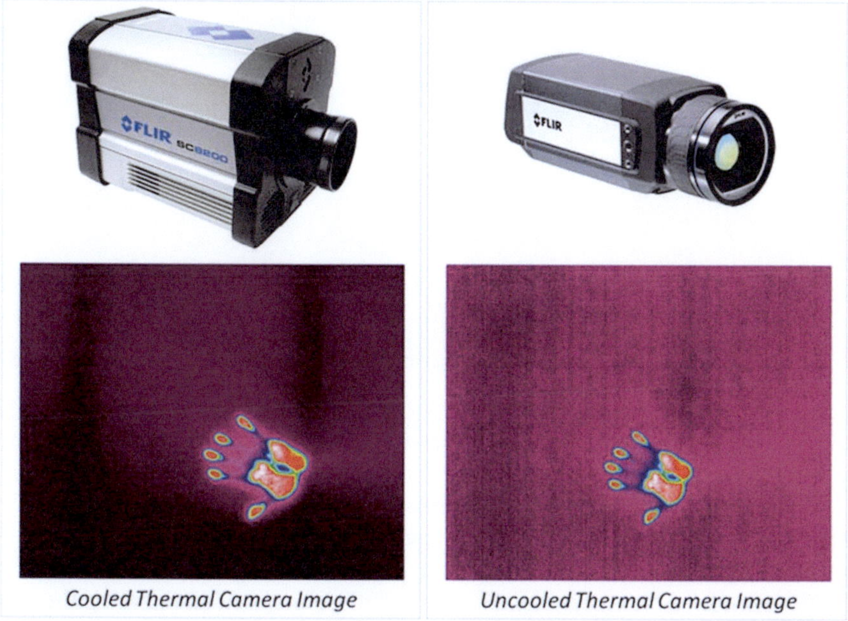

Fig. 4.3 Examples of cooled and uncooled cameras and respective image record. (Figure modified using images from the FLIR website)

down the camera, and it is bulky and expensive [47, 59]. However, the cooled detectors provide a superior image quality and a greater sensitivity to small differences in the scene temperatures than the uncooled detectors [47, 59]. The more recent generation of uncooled cameras can be used without cooling but utilizing a microbolometer [61]. These cameras will be stabilized at ambient temperatures using temperature control elements (based on changes in resistance and voltage) to reduce the image noise [59]. Although these cameras provide a lower quality of image and have lower sensitivity, they are less expensive in price and maintenance, smaller, and easier to use [59, 60].

Also related to the choice of the camera, infrared resolution is an important technical aspect to be considered (Fig. 4.4). Although infrared resolution is not the most important factor in the accuracy of the calculation of the temperature [62], it is always preferable to have a camera with the best resolution that you can afford [34]. Each pixel in the thermal image represents one thermal data, and a higher number of pixels will result in a more robust measure when the mean temperature is calculated in the ROI. One example of this is that 25 pixels is considered to be the minimum number of pixels to determine the average temperature for a region of interest for the ISO normative [45, 46]. In this regard, an infrared resolution of 320 × 240 pixels is commonly considered to be the minimum resolution necessary for satisfactory data in human assessments [19, 34].

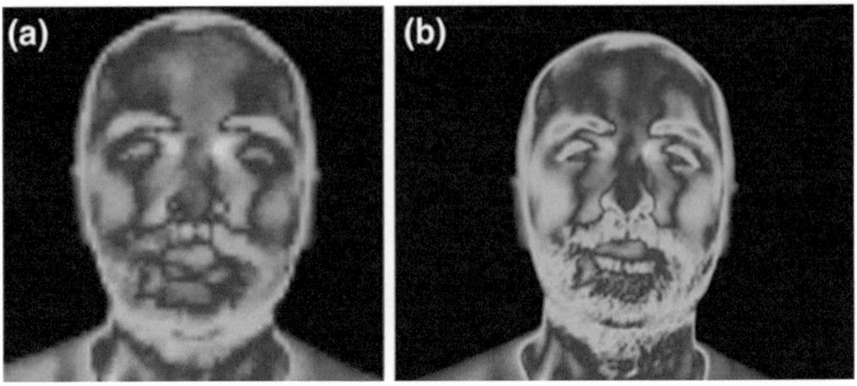

Fig. 4.4 Example of the same image recorded with an infrared camera with a resolution of 140 × 140 pixels (**a**) and with an infrared resolution of 320 × 240 pixels (**b**)

Furthermore, another critical parameter to consider is the Instantaneous Field of View (IFOV), which defines the spatial resolution of the camera by determining the smallest object size that can be accurately measured at a given distance [63]. The IFOV depends on the detector pixel pitch and the lens system, influencing the ability of the camera to distinguish fine details in the thermal image. A smaller IFOV results in better spatial resolution, allowing for more precise temperature measurements and reducing errors due to mixed pixels, particularly in small ROIs or when analyzing fine anatomical structures [40]. When selecting a camera for human assessments, ensuring that the IFOV is adequate for the intended measurement distance is essential to maintain measurement reliability and accuracy [40].

The emergence of low-cost infrared cameras has significantly expanded the accessibility of thermographic imaging across various domains, including medical applications [64–66]. These portable and affordable systems facilitate noncontact, real-time temperature assessment, proving particularly useful in resource-limited settings where high-end thermographic devices are not feasible [66]. However, these systems present significant technical limitations, such as lower spatial and thermal resolution, susceptibility to environmental interference, and calibration drift over time [40, 67]. Studies indicate that low-cost devices often overestimate temperature readings and require extended stabilization times, making them less reliable for precise clinical measurements [67]. Additionally, factors such as camera distance, emissivity settings, and ambient conditions must be carefully controlled to ensure accurate readings [40]. Therefore, while low-cost infrared cameras offer practical and scalable solutions for thermal imaging, users must implement rigorous validation protocols and consider their inherent limitations when interpreting results.

For sport science applications in particular, the capacity to record thermographic video is an important feature in the thermographic camera. There are a fewstudies using this feature in sport assessment [68–71]. In the past, it was suggested that the calculation of the temperature in a sports assessment could be disturbed by nearby electronic equipment (e.g., ergometer and computers) [60], explaining why researchers usually try to analyze static images recorded before and after exercise [12, 72]. However, we have observed that this kind of electronic device has a minimum influence on reflected temperature (max 1 °C), and this has practically no impact on skin temperature due to its high emissivity.

As observed in many other instruments, a proper calibration is also necessary for thermographic assessment. Infrared cameras are calibrated by the manufacturers with a high number of black body sources at different temperatures when the signal of each pixel is determined [73]. A black body source is an instrument with a stable and accurate temperature surface, with a high emissivity, configured with a specific temperature emission (Fig. 4.5). These black body sources usually have an emissivity as high as 0.98 [73]. In addition to the manufacturer's calibration, in research studies, the camera usually is calibrated again before every experiment starts. To do this, the black body sources need time to stabilize the temperature before the calibration. The calibration is more precise when the black body source is set at a higher temperature than the room temperature, and it is important to know the accuracy and the emissivity of the black body source in order to perform the best calibration possible. For example, the black body source can be configured to some specific temperature emission (e.g., 50 °C). When the black body is stabilized, a thermography camera can record a thermal image of where the black body is. After that, we can analyze the difference between the temperature of the black body source in the thermography and the set temperature of the black body source. If the difference is within the accuracy of the black body source, the camera is measuring correctly, but if we have a higher difference, we can compensate for this difference in the data analysis. However, if the difference is very high, it is recommended to send the camera to the manufacturer for calibration. In addition, the black body (and also a black plate thermistor) can be placed in the space of measurement to have this reference in the image, which also permits any drift in the temperature sensitivity during the laboratory daily routine to be checked helping to increase reliability [23, 60] (Fig. 4.6).

Another important aspect to consider is that the camera should be turned on at least 10 min before the measurements start. Infrared thermography cameras have a condenser to stabilize the electronics and compensate their temperature variations.

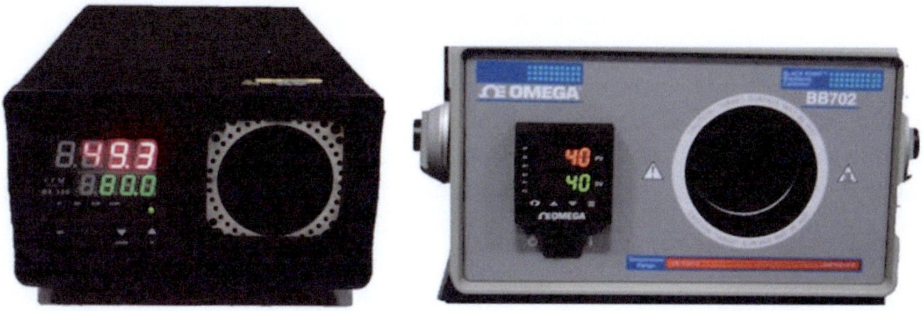

Fig. 4.5 Examples of black body systems

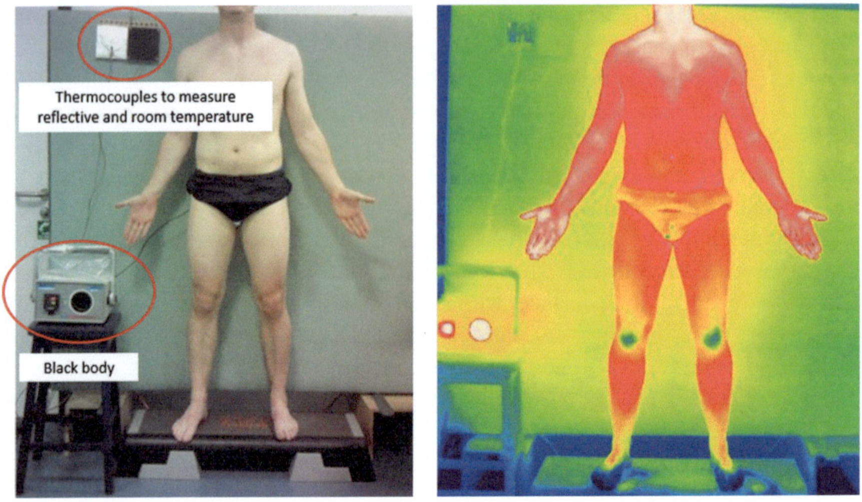

Fig. 4.6 Examples of different references (black body and thermocouples) placed in the space of measurement to increase the reliability

However, it needs time to adjust to the environmental conditions and to work properly. The time necessary for this stabilization may differ between camera models. For example, the minimum time determined for the uncooled camera used by the authors of this chapter was 5 min [44]. However, since their camera measurements became stable after 5 min of turning on the camera, a 10-min stabilization period has been suggested to ensure proper camera operation [44]. It is important to note that some cameras may require being turned on for a minimum of 10 min to stabilize [73]. The most common procedure to ensure camera stabilization time is to set some system with a stable temperature (black body source or hot plate system), turn on the camera, and record thermal images continuously (Fig. 4.7). When there is no variation in the temperature data with the time, it means that the camera operation is stable.

The self-calibration of infrared cameras is another crucial process to correct for thermal drift, a phenomenon where changes in the camera's internal temperature affect the accuracy of temperature measurements [74]. This drift occurs because infrared detectors not only capture radiation from the observed scene but also from the internal components of the camera, including the optics and sensor housing, which can introduce measurement bias over time [75]. To mitigate these effects, radiometric infrared cameras periodically perform an internal recalibration, often using a mechanical shutter to momentarily block external radiation and measure the camera's thermal signature [76]. This mechanical shutter can be heard while using the camera, and this allows the system to correct for temperature-dependent variations in pixel response, ensuring more stable and accurate readings [77]. However, this recalibration process has a potential drawback: at the moment of recalibration, the measured temperature may shift, leading to inconsistencies if measurements are

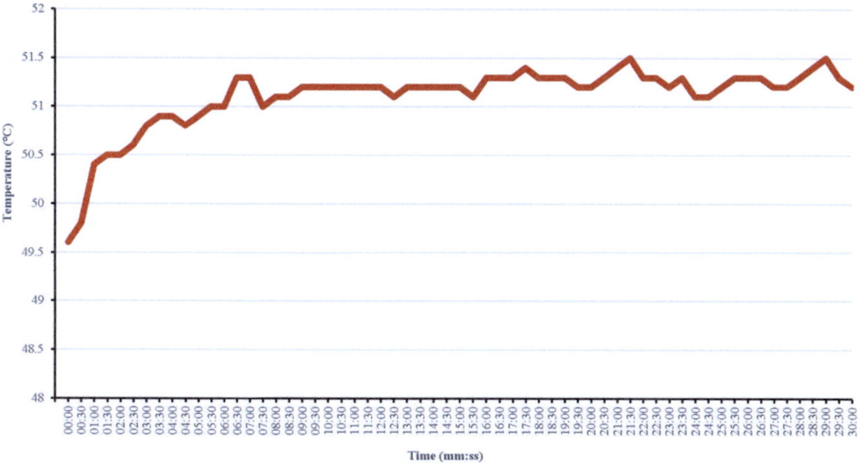

Fig. 4.7 Example of how to measure the time necessary for the stabilization of an infrared camera by the registration of the temperature of a black body at 50 °C

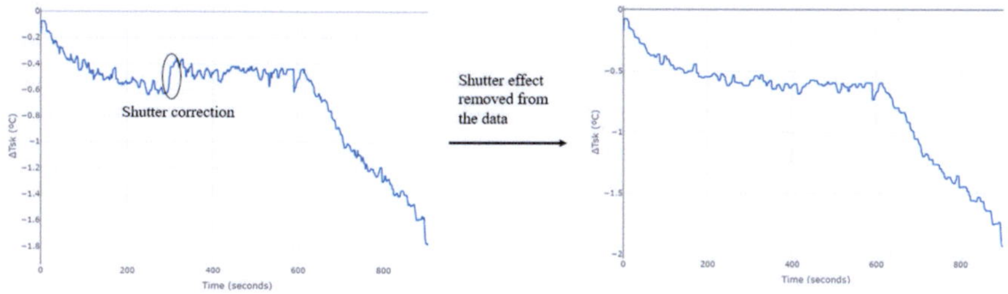

Fig. 4.8 Example of the effect of shutter correction in a thermography video data (variation of skin temperature), and how to remove this effect by calculating the measured temperature difference produced at that specific moment and subtracting it from subsequent data

taken immediately before or after the event [78]. Consequently, best practices for thermographic data collection include monitoring when self-calibration occurs and ensuring that critical measurements are not performed during or immediately after this process [79]. Understanding and accounting for these temporal variations are essential to maintain the reliability of temperature assessments in both controlled and field-based applications [74]. Furthermore, during thermal video measurements, this recalibration can be corrected by calculating the measured temperature difference produced at that specific moment and subtracting it from subsequent data (Fig. 4.8).

The position of the camera (distance and angle to the ROI) can affect the measurements and needs to be standardized [41, 80–82]. The camera lens should be placed perpendicularly to the ROI. Angles higher than 60° between the camera and the ROI considerably modify the data [82]. Although it has been observed that the angle effect (0° vs. 39°) is higher than the effect of the distance (1.2 vs. 3 m) [83], it is important to take all the thermal images of one study at the same distance to minimize these errors. In this sense, Tkáčová et al. [82] observed differences of 0.2 °C between measurements performed between 0.2 and 2.5 m between the camera and the ROI. Moreover, Machado et al. observed that the percentage of false positives and false negatives when determining thermal asymmetry increases with the distance (0.7 vs. 1.5 m) [41]. Although the distance is an important parameter to control, it also depends on the infrared resolution of the camera and the ROI [41]. Most of the studies with humans registered the measurements at distances between 1 and 3 m [11, 12, 29, 31, 44, 55, 72]. The camera must be placed as close as possible to the ROI and record the highest number of pixels possible in that region [47]. In order to have some reference values, a distance of 1–1.5 m can be enough for an ROI that is a body segment (Fig. 4.9a), and a distance of 3–4 m can be necessary when the ROI is involving all the whole surface of the subject body (Fig. 4.9b). Similarly, when taking thermographic images, the focus of the camera must be adjusted. Errors in the focus adjustment consistently produce errors in the temperature calculation of the ROI (see details in Fig. 4.10).

Although currently the application of infrared thermography for assessing skin temperature in real-world conditions (including field-based sports settings) is anecdotal, it is important to mention and consider for future applications. Aylwin et al. [84] investigated the use of thermography for monitoring skin temperature in athletes during competition and highlighted key methodological considerations. One of the main challenges is motion blur, which affects the accuracy of thermal measurements, particularly for lower-cost infrared cameras that have slower response times compared to better cameras. To mitigate these effects, high-speed cameras with short exposure times and frame rates suitable for dynamic conditions should be prioritized. Proper selection of the ROI is also crucial, as smaller ROIs can lead to less representative measurements, and missing data from certain body areas can alter the accuracy of mean skin temperature calculations. Furthermore, according to

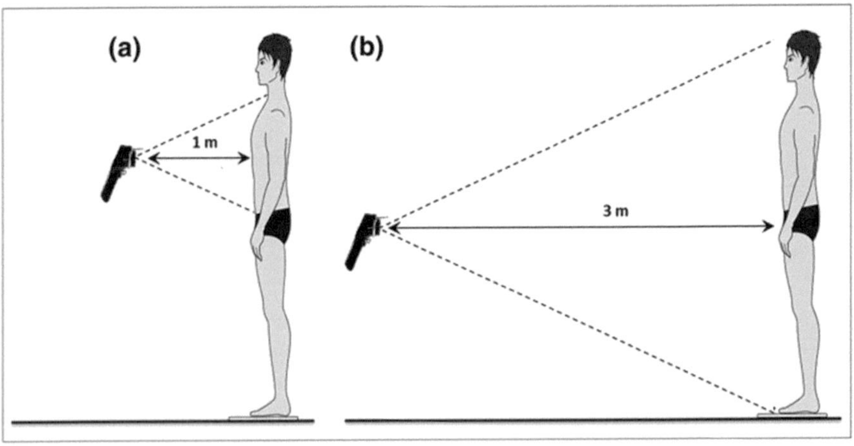

Fig. 4.9 Example of the position of the camera in relation to the ROI

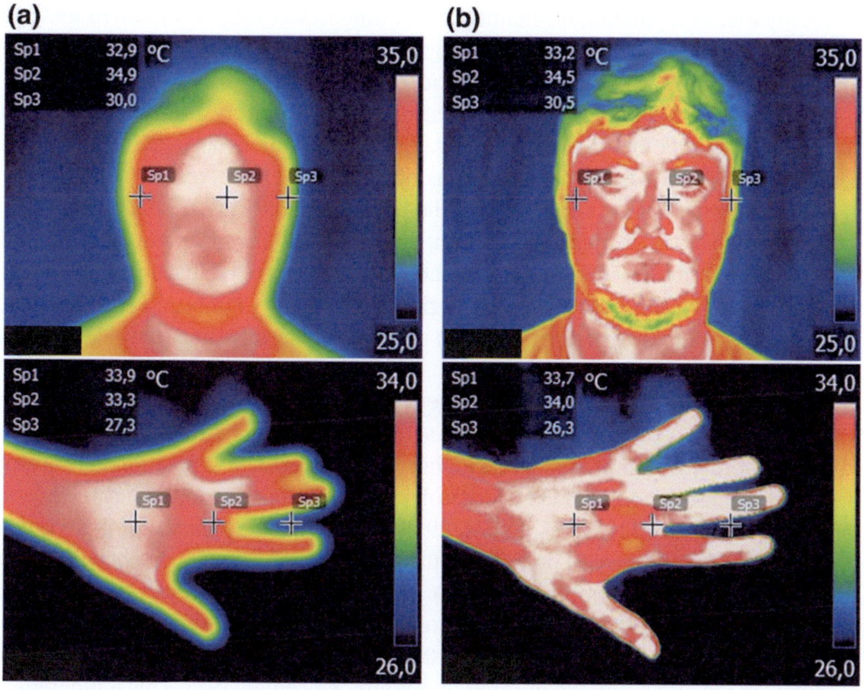

Fig. 4.10 Example of error calculation in images with an incorrect (**a**) and correct (**b**) adjustment of the focus

the application, the fact that other athletes may hide the ROI during the measurement, or that the athlete passes at a distance where the image is not in focus, are aspects that complicate its use in the field.

4.2.3 Methodological Aspects Related to the Space Where the Measurement Will Be Conducted

Considering the space measurement, the main methodological aspects to take into account are the room space, the room temperature, the relative humidity, and all sources of infrared radiation that might affect the thermographic measures by standing around the subject. The parameters of the room temperature and its relative humidity must be informed in the camera setup. The minimum space dedicated to the measurement is 6–9 m^2, but spaces of 12 m^2 are preferable [60, 73]. The room temperature is one of the most important factors affecting skin temperature because both are directly related [85–87]. Different authors have suggested that room temperature should be between 18 and 25 °C during thermography assessment in humans [61, 81]. This is because the room temperatures outside of this range could affect the thermoregulation mechanisms, resulting in sweating when the room temperature is higher than 25 °C, or shivering when the room temperature is lower than 18 °C [61]. Moreover, lower accuracy of the measurements was observed at low room temperatures (12 °C) [83]. Furthermore, temperature variation in the room should be no higher than 1–2 °C. When data from measurements made in the trunks of 64 male cyclists were analyzed, it was observed that an increase of 1 °C in the room temperature resulted in 0.35 °C higher skin temperature ($r = 0.5$ and $p < 0.01$). This influence was minimized when temperature variations (difference before and after) were determined (Fig. 4.11). Ring and Ammer suggested that an environmental temperature of 20° is better for visualizing inflammatory injuries [60]. Therefore, the ideal environmental temperature for thermographic measures may depend on the purpose of the experiment.

The relative humidity in the room where the measurements are taken affects the sweat evaporation rate during physical activity and sport performance [88]. The influence of relative humidity can be higher when the camera is positioned far from the ROI [73]. It is recommended to control relative humidity to values between 40% and 70% [11, 12, 29, 31, 32, 54, 68, 72, 89, 90]. It is also important to avoid or minimize, at best, the presence of particles in the air, such as dust or vapor, because these can absorb the radiation emitted by the participant and therefore cause measurement errors because not all the infrared radiation of the participant reaches the camera.

Reflected temperature is defined as the temperature of other objects that are reflected by the target into the infrared thermography camera [91]. This temperature should be measured and informed during camera setup configuration. The software of the camera will disregard this parameter from the calculation of the temperature of the ROI. It is recommended to measure the temperature reflected in each test. The reflected temperature can change when the environmental conditions change (e.g., variation in the room temperature), the conditions of the evaluator (e.g., changes in clothing), or any feature of the space changes. One of the most used methods to measure the reflected temperature is the "reflector method" described in the international normative ISO 18434-1:2008 [91]. This method consists of

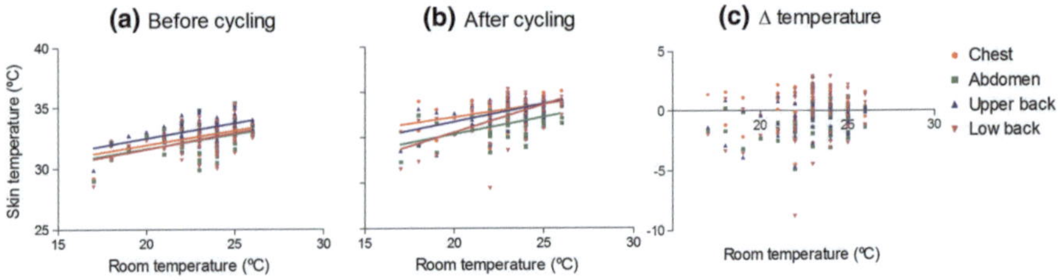

Fig. 4.11 Influence of the room temperature on the skin temperature measured before (**a**) and after cycling (**b**). The influence was removed when skin temperature variations (difference between temperature after and before cycling) were assessed (**c**)

1. Informing the following parameters to the camera: distance with a value of 0 and an emissivity with a value of 1.
2. Positioning cardboard with an aluminum foil at the same level as the participant (Fig. 4.12a).
3. Measuring the average temperature of the aluminum foil (using a rectangle ROI) (Fig. 4.12b).
4. Introducing the thermal average obtained, such as the reflected temperature.

The presence of infrared radiation sources close to the measurement space should be avoided. Infrared radiation sources include the incidence of sunlight in the space of measurement, the presence of electronic devices, and the number of people in the measurement space (evaluator and participant). Other sources of radiation are heating ducts, water pipes, heating lights, and airflow. Likewise, reflections of infrared radiation around the participant should be avoided, so we recommend placing anti-reflective material behind the plane of interest [19, 73].

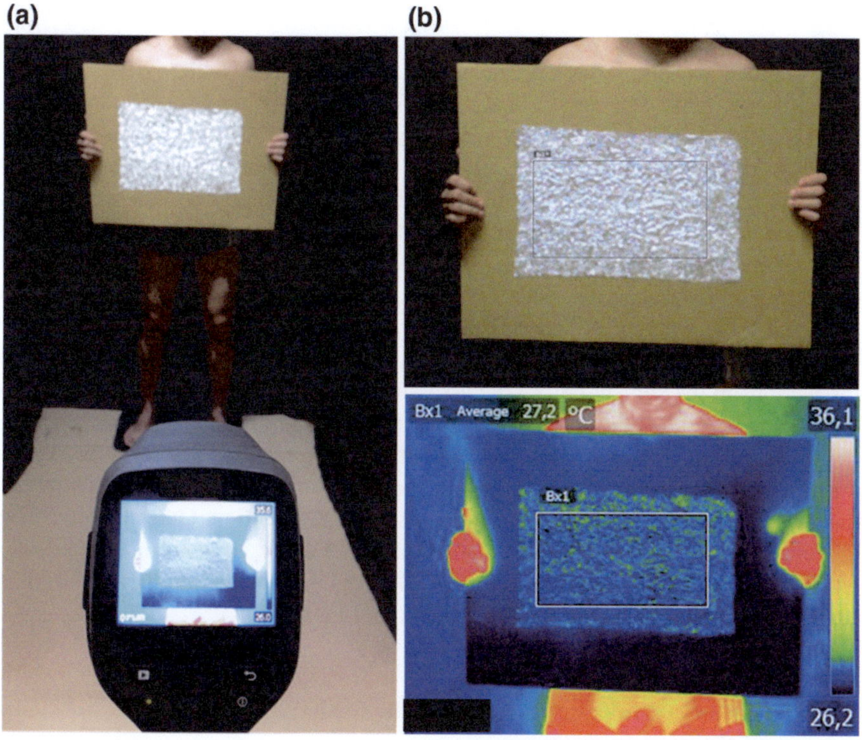

Fig. 4.12 Reflector method

4.2.4 Methodological Aspects Related to the Measuring Subject

Considering the characteristics of the subject being evaluated, one of the factors to take into account is the emissivity of human skin. Different studies determined the human skin emissivity and described values between 0.97 and 0.99 [92–94]. Thermographic studies with human subjects commonly consider an emissivity of 0.98 [11, 19, 27, 29, 31, 37, 95]. These studies considered this emissivity value for all the participants due to the lower variability observed—a standard deviation of 0.01 [92, 93]. This standard deviation in the emissivity resulted in approximately ±0.05 °C or ±0.12% of change in the absolute temperature, and such error will be smaller than the thermal sensitivity of the camera or the variability of the skin temperature between participants and trials. These values of emissivity of the skin mean that humans are a good emitter of infrared radiation, and therefore, it is possible to obtain accurate surface temperatures [94, 96]. The emissivity value must be introduced into the camera or thermographic software for the correct calculation of the temperature.

The thermal variability of the human skin temperature can be very high due to many factors, such as individual metabolism, blood flow, and adipose tissue, among others [34]. The present chapter has an applied orientation, and for this reason, the factors will be classified considering the experimental process and when they will be taken into account: (1) in the design of the experiment; (2) in the criteria for selection of the participants; and (3) to elaborate the instruction that participants need to follow (Fig. 4.13). Some of the points related to the selection of the participants or instructions they will receive can be interpreted in a different classification depending on the researcher. For example, some researchers will consider it important to classify the participant as a smoker or nonsmoker in the inclusion criteria for the research, while for others, such information will be pertinent when sending the instructions to the subject before the assessment. For this reason, it is important to understand that the authors of the present chapter do not think that their classification is the only one to be considered, but rather is the classification considered suitable for this chapter. On the other hand, the most important aspects in which the thermography evaluator can act were addressed in this section. Other factors, such as genetics, were not addressed in this chapter. Finally, as we commented, the TISEM checklist [37] is very useful to check if the most important methodological aspects are followed. This checklist is shown in Fig. 4.14.

4.2.4.1 Factors to Take into Account in the Design of the Assessment Protocol

Some factors are essential to take into account to obtain the maximum reproducibility in the thermographic measurements. The temperature of the skin usually presents high variability, which may limit the comparison of results between and

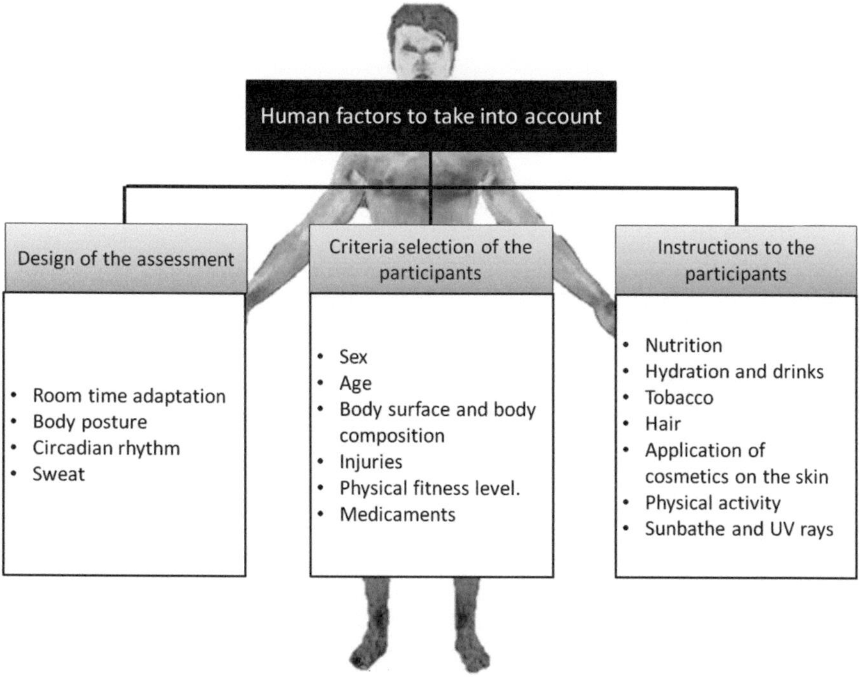

Fig. 4.13 Classification of the methodological aspects related to the person following the experimental process

within subjects. Therefore, some general recommendations can be followed to reduce these intervening factors.

Room Time Adaptation One of the most important methodological aspects is the time required to achieve a stable measurement of skin temperature in a controlled environment [43, 60]. When the individual arrives at the measurement space (typically, he/she comes from a different environmental temperature) and prepares for the thermographic measurements (e.g., undressing), it will produce a change in thermal conditions, and hence a variation in the skin temperature. Therefore, an adaptation period is necessary; otherwise, the repeatability of the measurements will be very low because the external conditions of each day and each subject are different. During this adaptation time, the participant must avoid all the movements and contact that can affect the skin temperature of the ROIs; for instance, actions such as folding or crossing the extremities, placing bare feet on a cold surface, etc. [60]. In this sense, the use of a carpet or rug on the ground is recommended. Studies commonly use room time adaptation ranging from 10 to 20 min [12, 27, 32, 54, 89, 97, 98]. Longer periods are not recommended because when the time adaptation exceeds 30 min, the temperature oscillation is higher and may result in thermal asymmetries [43, 60]. Marins and colleagues concluded that a minimum time of 10 min is suitable for temperature stabilization, but they suggested that longer times

1) The relevant individual data of the participants must be provided. Note: These could include, but are not limited to, age, sex, body mass, height, body mass index, ethnicity and whether they are smokers or not. An indication of physical activity profile (e.g. frequency, duration, intensity, and activity description) should be reported.
❑ Yes ❑ No ❑ Unclear
2) Participants should be instructed to avoid alcohol beverages, smoking, caffeine, large meals, ointments, cosmetics and showering for four hours before the assessment. Also, sunbathing (e.g. UV sessions or direct sun without protection) should be avoided before the assessment. Note: This should be confirmed verbally before the assessment. The use of any medicinal treatments or drugs should be recorded. Any condition that could not be avoided should be reported.
❑ Yes ❑ No ❑ Unclear
3) Extrinsic factors affecting skin temperature (e.g. physical activity prior to the assessment, massage, electrotherapy, ultrasound, heat or cold exposure, cryotherapy) should be clearly described.
❑ Yes ❑ No ❑ Unclear
4) Ambient temperature and relative humidity of the location where the assessment took place must be recorded and reported as mean ± standard deviation.
❑ Yes ❑ No ❑ Unclear
5) The assessment should be completed away from any source of infrared radiation (e.g. electronic devices, lightning) or airflow (e.g. under an air conditioning unit). Note: Any condition that could not be controlled should be reported.
❑ Yes ❑ No ❑ Unclear
6) The manufacturer, model and accuracy of the camera used should be provided. Note: When available it is recommended to provide the maintenance information of the equipment (e.g. when and where it was completed the last calibration).
❑ Yes ❑ No ❑ Unclear
7) An acclimation period in the examination room should be completed. Note: This item is only applicable for initial baseline measurements or basal analysis.
❑ Yes ❑ No ❑ Unclear
8) If necessary the camera should be turned on for some time prior to the test to allow sensor stabilization following the manufacturer's guidelines.
❑ Yes ❑ No ❑ Unclear
9) Conditions of image recording such as mean distance between object and camera, percentage of the region of interest within the image should be detailed.
❑ Yes ❑ No ❑ Unclear
10) The camera should be positioned perpendicular to the region of interest.
❑ Yes ❑ No ❑ Unclear
11) Emissivity settings of the camera must be reported. Note: 0.98 of emissivity is suggested for a dry clean skin surface.
❑ Yes ❑ No ❑ Unclear
12) The time of day at which the images were taken should be reported.
❑ Yes ❑ No ❑ Unclear
13) The standard body position of the subject and the regions of interest must be well described and appropriately selected. A visual example (with temperature scale presented and scale of colors properly configured) is recommended.
❑ Yes ❑ No ❑ Unclear
14) If the skin is dried (e.g. to remove surface water), the drying method should be clearly described.
❑ Yes ❑ No ❑ Unclear
15) The evaluation of thermograms and collection of temperature from the software should be clearly described.
❑ Yes ❑ No ❑ Unclear

Fig. 4.14 "Thermographic Imaging in Sports and Exercise Medicine" (TISEM) checklist. (Figure obtained from Moreira et al. [37])

can be required when participants come from extreme environments (cold or hot) [43].

Body posture during the image acquisition will be dependent on the ROI defined. Some examples and guidelines are present in the catalog of ROIs described in the

Glamorgan Protocol [33]. In this sense, the anatomical position is usually a very common position for thermal measurement using infrared thermography. However, there will be some cases not considered in that protocol and, for these cases, special attention should be given. The body posture needs to be the same during the whole room time adaptation [60, 99]. In some exercises, it is possible to take the thermal images during the moments of force production or along the movement (e.g., in a static contraction or during a body segmental movement) [32, 97, 100]. We do not recommend comparison of thermal images taken in different body postures because the posture changes the body surface exposed to the environment, and then the skin temperature may also be changed [23, 60, 99, 101] (Fig. 4.15).

Circadian Rhythm The circadian rhythm affects the core temperature [102, 103], skin blood flow [103], and skin temperature [104–107]. Skin temperatures are lower in the morning and increase during the day [104–107]. The hands are the body region with the highest increase in skin temperature during the day [104, 106]. These effects of skin temperature could be related to the same trend observed by Smolander et al. about the relationship between skin blood flow and the circadian rhythm [103]. In order to minimize this effect, some studies have performed all measurements in the same period of time [11, 31, 32, 39, 54]. Another possible strategy to reduce the intrasubject effect of the circadian rhythm, when the study has a cross-over experimental design, is to ensure that measurements of each participant are always performed at the same time of the day [89, 90].

Sweat Moistening of the skin surface due to perspiration may influence thermographic data [12]. Ammer [108] suggested that a film of water on the skin may act as a filter for infrared radiation, and that could lead to an error in the estimation of the skin temperature. Removing the sweat or water from the skin has been tried as

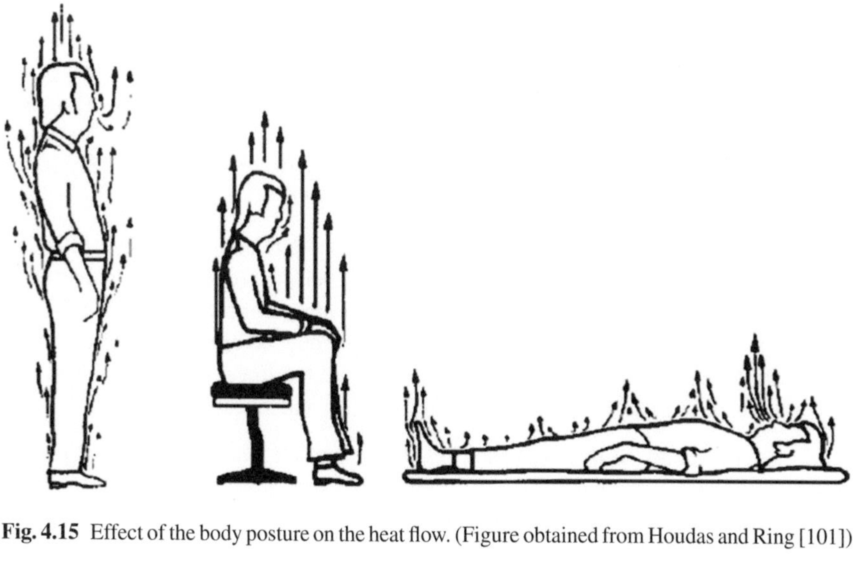

Fig. 4.15 Effect of the body posture on the heat flow. (Figure obtained from Houdas and Ring [101])

a solution [109, 110]. These studies [109, 110] tried to remove the sweat without friction from the skin. However, in other studies, temperature could increase as a consequence of rubbing the skin surface and also as a result of the reduction of the natural process of sweat evaporation [12]. On the other hand, the results of one study indicated that the accumulation of sweat during cycling is not enough to form a film of water and that sweat produced under these conditions does not affect the thermographic data [12]. Moreover, for the alternative to measure skin temperature (thermal contact sensors), sweat also affects the data, as it is accumulated in the sensor and its attachment [17]. However, the influence of the sweat on the skin emissivity in infrared measurements, as well as the possible strategies to deal with the effect of sweating on skin emissivity during thermographic examinations, is unclear to date. Hence, future studies are necessary to explore this methodological aspect in more detail.

4.2.4.2 Factors to Take into Account in the Criteria Selection of the Participants

In order to reduce the variability of the skin temperature results, many factors should be considered. It starts with the definition of the criteria for the selection of the participants. Some of the most relevant factors are commented.

Sex Women present higher core temperature [102, 111], lower whole-body sweat rate [112] and higher percentage of body fat than men, which influences heat loss and results in lower skin temperature in most of the body regions [52, 53, 113]. The lower heat production during exercise in women is another explanation for these lower skin temperatures [114]. These gender effects are more evident in the morning than in the evening, possibly because the blood flow is lower and then the thermal insulation effect of the body fat is higher [106]. It is also important to consider the influence of the menstrual cycle. Different studies have shown that core temperature [115, 116] and skin blood flow [117] were higher during the luteal phase than in the follicular phase, suggesting a possible higher skin temperature in women. However, a recent investigation suggested that the menstrual cycle does not affect skin temperature, but menopause results in lower skin temperature [118], which encourages further research on this topic.

Age Older people present lower basal values of skin temperature [87, 97, 119], and after an intervention aimed at increasing body temperature (e.g., physical exercise), a lower response of skin temperature was found among the elderly [87, 97]. The reasons for the skin temperature decrease with age may be related to a lower metabolic rate [120], lower core temperature [102], and the lower capacity of heat dissipation via vasodilation/vasoconstriction [87, 121] and sweat rate [120, 122].

Body Surface and Body Composition Body surface has an important role in whole-body heat exchange [123, 124]. Individuals with a greater body surface have a greater capacity for heat dissipation [124]. This happens because they have a greater skin surface and then greater absolute rates of convection, radiation, and sweat evaporation [123]. For this reason, to compare different groups or participants, normalization of body surface variables is recommended (e.g., sweat rate; mg cm^{-2} min^{-1}). Body surface is usually measured using predictive equations considering the height and body mass, and one of the most used equations is the DuBois and DuBois' equation [125]. Concerning body composition, the level of body fat influences the capacity for heat loss, resulting in lower skin temperatures [126, 118, 127, 128]. Body fat tissue had an insulation capacity, resulting in impairment in heat dissipation between the core and the skin [72, 126, 127, 129].

Injuries An injury is often related to variations in the regional blood flow. Changes in blood flow affect skin temperature, which can increase in the case of an inflammation, or decrease in the case of tissues with poor perfusion, degeneration, or reduced muscular activity [19, 61, 130]. Individuals with overuse and traumatic injuries could present with an alteration of skin temperature and thermal symmetry [19, 61] (Fig. 4.16a). In addition, other disorders such as diabetic neuropathy, fever, and vascular disease, among others, can influence skin temperature [20] (Fig. 4.16b). For this reason, studies selected participants without a history of injuries in the previous months [11, 89, 131]. This is certainly a fertile topic for further investigation, since the medical application of thermography, as mentioned before, has many advantages over other imaging methods.

Physical Fitness Level Physical fitness influences skin temperature [7, 27, 31, 32, 68, 72]. Larger heat loss is observed among participants with better physical fitness, which could be explained mainly through the higher capacity of evaporation of sweat during exercise [72, 132]. Moreover, trained people usually have greater capacity for heat transference between the core and the skin due to higher blood flow and lower body fat [7, 27, 72, 127, 129, 133]. The variability of fitness level will result in high variability of the skin temperature if considering a heterogeneous group.

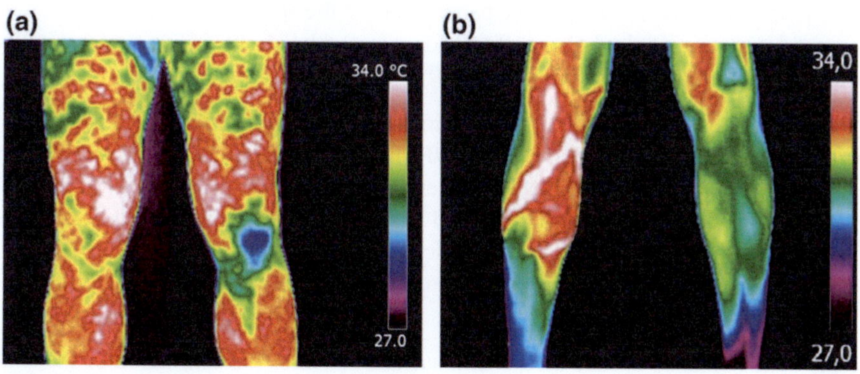

Fig. 4.16 Example of the presence of an undiagnosed injury in the knee (**a**) and a varicose vein in the leg (**b**)

Medicaments Medicaments can alter skin blood flow and then skin temperature. It is known that different kinds of medicaments, such as analgesics, anti-inflammatories, vasoactive, hormonal medications, prophylactics, and anesthetics, can affect normal skin temperature [34]. Furthermore, infrared thermography was used in different studies as a method to determine the response of a medicament [134–137]. In human subjects assessment, when the objective of a study is not to assess the medicament effects, it is common to recruit participants without a medicament prescription or to avoid the intake of medicaments before a study to remove this factor [31, 32, 54, 131].

4.2.4.3 Factors to Take into Account in Instructing Participants

It is suggested that participants be instructed through a previous session or using a written letter or email containing the information and procedures necessary to avoid possible measurement errors. When possible, a familiarization session can be very valuable to fully address all the factors that can influence measurement. Some of the most important factors are described below.

Nutrition It is commonly established that after food intake, blood supplies the stomach to the detriment of the skin, resulting in a lower skin temperature in the extremities [138, 139]. Heavy meals approximately 4 h before the measurement should be avoided for the human thermography assessments [27, 31, 44].

Hydration and Drinks Before the thermographic assessment, it is important to ensure that the subject is adequately hydrated. Hypohydration can affect the sweat rate response and skin temperature, as detected by infrared thermography [140]. On the other hand, the intake of different liquids has been associated with an alteration of skin temperature. For example, alcohol intake increases skin temperature due to the increase in the skin blood flow [141–144]. Liquids with caffeine and other stimulants, such as teas, stimulate the physiological function, increasing skin temperature [145–147]. For these reasons, participants of the studies are usually requested to avoid ingestion of alcohol or caffeine for at least 4 h [31, 39, 54] or, preferentially, 12 h before the measurements [27, 44, 89].

Tobacco It is well established that nicotine has a vasoconstrictive effect. Therefore, it may reduce skin temperature [148–150] (Fig. 4.17). In addition, it is known that smokers have a reduced circulatory system compared with nonsmokers (e.g., they have impaired capillary recruitment and impaired vasodilation) [150]. Thus, it is usually recommended that thermographic studies do not include smokers [31, 54], or that smokers are instructed to avoid smoking for 12 h before the measurement [27, 44, 89].

Hair The presence of hair on the skin surface can alter the estimation of the skin temperature using infrared thermography [34]. Hair is an avascular tissue [151] with a low heat capacity and then with a low temperature [152] (very similar to room temperature). For this reason, if there is a high hair density in the skin, this

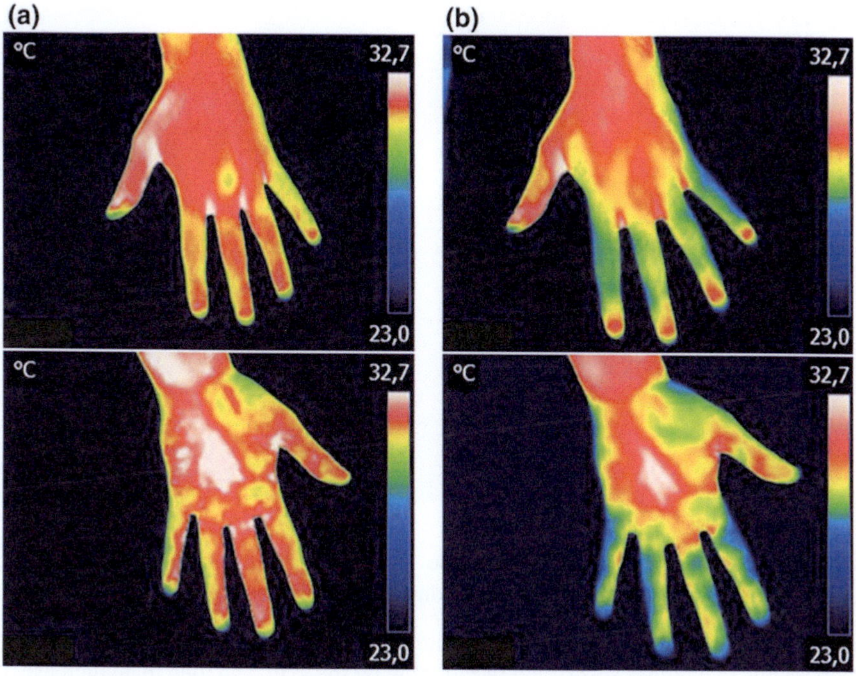

Fig. 4.17 Example of the effect of smoking on skin blood flow and then in skin temperature. Thermal images were taken before (**a**) and after smoking (**b**)

hair could be considered to be an artifact resulting in an underestimation of the skin temperature (Fig. 4.18). For this reason, some studies asked for participants to remove their body hair before the study [31, 32, 39, 54]. However, this could be inconvenient during participant recruitment.

Application of Cosmetics on the Skin This group of variables is especially important since they influence the radiation emitted by the skin and are difficult to detect. The use of cosmetics, creams, gels, or any other type of skin treatment is relevant to the analysis of thermographic images since they will act as a filter on the skin, distorting its radiation, influencing the emissivity, and resulting in an underestimation of the skin temperature [96]. Based on different studies, it is recommended to request subjects to avoid the use of skin lotions or other cosmetic products on the day of the measurement [27, 31, 32, 39].

Physical Activity Exercise recovery is a process in which the body works to restore the different physiological processes that occurred in the body during exercise, such as muscle damage, DNA damage, and oxidative stress, among others [153, 154]. In relation to the skin temperature, different studies observed that, after a training session, there is a progressive increase in the skin temperature up to 6-9 h after exercise [50, 105, 107]. This idea that exercise affects skin temperature a posteriori is an important factor to take into account, and for this reason the studies should request

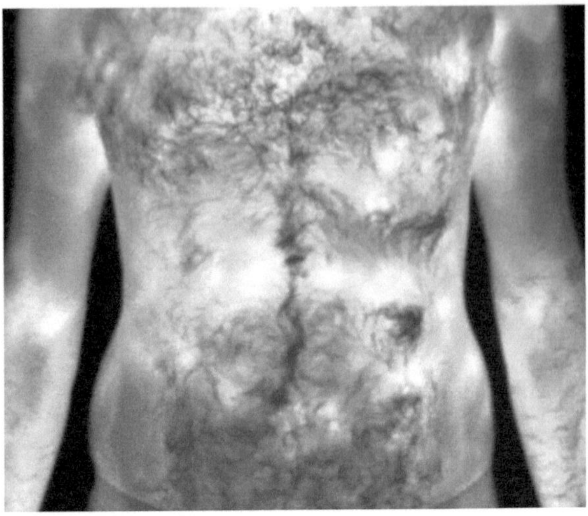

Fig. 4.18 Example of thermography of the chest of a subject with hair

that participants avoid physical exercise on the day of the study [131] and also refrain from high-intensity exercise the day before the study [27, 31, 90].

Sunbathing and UV Rays The exposure of the skin to sunbathing and UV radiation results in a higher absorption of radiation by the skin, and then in an increase of skin temperature [155]. A good recommendation for the participants in a human thermography assessment is to refrain from sunbathing or from being exposed to UV rays 12 h before the test [27, 44, 89].

4.3 Analysis of the Thermal Data

The analysis process in thermography data is still very conservative in some areas. That is the case for measurement in the field of exercise science, where thermography analysis is recent. It is attractive due to the facilitated access, but it is extremely important to ensure validity not only for the technical measure but for outcome interpretation. Before presenting conclusions from thermal data analyses, there are some important and mandatory steps for proper data analysis.

4.3.1 Qualitative Representation

The thermographic analysis can involve qualitative and quantitative approaches. Qualitative analysis consists of the representation and description of the images obtained. Although a quantitative analysis of the data is fundamental to drawing consistent conclusions, the qualitative analysis is very important for checking whether there is any factor that should be considered in the quantitative analysis. For example, in the qualitative analysis, we can see that the assessed person has a varicose vein in their legs, and therefore, there may be thermal asymmetry. Because of this, it is necessary to always perform a qualitative analysis to ensure the correct interpretation of the numerical data. In addition, this analysis can provide much faster feedback on the concerned subjects: the athlete, the coach, or the medical staff.

Another utility of qualitative analysis is to obtain a graphical representation of what is being evaluated. In this sense, it is important to know the different types of representation currently offered by the thermographic software. Infrared cameras usually have an incorporated digital visual lens that provides interesting visualization tools. Besides the infrared image (Fig. 4.19a), infrared images superimposed on the visual can also be obtained (Fig. 4.19b) as well as an image that integrates infrared and the visual (Fig. 4.19c). According to the camera, the software, and how both correct the differences between the lens in position and other characteristics, the overlap is more or less accurate.

Some of the recommendations usually performed for the qualitative representation of images with humans are listed below [61, 73]:

- Use of a rainbow palette, with white-red as hot and blue-black as cold.
- Use of an image range defined with a minimum of 23–24 °C and a maximum of 36–38 °C. In addition, the use of the same image field for all the comparable images makes them comparable.

4.3.2 *Quantitative Analysis*

Different variables can be obtained from the analysis of the ROI in the thermographic analysis, such as the mean temperature, the maximum temperature, the minimum temperature, and the standard deviation. In addition, it is possible to calculate other variables such as the thermal differences between ROIs (to express gradients or temperatures normalized with respect to an ROI), variation of the skin

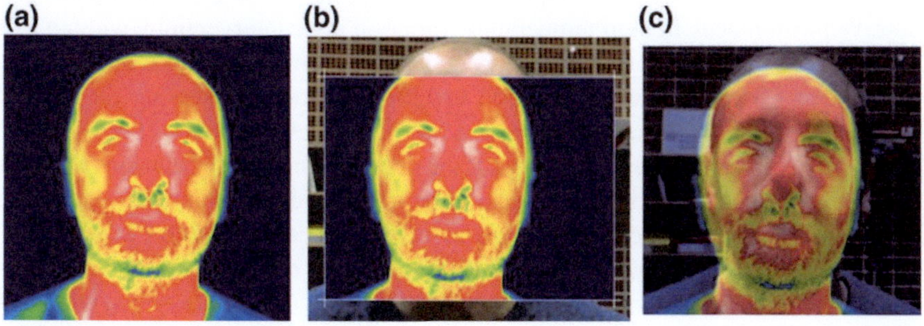

Fig. 4.19 Different thermal image representations

temperature (Δ: difference between before and after an intervention), the thermal symmetry (difference between both body hemispheres), and the amplitude temperature of the ROI (difference between the maximum and minimum values). Different studies used different variables for the analysis of thermal images, and there are still many open questions in this regard. The more common approaches to quantify thermal data are discussed below.

Mean temperature is the most used skin temperature variable in the literature [2]. The advantage of using the mean of skin temperature in large ROIs is that it can be a representative value and removes the effects of punctual higher values that can be a source of errors. Moreover, different studies showed that it is the most reliable parameter [38, 40]. For this reason, it is considered that it is the parameterless influenced by the experience of the evaluator [38]. However, when variations are important, the mean temperature may not provide the experiment with significant information [44, 89, 90].

Maximum temperature can be considered the second most used thermal variable in thermography studies [2]. Some previous studies with animals suggest associations between maximal temperature and injury [156, 157], and in humans, it was suggested that the maximum temperature can be more sensitive to skin blood flow and inflammation [158]. Although the maximum temperature presents lower reliability than the mean temperature, its reliability is also between very good and excellent [38, 40]. Related to the maximum temperature is the Tmax method suggested by Ludwig et al. [39], which is described in Sect. 4.2.1 of this chapter. A positive feature of this method is that, using it, all the ROIs will have the same number of pixels for the calculation, which minimizes the problem of the differences in the anatomical size between participants.

Minimum temperature is the minimum value obtained in an ROI. This parameter was suggested as more related to vasoconstriction of skin blood flow [158]. However, the reliability of minimum temperature is very low as it can be more affected by small differences in the ROIs determination [38, 40] and therefore be more affected by the evaluator's experience [38]. Further studies could explore the application of similar methodologies, such as the Tmax method, to the minimum value assessment.

Temperature variation (Δ), as produced by exercise, has been suggested as a valid measure for determining the effects of different interventions in studies using repeated measures design [44, 89, 90]. This idea is supported by some studies that observed differences in skin temperature variation after interventions without differences when looking at the absolute magnitudes of temperature [32, 89].

Thermal symmetries were explored in different studies assessing the normal thermal behavior of the participants or the risk of injuries [100, 159, 160]. Vardasca et al. [160] defined thermal symmetry as "the degree of similarity between two areas of interest, mirrored across the human body's longitudinal main axes which are identical in shape, identical in size and as near identical in position as possible." Thermal symmetry assessment is considered a valuable method to assess the physiological normality/abnormality in sport medicine [19, 160] because asymmetries higher than 0.5–0.7 °C are usually associated with a dysfunction in the musculoskeletal system [160–162].

Difference in temperature between ROIs can be a useful measurement for exploring thermal gradients [163, 164]. For example, the forearm-finger skin temperature gradient is usually considered to be an index of peripheral circulation and vasomotor tone [164]. Finally, the mean skin temperature is the most common variable used in thermal studies with thermal contact sensors [165]. This variable is determined using different equations and including temperatures from different body regions [165, 166]. There are equations that make the calculations with temperature data from only three regions, whereas other equations that are considered more accurate will need temperature data from up to 15 regions [165, 166]. There is no clear conclusion about how many measurements represent the ideal condition. There are studies showing good results with different numbers of measurements, with data from, for example, seven [166] or four [42] measurements. Although this variable is not commonly used in thermography studies, some studies measured infrared thermography in the same way, and it can be useful in studies when the main purpose is to obtain an overall value of skin temperature for the whole body [167, 168].

Other authors suggested that skin temperature should be normalized using the difference between each body region with one specific body region, rather than considering only absolute magnitudes of temperature [169, 170]. This normalization aims to reduce the variability of the absolute temperatures [170]. Forehead [170] and chin [169] were some of the body regions used as references, probably due to the consideration of these regions as more temperature-stable regions.

Alternative metrics have appeared in the last years, but they still need more research [171]. Entropy parameter measures the average number of bits of an event with respect to the probability distribution for the random variable, and it was suggested that it is related to exercise workload, exercise duration, and pulmonary ventilation during exercise [172]. Moreover, some studies have defined the pixelgraphy method as the quantification of the percentage of pixels in different temperature ranges, assuming that the percentage of pixels above 33° can be more related to muscle damage or physiological stress [173, 174]. Finally, as we commented in Sect. 4.2.1, machine learning methods are becoming more common where the algorithm can analyze the image and give a result based on a categorization without having to provide traditional thermal parameters.

All of these methods described, in addition to other methods, such as the spectral analysis of the image, need to be investigated in order to elucidate which method is the most suitable for each situation.

4.4 Conclusions

In this chapter, we present the major concerns and implications of selecting a methodology for determining human skin temperature from measurements using infrared thermography. It is possible to note that while thermography has many applications in the context of human movement, there is still a need for more research to determine the adequate procedures for data acquisition and analysis.

Furthermore, unlike other techniques for the assessment of human movement, there are significant efforts prior to data acquisition to ensure that the participant will arrive in a condition that minimizes the influence of intervening factors in the assessment.

References

1. Hillen B, Pfirrmann D, Nägele M, Simon P (2020) Infrared thermography in exercise physiology: the dawning of exercise radiomics. Sports Med (Auckland, N.Z.) 50(2):263–282. https://doi.org/10.1007/s40279-019-01210-w
2. Perpetuini D, Formenti D, Cardone D, Filippini C, Merla A (2021) Regions of interest selection and thermal imaging data analysis in sports and exercise science: a narrative review. Physiol Meas 42(8):08TR01. https://doi.org/10.1088/1361-6579/ac0fbd
3. Escamilla-Galindo V-L, Felipe JL, Alonso-Callejo A, Van-der-Horst R, de la Torre-Combarros A, Minafra P, Fernández-Muñoz D, Fernández-Cuevas I (2024) Return-to-play criteria based on infrared thermography during anterior cruciate ligament rehabilitation in football players. Biol Sport 42(2):161–167. https://doi.org/10.5114/biolsport.2025.144295
4. Gómez-Carmona P, Fernández-Cuevas I, Sillero-Quintana M, Arnaiz-Lastras J, Navandar A (2020) Infrared thermography protocol on reducing the incidence of soccer injuries. J Sport Rehabil:1–6. https://doi.org/10.1123/jsr.2019-0056
5. Hillen B, Andrés López D, Marzano-Felisatti JM, Sanchez-Jimenez JL, Ortiz C, de Anda RM, Nägele M, Salvador-Palmer MR, Pérez-Soriano P, Schömer E, Simon P, Priego-Quesada JI (2023) Acute physiological responses to a pyramidal exercise protocol and the associations with skin temperature variation in different body areas. J Therm Biol 115:103605. https://doi.org/10.1016/j.jtherbio.2023.103605
6. Machado ÁS, da Silva W, Priego-Quesada JI, Carpes FP (2023) Can infrared thermography serve as an alternative to assess cumulative fatigue in women? J Therm Biol 115:103612. https://doi.org/10.1016/j.jtherbio.2023.103612
7. Zarębska EA, Kusy K, Korman P, Słomińska EM, Zieliński J (2024) Exercise-induced changes in lower limbs skin temperature against plasma ATP among individuals with various type and level of physical activity. J Therm Biol 122:103877
8. Cuddy JS, Hailes WS, Ruby BC (2014) A reduced core to skin temperature gradient, not a critical core temperature, affects aerobic capacity in the heat. J Therm Biol 43:7–12. https://doi.org/10.1016/j.jtherbio.2014.04.002
9. Racinais S, Havenith G, Aylwin P, Ihsan M, Taylor L, Adami PE, Adamuz M-C, Alhammoud M, Alonso JM, Bouscaren N, Buitrago S, Cardinale M, van Dyk N, Esh CJ, Gomez-Ezeiza J, Garrandes F, Holtzhausen L, Labidi M, Lange G et al (2022) Association between thermal responses, medical events, performance, heat acclimation and health status in male and female elite athletes during the 2019 Doha World Athletics Championships. Br J Sports Med 56(8):439–445. https://doi.org/10.1136/bjsports-2021-104569
10. Sawka MN, Cheuvront SN, Kenefick RW (2012) High skin temperature and hypohydration impair aerobic performance. Exp Physiol 97:327–332. https://doi.org/10.1113/expphysiol.2011.061002
11. de Andrade Fernandes A, dos Santos Amorim PR, Brito CJ et al (2014) Measuring skin temperature before, during and after exercise: a comparison of thermocouples and infrared thermography. Physiol Meas 35:189
12. Priego Quesada JI, Martínez Guillamón N, Cibrián O, de Anda RM et al (2015a) Effect of perspiration on skin temperature measurements by infrared thermography and contact thermometry during aerobic cycling. Infrared Phys Technol 72:68–76. https://doi.org/10.1016/j.infrared.2015.07.008

13. van den Heuvel CJ, Ferguson SA, Dawson D, Gilbert SS (2003) Comparison of digital infrared thermal imaging (DITI) with contact thermometry: pilot data from a sleep research laboratory. Physiol Meas 24:717. https://doi.org/10.1088/0967-3334/24/3/308
14. Smith ADH, Crabtree DR, Bilzon JLJ, Walsh NP (2010) The validity of wireless iButtons and thermistors for human skin temperature measurement. Physiol Meas 31:95–114. https://doi.org/10.1088/0967-3334/31/1/007
15. Buono MJ, Ulrich RL (1998) Comparison of mean skin temperature using "covered" versus "uncovered" contact thermistors. Physiol Meas 19:297–300
16. Psikuta A, Niedermann R, Rossi RM (2013) Effect of ambient temperature and attachment method on surface temperature measurements. Int J Biometeorol:1–9
17. Priego-Quesada JI, Machado AS, Gil-Calvo M, Jimenez-Perez I, de Anda RMCO, Salvador Palmer R, Perez-Soriano P (2020b) A methodology to assess the effect of sweat on infrared thermography data after running: preliminary study. Infrared Phys Technol 109:103382. https://doi.org/10.1016/j.infrared.2020.103382
18. Tyler CJ (2011) The effect of skin thermistor fixation method on weighted mean skin temperature. Physiol Meas 32:1541–1547. https://doi.org/10.1088/0967-3334/32/10/003
19. Hildebrandt C, Raschner C, Ammer K (2010) An overview of recent application of medical infrared thermography in sports medicine in Austria. Sensors 10:4700–4715
20. Lahiri BB, Bagavathiappan S, Jayakumar T, Philip J (2012) Medical applications of infrared thermography: a review. Infrared Phys Technol 55:221–235
21. dos Santos Bunn P, Elisa Koppke Miranda M, Inoue Rodrigues A, de Souza Sodré R, Borba Neves E, Bezerra da Silva E (2020) Infrared thermography and musculoskeletal injuries: a systematic review with meta-analysis. Infrared Phys Technol 103435. https://doi.org/10.1016/j.infrared.2020.103435
22. Fitzgerald A, Berentson-Shaw J (2012) Thermography as a screening and diagnostic tool: a systematic review. N Z Med J 125:80
23. Ng E-K (2009) A review of thermography as promising non-invasive detection modality for breast tumor. Int J Therm Sci 48:849–859
24. Ramirez-GarciaLuna JL, Bartlett R, Arriaga-Caballero JE, Fraser RDJ, Saiko G (2022) Infrared thermography in wound care, surgery, and sports medicine: a review. Front Physiol 13. https://doi.org/10.3389/fphys.2022.838528
25. Bouzas Marins JC, de Andrade Fernandes A, Gomes Moreira D et al (2014) Thermographic profile of soccer players' lower limbs. Rev Andal Med Deporte 7:1–6. https://doi.org/10.1016/S1888-7546(14)70053-X
26. Priego Quesada JI, Martínez N, Salvador Palmer R et al (2016a) Effects of the cycling workload on core and local skin temperatures. Exp Thermal Fluid Sci 77:91–99. https://doi.org/10.1016/j.expthermflusci.2016.04.008
27. Priego Quesada JI, Carpes FP, Bini RR et al (2015b) Relationship between skin temperature and muscle activation during incremental cycle exercise. J Therm Biol 48:28–35. https://doi.org/10.1016/j.jtherbio.2014.12.005
28. Coull NA, West AM, Hodder SG, Wheeler P, Havenith G (2021) Body mapping of regional sweat distribution in young and older males. Eur J Appl Physiol 121(1):109–125. https://doi.org/10.1007/s00421-020-04503-5
29. Fournet D, Ross L, Voelcker T et al (2013) Body mapping of thermoregulatory and perceptual responses of males and females running in the cold. J Therm Biol 38:339–344. https://doi.org/10.1016/j.jtherbio.2013.04.005
30. Aylwin P, Havenith G, Cardinale M, Lloyd A, Ihsan M, Taylor L, Adami PE, Alhammoud M, Alonso J-M, Bouscaren N, Buitrago S, Esh C, Gomez-Ezeiza J, Garrandes F, Labidi M, Lange G, Moussay S, Mtibaa K, Townsend N et al (2023) Thermoregulatory responses during road races in hot-humid conditions at the 2019 Athletics World Championships. J Appl Physiol 134(5):1300–1311. https://doi.org/10.1152/japplphysiol.00348.2022

31. Abate M, Di Carlo L, Di Donato L et al (2013) Comparison of cutaneous termic response to a standardised warm up in trained and untrained individuals. J Sports Med Phys Fitness 53:209–215
32. Formenti D, Ludwig N, Gargano M et al (2013) Thermal imaging of exercise-associated skin temperature changes in trained and untrained female subjects. Ann Biomed Eng 41:863–871. https://doi.org/10.1007/s10439-012-0718-x
33. Ammer K (2008) The Glamorgan Protocol for recording and evaluation of thermal images of the human body. Thermol Int 18:125–144
34. Fernández-Cuevas I, Bouzas Marins JC, Arnáiz Lastras J et al (2015) Classification of factors influencing the use of infrared thermography in humans: a review. Infrared Phys Technol 71:28–55. https://doi.org/10.1016/j.infrared.2015.02.007
35. A Committee on Quality Control and Qualifications of the American Academy of Thermology (1986) Technical guidelines, Edition 2. Thermology 2:108–112
36. Schwartz RG (2006) Guidelines for neuromusculoskeletal thermography. Thermol Int 16:5–9
37. Moreira DG, Costello JT, Brito CJ, Adamczyk JG, Ammer K, Bach AJE, Costa CMA, Eglin C, Fernandes AA, Fernández-Cuevas I, Ferreira JJA, Formenti D, Fournet D, Havenith G, Howell K, Jung A, Kenny GP, Kolosovas-Machuca ES, Maley MJ et al (2017) Thermographic imaging in sports and exercise medicine: a Delphi study and consensus statement on the measurement of human skin temperature. J Therm Biol 69:155–162. https://doi.org/10.1016/j.jtherbio.2017.07.006
38. da Silva W, Machado ÁS, Kunzler MR, Jimenez-Perez I, Gil-Calvo M, Priego-Quesada JI, Carpes FP (2022) Reproducibility of skin temperature analyses by novice and experienced evaluators using infrared thermography. J Therm Biol 110:103345. https://doi.org/10.1016/j.jtherbio.2022.103345
39. Ludwig N, Formenti D, Gargano M, Alberti G (2014) Skin temperature evaluation by infrared thermography: comparison of image analysis methods. Infrared Phys Technol 62:1–6
40. Machado ÁS, Priego-Quesada JI, Jimenez-Perez I, Gil-Calvo M, Carpes FP, Perez-Soriano P (2021) Influence of infrared camera model and evaluator reproducibility in the assessment of skin temperature responses to physical exercise. J Therm Biol 98:102913. https://doi.org/10.1016/j.jtherbio.2021.102913
41. Machado ÁS, Cañada-Soriano M, Jimenez-Perez I, Gil-Calvo M, Carpes FP, Perez-Soriano P, Priego-Quesada JI (2024) Distance and camera features measurements affect the detection of temperature asymmetries using infrared thermography. Quant InfraRed Thermogr J 21(2):69–81. https://doi.org/10.1080/17686733.2022.2143227
42. Maniar N, Bach AJE, Stewart IB, Costello JT (2015) The effect of using different regions of interest on local and mean skin temperature. J Therm Biol 49–50:33–38. https://doi.org/10.1016/j.jtherbio.2015.01.008
43. Marins JCB, Moreira DG, Cano SP et al (2014) Time required to stabilize thermographic images at rest. Infrared Phys Technol 65:30–35. https://doi.org/10.1016/j.infrared.2014.02.008
44. Priego Quesada JI, Lucas-Cuevas AG, Salvador Palmer R et al (2016b) Definition of the thermographic regions of interest in cycling by using a factor analysis. Infrared Phys Technol 75:180–186. https://doi.org/10.1016/j.infrared.2016.01.014
45. ISO (2008a) Particular requirements for the basic safety and essential performance of screening thermographs for human febrile temperature screening. TC121/SC3-IEC SC62D
46. ISO (2009) Medical electrical equipment-deployment, implementation and operational guidelines for identifying febrile humans using a screening thermograph. TR 13154:2009 ISO/TR 8-600
47. Bach AJ, Stewart IB, Minett GM, Costello JT (2015) Does the technique employed for skin temperature assessment alter outcomes? A systematic review. Physiol Meas 36:R27
48. Smith CJ, Havenith G (2011) Body mapping of sweating patterns in male athletes in mild exercise-induced hyperthermia. Eur J Appl Physiol 111:1391–1404. https://doi.org/10.1007/s00421-010-1744-8

49. Barcelos EZ, Caminhas WM, Ribeiro E et al (2014) A combined method for segmentation and registration for an advanced and progressive evaluation of thermal images. Sensors 14:21950–21967. https://doi.org/10.3390/s141121950
50. Fernández-Cuevas I, Sillero-Quintana M, Garcia-Concepcion MA et al (2014) Monitoring skin thermal response to training with infrared thermography. New Stud Athl 29:57–71
51. Requena-Bueno L, Priego-Quesada JI, Jimenez-Perez I, Gil-Calvo M, Perez-Soriano P (2020) Validation of ThermoHuman automatic thermographic software for assessing foot temperature before and after running. J Therm Biol 102639. https://doi.org/10.1016/j.jtherbio.2020.102639
52. Jimenez-Perez I, Gil-Calvo M, Priego-Quesada JI, Aparicio I, Pérez-Soriano P, Ortiz de Anda RMC (2020) Effect of prefabricated thermoformable foot orthoses on plantar surface temperature after running: a gender comparison. J Therm Biol 91:102612. https://doi.org/10.1016/j.jtherbio.2020.102612
53. Chudecka M, Lubkowska A (2015) Thermal maps of young women and men. Infrared Phys Technol 69:81–87. https://doi.org/10.1016/j.infrared.2015.01.012
54. Merla A, Mattei PA, Di Donato L, Romani GL (2010) Thermal imaging of cutaneous temperature modifications in runners during graded exercise. Ann Biomed Eng 38:158–163. https://doi.org/10.1007/s10439-009-9809-8
55. Paolillo FR, Lins EC, Corazza AV et al (2013) Thermography applied during exercises with or without infrared light-emitting diode irradiation: individual and comparative analysis. Photomed Laser Surg 31:349–355. https://doi.org/10.1089/pho.2013.3505
56. Priego-Quesada JI, Pérez-Guarner A, Gandia-Soriano A, Oficial-Casado F, Galindo C, de Anda RMCO, Piñeiro-Ramos JD, Sánchez-Illana Á, Kuligowski J, Barbosa MAG, Vento M, Palmer RS (2020c) Effect of a Marathon on Skin Temperature Response After a Cold-Stress Test and Its Relationship With Perceptive, Performance, and Oxidative-Stress Biomarkers. Int J Sports Physiol Perform 1(aop):1–9. https://doi.org/10.1123/ijspp.2019-0963
57. Priego Quesada JI, Gil-Calvo M, Lucas-Cuevas AG, Aparicio I, Pérez-Soriano P (2017) Assessment of a mattress with phase change materials using a thermal and perception test. Exp Thermal Fluid Sci 81:358–363. https://doi.org/10.1016/j.expthermflusci.2016.10.024
58. Hillen B, López DA, Schömer E, Nägele M, Simon P (2022) Towards exercise radiomics: deep neural network-based automatic analysis of thermal images captured during exercise. IEEE J Biomed Health Inform 26(9):4530–4540
59. Vardasca R, Simoes R (2013) Current issues in medical thermography. In: Topics in medical image processing and computational vision. Springer, Berlin, pp 223–237
60. Ring EFJ, Ammer K (2000) The technique of infrared imaging in medicine. Thermol Int 10:7–14
61. Ring EFJ, Ammer K (2012) Infrared thermal imaging in medicine. Physiol Meas 33:R33–R46. https://doi.org/10.1088/0967-3334/33/3/R33
62. Ammer K (2005) Temperature readings from thermal images are less dependent on the number of pixels of the measurement area than on variation of room temperature. Thermol Int 15:131–133
63. Playà-Montmany N, Tattersall GJ (2021) Spot size, distance and emissivity errors in field applications of infrared thermography. Methods Ecol Evol 12(5):828–840. https://doi.org/10.1111/2041-210X.13563
64. de Deus Passos M, da Rocha AF (2022) Evaluation of infrared thermography with a portable camera as a diagnostic tool for peripheral arterial disease of the lower limbs compared with color Doppler ultrasonography. Arch Med Sci Atheroscler Dis 7:e66. https://doi.org/10.5114/amsad/150716
65. Li F, Wang M, Wang T, Wang X, Ma X, He H, Ma G, Zhao D, Yue Q, Wang P, Ma M (2023) Smartphone-based infrared thermography to assess progress in thoracic surgical incision healing: a preliminary study. Int Wound J 20(6):2000–2009. https://doi.org/10.1111/iwj.14063

66. Qu Y, Meng Y, Fan H, Xu RX (2022) Low-cost thermal imaging with machine learning for non-invasive diagnosis and therapeutic monitoring of pneumonia. Infrared Phys Technol 123:104201. https://doi.org/10.1016/j.infrared.2022.104201
67. Vardasca R (2019) Are the IR cameras FLIR ONE suitable for clinical applications? Thermol Int 23(3):95–102
68. Akimov EB, Son'kin VD (2011) Skin temperature and lactate threshold during muscle work in athletes. Hum Physiol 37:621–628
69. Duc S, Arfaoui A, Polidori G, Bertucci W (2015) Efficiency and thermography in cycling during a graded exercise test. J Exerc Sports Orthop 2:1–8
70. Merla A, Iodice P, Tangherlini A et al (2005) Monitoring skin temperature in trained and untrained subjects throughout thermal video. In: Conference proceedings: annual international conference of the IEEE, Engineering in Medicine and Biology Society, vol 2, pp 1684–1686. https://doi.org/10.1109/IEMBS.2005.1616767
71. Priego-Quesada JI, De la Fuente C, Kunzler MR, Perez-Soriano P, Hervás-Marín D, Carpes FP (2020a) Relationship between skin temperature, electrical manifestations of muscle fatigue, and exercise-induced delayed onset muscle soreness for dynamic contractions: a preliminary study. Int J Environ Res Public Health 17(18). https://doi.org/10.3390/ijerph17186817
72. Chudecka M, Lubkowska A (2010) Temperature changes of selected body's surfaces of handball players in the course of training estimated by thermovision, and the study of the impact of physiological and morphological factors on the skin temperature. J Therm Biol 35:379–385
73. Vollmer M, Möllmann K-P (2011) Infrared thermal imaging: fundamentals, research and applications. Wiley, Weinheim
74. Riou O, Berrebi S, Bremond P (2004) Nonuniformity correction and thermal drift compensation of thermal infrared camera. Thermosense XXVI 5405:294–302. https://www.spiedigitallibrary.org/conference-proceedings-of-spie/5405/1/Nonuniformity-correction-and-thermal-drift-compensation-of-thermal-infrared-camera/10.1117/12.547807.short
75. Budzier H, Gerlach G (2015) Calibration of uncooled thermal infrared cameras. J Sens Sens Syst 4(1):187–197. https://doi.org/10.5194/jsss-4-187-2015
76. Tempelhahn A, Budzier H, Krause V, Gerlach G (2016) Shutter-less calibration of uncooled infrared cameras. J Sens Sens Syst 5(1):9–16. https://doi.org/10.5194/jsss-5-9-2016
77. Bagavathiappan S, Lahiri BB, Saravanan T, Philip J, Jayakumar T (2013) Infrared thermography for condition monitoring – A review. Infrared Phys Technol 60:35–55. https://doi.org/10.1016/j.infrared.2013.03.006
78. Gupta R, Breitenstein O (2006) Temperature drift correction for fast lock-in infrared thermography. In: Proceedings of the 21st European Photovoltaic Solar Energy Conference, pp 4–8. http://www-old.mpi-halle.mpg.de/mpi/publi/pdf/7199_06.pdf
79. Peeters J, Louarroudi E, De Greef D, Vanlanduit S, Dirckx JJJ, Steenackers G (2017) Time calibration of thermal rolling shutter infrared cameras. Infrared Phys Technol 80:145–152. https://doi.org/10.1016/j.infrared.2016.12.001
80. Ammer K (2006) Influence of imaging and object conditions on temperature readings from medical infrared images. Pol J Environ Stud (Submitted)
81. Ivanitsky GR, Khizhnyak EP, Deev AA, Khizhnyak LN (2006) Thermal imaging in medicine: a comparative study of infrared systems operating in wavelength ranges of 3–5 and 8–12 microm as applied to diagnosis. Dokl Biochem Biophys 407:59–63
82. Tkáčová M, Hudák R, Foffová P, Živčák J (2010) An importance of camera subject distance and angle in musculoskeletal applications of medical thermography. Acta Electrotech Inform 10:57–60
83. Tian X, Fang L, Liu W (2023) The influencing factors and an error correction method of the use of infrared thermography in human facial skin temperature. Build Environ 244:110736. https://doi.org/10.1016/j.buildenv.2023.110736
84. Aylwin PE, Racinais S, Bermon S, Lloyd A, Hodder S, Havenith G (2021) The use of infrared thermography for the dynamic measurement of skin temperature of moving ath-

letes during competition; methodological issues. Physiol Meas 42(8):084004. https://doi.org/10.1088/1361-6579/ac1872
85. Dumke CL, Slivka DR, Cuddy JS et al (2015) The effect of environmental temperature on glucose and insulin after an oral glucose tolerance test in healthy young men. Wilderness Environ Med 26:335–342. https://doi.org/10.1016/j.wem.2015.03.002
86. Guéritée J, Tipton MJ (2015) The relationship between radiant heat, air temperature and thermal comfort at rest and exercise. Physiol Behav 139:378–385. https://doi.org/10.1016/j.physbeh.2014.11.064
87. Petrofsky JS, Lohman E, Suh HJ et al (2006) The effect of aging on conductive heat exchange in the skin at two environmental temperatures. Med Sci Monit 12:CR400–CR408
88. Maughan RJ, Otani H, Watson P (2011) Influence of relative humidity on prolonged exercise capacity in a warm environment. Eur J Appl Physiol 112:2313–2321. https://doi.org/10.1007/s00421-011-2206-7
89. Priego Quesada JI, Lucas-Cuevas AG, Gil-Calvo M et al (2015c) Effects of graduated compression stockings on skin temperature after running. J Therm Biol 52:130–136. https://doi.org/10.1016/j.jtherbio.2015.06.005
90. Priego Quesada JI, Carpes FP, Salvador Palmer R et al (2016c) Effect of saddle height on skin temperature measured in different days of cycling. Springerplus 5:205–214. https://doi.org/10.1186/s40064-016-1843-z
91. ISO (2008b) 18434-1:2008: condition monitoring and diagnostics of machines—thermoraphy—part 1: general procedures
92. Sanchez-Marin FJ, Calixto-Carrera S, Villaseñor-Mora C (2009) Novel approach to assess the emissivity of the human skin. J Biomed Opt 14:24006–24006–6. https://doi.org/10.1117/1.3086612
93. Steketee J (1973) Spectral emissivity of skin and pericardium. Phys Med Biol 18:686
94. Togawa T (1989) Non-contact skin emissivity: measurement from reflectance using step change in ambient radiation temperature. Clin Phys Physiol Meas 10:39
95. James CA, Richardson AJ, Watt PW, Maxwell NS (2014) Reliability and validity of skin temperature measurement by telemetry thermistors and a thermal camera during exercise in the heat. J Therm Biol 45:141–149. https://doi.org/10.1016/j.jtherbio.2014.08.010
96. Bernard V, Staffa E, Mornstein V, Bourek A (2013) Infrared camera assessment of skin surface temperature—effect of emissivity. Phys Med 29:583–591. https://doi.org/10.1016/j.ejmp.2012.09.003
97. Ferreira JJA, Mendonça LCS, Nunes LAO et al (2008) Exercise-associated thermographic changes in young and elderly subjects. Ann Biomed Eng 36:1420–1427. https://doi.org/10.1007/s10439-008-9512-1
98. Zontak A, Sideman S, Verbitsky O, Beyar R (1998) Dynamic thermography: analysis of hand temperature during exercise. Ann Biomed Eng 26:988–993
99. Pereira Franco VH, Vasquez-Bonilla AA, Sillero-Quintana M (2025) Influence of body position on skin temperature, heart rate, and blood pressure in active men. J Therm Biol 127:104009. https://doi.org/10.1016/j.jtherbio.2024.104009
100. Bertmaring I, Babski-Reeves K, Nussbaum MA (2008) Infrared imaging of the anterior deltoid during overhead static exertions. Ergonomics 51:1606–1619. https://doi.org/10.1080/00140130802216933
101. Houdas Y, Ring EFJ (1982) Temperature distribution. In: Human body temperature. Springer, berlin, pp 81–103
102. Chamberlain JM, Terndrup TE, Alexander DT et al (1995) Determination of normal ear temperature with an infrared emission detection thermometer. Ann Emerg Med 25:15–20. https://doi.org/10.1016/S0196-0644(95)70349-7
103. Smolander J, Härmä M, Lindgvist A et al (1993) Circadian variation in peripheral blood flow in relation to core temperature at rest. Eur J Appl Physiol 67:192–196. https://doi.org/10.1007/BF00376666

104. Costa CMA, Sillero-Quintana M, Piñonosa Cano S et al (2015) Daily oscillations of skin temperature in military personnel using thermography. J R Army Med Corps. https://doi.org/10.1136/jramc-2015-000462
105. Da Silva W, Godoy-López JR, Machado ÁS, Lemos AL, Sendra-Pérez C, Brejano MG, Carpes FP, Priego-Quesada JI (2024) Effect of different Volumes of exercise on skin temperature responses over the following 24 hours. J Therm Biol 123:103923
106. Marins JCB, Formenti D, Costa CMA et al (2015) Circadian and gender differences in skin temperature in militaries by thermography. Infrared Phys Technol 71:322–328. https://doi.org/10.1016/j.infrared.2015.05.008
107. Priego-Quesada JI, Catalá-Vilaplana I, Bermejo-Ruiz JL, Gandia-Soriano A, Pellicer-Chenoll MT, Encarnación-Martínez A, Ortiz C, de Anda R, Salvador-Palmer R (2022) Effect of 10 km run on lower limb skin temperature and thermal response after a cold-stress test over the following 24 h. J Therm Biol 105:103225. https://doi.org/10.1016/j.jtherbio.2022.103225
108. Ammer K (2009) Does neuromuscular thermography record nothing else but an infrared sympathetic skin response? Thermol Int 19:107–108
109. Novotny J, Rybarova S, Zacha D et al (2015) The influence of breaststroke swimming on the muscle activity of young men in thermographic imaging. Acta Bioeng Biomech 17:121
110. Zaidi H, Fohanno S, Polidori G, Taiar R (2007) The influence of swimming type on the skin-temperature maps of a competitive swimmer from infrared thermography. Acta Bioeng Biomech 9:47
111. van Ooijen AMJ, van Marken Lichtenbelt WD, Westerterp KR (2001) Individual differences in body temperature and the relation to energy expenditure: the influence of mild cold. J Therm Biol 26:455–459. https://doi.org/10.1016/S0306-4565(01)00060-2
112. Gagnon D, Kenny GP (2011) Sex modulates whole-body sudomotor thermosensitivity during exercise. J Physiol 589:6205–6217. https://doi.org/10.1113/jphysiol.2011.219220
113. Karki A, Karppi P, Ekberg J, Selfe J (2004) A thermographic investigation of skin temperature changes in response to a thermal washout of the knee in healthy young adults. Thermol Int 14:137–141
114. Gagnon D, Jay O, Lemire B, Kenny GP (2008) Sex-related differences in evaporative heat loss: the importance of metabolic heat production. Eur J Appl Physiol 104:821–829. https://doi.org/10.1007/s00421-008-0837-0
115. Gonzalez RR (1985) Blanchard LA (1998) Thermoregulatory responses to cold transients: effects of menstrual cycle in resting women. J Appl Physiol Bethesda Md 85:543–553
116. Stachenfeld NS, Silva C, Keefe DL (2000) Estrogen modifies the temperature effects of progesterone. J Appl Physiol Bethesda Md (1985) 88:1643–1649
117. Bartelink ML, Wollersheim H, Theeuwes A et al (1990) Changes in skin blood flow during the menstrual cycle: the influence of the menstrual cycle on the peripheral circulation in healthy female volunteers. Clin Sci 78:527–532
118. do Nascimento Silva RK, Matias FL, Gonçalves AF, de Almeida Ferreira JJ, de Andrade PR (2023) Skin temperature of women: a prospective longitudinal study. J Therm Biol 118:103741. https://doi.org/10.1016/j.jtherbio.2023.103741
119. Niu HH, Lui PW, Hu JS et al (2001) Thermal symmetry of skin temperature: normative data of normal subjects in Taiwan. Zhonghua Yi Xue Za Zhi Chin Med J Free China Ed 64:459–468
120. Inbar O, Morris N, Epstein Y, Gass G (2004) Comparison of thermoregulatory responses to exercise in dry heat among prepubertal boys, young adults and older males. Exp Physiol 89:691–700. https://doi.org/10.1113/expphysiol.2004.027979
121. Ho CW, Beard JL, Farrell PA et al (1997) Age, fitness, and regional blood flow during exercise in the heat. J Appl Physiol Bethesda Md 82:1126–1135
122. Stapleton JM, Poirier MP, Flouris AD et al (2015) Aging impairs heat loss, but when does it matter? J Appl Physiol Bethesda Md 1985 118:299–309. https://doi.org/10.1152/japplphysiol.00722.2014

123. Cramer MN, Jay O (2016) Biophysical aspects of human thermoregulation during heat stress. Auton Neurosci Basic Clin 196:3–13. https://doi.org/10.1016/j.autneu.2016.03.001
124. Havenith G (2001) Human surface to mass ratio and body core temperature in exercise heat stress—a concept revisited. J Therm Biol 26:387–393. https://doi.org/10.1016/S0306-4565(01)00049-3
125. Du Bois D, Du Bois EF (1916) Clinical calorimetry: a formula to estimate the appropriate surface area if height and weight be known. Arch Intern Med 17:863–871
126. Chudecka M, Lubkowska A, Kempińska-Podhorodecka A (2014) Body surface temperature distribution in relation to body composition in obese women. J Therm Biol 43:1–6. https://doi.org/10.1016/j.jtherbio.2014.03.001
127. Korman P, Kusy K, Straburzyńska-Lupa A, Sillero-Quintana M, Zarębska E, Zieliński J (2024) Exploring the correlation of skin temperature and body composition in athletes undergoing exhaustive physical exercise. J Therm Biol 123:103918
128. Salamunes ACC, Stadnik AMW, Neves EB (2017) The effect of body fat percentage and body fat distribution on skin surface temperature with infrared thermography. J Therm Biol 66:1–9. https://doi.org/10.1016/j.jtherbio.2017.03.006
129. Savastano DM, Gorbach AM, Eden HS et al (2009) Adiposity and human regional body temperature. Am J Clin Nutr 90:1124–1131. https://doi.org/10.3945/ajcn.2009.27567
130. Kumar P, Gaurav A, Rajnish RK, Sharma S, Kumar V, Aggarwal S, Patel S (2022) Applications of thermal imaging with infrared thermography in Orthopaedics. J Clin Orthop Trauma 24:101722. https://doi.org/10.1016/j.jcot.2021.101722
131. Zaproudina N, Varmavuo V, Airaksinen O, Närhi M (2008) Reproducibility of infrared thermography measurements in healthy individuals. Physiol Meas 29:515. https://doi.org/10.1088/0967-3334/29/4/007
132. Ichinose-Kuwahara T, Inoue Y, Iseki Y et al (2010) Sex differences in the effects of physical training on sweat gland responses during a graded exercise. Exp Physiol 95:1026–1032. https://doi.org/10.1113/expphysiol.2010.053710
133. Simmons GH, Wong BJ, Holowatz LA, Kenney WL (2011) Changes in the control of skin blood flow with exercise training: where do cutaneous vascular adaptations fit in? Exp Physiol 96:822–828. https://doi.org/10.1113/expphysiol.2010.056176
134. Ring EF, Collins AJ, Bacon PA, Cosh JA (1974) Quantitation of thermography in arthritis using multi-isothermal analysis. II. Effect of nonsteroidal anti-inflammatory therapy on the thermographic index. Ann Rheum Dis 33:353
135. Ring EF, Engel J, Page-Thomas DP (1984) Thermologic methods in clinical pharmacology-skin temperature measurement in drug trials. Int J Clin Pharmacol Ther 22:20–24
136. Toshihiko U, Yoshiharu T, Atsuhiro M et al (1987) Application of thermography to the evaluation of the histamine skin test in man. J Pharmacol Methods 18:103–110
137. Wilczyński S (2015) The use of dynamic thermal analysis to distinguish between genuine and counterfeit drugs. Int J Pharm 490:16–21. https://doi.org/10.1016/j.ijpharm.2015.04.077
138. Hirai A, Tanabe M, Shido O (1991) Enhancement of finger blood flow response of postprandial human subjects to the increase in body temperature during exercise. Eur J Appl Physiol 62:221–227
139. Westerterp-Plantenga MS, Wouters L, ten Hoor F (1990) Deceleration in cumulative food intake curves, changes in body temperature and diet-induced thermogenesis. Physiol Behav 48:831–836
140. Kieser E, Dellimore K, Scheffer C et al (2015) Development of diagnostic sensors for infant dehydration assessment using optical methods. In: Conference proceedings: annual international conference of the IEEE, Engineering in Medicine and Biology Society, vol 2015, pp 5537–5540. https://doi.org/10.1109/EMBC.2015.7319646
141. Ammer K, Melnizky P, Rathkolb O (2003) Skin temperature after intake of sparkling wine, still wine or sparkling water. Thermol Int 13:99–102
142. Hughes JH, Henry RE, Daly MJ (1984) Influence of ethanol and ambient temperature on skin blood flow. Ann Emerg Med 13:597–600. https://doi.org/10.1016/S0196-0644(84)80282-6

143. Mannara G, Salvatori GC, Pizzuti GP (1993) Ethyl alcohol induced skin temperature changes evaluated by thermography. Preliminary results. Boll Della Soc Ital Biol Sper 69:587
144. Wolf R, Tüzün B, Tüzün Y (1999) Alcohol ingestion and the cutaneous vasculature1. Clin Dermatol 17:395–403. https://doi.org/10.1016/S0738-081X(99)00023-1
145. Koot P, Deurenberg P (1995) Comparison of changes in energy expenditure and body temperatures after caffeine consumption. Ann Nutr Metab 39:135–142. https://doi.org/10.1159/000177854
146. Quinlan P, Lane J, Aspinall L (1997) Effects of hot tea, coffee and water ingestion on physiological responses and mood: the role of caffeine, water and beverage type. Psychopharmacology 134:164–173. https://doi.org/10.1007/s002130050438
147. Quinlan PT, Lane J, Moore KL et al (2000) The acute physiological and mood effects of tea and coffee: the role of caffeine level. Pharmacol Biochem Behav 66:19–28
148. Bornmyr S, Svensson H (1991) Thermography and laser-Doppler flowmetry for monitoring changes in finger skin blood flow upon cigarette smoking. Clin Physiol 11:135–141
149. Gershon-Cohen J, Borden AGB, Hermel MB (1969) Thermography of extremities after smoking. Br J Radiol 42:189–191
150. Ijzerman RG, Serne EH, van Weissenbruch MM et al (2003) Cigarette smoking is associated with an acute impairment of microvascular function in humans. Clin Sci 104:247–252
151. Barnes RB (1963) Thermography of the human body. Science 140:870–877
152. Togawa T, Saito H (1994) Non-contact imaging of thermal properties of the skin. Physiol Meas 15:291. https://doi.org/10.1088/0967-3334/15/3/007
153. Connolly DAJ, Sayers SP, McHugh MP (2003) Treatment and prevention of delayed onset muscle soreness. J Strength Cond Res 17:197–208
154. Ryu JH, Paik IY, Woo JH et al (2016) Impact of different running distances on muscle and lymphocyte DNA damage in amateur marathon runners. J Phys Ther Sci 28:450–455. https://doi.org/10.1589/jpts.28.450
155. Petersen B, Philipsen PA, Wulf HC (2014) Skin temperature during sunbathing–relevance for skin cancer. Photochem Photobiol Sci 13:1123–1125. https://doi.org/10.1039/c4pp00066h
156. Stokes JE, Leach KA, Main DCJ, Whay HR (2012) An investigation into the use of infrared thermography (IRT) as a rapid diagnostic tool for foot lesions in dairy cattle. Vet J 193:674–678
157. Whay HR, Bell MJ, Main DCJ (2004) Validation of lame limb identification through thermal imaging. In: Proceedings of 13th international symposium and 5th conference of lameness rumen. Maribor Slovenija, pp 11–15
158. Formenti D, Ludwig N, Rossi A, Trecroci A, Alberti G, Gargano M, Merla A, Ammer K, Caumo A (2018) Is the maximum value in the region of interest a reliable indicator of skin temperature? Infrared Phys Technol 94:299–304
159. Sillero-Quintana M, Fernández-Jaén T, Fernández-Cuevas I et al (2015) Infrared thermography as a support tool for screening and early diagnosis in emergencies. J Med Imaging Health Inform 5:1223–1228
160. Vardasca R, Ring F, Plassmann P, Jones C (2012) Thermal symmetry of the upper and lower extremities in healthy subjects. Thermol Int 22:53–60
161. Nowakowski AZ (2006) Advances of quantitative IR-thermal imaging in medical diagnostics. Brain 10:3
162. Vainionpaa M, Tienhaara E-P, Raekallio M et al (2012) Thermographic imaging of the superficial temperature in racing greyhounds before and after the race. Sci World J. https://doi.org/10.1100/2012/182749
163. Haddad DS, Brioschi ML, Baladi MG, Arita ES (2016) A new evaluation of heat distribution on facial skin surface by infrared thermography. Dentomaxillofac Radiol 45:20150264. https://doi.org/10.1259/dmfr.20150264
164. Keramidas ME, Geladas ND, Mekjavic IB, Kounalakis SN (2013) Forearm-finger skin temperature gradient as an index of cutaneous perfusion during steady-state exercise. Clin Physiol Funct Imaging 33:400–404. https://doi.org/10.1111/cpf.12043

165. Taylor NAS, Tipton MJ, Kenny GP (2014) Considerations for the measurement of core, skin and mean body temperatures. J Therm Biol 46:72–101. https://doi.org/10.1016/j.jtherbio.2014.10.006
166. Choi JK, Miki K, Sagawa S, Shiraki K (1997) Evaluation of mean skin temperature formulas by infrared thermography. Int J Biometeorol 41:68–75. https://doi.org/10.1007/s004840050056
167. Jimenez-Perez I, Gil-Calvo M, Vardasca R, Fernandes RJ, Vilas-Boas JP (2021) Pre-exercise skin temperature evolution is not related with 100 m front crawl performance. J Therm Biol 98:102926. https://doi.org/10.1016/j.jtherbio.2021.102926
168. Pérez-Guarner A, Priego-Quesada JI, Oficial-Casado F, Cibrián Ortiz de Anda RM, Carpes FP, Palmer RS (2019) Association between physiological stress and skin temperature response after a half marathon. Physiol Meas 40(3):034009–034019. https://doi.org/10.1088/1361-6579/ab0fdc
169. Haddad DS, Brioschi ML, Vardasca R et al (2014) Thermographic characterization of masticatory muscle regions in volunteers with and without myogenous temporomandibular disorder: preliminary results. Dentomaxillofac Radiol 43:20130440. https://doi.org/10.1259/dmfr.20130440
170. Sun P-C, Jao S-HE, Cheng C-K (2005) Assessing foot temperature using infrared thermography. Foot Ankle Int 26:847–853. https://doi.org/10.1177/107110070502601010
171. Verderber L, da Silva W, Aparicio-Aparicio I, Germano AMC, Carpes FP, Priego-Quesada JI (2024) Assessment of alternative metrics in the application of infrared thermography to detect muscle damage in sports. Physiol Meas 45(9):095014. https://doi.org/10.1088/1361-6579/ad7ad3
172. Bogomilsky S, Hoffer O, Shalmon G, Scheinowitz M (2022) Preliminary study of thermal density distribution and entropy analysis during cycling exercise stress test using infrared thermography. Sci Rep 12(1):14018. https://doi.org/10.1038/s41598-022-18233-5
173. Alburquerque Santana PV, Alvarez PD, Felipe da Costa Sena A, Serpa TK, de Assis MG, Pimenta EM, Costa HA, Sevilio de Oliveira Junior MN, Torres Cabido CE, Veneroso CE (2022) Relationship between infrared thermography and muscle damage markers in physically active men after plyometric exercise. J Therm Biol 104:103187. https://doi.org/10.1016/j.jtherbio.2022.103187
174. de Andrade Fernandes A, Marins JCB, de Andrade AGP, Albuquerque MR, Brito CJ, da Silva CD, do Valle MAAN, de Assis MG, Garcia ES, Pimenta EM (2024) Thermopixelgraphy: a new method for analyzing the skin temperature of the lower limbs through infrared thermography. J Therm Anal Calorim 149(22):12737–12745. https://doi.org/10.1007/s10973-024-13799-w

Chapter 5
Infrared Thermography for Injury Risk Detection, Return to Play and Recovery Individualization in Sports

Ismael Fernández-Cuevas, Javier Arnáiz Lastras, Víctor Escamilla Galindo, and Pedro Gómez Carmona

Abstract This chapter examines the evolving role of infrared thermography (IRT) in elite sports, integrating both medical and performance perspectives. We explore the physiological principles behind IRT and its key applications, including injury prevention (initial evaluation, injury risk assessment, and treatment adaptation), injury management (diagnostic support and return to play), and fatigue identification (individualization of recovery strategies). Drawing from extensive research and professional experience with elite athletes and teams across various sports, we demonstrate how IRT, when applied with standardized protocols, provides a non-invasive, rapid, and objective method to monitor and manage injury risk, optimize recovery, and enhance performance. These insights are further strengthened by a growing body of evidence, validating IRT's effectiveness in sports science.

5.1 Introduction: How Can Infrared Thermography Help Us to Prevent/Monitor Injuries and Improve Performance?

Despite the improvement of technology and training methodologies in recent decades, injuries are still a huge problem for professional and amateur athletes and sport institutions. In addition to that, the increase in the number of competitions and

I. Fernández-Cuevas (✉)
Sports Department, Faculty of Physical Activity and Sport Sciences—INEF, Universidad Politécnica de Madrid, Madrid, Spain

PEMA THERMO GROUP S.L., ThermoHuman, Madrid, Spain
e-mail: ismael.fernandez@thermohuman.com

J. Arnáiz Lastras · V. Escamilla Galindo · P. Gómez Carmona
Sports Department, Faculty of Physical Activity and Sport Sciences—INEF, Universidad Politécnica de Madrid, Madrid, Spain

PEMA THERMO GROUP S.L., ThermoHuman, Madrid, Spain
e-mail: javier.arnaiz@thermohuman.com; pedro.gomez.carmona@thermohuman.com

J. I. Priego-Quesada (ed.), *Application of Infrared Thermography in Sports Science*,
https://doi.org/10.1007/978-3-031-93311-0_5

travels is prioritizing recovery over training due to a congested schedule and the lack of time to plan normalized training periods [1]. Moreover, injuries are not just a health matter, but also an economic problem for the team or athlete [2], and obviously a factor influencing individual and team performance.

Nevertheless, we cannot deny the significant and useful changes that new technologies are bringing to sport professionals to reduce injury incidence. One of the key factors is related to the quantification of training loads: internal and external. We speak about a new perspective that has arisen since the advent of tools that help to control the training loads. On one hand, we have the external loads, such as encoders, accelerometers or the most popular global positioning systems (GPS) that calculate impact, accelerations, high-speed distances, and so on. On the other hand, we have the internal loads to quantify the biological responses of each individual according to the type of training. In both cases, we have seen significant advancements, including the implementation of new methodologies and a deeper understanding of the most relevant metrics that can assist professionals in better managing both competition and training loads. These developments aim to provide sports professionals with more objective data, both on and off the field (including "invisible training"), enabling them to make more informed decisions based on science, metrics, and individual needs.

Tim Gabbett's work on the acute:chronic workload ratio has been foundational in understanding the relationship between training load and injury risk [3]. However, recent research has highlighted the complexity of this relationship. For instance, athletes with higher chronic workloads may exhibit greater resilience to acute spikes in training load [4]. Additionally, individual factors—such as strength, fitness, and injury history—are crucial in determining how athletes respond to workload variations. On top of that, we observe the growing role of technology and artificial intelligence (AI) in personalized training load monitoring, promoting more accurate and individualized strategies for injury prevention and performance optimization in elite sports with different technologies while respecting traditional training principles, which has been denominated as Sport Sciences 3.0 [5]

In this context, we found different technologies, tools, and methods focused on injury prevention. Among all of them, we identify infrared thermography (IRT) as one of the most interesting technologies for measuring internal load and, consequently, helping to identify injury risk in athletes. In recent years, some authors have described how this old technology can be used as a tool for injury prevention [6–11]. IRT stands out as a highly attractive alternative for measuring internal load due to three key advantages: it is fast (requiring only a photo), noninvasive (no radiation is emitted, and the subject is not manipulated), and it provides objective data (in addition to the image, metrics are extracted).

IRT is a safe, noninvasive and low-cost technique that allows for the rapid and noninvasive recording of radiating energy that is released from the body. IRT measures skin temperature (Tsk) and has been widely used since the early 1960s in different sectors. During the first decades after its development, research into the use of IRT in humans was mainly focused on its applications as a diagnostic tool. However, IRT was replaced by newer and more accurate technologies (such as X-rays and magnetic resonance imaging). Recent technical advances in infrared cameras have made new human applications of IRT (beyond diagnostic techniques) possible. Among its various applications, injury prevention and monitoring have emerged as some of the most valuable uses of IRT, given its ability to detect changes in body surface temperature. These temperature variations can indicate physiological imbalances associated with injury risk or the healing process of damaged tissue [9]. More recently, a growing body of research has explored the relationship between fatigue and Tsk. In this context, IRT has been shown to help identify athletes exhibiting abnormal Tsk patterns, which may correspond to different types of fatigue [12].

It is widely known that there is a correlation between muscle activation and the variation of skin temperature [13]. Depending on the type of exercise and intensity, the area adjacent to the muscles involved in the exercise could experience a decrease or increase of the skin temperature, as result of a combination of factors, such as metabolism, muscle contraction, sweating or skin blood flow [14].

Ideally, our skin is supposed to maintain a constant thermal pattern over time, with the constant aim of keeping the body in thermal balance or "homeothermy" [15]. However, a lot of factors can affect the thermal asymmetry of a person during their lifetime, and most of them are related to pathological reasons such as inflammation or nerve dysfunctions [16].

In this sense, IRT can help us to identify thermal asymmetries by comparing bilateral body areas (e.g., left and right knee, or dominant and nondominant calf) [17, 18]. Thermal asymmetries on those regions of interest (ROIs) may indicate a nonpathological pattern (related to dominance factors or ancient injuries or surgeries) [19], but it could also show us a potential injury risk due to an incorrect workload assimilation, excessive activity, asymmetrical exercise execution or muscle overload. Those asymmetries could appear before other indicators as pain, which is extremely useful for implementing preventive strategies before injury occurs. More importantly, thermal asymmetries are significant regardless of whether they present as hyperthermic (hot spots) or hypothermic (cold patterns). In this context, IRT is not only useful for detecting these asymmetries but also for understanding the underlying physiological changes in the tissue, whether or not they are associated with pain or discomfort. This allows for the development of personalized treatment and training interventions. For example, if an athlete reports stiffness in the right hamstring and IRT reveals a hypothermic pattern, treatments aimed at increasing blood flow and Tsk, along with activation-focused exercises, may help restore thermal balance.

Moreover, IRT can also be extremely useful once an injury occurs because we can monitor the thermal asymmetry evolution in order to check if the rehabilitation process is correctly stimulating the areas involved and taking as reference the thermal pattern before the injury [20, 21].

Obviously, there are a lot of factors affecting Tsk [22], which makes it, in some cases, impossible to prevent certain injuries. Nevertheless, the frequent record of thermal images allows for an individual comparison of thermal asymmetries over time, providing a reliable database evolution (based on what we call neutralized asymmetry) that is a key factor in determining the normal thermal values of an athlete (based on the evaluations performed in baseline conditions and with no pain, injury or influence factors), and therefore for individualizing and better understanding any changes to the thermal pattern, and therefore the reasons that may lead to a concerning thermal asymmetry.

To summarize, IRT represents a fast, cheap, and useful technique that allows the sport and health professionals to better understand the internal load and the physiology and health status of a subject. Indeed, if IRT is used following a standardized protocol, with a reliable analysis method or software, and combined with other technologies and the collaboration of different sport professionals, it could really help to decrease the injury incidence [7–10, 23] improving the return to play [20, 21, 24] and helping identifying the type of fatigue after competition [12, 25].

5.2 The Protocol: Methodological Aspects and Equipment Requirements

IRT is not a perfect technique. Therefore, the use of a standardized protocol seems to be absolutely essential for ensuring the quality of the thermal results [26].

We are also aware of the differences between using IRT in a scientific context or facing the reality outside of a laboratory. As researchers, we must always try to maintain the protocol quality; nevertheless, we should also be able to manage real situations with top elite athletes and institutions where not all the elements are under control (not enough time to acclimatize, lack of information, etc.), but the results must be still enough reliable to provide trustable insights.

In addition to Chap. 4 of this book, we will enumerate some of the more recognized protocols. Then, we will show a synopsis from the TISEM consensus, which is the main consensus for the use of IRT in sports and exercise medicine [27]. Finally, we will describe the different approaches to analyzing thermal images: qualitative and quantitative analysis.

5.2.1 Thermographic Protocols in the Literature

In recent years, several organizations from the health sector have generated and published their own protocols and quality assurance guidelines. For example, below we list some of the most relevant publications in the last several years:

(a) In 2002, International Academy of Clinical Thermology (IACT) published Standards and Protocols in Clinical Thermographic Imaging, which describes some of the basic requirements for performing an accurate clinical assessment with IRT [28].
(b) In 2004, the International Organisation for Standardisation (ISO) published ISO 9886 focusing on the ergonomic evaluation of thermal strain by physiological measurements [29].

(c) In 2006, Schwartz published a complete guideline prepared by members of the American Academy of Thermology [30]. The guide was aimed at neuromuscular thermologists and other interested parties.
(d) Among the large number of academies, associations, and societies, the European Association of Thermology (EAT) has been one of the most active institutions in recent years in publishing IRT-related studies. The University of Glamorgan group (recently renamed South Wales University) has worked to better understand the technical factors that affect IRT measurements and to create a strict protocol for reducing errors and increasing the accuracy and the precision of temperature measurements [26, 31–40]. Ring and Ammer published one of the most cited works regarding the requirements for using IRT in the medical sector [38]. Other authors from this active group have noted important factors in quality assurance [37] or specific applications in fever screening, updated after COVID-19 pandemic [35, 36]. Most of the works are summarized in the Glamorgan Protocol [33].
(e) The research group from the Technical University of Madrid presented during the XIII EAT Congress in 2015 a "Protocol for thermographic assessment in humans" [41], gathering the most recent updates and results from the research and professional experience of the previous years. The report was reviewed by Kurt Ammer and the EAT scientific board.
(f) In 2017, a prominent group of researchers and experts in the application of IRT in health and sports sciences, led by Danilo Gomes Moreira and Manuel Sillero-Quintana, published the first major consensus on the use of IRT in sports and exercise medicine [27]. Today, these guidelines known as TISEM remain the most widely recognized and cited reference in the field, with hundreds of citations and continued relevance in both research and applied practice.

To summarize, most thermographic guidelines agree on the following items:

1. Location requirements
2. Equipment characteristics
3. Subject/patient information and preparation
4. Subject/patient assessment
5. Processing and presentation of results

Regardless of the protocol followed, it is clear that it is absolutely necessary to follow standardized guidelines when using IRT in humans to avoid risking the quality and objectivity of our research or professional work on IRT.

5.2.2 *Thermographic Protocol for Elite Sports Institutions*

This section is based on both the TERMOINEF [41] and Thermographic Imaging in Sports and Exercise Medicine (TISEM) consensus [27]. In addition to the methodological aspects outlined in Chap. 4, this section integrates practical considerations derived from experience working with elite-level teams. Unlike controlled laboratory environments, real-world settings in high-performance sports often present challenges such as limited acclimatization time, incomplete data, and suboptimal measurement conditions. Despite these constraints, it is possible to ensure a sufficient level of methodological rigor to obtain valid and reliable results. However, it is essential to acknowledge the distinction between research-oriented applications, which adhere to the highest scientific standards, and field-based implementation, where balancing data quality and efficiency is a key priority.

Considerations Prior to the Assessment
Before the assessment, it is highly recommended to have an informative session with the athlete being assessed or the medical staff and professionals involved if they are not familiar with IRT. In this session, which may be physical or virtual, the subjects will be provided with written information, including information about IRT, the processes to be followed during data collection, the type of information to be provided in the final report, and recommendations to be followed before the thermographic session.

We all know that in some institutions and teams, the access to top elite athletes is restricted, but this information should be provided at least to the medical staff so that they could inform them or at least take into account the following factors before the session:

- All evaluations of the same subject should be conducted at the same time.
- No physical activity should be performed for at least 6 h prior to the evaluation (so, ideally before the first training session).
- Avoid the application of creams, gels or sprays on the skin area to be evaluated
- Avoid coffee, alcohol or stimulants for 6 h prior to the evaluation.
- Do not smoke in the 6 h prior to the assessment.
- Do not receive any treatment or massage (it is especially important to coordinate the session with the physiotherapists, so they do not proceed before the evaluation).
- Avoid direct sun or UV sessions before evaluation.
- The previous day (when possible), avoid any drugs or treatment with any substance that may alter the normal thermogram.
- Avoid taking a shower or bath before the evaluation.
- Do not alter rest or meal habits.

In addition to that, and since not all facilities are perfectly adapted to data collection, it is highly advisable to visit the location before the day, or with enough time to get it adapted. Sometimes, the only place available is some space in the changing room, the gym or near to the place the athletes prepare themselves before the training session (Fig. 5.1). We suggest that it should not occur in a place where the humidity of the showers could change the room temperature and humidity.

In certain situations, the acclimatization period may need to be shortened to maximize the number of athletes assessed. While the importance of this preparatory phase is well established, studies have shown that in elite football teams, Tsk values

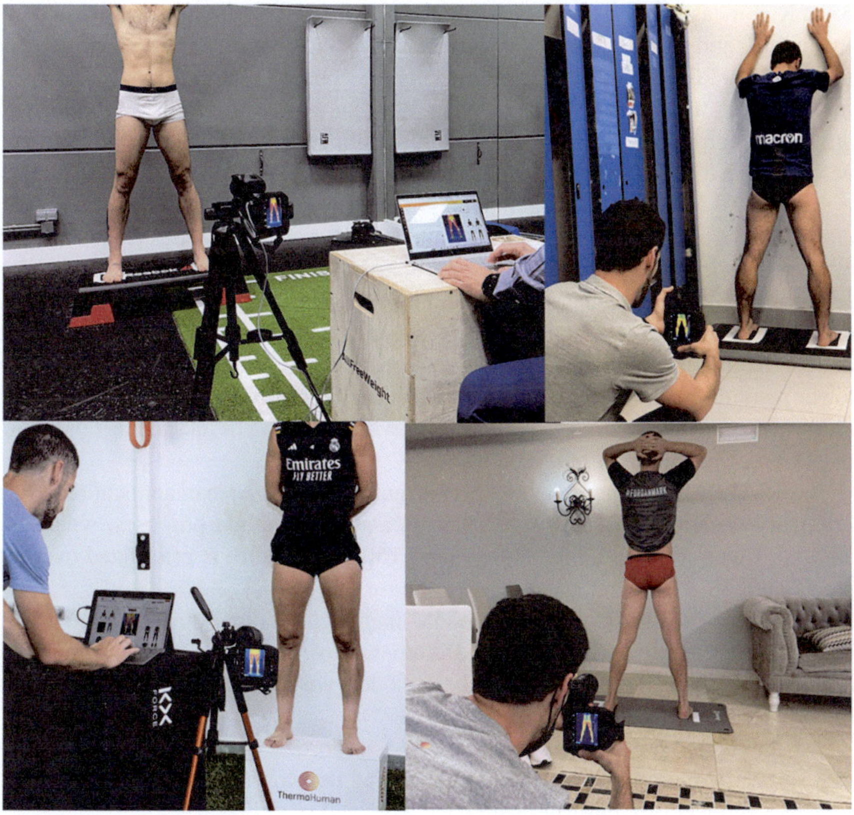

Fig. 5.1 Different examples of data collection locations performed with top sports institutions

remain largely consistent even with 0 minutes of acclimatization, compared to 5 or 10 minutes, with the exception of the hamstring region [42].

Although the recommended acclimatization period is 15 minutes, if this is not feasible, we strongly advise ensuring that athletes have been in the facility for at least 10 minutes before assessment—whether in the locker room or having breakfast—to minimize the impact of external environmental conditions. For evaluations conducted immediately after waking, such as during training camps, we recommend a minimum of 15 minutes post-waking to prevent false positives caused by body positioning during sleep.

Standardized protocols are crucial and should be followed whenever possible. However, in high-performance settings, where logistical constraints are common, it is often preferable to collect thermal imaging data with some limitations rather than forgoing the assessment entirely.

Selection of the Camera

We strongly recommend using a thermal imaging camera with a resolution equal to or greater than 320 (horizontal) x 240 (vertical) pixels, a minimum sensitivity of

65 m K, a measurement range covering the possibilities of the human environment (between 0 and 50 °C), a thermal sensitivity of 0.02 °C or lower, and standard data outputs for a PC. The resolution of the camera is one of the most important factors for obtaining high-quality thermal images. A camera with a high resolution will provide a sharper image and a greater number of pixels of the considered ROI, and will improve the quality and reliability of the results.

In recent years, several manufacturers have introduced low-cost thermal cameras, offering attractive pricing but limited resolution capabilities. Research has evaluated their performance, concluding that image quality improves with higher-resolution cameras and closer assessment distances [43]. Regarding practical applications, low-cost cameras can provide reliable asymmetry analysis within a single image, making them suitable for injury prevention and monitoring. However, their lower resolution and potential variability render them insufficient for longitudinal assessments, such as fatigue identification, where consistent and precise measurements over time are required.

Surprisingly, most thermal cameras currently used for human applications were originally designed for industrial, security, and construction purposes, where the ability to measure high temperatures (e.g., 150 °C or more) is prioritized over the precision required for human skin assessment.

During the COVID-19 pandemic, some manufacturers introduced Elevated Skin Temperature cameras with optimized temperature ranges (15–40 °C), improved accuracy, and lower costs. However, these models were discontinued in 2023 due to reduced demand.

Despite this, the growing interest in IRT for sports science and human physiology is driving manufacturers to reintroduce and develop thermal cameras tailored for these applications, ensuring better accuracy and usability.

Considerations on the Evaluation Day

On the day of data collection, it is necessary to carry out the following steps before starting:

- On the first day, the subject should be briefly informed about the characteristics of IRT, the purpose of the data collection, and the data details (including anonymity and data processing).
- On the first day, subjects have to sign the informed consent.
- The subject has to undress, staying in underwear (women preferably without bra) for around 10–15 minutes to acclimate to the temperature of the room, which should be between 18 and 25 °C (ideal: 21 °C). If this is not possible, at least 10 minutes indoors.
- During the acclimatization time, we recommend performing a survey with questions concerning the existence of possible influencing factors such as the previous training session, the intake of medicaments, the application of creams or ice on the skin and other influencing factors gathered in the literature [19].

The technician/researcher. The person charged with carrying out the evaluation must perform the following tasks before starting:

- To prepare the room at a suitable temperature (between 18 and 25 °C).
- To turn on the IRT camera at least 20′ before (to avoid calibration problems).
- To prepare the tripod (if necessary).
- To locate a step to elevate the subject and isolate the feet from the cold floor. Placing marks on the step will help to standardize the location of the subject's feet. (Fig. 5.1)
- To prepare roll up (if necessary) or make sure to have a proper and non-reflecting background.
- To turn on the weather station.
- To prepare the "black body" or calibration system (if necessary).
- To check with the IRT camera any radiation source in the room (hot pipes, windows, etc.).
- To prepare the notebook with questionnaire data collection for each subject or have a software ready to take those data.
- To turn on the computer, which will be used for transfer and analysis of the photos.
- To check the card or USB drive (or the connection of the camera to the computer) for the data acquisition.
- To turn on the computer, which will be used to transfer the photos and to analyze them.

Camera position. This depends on the subject's height and camera lens. Normally, it is placed between 2.5 and 3 m (for a subject of 1.70 m height and 42° lens). Regarding the height of the camera, it will be located so that the middle of the image matches the geometric center of the area to be evaluated (about 65 cm for the lower limbs and 125 cm for the upper body). It is important to keep the IRT camera at a perpendicular angle in relation to the subject. We recommend using a transparent template on the screen of the thermal camera to guide the technician on the placement of the subject (see Fig. 5.2). Currently, software solutions such as ThermoHuman

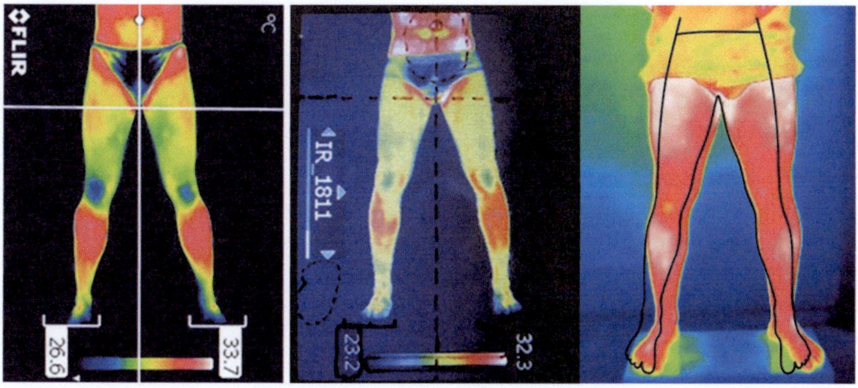

Fig. 5.2 Lower body reference (left image), transparent template on the thermal camera screen (middle image) and automatic silhouette for guiding the data collection in ThermoHuman software (right image)

software offers a similar guidance showing a silhouette during "plug-and-play" feature in order to make sure that distance and subject position is correct for all protocols.

Alternative guidelines, such as the Glamorgan protocol [33], have suggested capturing thermal images from closer perspectives to enhance the resolution of ROIs. While higher pixel density improves the precision of thermal assessments, practical constraints in applied settings—such as time limitations and the need for rapid evaluations—must also be considered.

ThermoHuman's imaging protocol consists of eight standardized thermal images, covering both the upper and lower body (anterior and posterior views), as well as specific ROIs including the face, chest, hands, and soles of the feet. This approach ensures a comprehensive assessment while maintaining efficiency.

For scenarios requiring high-throughput evaluations, a full-body protocol enables the analysis of an athlete using just two thermal images, optimizing the balance between accuracy and speed, particularly when assessing large groups (e.g., 20 athletes in a limited time frame). Additionally, ThermoHuman software incorporates an automated quality control system, which detects potential protocol deviations, including incorrect positioning, inadequate segmentation, improper distance, or other inconsistencies, ensuring the reliability of the thermal analysis.

During the evaluation. Subjects should follow the following instructions in lower body images (Fig. 5.2):

- The subject stands on a "step" platform to raise his feet (about 10 cm).
- Maintaining an anatomical position (upright and symmetrical).
- Slightly open legs (about 40 cm).
- Feet straight toward the camera.
- Looking ahead.
- Arms crossed or hands behind the head in lower body images.
- Feet must be clearly distinguishable from the background.
- Use of the transparent template or silhouette as a reference:
  - The lower brands are the places where feet should appear.
  - The crossing of the two lines coincides with the abductors area.
  - The highest point on the vertical line must match the navel. In the case of upper body images.

- The subject stands on a "step" platform to raise his feet (about 10 cm).
- Maintaining an anatomical position (upright and symmetrical).
- Same leg position.
- Looking ahead.
- Arms relaxed, stretched and slightly supinated so you can see the front or back of the forearm.
- Both hands with fists clenched.
- The head and fists must be clearly distinguishable from the background.

- The transparent template might be used as reference.
  - The temperature bar is just above the knees.
  - The body shape must fit on the image (see Fig. 5.3). That is, the head should be very close to the top edge and the hands to the lateral edges of the image.

– To facilitate adherence to standardized procedures, the TISEM consensus introduced a 15-item checklist highlighting the key considerations for conducting IRT evaluations in sports and exercise settings. This checklist serves as a practical reference to ensure methodological consistency, minimize potential errors, and enhance the reliability of thermal assessments. Figure 5.3 provides a visual representation of these 15 critical items, which should be carefully considered in every IRT evaluation.

Fig. 5.3 Adaptation of the checklist suggested by Gomes Moreira and collaborators in TISEM consensus [27]

5.2.3 Thermal Image Analysis

Once the thermal image is taken, the professional has two main ways to analyze the thermograms: qualitative and quantitative methods (Fig. 5.4).

The quality of the qualitative analysis depends on the experience of the thermographer. The image is interpreted by analyzing the color patterns of the thermograms, and a correct analysis depends on the accurate settings of the camera's temperature ranges. This method can help to identify the most evident asymmetries, but it could be that significant asymmetries are not seen.

Although qualitative analysis is valuable and can be highly useful for both thermographers (in detecting abnormalities in thermograms) and athletes (in enhancing their understanding of the results), it has often been misused. Some professionals rely solely on color visualization for IRT analysis, despite the fact that these color ranges are typically automatically distributed and, in some cases, manually manipulated. The following example illustrates how the same image can be altered by adjusting the color range, resulting in entirely different contrasts and potentially leading to opposing qualitative interpretations (see Fig. 5.5).

This is why qualitative analysis should always be complemented by quantitative analysis, which enhances the objectivity and reliability of IRT. Quantitative analysis provides thermal data for each ROI through manual or automated procedures,

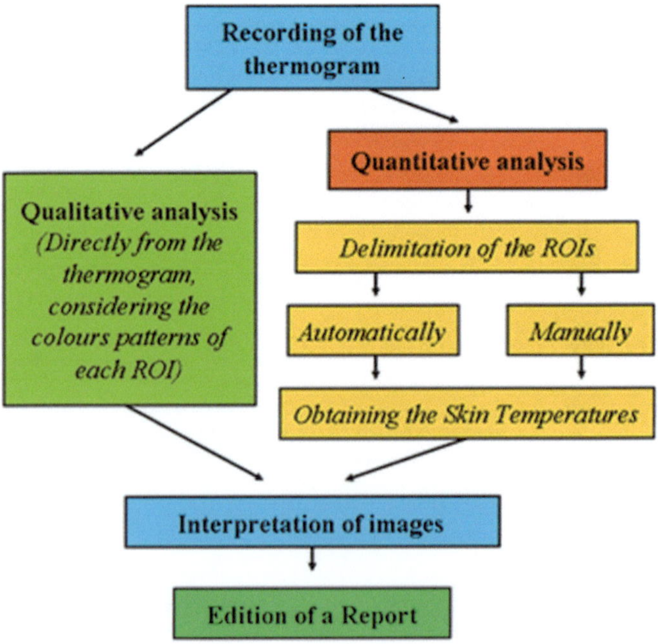

Fig. 5.4 Flowchart of the analysis of thermograms. (Figure adapted from Sillero-Quintana et al. [41])

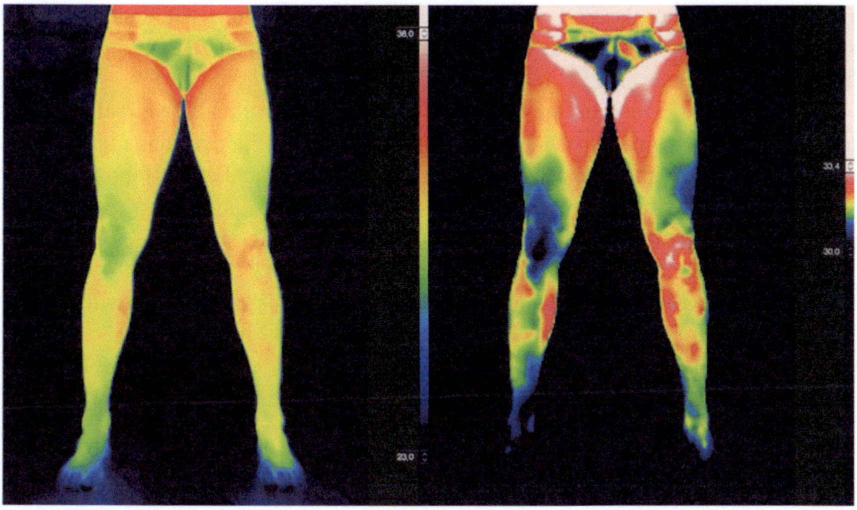

Fig. 5.5 The same thermogram is displayed using two different color ranges: the left thermogram goes from 23 to 38 °C, while the right image ranges from 30 to 33.4 °C

enabling objective measurements and metrics such as thermal asymmetries, standard deviation, pixel count, and coefficient of variation (among others) [44]. The thermal results of each ROI can be compared with contralateral or adjacent ROIs within the same subject or with corresponding ROIs in other subjects. Furthermore, thermal data can be compared across previous thermograms regardless of temperature range settings, ensuring that palette or color scale manipulation does not affect quantitative analysis (Fig. 5.5).

Manual

One of the most controversial points regarding IRT applications in humans is the selection of the ROIs. Many IRT studies have developed their own criteria for creating and selecting ROIs.

The controversy surrounding ROI selection stems from the manual process required to define ROIs. One of the key concerns is the suboptimal reliability of intra- and inter-examiner intraclass correlation coefficient results, largely influenced by the observer's ability to accurately select the ROI [39, 45]. Additionally, compared to automated processes, manual segmentation is time-consuming and prone to human errors such as the Narcissus effect, negligence, or visual fatigue, among others [46]. While manual selection may be appropriate in specific cases, technological advancements increasingly support the superiority of semiautomated and automated ROI segmentation methods, as they offer greater reliability and efficiency in IRT analysis [47].

Automatic

To improve reliability and to open up the possibility of comparing IRT results among studies, we recommend semiautomatic or automatic and objective procedures to select the ROI. In this sense, software solutions with automatic ROI selection features would be a first step, such as the one analyzed and proposed by different research groups [7, 48–53].

Among these, we highlight ThermoHuman, a scientifically validated online Software as a Service (SaaS) [53]. Since its launch in 2015, it has continuously evolved and is now the leading solution for the automatic analysis of thermal images in humans. The software is updated monthly, providing enhancements for researchers, healthcare professionals, and sports specialists in three key areas: data collection, thermal data analysis, and result interpretation.

From a data collection perspective, ThermoHuman is compatible with various thermal camera brands (FLIR, Hikmicro, ICI, etc.) and offers a "plug-and-play" feature, allowing the thermographer to connect the camera to a computer or laptop and perform thermogram analysis in real time. This capability is particularly useful in time-sensitive contexts, such as sports teams with limited time to capture images before training. In a standard setup, a thermographic evaluation involving two thermograms (front and back), along with software-generated results and a report, can be completed in less than 15 seconds per subject.

Thermogram processing has been one of ThermoHuman's most significant advancements. Once a thermogram is captured and uploaded to the software, automated computer vision algorithms can identify more than 150 ROIs across the human body. Currently, ThermoHuman provides dozens of metrics, including average, maximum, and minimum Tsk, standard deviation, thermal asymmetry, neutralized asymmetry, coefficient of variation, number of pixels, and even histograms for each ROI. The software supports the analysis of eight body protocols: upper anterior, upper posterior, lower anterior, lower posterior, face, hands, chest, and feet. These protocols are represented with avatars displaying various metrics and alerts, both at the individual and group levels (Fig. 5.6).

The most significant and exciting innovation lies in the interpretation of results. As researchers and experts in IRT, we must recognize both the power and the potential risk of this technique: it is so intuitive and easy to use that anyone might attempt to interpret the results without proper expertise. However, we know that Tsk is complex, and IRT findings must be analyzed correctly.

The rapid advancements in artificial intelligence (AI) have now reached this field, enabling the use of neural networks and other techniques for automatic ROI segmentation and even pathology detection [52]. In terms of AI-driven thermographic interpretation, ThermoHuman provides valuable insights in three key areas: asymmetry detection for injury prevention, fatigue identification, and Return to Play (RTP) assessments. While it offers insights from the very first thermogram, ThermoHuman becomes significantly more precise by the third scan of an individual, leveraging AI to generate natural language interpretations. These insights help professionals better understand prevention alerts, fatigue trends, and injury progression charts. Besides that, a PRE-POST mode allows researchers to further analyze IRT images from individuals and groups after an intervention, enabling a better understanding from an academic perspective. Undoubtedly, these capabilities will

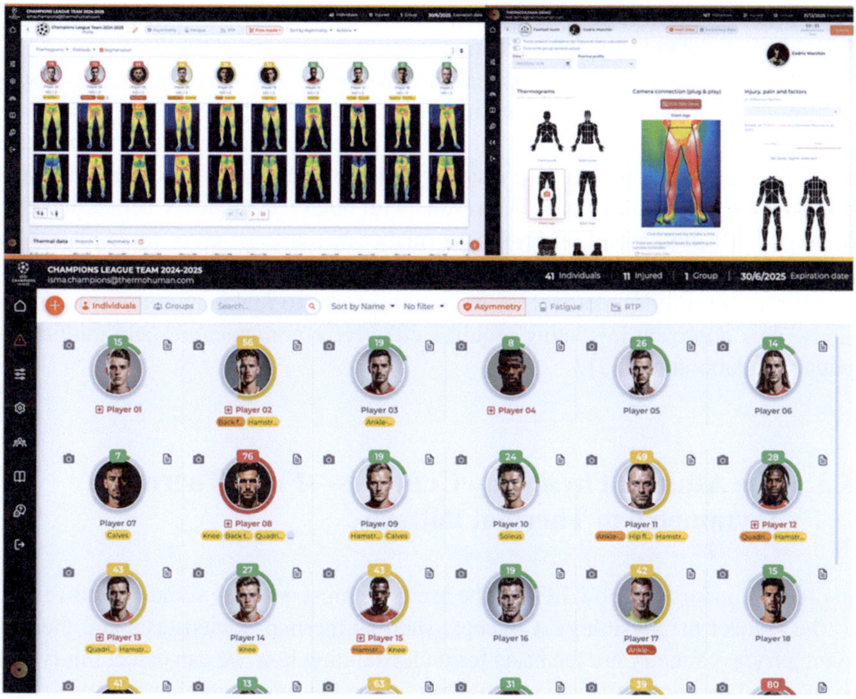

Fig. 5.6 Different interfaces of ThermoHuman 3.0 software

continue to evolve, further enhancing decision-making for healthcare and sports professionals.

These features, along with the ability to export data and reports in PDF, Excel, or via an API, enable medical and technical staff to identify players and body regions at risk of injury. They also help detect athletes with abnormal thermal patterns associated with various types of fatigue or Tsk values that deviate from the expected recovery progression in common injuries (e.g., anterior cruciate ligament (ACL) injuries, muscle injuries, fractures, or sprains).

Another factor contributing to the increasing popularity of IRT in the sports and healthcare sectors is the emergence of solutions similar to ThermoHuman. One such solution is Thermalytix, developed by Niramai, which has a strong scientific foundation and validation [52, 53]. Thermidas, another scientifically driven company, holds Conformité Européenne (CE) marking and Food and Drug Administration (FDA) clearance for solutions primarily focused on diabetic foot care and peripheral arterial disease [55]. In the sports domain, alongside ThermoHuman, we find notable researchers working on projects that combine IRT and biomarkers to offer alternative methods for injury prevention [54, 56]. However, it is important to note that there are also software solutions available that lack a solid scientific foundation, validation, or evidence of quality assurance. Since we want IRT to be used appropriately, we strongly recommend evaluating IRT software solutions based on their scientific credibility and validation evidence.

Despite these impressive technical advancements and scientific evidence supporting the role of IRT in reducing injury incidence [8, 9], monitoring injury progression [24], and measuring internal load [25], we must recognize the complex and multifactorial nature of human physiology and epidemiology. Therefore, our clear recommendation is to combine IRT results with other techniques, whether diagnostic (e.g., MRI, ultrasound, X-ray), performance-based (e.g., GPS, force plates), or other physiological metrics (e.g., heart rate, creatine kinase (CK), myoglobin, heart rate variability (HRV), cortisol, or hydration levels) to make more informed decisions. This approach inevitably requires effective communication and multidisciplinary collaboration [57].

5.3 The Athlete: Physiology Concepts—From Thermal Asymmetry to Thermal Pattern

In order to understand how IRT can be used, we must describe some aspects related to the subject of physiology. Concepts such as thermal homeostasis and thermal symmetry/asymmetry are the basis for understanding how we can detect injury risk areas. Another concept worth highlighting is the thermal profile, or the specialization thermal characteristics derived from a sport discipline or concrete performance. Finally, it is also worth summarizing the factors affecting human Tsk.

5.3.1 Biological Principles

Human skin, with an emissivity (an object's ability to emit radiation) of 0.98, is almost equal to a black body radiator [58]. The physics of heat radiation and the physiology of thermoregulation in the human body make the reliable and valid interpretation of thermal images difficult. Tsk regulation is a complex system that depends on blood-flow rate, local structures of subcutaneous tissues and the activity of the sympathetic nervous system [59]. However, there is evidence that the sympathetic nervous system is the primary regulator of blood circulation in the skin and is, therefore, the primary regulator of thermal emission [60]. Vasoconstriction and vasodilation of the blood vessels function to regulate blood flow in the skin. Thermoreceptors in the skin, also known as Ruffini corpuscles, recognize the ambient temperature. An increased temperature results in vasodilation, leading to increased blood flow to the skin, whereas vasoconstriction occurs by a decrease in temperature and results in reduced blood flow to the skin [61]. These physiological processes combine with heat transfer and thermoregulation in convection, conduction, radiation and sweat evaporation. Heat transfer by radiation is of great value in medicine [62]. To date, the mechanism of thermoregulatory adaptation to exercise is complex and not entirely understood. In addition to that, although it is clear that Tsk can be influenced by vascular, neurological, and biomechanical factors, IRT outcomes effectively highlight which aspect is predominant. This is similar to how a thermometer can indicate a pathological condition by showing a temperature of 38 °C, but it does not explain the underlying causes of the fever.

Heat Exchange Between the Human Body and Environment

In the case of the human body, the high metabolic activity generates a great amount of thermal energy, which must be dissipated to maintain the stability of the physiological processes of the body [63]. The mechanisms used by the human body to transfer the calorific energy are: conduction, convection, radiation, and evaporation (see Fig. 5.7).

The heat transfer by emission and absorption of radiation is called thermal transfer by radiation. We invite the reader to consult Chap. 2 of this book and the work of Incropera [48] or any other literature reference belonging to the field of physics in order to deepen their knowledge about the topic of energy transfer by radiation.

Bodies emit thermal radiation due to their temperature: the higher the temperature, the more thermal radiation that is emitted. However, each material has a different emissivity (see Chap. 2 of this book for emissivity values of different materials). Emissivity is the ability of each object to emit infrared radiation [65]. Not all objects emit 100% of what they should; if they did so, the infrared signal that would be able to pick up a camera would be directly related to the temperature, according to the Stefan-Boltzmann Law, which relates directly to the energy temperature irradiated.

From the amount of energy dissipated by the human skin, approximately 60% is in the form of infrared radiation [63], and the remaining 40% corresponds to convection. Normally, when working with opaque bodies (as human bodies), the transmissivity is equal to "0", because the radiation is unable to pass through the body. Consequently, the main concepts in working with radiation and human bodies are the emissivity and reflectivity.

Fig. 5.7 Mechanisms of heat excess release in the human body

5.3.2 Thermal Homeostasis, Thermal Symmetry/Asymmetry and Thermal Profile

Given that certain physiological disorders may influence the local exchange thermal response [23, 67–69], the use of IRT in sports can help to identify risk of injury in muscle, joint, bone or tendon areas since, in cases of overuse, overtraining, compensation, inflammation, inhibition, weakness, local fatigue or other risk factors the local and global Tsk show increasing levels of asymmetry.

This point is critical, as injury risk has been associated with hot spots and hyperthermic asymmetries. Based on our experience, thermal asymmetries are the most reliable metric for identifying injury risk. However, the following considerations should be taken into account:

1. Hot and cold regions: both hyperthermic and hypothermic areas can be equally important, particularly when analyzing muscle regions.
2. IRT is more effective at identifying "who" rather than "where": global asymmetries are a stronger indicator of risk than localized temperature variations.
3. Not all asymmetries lead to injuries, and not all injuries present asymmetries: IRT inherently includes some degree of noise, and unfortunately, certain injuries may not display thermal abnormalities before they occur.

Additionally, conducting recurrent assessments (e.g., on a weekly basis) is crucial for enhancing the accuracy of IRT in detecting anomalies across its various applications. Through periodic thermographic monitoring, a specific thermal profile can be established for each athlete. We essentially build a "map" representing the normal thermal characteristics of each ROI under neutral conditions (excluding assessments conducted with pain, injury, or influencing factors).

This approach enables the individualization of alarm thresholds for each ROI based on the athlete's unique thermal profile. In other words, if an individual consistently exhibits asymmetries due to non-injury-related factors (e.g., varicose veins), these patterns can be accounted for, effectively reducing irrelevant "noise". This process is achieved by calculating what we term *neutralized asymmetry*.

For instance, if an athlete with varicose veins consistently displays a 0.40 °C asymmetry in the right calf across multiple assessments—an initially concerning finding—we can neutralize this chronic asymmetry. In this case, 0.40 °C would be considered the individual's baseline, whereas for others, the expected Tsk asymmetry might be 0 °C. As a result, alarm thresholds would only be triggered if the asymmetry exceeds 0.70 °C or falls below 0.10 °C (see Fig. 5.8).

Considering the concept of anatomical proportionality, the thermal response between two contralateral body parts is expected to be symmetrical [18, 66, 70]. Thermal monitoring comparing bilateral body parts indicates that differences up to 0.1 °C [66], 0.2 °C [70], 0.3 °C [71], 0.4 °C [72, 73], 0.5 °C [17, 74], or 0.62 °C [75]

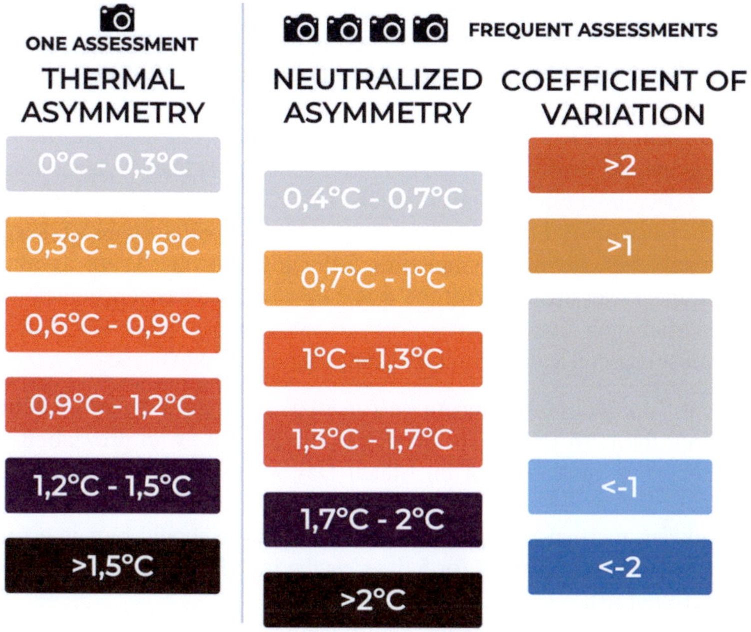

Fig. 5.8 Description of the different levels of alarms for the three main metrics used by ThermoHuman for prevention and fatigue identification: thermal asymmetry based on the comparison of contralateral ROI, neutralized asymmetry once we have more than 2 evaluations of the same subject, and coefficient of variation for each ROI when we have more than 3 evaluations (expressed in size effect)

are considered acceptable. Results depend on the body location, as normally distal regions show bigger asymmetries than proximal regions [70]. However, differences above these values may indicate that the ROI with a higher or lower temperature, contrasted with the individual's usual thermal profile settings, might have some inflammatory problem (hyperthermia) or degenerative/inhibition (hypothermia) [23, 68, 73, 76, 77].

According to our research and practical experience, we propose a "level of attention" scale to be considered depending on the bilateral temperature differences recorded and the coefficient of variation for the fatigue application (Fig. 5.8).

Based on the figure above showing indications for the level of attention, it should be borne in mind that a difference ≤0.3 °C would be considered normal [66, 70, 71]. For levels higher than 0.3 °C, the following would be advised: (a) confirming that the images have been correctly taken and segmented; (b) verifying if some external factor could be influencing the result; (c) monitoring the athlete to assess his/her environmental and training conditions. We do not recommend making drastic decisions based solely on a single abnormal assessment. Instead, we advise integrating

IRT findings with other technologies to confirm whether the detected thermal asymmetries warrant concern. As previously suggested, conducting regular thermographic assessments—ideally once a week or between competitions—enhances the reliability of the data.

Furthermore, in cases where significant thermal asymmetries are detected in muscle regions, we recommend remaining the alert for at least 7 days, even if the asymmetry decreases throughout the week. This recommendation is based on our experience, particularly in team sports—especially football—where we frequently observe significant thermal asymmetries following matches. These asymmetries tend to diminish over the course of the week; however, we have encountered injury cases even after such reductions.

These observations lead to two interesting take-home messages:

- The first 72 hours post competition (or after the highest training load of the week) are the most critical for identifying injury risk, with prevention alarms typically being most evident on Match Day +1 (MD +1) or Match Day +2 (MD +2).
- A significant thermal asymmetry in a muscle region should remain a red flag, even if it decreases after 3, 4, or 5 days, as the initial warning may still indicate an underlying risk.

In some cases, IRT results may appear unexpectedly extreme. If the assessment protocol was not strictly followed or if the thermogram suggests a potential false alarm, several alternatives can be considered. The most common approach is to disregard the result, invalidate it in the subject's historical data, and reassess during the next evaluation. However, if the athlete is still available, repeating the assessment after 15 min is recommended to confirm whether this difference is maintained [73].

Another possibility is to spray the area with an alcohol or convection cooling and wait for five minutes to see the thermal response of the area [73]. This technique, known as dynamic thermography [78], provides a valuable approach that, while requiring additional time, helps mitigate potential external influencing factors. By temporarily lowering Tsk through convection cooling, spray, or cold water, this method effectively resets the Tsk, allowing for the observation of its natural recovery pattern, which more accurately reflects internal physiological conditions.

IRT also enables the assessment of metabolic activity levels following trauma [23] or post surgery [73]. Regular monitoring throughout the rehabilitation process allows for tracking the athlete's thermal profile as it progresses toward preinjury baseline conditions. Additionally, thermographic monitoring provides valuable insights into whether medical and physiotherapy interventions are progressing as expected.

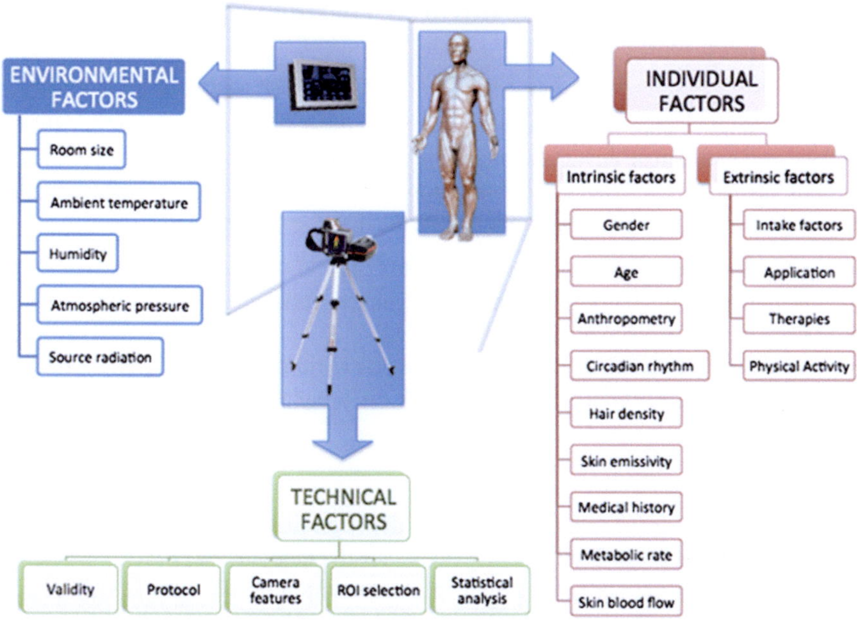

Fig. 5.9 Representation of IRT influence factors classification in humans. (Figure obtained from Fernández-Cuevas et al. [22])

5.3.3 Factors Influencing the Application of Infrared Thermograhy on Humans

Working with IRT requires taking into account many factors, which could influence either the evaluation or the interpretation of thermal images [45]. Pretending to control such a quantity of factors could entail an "impossible mission", but in many cases, just the knowledge of these factors is a first but important step. Fernández-Cuevas and collaborators published a comprehensive classification of factors that could affect the use of IRT in humans [22]. Even if this review was published some years ago, the classification and factors are still valid. They divided them into three main groups (see Fig. 5.9).

- Environmental factors: these are related to the place where the evaluation is made.
- Individual factors: these are related to the subject assessed and his/her individual characteristics, which could influence Tsk. These will be divided into intrinsic and extrinsic factors.
- Technical factors: linked to the equipment used during the IRT evaluation.

As we described in the previous subchapter about protocols, the best way to be aware of the existence of such factors is to use standardized protocols and surveys, so we can note the existence of some of those factors.

5.4 Main Applications of Infrared Thermography in Sport Injury Detection

Physical activity and exercise are some of the most promising IRT applications. Indeed, technological advances in infrared cameras, software, and new research groups opening new lines of investigation have allowed for a resurgence of this topic, enhancing new and old applications, such as the following:

- The quantification of training workload [11, 14, 79–82]
- The detection of anatomical and biomechanical imbalances [83–87]
- The warm-up efficiency evaluation [88–90]
- The evaluation of fitness and performance conditions [79, 91–94]
- The detection of high temperature risk in pregnant women during exercise [95, 96]
- The detection of muscle damage and/or delayed onset muscle soreness (DOMS) [67, 97–102]
- The identification of fatigue during and after exercise [25, 103–107]
- The support in screening and early diagnosis in emergencies [108]
- The evaluation of energy efficiency in some disciplines [92, 109–111]
- The detection of the lactate threshold [79, 112–114]
- The monitoring of the respiration rate [115]
- Clothing design and thermal comfort [116–122]
- The prevention of injuries [6, 8–10, 23]
- The follow-up of injuries or RTP [10, 20, 21, 23, 24]
- The combination with other performance technologies such as GPS, force plates, etc. [25, 54, 123, 124]
- The correlation with internal load parameters [25, 125–128]
- The relationship with the immune system [129]

From all these applications, we highlight three main ones with subapplications:

1. Prevention
    (a) Initial evaluation
    (b) Injury prevention
    (c) Treatment/training individualization
2. Return to Play
    (a) Diagnosis support
    (b) Injury evolution
3. Internal load assimilation
    (a) Fatigue identification
    (b) Recovery strategy individualization

5.4.1 Prevention

A professional football team has an average of 50 injuries per season [130] and more than 30% of them are expected to be muscle injuries [131]. Of those muscle injuries, 92% will impact four areas: hamstring (37%), adductors (23%), quadriceps (19%) and calf muscle (13%). More than 50% of injuries occur in the muscle and tendon, while two-thirds of them are from muscle fatigue, and the remaining third are from trauma [132]. Nowadays, the main concern is focused on the increasing incidence of hamstring injuries [133], due, among other factors, to the increasing demands and congested schedule [1]. Furthermore, 16% of muscle injuries were reinjuries, which resulted in significantly longer absences compared to initial injuries [132]

The implementation of IRT has the potential to significantly reduce muscle injuries, which can positively impact team performance and result in substantial cost savings. Corte et al. [8] reported a 63% reduction in muscle injuries when comparing two consecutive seasons—one without and one with the implementation of an IRT protocol—in a Brazilian team. Similarly, Gómez-Carmona et al. [9] observed a 70% decrease in muscle injuries in a Spanish professional team following the adoption of IRT. In the case of the Brazilian team, the estimated cost savings amounted to $317,500 [8].

While it is evident that IRT becomes more accurate and reliable for injury risk detection when athletes undergo frequent assessments—allowing for the establishment of individual thermal profiles and neutralized asymmetries for each ROI—it can also provide valuable insights from the very first thermogram. Additionally, general asymmetry indices, such as the Thermal Risk Index (TRI), localized alarms based on neutralized asymmetry, and the coefficient of variation for each region, can help address the three main applications of IRT in injury prevention:

- Initial evaluation: the first thermogram can help identify asymmetries that may be linked to underlying issues when a new player joins a team or during the initial assessment of an athlete in a national team setup.
- Injury prevention: with regular assessments (ideally conducted weekly), IRT can detect global and localized imbalances potentially associated with overuse, inflammation, compensation, inhibition, fatigue, or other injury risk factors.
- Treatment and training individualization: IRT can assist health and performance professionals in understanding the physiological processes underlying pain or discomfort, guiding appropriate treatment and training load adjustments based on the thermal response of the affected tissue—whether it exhibits a cold or hot pattern.

5.4.1.1 Initial Evaluation

When a new player joins a team, the medical staff conducts a series of assessments to ensure the athlete is fit for competition and meets the club's physical requirements. These medical evaluations typically include electrocardiograms, stress tests, and other diagnostic procedures. However, these assessments are often performed under significant time constraints, with limited access to prior medical history, and in high-pressure situations where closing the transfer deal is a priority. The ability of medical professionals to identify potential physical concerns is crucial, as history has shown that undetected pre-existing conditions can prevent players from reaching their expected performance levels. In this context, IRT emerges as a valuable tool, offering an objective, non-invasive method to detect physiological anomalies that may warrant further investigation.

In the following example (Fig. 5.10), we analyze the first thermogram of a new player. At the center of the figure, a chart displays normative data and the standard deviation of Tsk values for young subjects across different ROIs in the anterior lower limbs [134], represented by the brown-shaded area. Using this application developed by ThermoHuman, a single thermographic image allows for the immediate identification of key thermal asymmetries, visualized through an avatar with color-coded ROIs.

Beyond these initial insights—based on intrasubject comparisons of Tsk values from right and left ROIs—the system also evaluates how "normal" the player's absolute Tsk values are in relation to a reference dataset of healthy individuals [134]. While absolute Tsk values should be interpreted with caution due to potential external influencing factors [22], this feature provides valuable clinical insights. For instance, in this specific case, a medical practitioner assessing the player for the first time would observe:

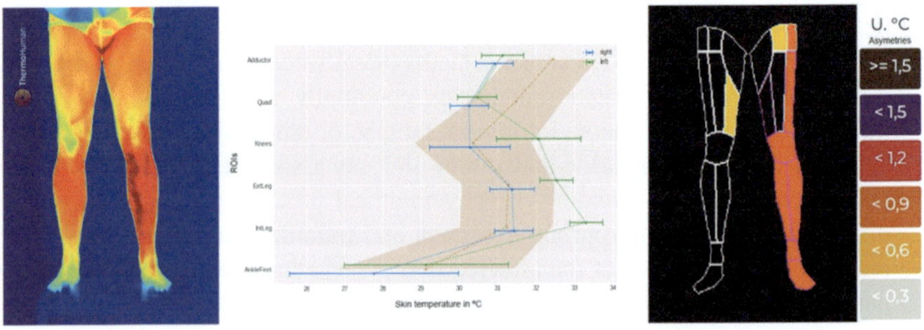

Fig. 5.10 Normative data, represented by the brown-shaded area, along with the standard deviation of Tsk values for young subjects across different ROIs of the anterior lower limbs [134]. The blue and green lines indicate the absolute Tsk values from the thermogram shown on the left. On the right side, the segmentation of the thermogram, processed by ThermoHuman software, highlights significant asymmetries using color-coded ROIs, with the intensity of asymmetry represented according to the scale on the right

(1) Significant Tsk asymmetries in the right leg, particularly at the knee, tibia, and ankle.
(2) Knee and Tibia Tsk values are warmer compared to normative data.

This information, captured and analyzed within seconds, can guide the medical team in conducting further assessments to identify potential risks related to previous injuries, overloaded regions, postural imbalances, vascular issues, and other concerns.

Speaking about vascular issues: here is an example of a player in whom a varicose was discovered in an initial evaluation in the right lower limb presenting an asymmetry of more than 2 °C. We informed the medical doctor, who transfer the case to the vascular specialist to assess the case and treat the player (Fig. 5.11).

Other thermal changes can also be signs of postural imbalances [87] which, combined with assessments such as tread analysis and manual scan, can increase the success rate in identifying potentially dangerous imbalances for the player.

Beyond the context of a new signing in a club, this application is also valuable for the initial evaluation of players joining a national team for international

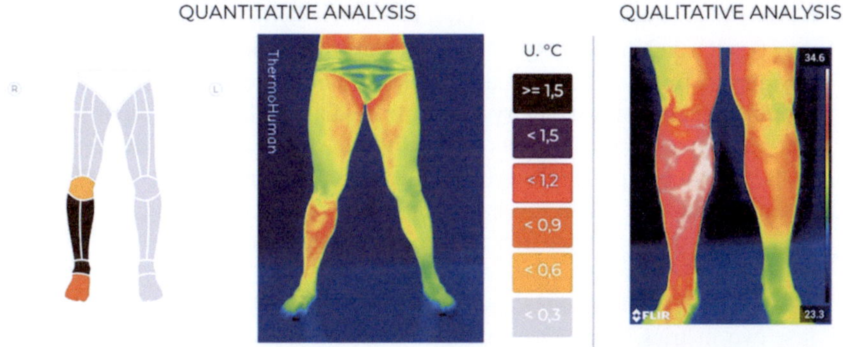

Body regions	Asymmetry	Avg temp. Left / Right	Std. deviation Left / Right
Front legs			
Outer upper thigh	-0,1 °C	31,81 °C / 31,71 °C	0,21 °C / 0,19 °C
Central upper thigh	-0,04 °C	32,4 °C / 32,36 °C	0,45 °C / 0,43 °C
Outer front thigh	-0,14 °C	31,04 °C / 30,9 °C	0,47 °C / 0,55 °C
Central front thigh	-0,06 °C	31,83 °C / 31,77 °C	0,5 °C / 0,53 °C
Front adductor	0,04 °C	32,73 °C / 32,77 °C	0,63 °C / 0,66 °C
Inner front thigh	-0,03 °C	32,16 °C / 32,13 °C	0,32 °C / 0,51 °C
Knee	0,31 °C	30,65 °C / 30,96 °C	0,59 °C / 0,74 °C
Outer front leg	2,11 °C	30,73 °C / 32,84 °C	0,6 °C / 0,57 °C
Inner front leg	2,15 °C	31,03 °C / 33,19 °C	0,84 °C / 0,71 °C
Ankle	2,08 °C	29,31 °C / 31,4 °C	0,38 °C / 0,45 °C
Anterior foot	0,77 °C	27,79 °C / 28,56 °C	1,96 °C / 2,03 °C

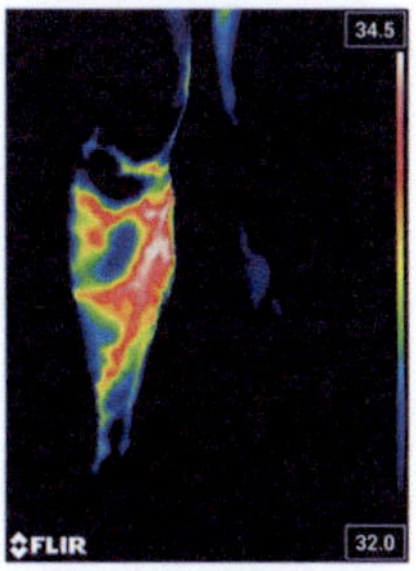

Fig. 5.11 Example of a varicose vein detected by IRT during the initial thermal evaluation of a football player. On the left, the quantitative analysis performed by ThermoHuman highlights areas with significant asymmetries using a color-coded representation, corresponding to the data in the table below. On the right, two close-up thermograms with different contrast settings provide a clearer visualization of the varicose pattern

competitions or international breaks. Given the limited preparation time and the potential lack of medical information from their respective clubs, IRT can offer valuable insights into the player's condition during the short training and competition periods typically associated with national team duty.

5.4.1.2 Injury Prevention

As mentioned in the previous subchapter, for an optimal tracking of the asymmetries, automatic solutions that quantify the temperature of different ROIs are needed. Using an algorithm for artificial vision, each ROI (muscle areas and/or joint) can automatically be identified and, therefore, comparative reports can be carried out easily between a left–right member or a dominant–nondominant in a scientific and reliable way. This allows us to accumulate and manage the data recorded, establishing the thermal evolution of each player, which is essential for assessing individual training load and match response.

The primary metrics we will use include:

- *Thermal Risk Index (TRI)*: a general index ranging from 0 to 100, based on the number of regions of interest (ROIs) exhibiting significant asymmetry (greater than 0.3 °C) and the severity of those asymmetries. This index allows us to classify players from least to most asymmetrical during group sessions and track the overall evolution of asymmetry over time in individual tracking reports.
- *Thermal asymmetry*: this refers to the difference in mean Tsk between bilateral ROIs, such as comparing the average Tsk of the right knee to that of the left knee. A classification of alarm levels is established (Fig. 5.8) starting from a 0.3 °C difference, with asymmetries above this threshold considered potentially significant.
- *Neutralized TRI and neutralized asymmetry*: these are adapted versions of the first two metrics, used when multiple assessments of the same individual are available and there are no influencing factors such as pain or injury. These global and localized asymmetry indices are individualized, with thresholds adjusted based on each individual's historical average for each ROI. This allows for a more precise and personalized assessment (Fig. 5.8).
- *Softened coefficient of variation*: this metric highlights the most relevant trends in the Tsk of a single ROI over time (e.g., hyperthermic or hypothermic patterns). By subtracting the general increase or decrease in Tsk, it emphasizes those ROIs that have experienced a significant change, which helps understanding if a thermal asymmetry alarm is warm or cold.
- *AI automatic alarms*—asymmetry alarms: besides the metrics, ThermoHuman creates, based on AI algorithms, automatic alarms ranging in three levels of severity, to automatically identify the ROIS with a more concerning asymmetry evolution. Those are the levels and its meaning:
  - Alarm level 1: review

No immediate action is required, but the player's thermal profile should continue to be monitored to track potential changes.

- Alarm level 2: review and monitoring

  Inform the medical/performance staff about the detected asymmetry.
  Implement closer monitoring to assess its progression.
  Consider adjustments in workload and the integration of complementary assessments if necessary.

- Alarm level 3: exhaustive review and monitoring

  Recommend a medical assessment and targeted treatment.
  Implement load reduction strategies and additional diagnostic tests to identify potential underlying issues.
  If the asymmetry continues to increase, suggest temporary training restrictions or rest, with close follow-up to confirm the resolution of the trend.

We always recommend combining IRT results with other information (such as wellness survey [25], clinical evaluations, etc.) and technologies such as GPS [54], force plates [124] or variables like CK values [125] or heart rate variability [128] before making any decision in regards training load. In other words, IRT allows a very fast assessment which might create alarms that should be confirmed with more objective data or diagnosis technologies.

Case Study: IRT Alarms Predict Foot Pain Onset 1 Week in Advance
Nonetheless, this is a great example of how IRT can aid in injury prevention just by being combined with the wellness survey. During a blind IRT study conducted with a professional Spanish LaLiga team, one player began reporting significant pain in his right foot in his daily wellness survey. Thermal images were captured twice a week, prior to training sessions, following TISEM guidelines [27].

Upon analyzing the asymmetry evolution in this case, we observed that the asymmetry in the affected ROI was significantly elevated (0.79 °C) even 1 week prior to the player's pain report. What is particularly interesting is not just the alarm triggered by the asymmetry, but the consistent local and global patterns of increase. As shown in Fig. 5.12, the foot's asymmetry rose from 0.79 to 0.90 °C just 4 days later, and reached 1.33 °C in the assessment immediately before the player reported pain. Additionally, we noted a significant and progressive increase in the Thermal Risk Index (TRI), indicating that the asymmetry was not confined to the foot but reflected an overall trend of increasing asymmetry. Fortunately, the player did not sustain an injury, but this case highlights the potential of IRT in detecting both localized and global alarms well before the onset of pain, demonstrating its value in injury prevention.

Dynamic thermography involves thermal monitoring after a physical stress (cold, heat, exercise) and can be ideal for monitoring possible alterations that are not visible on a daily basis [78]. This technique can often used on the MD +1, involving immersion in cold water which many professional teams perform. In the example of Fig. 5.13, we can observe how the cold stress on the Tsk is not symmetrical, and

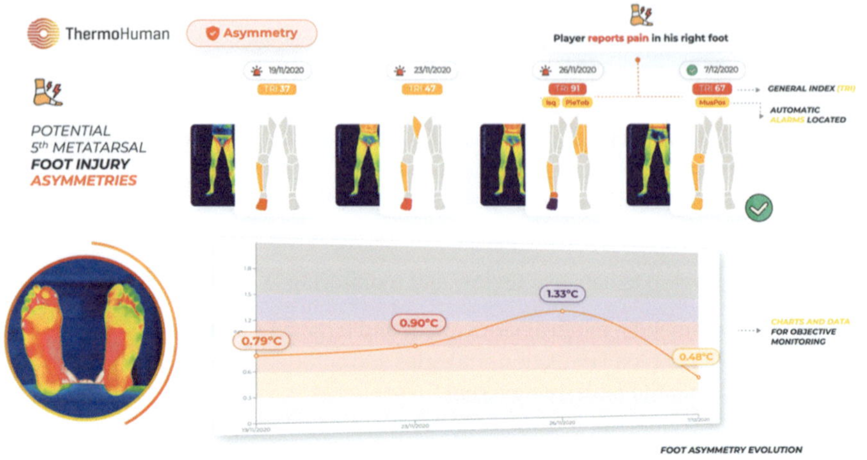

Fig. 5.12 Thermal images, chart and ThermoHuman avatars evolution of foot thermal asymmetries on a professional football player before and after reporting pain in his right foot. (Designed by Maria Sambeat)

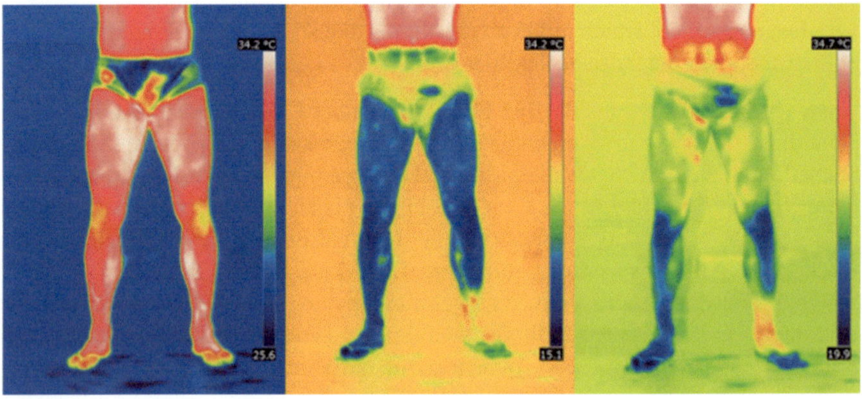

Fig. 5.13 Dynamic thermography after immersion in cold water, where a different thermal response can be seen in an ankle with an old injury that permits a clear recognition of the affected area

therefore we can identify a warmer area around the individual's left ankle. This warmer area could be related to an old injury or area that has been abnormally activated or irrigated. In this case, this thermal pattern was reported to the medical staff in order to follow the evolution and to determine if there could be any dysfunction or problem.

5.4.1.3 Treatment/Training Individualization

When applying IRT to a group of athletes, we typically categorize them into two main groups: (1) healthy athletes and (2) athletes experiencing pain and/or injury.

Healthy athletes: for this group, initial evaluations and injury prevention metrics are crucial for identifying potential risks. Since IRT is often one of the first assessments conducted during the morning routine -and its results are generated automatically—it serves as an effective screening tool. This allows the performance and medical staff to flag cases that may require further investigation.

When IRT alarms are detected, the integration of additional data becomes essential. Many teams use Athlete Monitoring Systems that aggregate data from various technologies via API, facilitating the combination of alerts and confirming potentially concerning cases. If these findings are validated through complementary assessments, staff can implement appropriate interventions based on the alarm levels described in the previous subchapter.

Athletes with pain and/or injury: in cases involving pain or injury, IRT plays a critical role not only in confirming thermal asymmetries in the affected region but also in assessing compensatory patterns throughout the body. More importantly, IRT allows differentiation between hyperthermic and hypothermic asymmetries, which has significant implications for diagnosis and intervention strategies.

It is commonly assumed that most injuries are associated with hyperthermia due to inflammation and overload [108]. This is why ThermoHuman's initial IRT evaluations highlight significant asymmetries by marking warmer ROIs on the avatars. However, experience has shown that hypothermic asymmetries—associated with neuromuscular inhibition, reduced strength, or internal fatigue—can be just as relevant, particularly in muscle regions.

Understanding whether the affected ROI presents a hyperthermic or hypothermic asymmetry can drastically alter the interpretation of the issue and, consequently, the intervention strategy. This distinction enables a more precise and individualized approach to both physiotherapy and rehabilitation training.

Case Study: Hypothermic Asymmetry in a Professional Football Player

A compelling example of this concept comes from a professional football player in the Chinese Super League who underwent weekly IRT assessments. Figure 5.14 illustrates the results of an evaluation conducted on MD +1 (the day after a match), when the player reported discomfort in his right posterior central thigh (hamstring region).

Interestingly, the ThermoHuman avatar displayed a significant thermal asymmetry in the opposite ROI, with the left hamstring being 0.37 °C warmer. Meanwhile, a purple marker highlighted the player's reported pain on the right side. At first glance, this discrepancy might seem confusing. However, by analyzing the softened coefficient of variation (explained in the previous subchapter) and comparing the current assessment with the player's historical thermal data, it became evident that the most significant change had occurred in the left hamstring, while the right hamstring exhibited a distinct hypothermic trend.

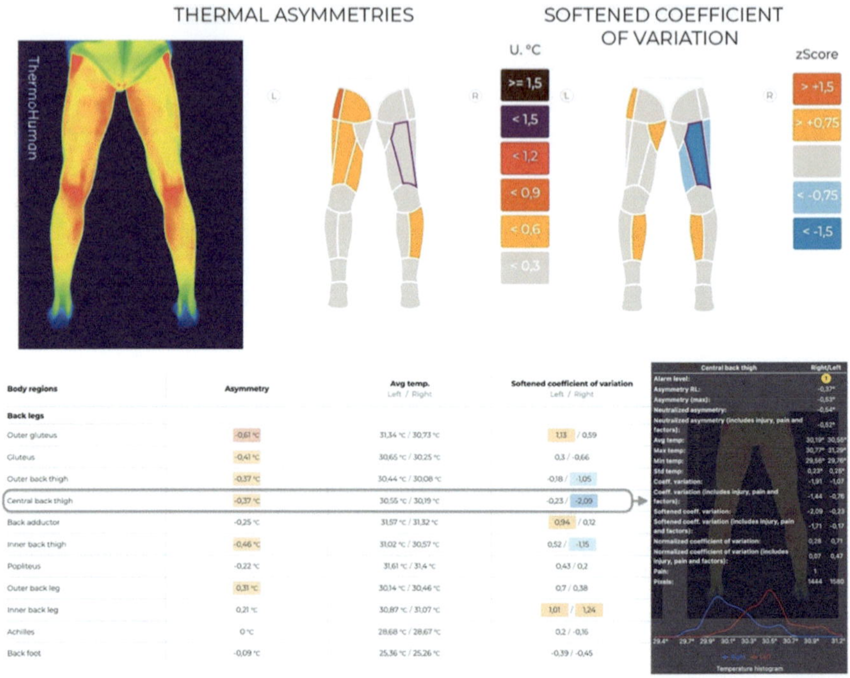

Fig. 5.14 IRT evaluation of a football player on MD +1 using ThermoHuman software. The figure presents the thermogram on the left, the thermal asymmetries avatar in the center, and the softened coefficient of variation on the right. Below, the table displays relevant data, while the lower right corner highlights all metrics for the central posterior thigh, including a histogram

In other words, the area of reported pain was associated with lower Tsk, suggesting that the underlying cause of discomfort could be related to neuromuscular inhibition, strength deficits, or fatigue rather than inflammation. This insight fundamentally shifted the interpretation of the injury and guided the intervention strategy accordingly.

Proposed Intervention for Hypothermic Alarms

Based on the hypothermic findings, the intervention plan should focus on restoring balance by increasing Tsk and neuromuscular activation in the affected region:

- From a physiotherapy perspective: address the asymmetry by promoting blood flow and increasing local Tsk through techniques such as diathermy, activation massage, or other interventions.
- From a training perspective:

– Assess potential strength and neuromuscular deficits using tests such as the Countermovement Jump, electromyography, or strength testing.
  – Introduce targeted activation and eccentric exercises to promote neuromuscular recovery and restore symmetrical thermal patterns by increasing Tsk.

This case demonstrates how IRT provides critical insights beyond simple injury detection, enabling a more nuanced and proactive approach to treatment and performance optimization.

5.4.2 *Return to Play*

While IRT and other technologies can contribute to reducing injury incidence, achieving a completely injury-free high-performance environment remains unrealistic. Therefore, it is essential to be prepared for the challenge of managing injuries effectively.

The first and most critical point to emphasize in injury management is that IRT is not a standalone diagnostic tool for sports injuries. Although it can be highly valuable in the assessment process and has demonstrated good sensitivity and specificity [135], more established imaging technologies—such as ultrasound, magnetic resonance imaging (MRI), and X-rays—are better suited for definitive musculoskeletal injury diagnosis.

However, IRT plays a particularly interesting role in the Return to Play (RTP) process, with two key applications:

1. *Diagnosis support and complex cases*: IRT can assist in the initial evaluation, helping to identify thermal asymmetries that may support clinical diagnosis.
2. *Monitoring rehabilitation progress*: IRT provides an objective measure of the injured area's thermal evolution while also tracking compensatory imbalances elsewhere in the body.

These applications contribute to a more *individualized rehabilitation approach*, enabling better load management and treatment adjustments. As a result, IRT may accelerate RTP timelines and reduce the risk of reinjury, optimizing both rehabilitation and performance outcomes.

5.4.2.1 Diagnosis Support

If someone had asked me 10 years ago about the typical thermal response following a sports injury, my answer would have been emphatic: *hot*. Hyperthermia is commonly associated with inflammation, stress, and overuse, making it logical to assume that most sports injuries present with increased temperature. And in many cases, this is indeed true. However, not all injuries follow this pattern.

A thorough review of the literature reveals studies reporting diverse thermal responses after sports injuries. Rochcongar and Schmitt were the first to describe different thermal asymmetry patterns following muscle injuries in 1979, identifying *normothermic, hyperthermic, and hypothermic responses* [136]. Years later, BenEliyahu published a comprehensive review of various sports injuries—including epicondylitis, patellofemoral syndrome, ankle injuries, stress fractures, and

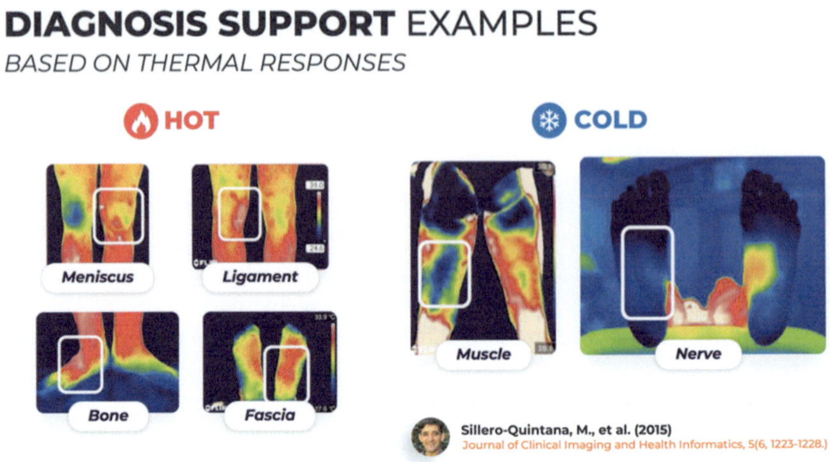

Fig. 5.15 Division of the two main thermal responses after sport injuries depending on the affected tissue: hyperthermic and hypothermic asymmetries. (Based on Siller-Quintana and collaborators [108])

shoulder injuries—reporting predominantly hyperthermic patterns in almost all cases [137].

More recently, Sillero-Quintana and collaborators conducted a *descriptive study* [108] on the thermal asymmetries of injuries treated at Clinica CEMTRO, a renowned FIFA medical center specializing in sports injuries. The study, which analyzed *201 cases*, yielded two key conclusions:

1. The average thermal asymmetry between injured and non-injured ROIs was 0.5 °C.
2. IRT can be considered a good support tool to correctly identify sports injuries [108].

Beyond these findings, Fig. 5.15 illustrates a crucial takeaway from Sillero-Quintana's work: thermal responses to injury vary depending on the affected tissue. Injuries may present as either *hyperthermic or hypothermic asymmetry*, a distinction that can be highly useful in supporting diagnosis and differentiating complex cases.

Case Study: Plantar Fasciitis vs Baxter's Neuropathy: A Case Study of Diagnosis Support

A 25-year-old female 800 m runner was diagnosed with plantar fasciitis in late 2020 during her preparation for the Olympic Games. She reported plantar heel pain (5/10 on the NRS scale), particularly in the mornings and after rest. Despite physiotherapy, shockwave therapy, cortisone injections, and insoles, her symptoms persisted, preventing her from running for over 2 months.

On March 10, 2021, an IRT evaluation was conducted following TISEM consensus [27]. Interestingly, the thermal image revealed a hypothermic pattern (−1.92 °C) in the painful area—more characteristic of nerve-related conditions than inflammatory processes. Given that not all plantar heel pain cases are plantar fasciitis [138], we recommended a differential diagnosis, considering possible nerve entrapment syndromes, such as Baxter's neuropathy (Inferior Calcaneal Nerve compression).

Following this insight, the athlete consulted a specialist who identified a blocked right ankle joint and initiated a treatment focused on joint mobilization and pressure relief in the medial foot. Within 2 weeks, she was able to run pain-free, experiencing only mild tingling potentially indicating nerve involvement. A follow-up IRT evaluation (March 24, 2021) showed a significant reduction in thermal asymmetry from −1.92 to −0.12 °C, supporting the clinical improvement (Fig. 5.16).

This case highlights IRT as a valuable tool for differential diagnosis and monitoring in sports medicine. In this instance, IRT helped reassess an initial misdiagnosis,

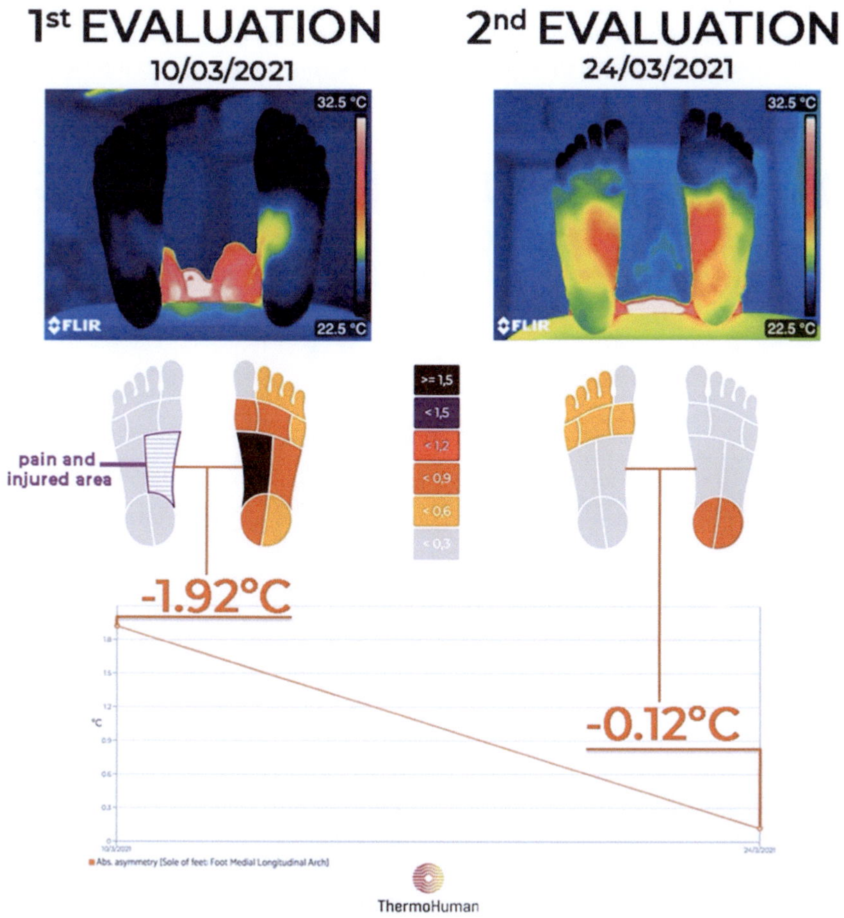

Fig. 5.16 Thermal images, ThermoHuman asymmetry avatars and chart representing the absolute asymmetry evolution of the foot medial longitudinal arch from the first evaluation to the second one, 14 days after

guiding a more effective treatment and confirming its success through objective thermal data.

5.4.2.2 Injury Monitoring

Once the diagnosis is confirmed, the primary challenge is to facilitate the athlete's return to training and competition as efficiently and safely as possible. From an IRT perspective, while the typical acute thermal asymmetries following an injury are well documented, there remains a gap in understanding the normative progression of thermal asymmetry across different injury types.

This continues to be a significant challenge for researchers and IRT specialists. However, recent case studies and descriptive research have provided valuable insights into the expected thermal evolution of various injuries, allowing for better comparison and assessment of an athlete's recovery trajectory (Fig. 5.17). This figure summarizes the primary thermal characteristics of common sports injuries based on findings from multiple studies, including anterior cruciate ligament (ACL)

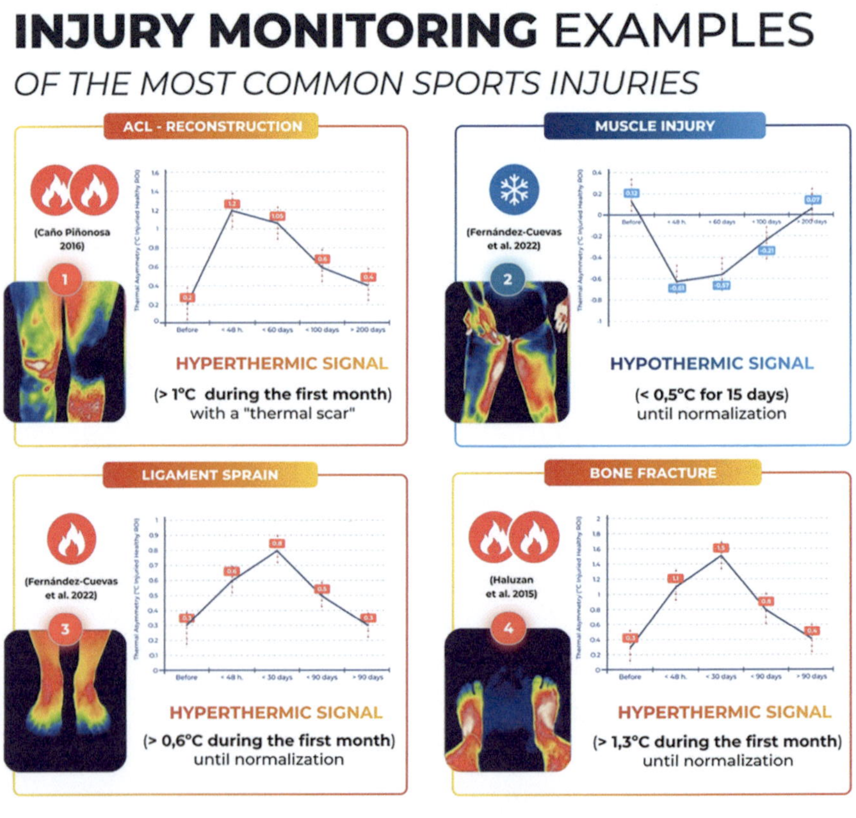

Fig. 5.17 Asymmetry thermal evolution graphics representing ACL, muscle injury, ligament sprain and bone fracture normative data based on different studies [20, 139–141]. (Designed by Maria Sambeat)

injuries [20, 24], muscle injuries [141], ligament sprains [140], and bone fractures [21, 139]. Notably, muscle injuries stand out due to their predominant hypothermic response [136, 140, 141].

Case Study: IRT for Hamstring Injury Follow-Up in a Premier League Football Player

This case study examines the rehabilitation process of a 26-year-old professional football player from a Premier League team who sustained a grade 3 biceps femoris injury in his right leg, requiring surgical intervention. IRT assessments were conducted monthly under controlled conditions before treatment or rehabilitation sessions.

Initial IRT evaluations revealed a hypothermic asymmetry (−0.41 °C) in the injured leg compared to the noninjured side, consistent with findings in the scientific literature [141]. However, a concerning trend emerged during follow-up: rather than progressing toward thermal symmetry, the asymmetry gradually worsened, deviating from the expected recovery pattern for similar injuries [140] (Fig. 5.18).

Nearly 5 months post injury, the athlete was deemed ready for RTP. He reported no symptoms, ultrasound imaging confirmed proper muscle healing, and the coaching staff was prepared for his reintegration. However, IRT continued to indicate a persistent hypothermic pattern. Given these findings, a strength assessment was conducted, revealing a significant strength deficit in the injured hamstring. Consequently, RTP was postponed.

One month later, following targeted rehabilitation, strength levels were restored, and IRT asymmetry had nearly resolved (0.04 °C) (Fig. 5.18). This case highlights the value of IRT in RTP decision-making by providing objective, functional, and comparative data. Unlike structural imaging modalities such as MRI or ultrasound,

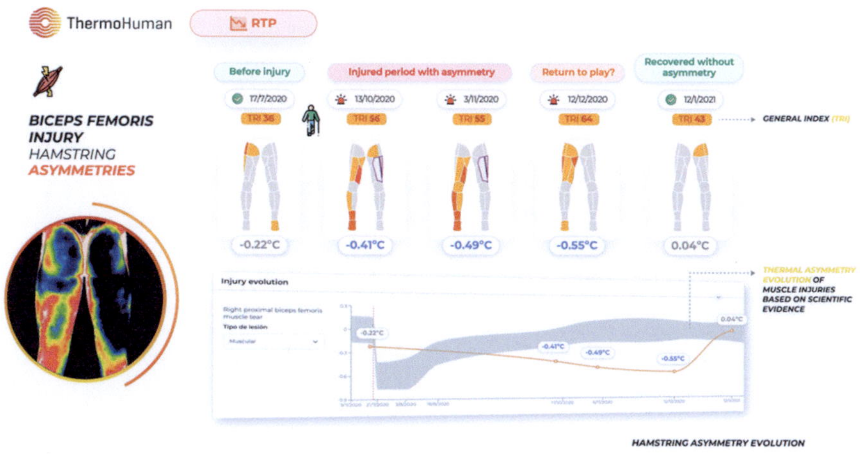

Fig. 5.18 ThermoHuman avatars and injury evolution chart of a left biceps femoris grade 3 thermal asymmetries on a professional football player. (Designed by Maria Sambeat)

which assess tissue morphology and structure, IRT evaluates physiological function and energy dynamics. Therefore, IRT should be integrated as a complementary tool in rehabilitation to enhance injury monitoring and optimize RTP strategies.

Case Study: ACL Rehabilitation in a Rugby Player

We present an anterior cruciate ligament (ACL) injury evolution in a rugby player: from the day before the injury until the return to play (Fig. 5.19). What makes this case study unique is that during the whole process, the rugby player was monitored by IRT.

Before the surgery (from day 12 to 94), the subject followed a progressive five days a week plan with isometric and concentric exercises to prepare the limb muscles and to maintain the physical condition. After the surgery (from day 111 to 242), the rehabilitation program consisted of a progressive adaptation to gain range of motion, to reduce the inflammation, to strengthen the muscles, and to gain ability and confidence.

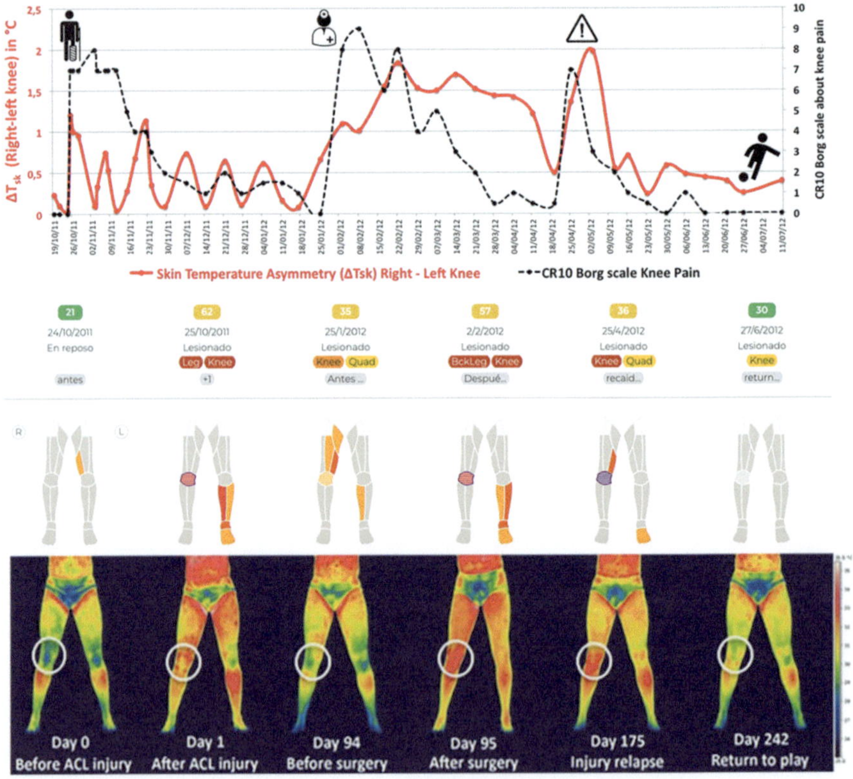

Fig. 5.19 Infrared images, ThermoHuman asymmetry avatars of the subject in specific moments from the day before the injury (day 0) to the return to play (day 242)

The results can be divided into three phases. First, from the injury to the surgery, we saw a decrease in pain perception and regular Tsk variations, from the biggest (+1.20 °C on day 1) to the smallest thermal asymmetry (+0.08 °C on day 82). Second, from the surgery (day 95), we found a similar decrease in the evolution of pain, but a different Tsk behavior, which was maintained during the first weeks of the rehabilitation due to the exercises performed. The third phase started some weeks before the return to play (day 175), when the subject again felt a "crack" on his right knee during his first game in a training session. Thermal asymmetry (+1.36 °C) and pain increased—some years after that, the subject was informed in a medical assessment that this crack was actually an injury relapse. However, the rehabilitation continued, ending on the June 27, 2012 (day 242), with his first game, and the last one of the season for the team.

In conclusion, IRT results were especially useful regarding the return to play: since Tsk is directly related to the inflammation and muscle activity, it helped both the medical and technical staff to adapt and optimize the rehabilitation process. Besides the knee evolution, IRT also helped measuring imbalances created in areas like the ankle as compensation or the quadriceps, which is crucial for the knee stability. Moreover, the subject described to us how IRT helped him to visualize the evolution, and somehow motivated him to move forward despite the relapse.

Recent studies have provided comparative data to better understand the typical recovery thermal evolution of ACL injuries in football players [24]. Notably, research has highlighted sex-based differences in thermal asymmetry trends, with male athletes requiring, on average, over 600 days to restore normal thermal balance, whereas female athletes achieve symmetrical knee values in approximately 275 days [141].

5.4.3 *Internal Load Assimilation*

The correlation between internal load and Tsk represents one of the most promising areas of IRT research. As discussed earlier in this chapter, both external and internal load parameters are crucial for understanding and optimizing athletic performance and adaptation. While technological advancements have significantly improved external load monitoring (e.g., GPS systems), the measurement of internal load remains more challenging due to its subjective nature, complexity, and, in some cases, invasive or time-consuming assessment methods.

IRT emerges as a compelling alternative for internal load monitoring, offering an objective, rapid, and noninvasive means of assessing Tsk—an often—overlooked physiological parameter linked to both health and performance. Although further research is needed to refine our understanding of the relationship between Tsk and internal load, existing evidence suggests that IRT holds great potential for identifying fatigue patterns post competition and guiding individualized recovery strategies.

From a methodological standpoint, it is important to distinguish the use of IRT for internal load and fatigue monitoring from its applications in injury prevention and RTP. While the latter primarily rely on thermal asymmetries, internal load and fatigue assessments are based on distinct thermal metrics:

- *Coefficient of variation (CV):* this metric quantifies temperature fluctuations by analyzing the historical mean and standard deviation of each ROI across multiple baseline assessments conducted in the absence of pain, injury, or external influencing factors. A minimum of three evaluations is required. CV values are visually represented using orange (hyperthermic) or blue (hypothermic) gradients, aiding in the identification of abnormal thermal patterns.
- *Thermal Status Index (TSI)*: a global index ranging from −100 to 100, TSI tracks global temperature variations over time to determine whether an athlete is exhibiting warming, cooling, or stable thermal trends. It requires at least three evaluations and utilizes Z-scores to classify individuals with significant hyperthermic or hypothermic tendencies.

5.4.3.1 Fatigue Identification

The relationship between Tsk and fatigue has been previously explored in the literature. Several studies have analyzed this connection during exercise [103], post-exercise [104], across different sports such as football [25] and volleyball [107], and even in female athletes exclusively [105]. Additionally, Tsk has been correlated with various internal load markers, including heart rate variability (HRV) [128], blood lactate levels [112], creatine kinase (CK) [56], and neuromuscular fatigue-related tests and metrics [97, 123].

A pivotal moment in this field was the publication of Robin Thorpe's study on the individualization of postexercise recovery strategies (cooling vs. heating) [12]. Thorpe, a former sports scientist at Manchester United, emphasized a key concept: fatigue is not a singular physiological phenomenon, but rather a multifaceted process with distinct underlying mechanisms.

Thorpe identified two primary types of fatigue: *metabolic fatigue (central fatigue)* and *structural damage (peripheral fatigue)*. While both contribute to what is generally referred to as "fatigue," they arise from fundamentally different physiological processes. This distinction is crucial because it allows for a more tailored approach to recovery. According to Thorpe's findings, *metabolic fatigue* should be counteracted with heating therapies, whereas *structural damage* requires cooling strategies (Fig. 5.20).

Interestingly, Thorpe did not mention IRT in his study [12]. However, when describing the characteristics of structural damage, he noted an increase in temperature, suggesting that cooling strategies might be the most effective recovery method on MD +1 or MD +2.

Building on this premise, we hypothesized that metabolic fatigue, which is more closely linked to central nervous system stress, sleep disturbances, and other

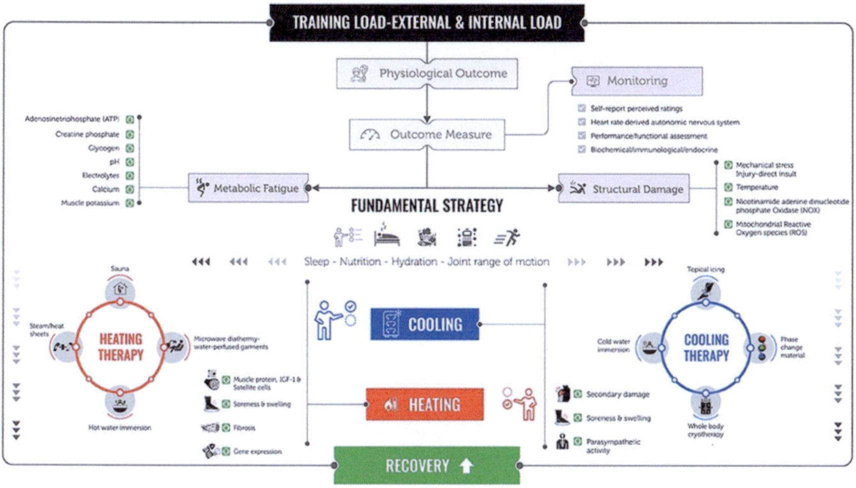

Fig. 5.20 Practical framework to enhance recovery in athletes from Thorpe (2021) Post-Exercise Recovery: Cooling & Heating, a Periodized Approach [12]

systemic factors, might be associated with a decrease in temperature. Consequently, heating therapies could be a more suitable recovery approach in such cases. This hypothesis was later confirmed [25], leading to a key conclusion: postcompetition fatigue can be identified by analyzing individual Tsk trends, with structural damage presenting as hyperthermia and metabolic or central fatigue manifesting as hypothermia.

5.4.3.2 Recovery Strategy Individualization

Considering the insights discussed in the previous subchapter and the increasing issue of fixture congestion in high-performance sports, IRT presents a valuable application for identifying the type of fatigue that athlete experience after competition or high training loads. This allows practitioners to tailor recovery strategies based on Tsk assessments.

Unfortunately, recovery strategies have historically been underprioritized by many institutions and performance practitioners. However, the growing impact of congested schedules, increased injury rates, more competitions, extended travel, reduced training time, sleep disturbances, and heightened stress has shifted the focus toward recovery. The traditional one-size-fits-all approach is no longer viable, and both performance experts and athletes are demanding more individualized recovery strategies.

Scientific evidence suggests that cold-based strategies are optimal immediately post competition. However, there is no clear consensus on whether cooling or heating strategies are more effective on MD +1 and MD +2. This is where IRT provides

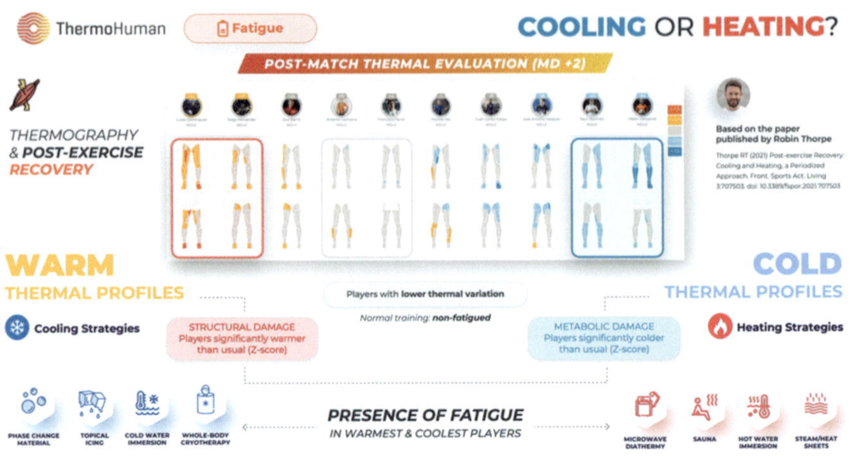

Fig. 5.21 Thermal responses from a football team on MD +2 based on TSI and coefficient of variation, leading to different profiles that allow individualizing the recovery strategy based on Thorpe's publication [12]. (Designed by Maria Sambeat)

an added value. As illustrated in Fig. 5.21, players who competed within the past 72 hours can be assessed using the coefficient of variation and TSI to determine:

- Neutral tendency → indicates good workload assimilation.
- Hyperthermic tendency → associated with structural damage and peripheral fatigue, suggesting that cooling strategies (e.g., Cold Water Immersion) are recommended.
- Hypothermic tendency → linked to central fatigue, often resulting from fixture congestion, sleep deprivation, stress, and overtraining, making heat-based recovery strategies more appropriate.

This case study from a professional football team demonstrates how IRT can be integrated to individualize recovery strategies. Currently, we are conducting data collection projects to further investigate the effectiveness of IRT-based recovery prescriptions, as well as their impact on athlete recovery perception and performance outcomes.

5.5 Conclusions

IRT has demonstrated significant potential as a valuable tool for sports professionals, with applications extending far beyond injury assessment. Its noninvasive nature, increasing accessibility, and rapid, objective data acquisition make it a promising technology for internal load monitoring, positioning it as a key asset for both health and performance practitioners.

However, to fully harness the benefits of IRT, it is crucial to implement standardized protocols that minimize the risk of false positives and enhance the reliability of data interpretation. Without rigorous methodological control and reliable software solutions, the effectiveness of IRT in decision-making processes may be compromised.

While IRT is not a standalone solution for injury prevention or a magic bullet for optimizing recovery, its scientific foundation and growing adoption in elite sports institutions confirm its value. Current evidence supports its role in injury prevention, RTP, and individualized postcompetition recovery strategies, particularly through the analysis of Tsk evolution and thermal asymmetries. As research continues to expand, IRT is poised to become an integral component of modern sports science, bridging the gap between physiological assessment and data-driven performance optimization.

References

1. Page RM, Field A, Langley B, Harper LD, Julian R (2023) The effects of fixture congestion on injury in professional male soccer: a systematic review. Sports Med 53(3):667–685
2. Pulici L, Certa D, Zago M, Volpi P, Esposito F (2023) Injury burden in Professional European Football (Soccer): systematic review, meta-analysis, and economic considerations. Clin J Sport Med 33(4):450–457
3. Gabbett TJ (2016) The training—injury prevention paradox: should athletes be training smarter and harder? Br J Sports Med 50:273–280. https://doi.org/10.1136/bjsports-2015-095788
4. Bowen L, Gross AS, Gimpel M, Bruce-Low S, Li F-X (2020) Spikes in acute:chronic workload ratio (ACWR) associated with a 5–7 times greater injury rate in English Premier League football players: a comprehensive 3-year study. Br J Sports Med 54(12):731
5. Buchheit M, Laursen PB (2024) Sports Science 3.0: integrating technology and AI with foundational knowledge. Available on: https://martin-buchheit.net/wp-content/uploads/2024/08/SPSR231_Buchheit-1.pdf
6. Tauchmannova H, Gabrhel J, Cibak M (1993) Thermographic findings in different sports, their value in the prevention of soft tissue injuries. Themol Osterr 3:91–95
7. Barcelos EZ, Caminhas WM, Ribeiro E et al (2014) A combined method for segmentation and registration for an advanced and progressive evaluation of thermal images. Sensors 14:21950–21967. https://doi.org/10.3390/s141121950
8. Côrte AC, Pedrinelli A, Marttos A, Souza IFG, Grava J, Hernandez AJ (2019) Infrared thermography study as a complementary method of screening and prevention of muscle injuries: pilot study. BMJ Open Sport Exerc Med 5(1)
9. Gomez-Carmona P, Fernández-Cuevas I, Sillero-Quintana M, Arnaiz-Lastras J, Navandar A (2020) Infrared thermography protocol on reducing the incidence of soccer injuries. J Sport Rehabil 29(8):1222–1227
10. Sillero-Quintana M, Gomez-Carmona PM, Fernández-Cuevas I (2021) Infrared thermography as a means of monitoring and preventing sports injuries. In: Research anthology on business strategies, health factors, and ethical implications in sports and eSports. IGI Global, pp 832–865
11. Čoh M, Širok B (2007) Use of the thermovision method in sport training. Facta Univ-Ser Phys Educ Sport 5:85–94
12. Thorpe R (2021) Post-exercise recovery: cooling and heating, a periodized approach. Front Sports Act Living 3(236). https://doi.org/10.3389/fspor.2021.707503

13. Kenny GP, Jay O (2007) Sex differences in postexercise esophageal and muscle tissue temperature response. Am J Physiol Regul Integr Comp Physiol 292:R1632–R1640. https://doi.org/10.1152/ajpregu.00638.2006
14. Fernández Cuevas I (2012) Effect of endurance, speed and strength training on skin temperature measured by infrared thermography = Efecto del entrenamiento de resistencia, velocidad y fuerza en la temperatura de la piel a través de la termografía infrarroja. Universidad Politécnica de Madrid
15. Frim J, Livingstone SD, Reed LD et al (1990) Body composition and skin temperature variation. J Appl Physiol Bethesda Md 68:540–543
16. Ring EFJ (1990) Quantitative thermal imaging. Clin Phys Physiol Meas 11:87. https://doi.org/10.1088/0143-0815/11/4A/310
17. Niu HH, Lui PW, Hu JS et al (2001) Thermal symmetry of skin temperature: normative data of normal subjects in Taiwan. Zhonghua Yi Xue Za Zhi Chin Med J Free China Ed 64:459–468
18. Uematsu S, Edwin DH, Jankel WR et al (1988) Quantification of thermal asymmetry: part 1: normal values and reproducibility. J Neurosurg 69:552–555
19. De León-Muñoz A, Priego-Quesada JI, Marzano-Felisatti JM, Sanchez-Jimenez JL, Sendra-Pérez C, Aparicio-Aparicio I (2024) Preliminary application of infrared thermography to monitoring of skin temperature asymmetries in professional padel players. Sensors 24(14). https://doi.org/10.3390/s24144534
20. Piñonosa Cano S, Sillero Quintana M, Milanovic L et al (2013) Thermal evolution of lower limbs during a rehabilitation process after anterior cruciate ligament surgery. Kinesiol Zagreb Croat 45:121–129
21. Castonguay T, Dover G (2023) Infrared thermography—a novel tool for monitoring fracture healing: a critically appraised topic with evidence-based recommendations for clinical practice. J Sport Rehabil:1–6. https://doi.org/10.1123/jsr.2022-0390
22. Fernández-Cuevas I, Bouzas Marins JC, Arnáiz Lastras J et al (2015) Classification of factors influencing the use of infrared thermography in humans: a review. Infrared Phys Technol 71:28–55. https://doi.org/10.1016/j.infrared.2015.02.007
23. Hildebrandt C, Raschner C, Ammer K (2010) An overview of recent application of medical infrared thermography in sports medicine in Austria. Sensors 10:4700–4715
24. Escamilla-Galindo V-L, Felipe JL, Alonso-Callejo A, Van-der-Horst R, de la Torre-Combarros A, Minafra P et al (2024) Return-to-play criteria based on infrared thermography during anterior cruciate ligament rehabilitation in football players. Biol Sport 42:161–167. https://doi.org/10.5114/biolsport.2025.144295
25. Majano C, Garcia-Unanue J, Fernández-Cuevas I, Escamilla-Galindo V, Alonso-Callejo A, Sanchez-Sanchez J et al (2023) Association between physical demands, skin temperature and wellbeing status in elite football players. Sci Rep 13(1):13780. https://doi.org/10.1038/s41598-023-40396-y
26. Ammer K (2006) Influence of imaging and object conditions on temperature readings from medical infrared images. Pol J Environ Stud (Submitt)
27. Gomes Moreira D, Costello JT, Brito CJ, Adamczyk JG, Ammer K, Bach AJE et al (2017) Thermographic imaging in sports and exercise medicine: a Delphi study and consensus statement on the measurement of human skin temperature. J Therm Biol 69:155–162. https://doi.org/10.1016/j.jtherbio.2017.07.006
28. IACT (2002) Thermology guidelines. Standards and protocolos in clinical thermography imaging
29. ISO (2004) Ergonomics—evaluation of thermal strain by physiological measurements. ISO 9886
30. Schwartz RG (2006) Guidelines for neuromusculoskeletal thermography. Thermol Int 16:5–9
31. Ammer K (2015) Do we need reference data of local skin temperatures? Thermol Int 25:45–47
32. Ammer K (2003) Need for standardisation of measurements in thermal imaging. In: Thermography and Lasers in Medicine. Akad Cent Graf-Mark Lodart SA Lodz, pp 13–18

33. Ammer K (2008) The Glamorgan Protocol for recording and evaluation of thermal images of the human body. Thermol Int 18:125–144
34. Ammer K, Ring EFJ (2006) Standard procedures for infrared imaging in medicine. In: Biomedical engineering handbook. CRC Press, Boca Raton
35. Mercer JB, Ring EFJ (2009) Fever screening and infrared thermal imaging: concerns and guidelines. Thermol Int 19:67–69
36. Howell KJ, Mercer JB, Smith RE (2020) Infrared thermography for mass fever screening: repeating the mistakes of the past. J Med Virol 30:5–6
37. Plassmann P, Ring EFJ, Jones CD (2006) Quality assurance of thermal imaging systems in medicine. Thermol Int 16:10–15
38. Ring EFJ, Ammer K (2012) Infrared thermal imaging in medicine. Physiol Meas 33:R33–R46. https://doi.org/10.1088/0967-3334/33/3/R33
39. Ring EFJ, Ammer K (2000) The technique of infrared imaging in medicine. Thermol Int 10:7–14
40. Ring EFJ, Ammer K, Wiecek B et al (2007) Quality assurance for thermal imaging systems in medicine. Thermol Int 17:103–106
41. Sillero-Quintana M, Fernández Cuevas I, Arnaiz Lastras J, Bouzas Marins JC (2015) TERMOINEF group: protocol for thermographic assessment in humans
42. Majano C, García-Unanue J, Fernández-Cuevas I, Escamilla-Galindo V, Alonso-Callejo A, Gallardo L, Felipe JL (2024) Assessment of the acclimatization influence on thermography patterns in youth football players: a pilot study. J Therm Anal Calorim. https://doi.org/10.1007/s10973-024-13432-w
43. Machado ÁS, Cañada-Soriano M, Jimenez-Perez I, Gil-Calvo M, Carpes FP, Perez-Soriano P, Priego-Quesada JI (2022) Distance and camera features measurements affect the detection of temperature asymmetries using infrared thermography. Quant InfraRed Thermogr J:1–13. https://doi.org/10.1080/17686733.2022.2143227
44. Perpetuini D, Formenti D, Cardone D, Filippini C, Merla A (2021) Regions of interest selection and thermal imaging data analysis in sports and exercise science: a narrative review. Physiol Meas 42(8):08TR01. https://doi.org/10.1088/1361-6579/ac0fbd
45. Zaproudina N, Varmavuo V, Airaksinen O, Närhi M (2008) Reproducibility of infrared thermography measurements in healthy individuals. Physiol Meas 29:515. https://doi.org/10.1088/0967-3334/29/4/007
46. Singh J, Arora AS (2020) Automated approaches for ROIs extraction in medical thermography: a review and future directions. Multimed Tools Appl 79(21):15273–15296. https://doi.org/10.1007/s11042-018-7113-z
47. Sathish D, Kamath S, Prasad K, Kadavigere R, Martis RJ (2017) Asymmetry analysis of breast thermograms using automated segmentation and texture features. SIViP 11(4):745–752. https://doi.org/10.1007/s11760-016-1018-y
48. Fernandez Cuevas I, Marins JC, Gomez Carmona PM et al (2012) Reliability and reproductibility of skin temperature of overweight subjects by an infrared thermograpy software designed for human beings. Thermol Int 22:130–137
49. Fournet D, Redortier B, Havenith G (2012) A method for whole-body skin temperature mapping in humans. Thermol Int 22:157–159
50. Varajão J, Cunha M, Bjørn-Andersen N et al (2014) Segmentation algorithms for thermal images. Proc Technol 16:1560–1569. https://doi.org/10.1016/j.protcy.2014.10.178
51. Requena-Bueno L, Priego-Quesada JI, Jimenez-Perez I, Gil-Calvo M, Pérez-Soriano P (2020) Validation of ThermoHuman automatic thermographic software for assessing foot temperature before and after running. J Therm Biol 92:102639. https://doi.org/10.1016/j.jtherbio.2020.102639
52. Kakileti ST, Sampangi S, Ramprakash HV, Manjunath G (2024) Evaluating clinical efficacy of thermalytix as a breast cancer screening tool in young women. J Young Womens Breast Cancer Health 1(1&2). Retrieved from https://journals.lww.com/ywbc/fulltext/2024/01000/evaluating_clinical_efficacy_of_thermalytix_as_a.10.aspx

53. Kakileti ST, Madhu HJ, Krishnan L, Manjunath G, Sampangi S, Ramprakash HV (2020) Observational study to evaluate the clinical efficacy of thermalytix for detecting breast cancer in symptomatic and asymptomatic women. JCO Glob Oncol (6):1472–1480. https://doi.org/10.1200/GO.20.00168
54. de Andrade Fernandes A, Bouzas Marins JC, Pereira de Andrade AG, Rodrigues Albuquerque M, Brito CJ, da Silva CD et al (2024) Thermopixelgraphy: a new method for analyzing the skin temperature of the lower limbs through infrared thermography. J Therm Anal Calorim. https://doi.org/10.1007/s10973-024-13799-w
55. De Deus Passos M, Ferreira Da Rocha A (2022) Evaluation of infrared thermography with a portable camera as a diagnostic tool for peripheral arterial disease of the lower limbs compared with color Doppler ultrasonography. Arch Med Sci Atheroscler Dis 7(1):66–72. https://doi.org/10.5114/amsad/150716
56. Rodrigues Júnior JL, Oliveira WHD, Falqueto H, Andrade AGP, Morandi RF, Albuquerque MR et al (2021) Correlation between strength and skin temperature asymmetries in the lower limbs of Brazilian elite soccer players before and after a competitive season. J Therm Biol 99:102919. https://doi.org/10.1016/j.jtherbio.2021.102919
57. Ekstrand J, Hägglund M, Waldén M, Gauffin H, Baudot C, Biosca P et al (2025) Higher level of communication between the medical staff and the performance staff is associated with a lower hamstring injury burden: a substudy on 14 teams from the UEFA Elite Club Injury Study. BMJ Open Sport Exerc Med 11(1):e002182. https://doi.org/10.1136/bmjsem-2024-002182
58. Kellogg DL, Pérgola P (2000) Skin responses to exercise and training. In: Garrett WE, Kirkendall DT (eds) Exercise and sport science. Lippincott-Raven Publishers, Philadelphia, pp 239–250
59. Charkoudian N (2003) Skin blood flow in adult human thermoregulation: how it works, when it does not, and why. Mayo Clin Proc 78:603–612. https://doi.org/10.4065/78.5.603
60. Wallin BG (1990) Neural control of human skin blood flow. J Auton Nerv Syst 30(Suppl):S185–S190
61. Blatteis CM (1998) Physiology and pathophysiology of temperature regulation. World Scientific, Singapore
62. Wilmore JH, Costill DL, Kenney WL (2008) Physiology of sport and exercise. Human Kinetics
63. Incropera FP (1999) Introduction to heat transfer: with brief fluid. Wiley, New York
64. Anbar M (1987) Computerized thermography. Int J Technol Assess Health Care 3:613–621
65. Escamilla-Galindo V, Fernández-Cuevas I, Estal-Martínez A (2022, 30 August – 2 September 2022) Description of the thermal pattern of 950 athletes using thermography to measure skin temperature. Paper presented at the 27th Annual Congress of the European College of Sport Sciences ECSS, Sevilla, Spain
66. Al-Nakhli HH, Petrofsky JS, Laymon MS, Berk LS (2012) The use of thermal infra-red imaging to detect delayed onset muscle soreness. J Vis Exp. https://doi.org/10.3791/3551
67. Hildebrandt C, Zeilberger K, Ring EFJ, Raschner C (2012) The application of medical infrared thermography in sports medicine. Ultrasound 10:2
68. Zaproudina N, Ming Z, Hänninen OO (2006) Plantar infrared thermography measurements and low back pain intensity. J Manip Physiol Ther 29:219–223
69. Zhang HY, Son S, Yoo BR, Youk T-M (2023) Reference Standard for Digital Infrared Thermography of the Surface Temperature of the Lower Limbs. Bioengineering 10(3). https://doi.org/10.3390/bioengineering10030283
70. Mara de Rezende C, Gomes da Silva A, Sillero-Quintana M, Bouzas Marins JC (2024) Patterns of thermal normality of the knee joint in professional athletes of different sports. J Therm Biol 125:103991. https://doi.org/10.1016/j.jtherbio.2024.103991
71. Vardasca R, Ring F, Plassmann P, Jones C (2012) Thermal symmetry of the upper and lower extremities in healthy subjects. Thermol Int 22:53–60

72. Pichot C (2001) Aplicación de la termografía en el dolor lumbar crónico. Rev Soc Esp Dolor 8:43–47
73. Macdonald A, Petrova N, Ainarkar S, Allen J, Plassmann P, Whittam A et al (2017) Thermal symmetry of healthy feet: a precursor to a thermal study of diabetic feet prior to skin breakdown. Physiol Meas 38(1):33. https://doi.org/10.1088/1361-6579/38/1/33
74. Feldman F, Nickoloff EL (1984) Normal thermographic standards for the cervical spine and upper extremities. Skeletal Radiol 12:235–249. https://doi.org/10.1007/BF00349505
75. Garagiola U, Giani E (1990) Use of telethermography in the management of sports injuries. Sports Med Auckl NZ 10:267–272
76. Catunda IS, de Lima Junior MO, da Costa Araujo FA, Fontan Soares C, Ferreira JCA, Kosminsky M et al (2024) Is muscle pain in temporomandibular disorder a hyperactivity problem? A clinical, thermographic, and electromyographic analysis. J Craniofac Surg 35(7). Retrieved from https://journals.lww.com/jcraniofacialsurgery/fulltext/2024/10000/is_muscle_pain_in_temporomandibular_disorder_a.66.aspx
77. de Weerd L, Mercer JB, Weum S (2011) Dynamic infrared thermography. Clin Plast Surg 38:277–292. https://doi.org/10.1016/j.cps.2011.03.013
78. Akimov EB, Son'kin VD (2011) Skin temperature and lactate threshold during muscle work in athletes. Hum Physiol 37:621–628
79. Chudecka M, Lubkowska A (2010) Temperature changes of selected body's surfaces of handball players in the course of training estimated by thermovision, and the study of the impact of physiological and morphological factors on the skin temperature. J Therm Biol 35:379–385
80. Fernández-Cuevas I, Sillero-Quintana M, Garcia-Concepcion MA et al (2014) Monitoring skin thermal response to training with infrared thermography. New Stud Athl 29:57–71
81. Fernández-Cuevas I, Torres G, Sillero-Quintana M, Navandar A (2023) Thermographic assessment of skin response to strength training in young participants. J Therm Anal Calorim. https://doi.org/10.1007/s10973-023-11978-9
82. Abate M, Di Carlo L, Di Romualdo S et al (2009) Postural adjustment in experimental leg length difference evaluated by means of thermal infrared imaging. Physiol Meas 31:35
83. Koprowski R (2015) Automatic analysis of the trunk thermal images from healthy subjects and patients with faulty posture. Comput Biol Med 62:110–118. https://doi.org/10.1016/j.compbiomed.2015.04.017
84. Kwok G, Yip J, Yick K-L, Cheung M-C, Tse C-Y, Ng S-P, Luximon A (2017) Postural Screening for Adolescent Idiopathic Scoliosis with Infrared Thermography. Sci Rep 7(1):14431. https://doi.org/10.1038/s41598-017-14556-w
85. Lasanen R, Malo MKH, Airaksinen O, Karhu J, Töyräs J, Julkunen P (2018) Infrared thermography reveals effect of working posture on skin temperature in office workers. Int J Occup Saf Ergon 24(3):457–463. https://doi.org/10.1080/10803548.2017.1336299
86. Roggio F, Petrigna L, Trovato B, Zanghì M, Sortino M, Vitale E et al (2023) Thermography and rasterstereography as a combined infrared method to assess the posture of healthy individuals. Sci Rep 13(1):4263. https://doi.org/10.1038/s41598-023-31491-1
87. Abate M, Di Carlo L, Di Donato L et al (2013) Comparison of cutaneous termic response to a standardised warm up in trained and untrained individuals. J Sports Med Phys Fitness 53:209–215
88. Adamczyk JG, Olszewska M, Boguszewski D, Białoszewski D, Reaburn P (2016) Is it possible to create a thermal model of warm-up? Monitoring of the training process in athletic decathlon. Infrared Phys Technol 76:555–559. https://doi.org/10.1016/j.infrared.2016.04.017
89. Trovato B, Sortino M, Petrigna L, Roggio F, Musumeci G (2024) The influence of static and dynamic warm-up on knee temperature: infrared thermography insights before and after a change of direction exercise. Front Physiol 15. https://doi.org/10.3389/fphys.2024.1393804
90. Formenti D, Ludwig N, Gargano M et al (2013) Thermal imaging of exercise-associated skin temperature changes in trained and untrained female subjects. Ann Biomed Eng 41:863–871. https://doi.org/10.1007/s10439-012-0718-x

91. Priego Quesada JI, Carpes FP, Bini RR et al (2015) Relationship between skin temperature and muscle activation during incremental cycle exercise. J Therm Biol 48:28–35. https://doi.org/10.1016/j.jtherbio.2014.12.005
92. Kapoor M, Vasdev V, Singh RK, Jaipurkar R, Sikri G (2022) Relationship between aerobic fitness and lower limb skin temperature during cycling exercise testing among well-trained athletes and nonathletes: a cross-sectional study. Med J Armed Forces India. https://doi.org/10.1016/j.mjafi.2022.04.017
93. Hillen B, Pfirrmann D, Nägele M, Simon P (2019) Infrared thermography in exercise physiology: the dawning of exercise radiomics. Sports Med. https://doi.org/10.1007/s40279-019-01210-w
94. Sillero Quintana M, Conde Pascual E, Gómez Carmona PM et al (2012) Effect of yoga and swimming on body temperature of pregnant women. Thermol Int 22:143–149
95. Dervis S, Dobson KL, Nagpal TS, Geurts C, Haman F, Adamo KB (2021) Heat loss responses at rest and during exercise in pregnancy: a scoping review. J Therm Biol 99:103011. https://doi.org/10.1016/j.jtherbio.2021.103011
96. da Silva W, Machado ÁS, Souza MA, Kunzler MR, Priego Quesada JI, Carpes FP (2018) Can exercise-induced muscle damage be related to changes in skin temperature? Physiol Meas 39(10):104007. Retrieved from http://stacks.iop.org/0967-3334/39/i=10/a=104007
97. Priego-Quesada JI, De la Fuente C, Kunzler MR, Perez-Soriano P, Hervás-Marín D, Carpes FP (2020) Relationship between skin temperature, electrical manifestations of muscle fatigue, and exercise-induced delayed onset muscle soreness for dynamic contractions: a preliminary study. Int J Environ Res Public Health 17(18):6817. https://doi.org/10.3390/ijerph17186817
98. Stewart IB, Moghadam P, Borg DN, Kung T, Sikka P, Minett GM (2020) Thermal infrared imaging can differentiate skin temperature changes associated with intense single leg exercise, but not with delayed onset of muscle soreness. J Sports Sci Med 19:469–477. Retrieved from https://www.jssm.org/hf.php?id=jssm-19-469.xml#7
99. Jones TW, Shillabeer BC, Cardinale M (2020) Skin temperature, training load, and subjective muscle soreness in junior endurance athletes: a case study. Int J Sports Physiol Perform 15(9):1349–1352. https://doi.org/10.1123/ijspp.2019-0748
100. Alburquerque Santana PV, Alvarez PD, da Costa F, Sena A, Serpa TK, de Assis MG, Pimenta EM et al (2022) Relationship between infrared thermography and muscle damage markers in physically active men after plyometric exercise. J Therm Biol 104:103187. https://doi.org/10.1016/j.jtherbio.2022.103187
101. Lamberty de Morais AC, Sosa Machado Á, Ferreira Pereira ME, da Silva W, Priego-Quesada JI, Carpes FP (2024) Intensity and volume of physical exercise influence DOMS and skin temperature differently in healthy adults. Sci Rep 14(1):30282. https://doi.org/10.1038/s41598-024-79785-2
102. Hadžić V, Širok B, Malneršič A, Čoh M (2019) Can infrared thermography be used to monitor fatigue during exercise? A case study. J Sport Health Sci. https://doi.org/10.1016/j.jshs.2015.08.002
103. Shakhih MFM, Ridzuan N, Wahab AA, Zainuddin NF, Delestri LFU, Rosslan AS, Kadir MRA (2021) Non-obstructive monitoring of muscle fatigue for low intensity dynamic exercise with infrared thermography technique. Med Biol Eng Comput 59(7):1447–1459. https://doi.org/10.1007/s11517-021-02387-x
104. Sosa Machado Á, da Silva W, Priego-Quesada JI, Carpes FP (2023) Can infrared thermography serve as an alternative to assess cumulative fatigue in women? J Therm Biol 115:103612. https://doi.org/10.1016/j.jtherbio.2023.103612
105. Perpetuini D, Formenti D, Cardone D, Trecroci A, Rossi A, Di Credico A et al (2023) Can data-driven supervised machine learning approaches applied to infrared thermal imaging data estimate muscular activity and fatigue? Sensors 23(2). https://doi.org/10.3390/s23020832
106. Seixas Duarte T, Ferreira-Júnior JB, Oliveira JCG, Amaral Costa CM, Coelho DB, Bouzas Marins JC, Bara-Filho MG (2025) Is skin temperature associated with training load during

preparatory period of a professional volleyball team? J Therm Biol 127:104038. https://doi.org/10.1016/j.jtherbio.2024.104038
107. Sillero-Quintana M, Fernández-Jaén T, Fernández-Cuevas I et al (2015) Infrared thermography as a support tool for screening and early diagnosis in emergencies. J Med Imaging Health Inform 5:1223–1228
108. Bertucci W, Arfaoui A, Janson L, Polidori G (2013) Relationship between the gross efficiency and muscular skin temperature of lower limb in cycling: a preliminary study. Comput Methods Biomech Biomed Engin 16(Suppl 1):114–115. https://doi.org/10.1080/10255842.2013.815902
109. Adamczyk JG, Mastej M, Boguszewski D, Białoszewski D (2014) Usage of thermography as indirect non-invasive method of evaluation of physical efficiency. Pilot study. Pedagogics Phsychol Med Biol Probl Phys Training Sports 3:90–95. https://doi.org/10.6084/m9.figshare.938187
110. Duc S, Arfaoui A, Polidori G, Bertucci W (2015) Efficiency and thermography in cycling during a graded exercise test. J Exerc Sports Orthop 2(3):1–8. Retrieved from https://hal.univ-reims.fr/hal-02314094/document
111. Adamczyk JG, Boguszewski D, Siewierski M (2014) Thermographic evaluation of lactate level in capillary blood during post-exercise recovery. Kinesiology 46(2):186–193
112. Gomes Moreira D, Brito CJ, de Almeida Ferreira JJ, Bouzas Marins JC, Díaz L, de Durana A, Couceiro Canalejo J et al (2021) Lactate concentration is related to skin temperature variation after a specific incremental Judo test. J Strength Cond Res 35(8). https://doi.org/10.1519/JSC.0000000000003095
113. Korman P, Kusy K, Straburzyńska-Lupa A, Kantanista A, Sillero-Quintana M, Zieliński J (2024) Response of skin temperature, blood ammonia and lactate during incremental exercise until exhaustion in elite athletes. Sci Rep 14(1):2237. https://doi.org/10.1038/s41598-024-52374-z
114. Lewis GF, Gatto RG, Porges SW (2011) A novel method for extracting respiration rate and relative tidal volume from infrared thermography. Psychophysiology 48:877–887. https://doi.org/10.1111/j.1469-8986.2010.01167.x
115. Dotti F, Ferri A, Moncalero M, Colonna M (2016) Thermo-physiological comfort of soft-shell back protectors under controlled environmental conditions. Appl Ergon 56:144–152. https://doi.org/10.1016/j.apergo.2016.04.002
116. Fournet D, Redortier B, Havenith G (2015) Can body-mapped garments improve thermal comfort for sport in the cold? Extreme Physiol Med 4:A74
117. Fournet D, Ross L, Voelcker T et al (2013) Body mapping of thermoregulatory and perceptual responses of males and females running in the cold. J Therm Biol 38:339–344. https://doi.org/10.1016/j.jtherbio.2013.04.005
118. Kasprzyk T, Cholewka A, Kucewicz M, Sieroń K, Sillero-Quintana M, Morawiec T, Stanek A (2018) A quantitative thermal analysis of cyclists' thermo-active base layers. J Therm Anal Calorim 136(4):1689–1699. https://doi.org/10.1007/s10973-018-7775-9
119. Raccuglia M, Heyde C, Lloyd A, Hodder S, Havenith G (2019) The use of infrared thermal imaging to measure spatial and temporal sweat retention in clothing. Int J Biometeorol. https://doi.org/10.1007/s00484-019-01701-5
120. Arlegui L, Smallcombe JW, Fournet D, Tolfrey K, Havenith G (2021) Body mapping of sweating patterns of pre-pubertal children during intermittent exercise in a warm environment. Eur J Appl Physiol 121(12):3561–3576. https://doi.org/10.1007/s00421-021-04811-4
121. Salopek Cubric I, Cubric G, Potocic Matkovic VM, Pavko Cuden A (2021) The comfort of knitted fabrics: interaction of sportswear and athlete's body. Commun Dev Assembling Textile Prod 2(1):70–79. https://doi.org/10.25367/cdatp.2021.2.p70-79
122. Menezes P, Rhea MR, Herdy C, Simão R (2018) Effects of strength training program and infrared thermography in soccer athletes injuries. Sports 6(4):148. https://doi.org/10.3390/sports6040148

123. Majano C, García-Unanue J, Hernandez-Martin A, Sánchez-Sánchez J, Gallardo L, Felipe JL (2023) Relationship between repeated sprint ability, countermovement jump and thermography in elite football players. Sensors 23(2). https://doi.org/10.3390/s23020631
124. de Carvalho G, Girasol CE, Gonçalves LGC, Guirro ECO, Guirro RR, d. J. (2021) Correlation between skin temperature in the lower limbs and biochemical marker, performance data, and clinical recovery scales. PLoS One 16(3):e0248653. https://doi.org/10.1371/journal.pone.0248653
125. Sillero-Quintana M, Jones-Rando J, Refoyo I, Bouzas Marins JC, Seixas A (2022) Effects of resistance training on skin temperature and its relationship with Central Nervous System (CNS) Activation. Healthcare 10(2). https://doi.org/10.3390/healthcare10020207
126. Masur L, Brand F, Düking P (2024) Response of infrared thermography related parameters to (non-)sport specific exercise and relationship with internal load parameters in individual and team sport athletes—a systematic review. Front Sports Act Living 6. https://doi.org/10.3389/fspor.2024.1479608
127. Brito Marques WK, Albuquerque Santana PV, Ferreira de Oliveira VF, de Azambuja Pussieldi G, Bernardes Leite L, Forte P et al (2024) Assessment of heart rate variability and infrared thermography in response to exercise-induced muscle damage. Sport Sci Health. https://doi.org/10.1007/s11332-024-01304-7
128. Brito CJ, Gomes Moreira D, Ferreira JJ, López Díaz-de-Durana A, Miarka B, Bouzas Marins JC, Sillero-Quintana M (2020) Immune response related with skin thermal pattern in judokas: a new application for infrared thermography? J Strength Cond Res 34(10). https://doi.org/10.1519/JSC.0000000000002672
129. Ekstrand J, Spreco A, Bengtsson H, Bahr R (2021) Injury rates decreased in men's professional football: an 18-year prospective cohort study of almost 12 000 injuries sustained during 1.8 million hours of play. Br J Sports Med 55(19):1084. https://doi.org/10.1136/bjsports-2020-103159
130. Gurau TV, Gurau G, Musat CL, Voinescu DC, Anghel L, Onose G et al (2023) Epidemiology of Injuries in Professional and Amateur Football Men (Part II). J Clin Med 12(19). https://doi.org/10.3390/jcm12196293
131. Ekstrand J, Hägglund M, Waldén M (2011) Epidemiology of muscle injuries in professional football (soccer). Am J Sports Med 39:1226–1232
132. Ekstrand J, Bengtsson H, Waldén M, Davison M, Khan KM, Hägglund M (2023) Hamstring injury rates have increased during recent seasons and now constitute 24% of all injuries in men's professional football: the UEFA Elite Club Injury Study from 2001/02 to 2021/22. Br J Sports Med 57(5):292. https://doi.org/10.1136/bjsports-2021-105407
133. Fernández-Cuevas I, Fernández-Muñoz D, Sillero-Quintana M (2024, 6th–8th September, 2024) Normative skin temperature values and gender differences in a healthy and young sample: infrared thermography analysis of over 3600 subjects. Paper presented at the XVI Congress of the European Association of Thermology, Wroclaw, Poland
134. dos Santos Bunn P, Elisa Koppke Miranda M, Inoue Rodrigues A, de Souza Sodré R, Borba Neves E, Bezerra da Silva E (2020) Infrared thermography and musculoskeletal injuries: a systematic review with meta-analysis. Infrared Phys Technol 109:103435. https://doi.org/10.1016/j.infrared.2020.103435
135. Rochcongar P, Schmitt M (1979) Thermographic study of muscular lesions in sport (author's transl). J Belge Med Phys Rehabil 2(4):335–342. Retrieved from http://www.ncbi.nlm.nih.gov/entrez/query.fcgi?cmd=Retrieve&db=PubMed&dopt=Citation&list_uids=549939
136. BenEliyahu DJ (1990) Infrared thermography in the diagnosis and management of sports injuries: a clinical study and literature review. Chiropractic Sports Med 4(2):46–53
137. Menz HB, Jordan KP, Roddy E, Croft PR (2010) Characteristics of primary care consultations for musculoskeletal foot and ankle problems in the UK. Rheumatology (Oxford, England) 49(7):1391–1398. https://doi.org/10.1093/rheumatology/keq092

138. Haluzan D, Davila S, Antabak A, Dobric I, Stipic J, Augustin G et al (2015) Thermal changes during healing of distal radius fractures—Preliminary findings. Injury 46:S103–S106. https://doi.org/10.1016/j.injury.2015.10.046
139. Fernández-Cuevas I, del Estal A (2020) Thermal profile description of most common soccer injuries by Infrared Thermography: case studies. In: Paper presented at the I International Congress on Application of Infrared Thermography in Sport Science, Valencia, Spain
140. Fernández-Cuevas I, Escamilla-Galindo V, Estal-Martínez A (2022, 30 August – 2 September 2022) Hamstring injuries in professional soccer players get colder. Infrared Thermography as an additional technology for return to play decisions. Paper presented at the 27th Annual Congress of the European College of Sport Sciences ECSS, Sevilla, Spain
141. Fernández-Cuevas I, Escamilla-Galindo V, Ceniza-Villacastín J, Fernández-Muñoz D (2024, 25th – 27th May) Gender differences in knee skin temperature among ACL injured football players. Paper presented at the XXXI Isokinetic Medical Group Conference Football Medicine Bridging Science and Practice, Civitas Metropolitano Stadium – Madrid, Spain

Chapter 6
Applications of Infrared Thermography to Monitor Muscle Soreness, Muscle Damage and Fatigue

Álvaro S. Machado, Willian da Silva, Jose Ignacio Priego-Quesada, and Felipe P. Carpes

Abstract Intense physical exercise leads to both acute and chronic physiological adaptations. Acute adaptations are often associated with transient tissue damage, resulting in delayed onset muscle soreness (DOMS). In recent years, there has been growing interest in strategies not only to minimize DOMS but also to better monitor recovery. Infrared thermography has emerged as a promising tool for assessing DOMS, typically by correlating physiological responses with changes in skin temperature. This chapter discusses recent evidence on these topics, highlighting the potential applications and limitations of using infrared thermography to monitor exercise adaptation and muscle soreness induced by exercise.

6.1 Introduction

Delayed onset muscle soreness (DOMS) is a physiological process resulting from strenuous exercise, which can typically be perceived shortly after its onset. It is also experienced when participants engage in exercises different from those they are

Á. S. Machado · F. P. Carpes
Applied Neuromechanics Research Group, Laboratory of Neuromechanics, Federal University of Pampa, Uruguaiana, RS, Brazil
e-mail: carpes@unipampa.edu.br

W. da Silva
Escuela de Kinesiologia, Facultad de Ciencias, Pontificia Universidad Católica de Valparaíso, Valparaíso, Chile
e-mail: willian.dasilva@pucv.cl

J. I. Priego-Quesada (✉)
Research Group in Sport Biomechanics (GIBD), Department of Physical Education and Sports, University of Valencia, Valencia, Spain

Biophysics and Medical Physics Group, Department of Physiology, University of Valencia, Valencia, Spain
e-mail: j.ignacio.priego@uv.es

J. I. Priego-Quesada (ed.), *Application of Infrared Thermography in Sports Science*,
https://doi.org/10.1007/978-3-031-93311-0_6

familiar with [1, 2], such as in terms of changing the typical range of articular motion [3] or adding acute increases in external workload [4]. Whether in athletes, sedentary individuals, patients, or special populations (such as older adults, obese individuals, or people with disabilities), intense physical activity can potentially induce mechanical tension in muscle fibers to the point of microtears, necessitating a subsequent adaptive response [4].

Following a high-intensity exercise session or involving unfamiliar movement patterns, the perception of DOMS typically begins the following day. The pain sensation is expected to peak at around 48 h post exercise and may persist, albeit at a diminished intensity, for up to 96 h [5]. This indicates that DOMS may not only extend over a prolonged period but also can vary in intensity over the recovery time [4].

A condition of DOMS frequently results in a temporary reduction in the articular range of motion, which in some cases may impair motor performance [2]. In clinical practice, DOMS may hinder adherence to and accuracy in executing an exercise protocol due to a self-regulated intensity adjustment aimed at avoiding discomfort [1]. It is noted that during postural tasks, the condition of DOMS in postural muscles may lead to some acute adaptations that cause transient postural control deficits [6], while DOMS in the main contributing muscles to manage impact forces will influence the kinematics strategies during landing [7]. In this context, monitoring DOMS can assist coaches and therapists in adjusting the training load or assessing the effectiveness of exercise programs and therapeutic interventions.

Although individuals can perceive the onset and progression of DOMS after exercise, the accuracy of self-reported intensity may be influenced by subjective factors in addition to depending on a difficult standardization of muscle stimulation [8, 9]. Therefore, the clinical assessment requires significant familiarity with measurement tools or access to invasive assessment protocols. Moreover, subjective instruments used to map individual sensations may limit the reliability of intersubject comparisons [10]. Considering physical exercise as a powerful intervention in many diseases, it is also relevant to consider that there will be cases in which the participants might not be able to accurately indicate their perceptions of soreness using common scales, which makes it more relevant to search for alternative tools to assess DOMS.

Even when subjective scales are validated and their application and interpretation are standardized, the reliance on self-reports from heterogeneous populations and protocols presents a challenge for researchers, coaches, and therapists. As DOMS is associated with muscle damage or motor function impairment, this process may also be monitored by muscle biopsies and/or magnetic resonance [11], blood markers (e.g., creatine kinase [CK]) [12], or decreases in jump performance [13, 14]. However, all these methods also present limitations, such as invasiveness, economic cost, high variability or the need for skill acquisition in some jumps (e.g., countermovement jump test), as well as the requirement for maximal effort to ensure valid results. In this context, infrared thermography emerges as an objective, noninvasive, and user-friendly alternative, which is hypothesized to enable the mapping of regions affected by DOMS.

While DOMS is primarily associated with pain perception—something a localized measurement alone may not directly detect—it also involves muscle damage [1, 4], which initiates the inflammatory process responsible for nociceptor sensitization [4, 15]. Therefore, the hypothesis underlying the use of infrared thermography in assessing DOMS is based on an indirect measurement approach, detecting localized heat production as a proxy for inflammation. In this hypothesis, the inflammation responsible for muscle regeneration following exercise-induced damage (and, consequently, the onset of DOMS) also generates localized heat. The heat produced by the muscles would then be conducted from the muscle tissue to more superficial layers, eventually warming the skin—an effect that could be detected using infrared thermography. In this chapter, we will examine the advancements and challenges in the literature regarding the objective detection of DOMS through infrared thermography.

6.2 Mechanisms of Delayed Onset Muscle Soreness

DOMS can be classified as resultant of a type I muscle strain and is assumed to result from exercise-induced muscle damage [16, 17]. The symptoms of DOMS include muscle tenderness, reduced articular range of motion, stiffness, swelling, and local pain [16]. Curiously, the sensation of pain can range from simple discomfort that disappears during daily activities or progress to severe debilitating pain that could restrict movement [2]. The etiology of DOMS has intrigued scientists for many years. Different theories have been proposed, such as the muscle spasm theory, lactic acid theory, connective tissue theory, ruptured tissue theory, enzyme influx theory, and fluid theory [18]. However, one isolated theory is unable to explain the etiology of DOMS. Instead, a sequence of unique events is proposed [2]. The most supported theory is that biochemical, thermal, and mechanical changes associated with the inflammation process sensitize small muscles afferent to types III and IV, leading to the DOMS sensation (see Fig. 6.1) [20].

The release of calcium ions (Ca^{2+}) into the cell, following exercise-induced muscle damage and consequently damage to the cell membrane, is responsible for triggering the activation of several proteases like phospholipase A2 [19]. This enzyme is responsible for cleaves arachidonic acid from the phospholipids of the cell membrane, resulting in free arachidonic acid into the cell, which in turn gives rise to inflammatory mediators how thromboxanes, prostaglandins, and leukotrienes [19]. The metabolism of arachidonic acid can follow 1 or 2 pathways: (I) The cyclooxygenase (COX) pathway, which leads to the production of prostaglandins E2 (PGE2) and thromboxanes or (II) lipoxygenase pathway, that potentialize the production of leukotrienes [19]. The PGE2 increases the vascular permeability and is responsible for the pain perception in the muscle by sensitizing the types III and IV afferent nerve fibers, whereas leukotrienes also increase vascular permeability and are responsible for attracting neutrophils to the site of structural damage potentializing oxidative damage [19, 21, 22].

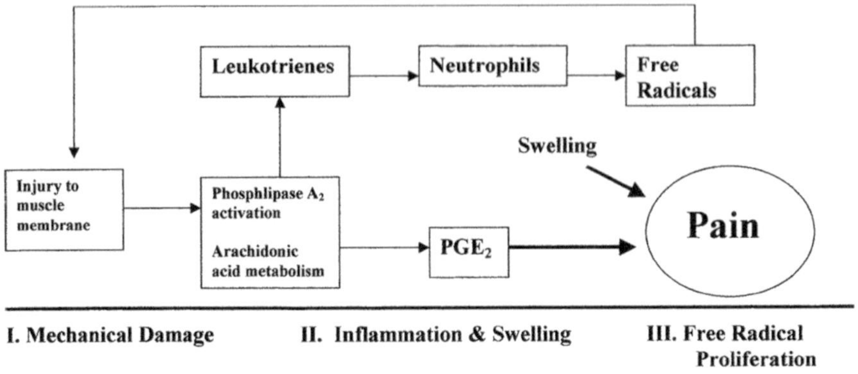

Fig. 6.1 The proposed sequence of events from delayed onset muscle soreness (DOMS). PGE2 prostaglandin E2. (Figure modified from Connolly et al. [19])

Additionally, the chemical response that occurs after exercise is accompanied by intramuscular edema, and it is thought to be one of the mechanisms responsible for nociceptor activation and subsequent DOMS sensations [23]. Recently, another pathway was proposed to be associated with DOMS [23]. This pathway involves the activation of B2-bradykinin receptors which are released during exercise, and the nerve growth factor (NGF). By this mechanism, bradykinin triggers the process associated with mechanical hyperalgesia by the upregulation of NGF through B2 receptors leading to DOMS [24, 25]. Also, the development of hyperalgesia may be associated with inflammation in the extracellular matrix [26].

Thus, in the complex DOMS etiology, the inflammation process appears to play an essential role. This process is responsible for recruiting several immune cells to the injury site, resulting in muscle edema and increased muscle temperature [23]. Also, the increase in temperature could activate nociceptors within muscle fibers and myotendinous junctions in a DOMS condition [2]. This process would reflect the rise of skin surface temperature in response to cutaneous vasodilatation by heat transfer from the heated blood flow to the skin by conduction and convection [27], which also influences the heat transfer by radiation from the skin to the environment [28].

The heat transference between muscle and the skin would be mainly mediated by the skin blood flow. For this reason, it is important to understand how is the literature about the effect of DOMS and muscle damage condition on skin blood flow. Sensitization of nociceptive (group III-IV afferents) and mechanosensitive afferents following muscle damage can lead to a reduction in skin blood flow due to increased sympathetic activity and vascular resistance [29]. Specifically, reduced resting femoral blood flow and leg vascular conductance have been observed when both nociceptive and mechanoreceptor pathways are activated, suggesting an autonomic-mediated vasoconstriction [29]. The increase in inflammatory mediators such as prostaglandins and bradykinin following muscle damage contributes to vasodilation; however, this response is often counterbalanced by sympathetic vasoconstriction, particularly in individuals experiencing significant muscle pain [30]. Moreover, intraindividual variability on resting blood flow was observed that can be associated with also the high DOMS variability after an intervention [30]. However, this topic is recent, and further investigation is necessary. All these physiological mechanisms exposed in the current section are considered in studies addressing associations between DOMS conditions and skin temperature changes.

6.3 Skin Temperature Responses and DOMS

The motivation to explore the possible applicability of infrared thermography to monitor muscle damage and muscle soreness in sports medicine is associated with the fact that the local inflammation response would increase the muscle temperature, and this heat generated will be transferred to the skin and captured by a thermal camera [28]. However, to date, there is no consensus regarding this infrared thermography applicability, and the studies present contradictory results.

In one of the first studies performed on this topic, the authors evaluated the changes in skin temperature measured in 20 soccer athletes submitted to two consecutive soccer matches separated by 3 days [31]. The authors aimed to determine whether the skin temperature would show different behavior in different regions of interest (ROIs) monitored with infrared thermography in response to the consecutive exercise sessions. The skin temperature was quantified in the anterior and posterior thighs and anterior and posterior legs, while the magnitude of muscle damage was estimated by quantifying the activity of CK. The results described an increase in skin temperature 24 h after the first match in all ROIs monitored with a return to baseline values before the second match. In addition, 24 h after the second match, the skin temperature increases even more, but skin temperature does not return to the baseline values 48 after the second match (see Fig. 6.2). The results suggest a cumulative effect of the exercise on skin temperature responses. However, when a correlation analysis was conducted, the association between skin temperature data and muscle damage estimated by the CK activity showed a small magnitude ($r = 0.3$–0.4). One possible explanation is that soccer engages large muscle groups simultaneously, with variability in the intensity of movements performed by athletes. This variation may influence the extent of muscle damage, making it challenging to establish a clear correlation in this context. Moreover, there is a lack of other variables that quantify the physiological state of the soccer players, such as pain perception or jump performance. The manuscript also presents relevant insights related to the need to consider the skin temperature responses to the training load and not only consider contralateral thermal asymmetry as most studies have considered to date. This study is considered one of the first on this topic and played a key role in sparking the interest of many researchers, who subsequently investigated more about the link between DOMS and skin temperature.

As we commented, the rationale of an association between muscle damage leading to DOMS and changes in skin temperature has attracted the interest of

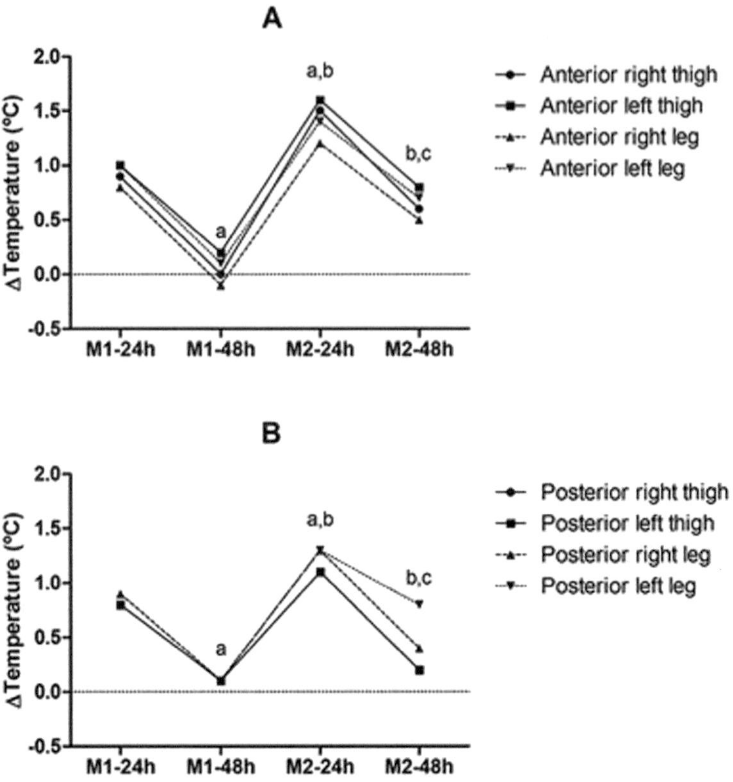

Fig. 6.2 Skin temperature variation of the anterior and posterior thighs and legs 24 h and 48 h after a first (M1) and second (M2) soccer match. (Figure obtained from de Andrade Fernandes et al. [31])

researchers. Under the premise that while for whole-body exercise the muscle damage cannot be sufficient to elicit changes for a specific body site, the exercise of a single muscle group could lead to more relevant conclusions. This distinction became evident when findings from whole-body exercises, such as marathon and half-marathon running [13, 14], were compared to those from a study in which 20 untrained participants were subjected to a localized exercise protocol to induce muscle damage in calf muscles [32]. The participants performed multiple sets of bilateral calf raises with body weight until fatigue [33]. Skin temperature was monitored considering the regions of the anterior and posterior legs in the moments: pre-exercise, immediately post, and 48 h post exercise. To monitor muscle damage and DOMS, blood samples were collected to quantify the CK activity and acetylcholinesterase (AChe). DOMS was evaluated using a numerical pain rate scale. The authors found that skin temperature increases in the posterior leg (0.9 ± 1.1 °C) and decreases in the anterior leg (−0.8 ± 1.4 °C) immediately post exercise in comparison with pre-exercise. However, 48 h post exercise, the skin temperature returned to values that did not differ from baseline values (see Fig. 6.3), even with the presence

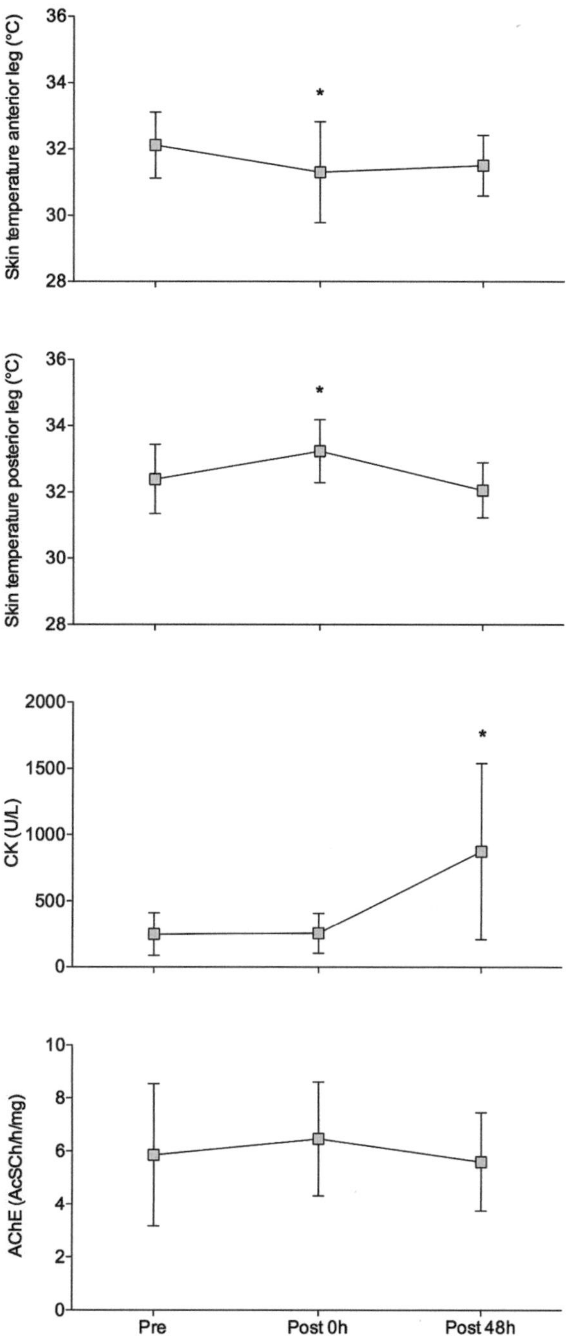

Fig. 6.3 Mean and standard deviation of absolute skin temperature in anterior and posterior legs, creatine kinase (CK), and acetylcholinesterase (AChE). (Figure reproduced from da Silva et al. [32])

of high muscle damage as quantified by CK activity and high sensation of DOMS (6.5 points in the numerical rate pain scale). Different from the study performed with soccer players, no significant correlations were observed between CK activity or DOMS and skin temperature variation post exercise and 48 h post exercise. Even though skin temperature can change less and more slowly in untrained participants compared with trained participants [34], these two studies [31, 32] have similar main outcomes but with opposite results.

Although it is possible to speculate that the lack of skin temperature responses may be due to the fact that the muscle damage generated in some studies is very localized [32, 35], this does not hold true as no clear skin temperature responses have been observed after a marathon [14] or half marathon either [13]. Therefore, more general muscle damage contextualized to real competitions has not resulted in increases in skin temperature either.

On the way to building a theory for the use of infrared thermography in monitoring DOMS, other points are worth mentioning. To date, most studies considered only men in their analyses, but the inferences were made for both men and women. When the relationship between DOMS, pain, and skin temperature was investigated in men and women [35], results indicated that skin temperature response after exercise is sex-dependent. Such conclusions were drawn based on measurements of 22 participants (10 men and 12 women) who were submitted to a squat exercise protocol to induce muscle damage. The protocol involved maximal squat repetitions with body weight at the highest velocity for 1 min. After 15 s of interval, the next set of 1 min began, until exhaustion. Skin temperature, DOMS and pressure pain thresholds (PPT) were evaluated pre-exercise, post and 48 h post exercise. In this study, three ROIs were considered. The anterior thigh, bilaterally, and biceps brachii as a control region. DOMS increased post exercise in comparison with pre-exercise (5.1 ± 2.2 points vs. 0.7 ± 1.1 points) and 48 h post exercise (4.0 ± 2.9 points vs. 0.7 ± 1.1 points, $p < 0.01$), without sex differences. However, the use of a mechanical pain evaluation performed through the PPT showed a sex difference. The PPT was lower in women compared with men for most of the moments. For men, the PPT increased post exercise but was lower 48 h post in both vastus lateralis and rectus femoris. As well as the PPT, the skin temperature response was sex dependent. The mean and maximum temperature increased post exercise for men and maximum skin temperature was lower 48 h post exercise in comparison the pre-exercise. The skin temperature pre-exercise did not differ with 48 h post exercise in any thermal parameters considered (maximum, minimum and mean of skin temperature). For women, the minimum skin temperature was higher 48 h post exercise in comparison with pre-exercise (pre-exercise: 28.9 ± 1.1 °C vs. 48 h post exercise: 30.3 ± 0.8 °C). Also, women showed lower skin temperature than men in most of the moments and thermal parameters considered. DOMS was not predicted by the skin temperature, but the PPT shows a direct relationship with the variation in the maximum skin temperature pre- to 48 h post exercise ($r = 0.15$). This seems to be one of the first studies to show that skin temperature response after exercise can be sex-dependent and that inferences not necessarily can be generalized for both sexes. It is important to understand that if peripheral blood flow can influence pain perception [30], and sex is known to affect both peripheral blood flow and skin temperature response which can be also affected by menstrual cycle [36–38], it seems logical to think that the possible use of skin temperature as an indicator of DOMS is even more difficult in the female sex.

The magnitude of muscle damage could be another factor influencing IRT evaluation outcomes. To test this hypothesis, 30 healthy men were submitted to two protocols of exercise performed at different intensities. Thermal responses to different magnitudes of exercise-induced muscle damage were investigated [39]. In this study, all participants were randomly divided into mild (10 sets of maximal eccentric elbow flexion at 60 degrees/s) and moderate (30 sets of maximal eccentric elbow flexion at 60 degrees/s) muscle damage groups. Skin temperature, DOMS, PPT, and CK activity, as well as neuromuscular performance, were evaluated in pre-, and immediately post exercise, 30 min after and 24 h, 48 h, and 72 h after exercise. The authors found that skin temperature increased immediately post exercise (~4%), for both groups and returned close to the baseline values 30 min after the exercise ended. After 30 min following the exercise, the skin temperature remained practically unchanged. In addition, the skin temperature did not correlate with DOMS and CK activity in any moment and group evaluated.

From this point on, some studies also initiate a debate about the need to establish a better time window to perform skin temperature measurements following exercise. In this sense, a study design considering measurements throughout a competition season, for instance, although quite challenging, could provide valuable information regarding the existence of a possible better measurement time window throughout the different phases of a training calendar. Considering this design, a study followed professional soccer players during a competitive season considering 19 matches with an interval of 7 days between matches [40]. Throughout the season, IRT measurements, as well as the perception of recovery, fatigue, and pain were evaluated 48 h after each match. To obtain IRT data, each lower limb was divided into 7 ROIs, totaling 14 anterior and posterior ROIs analyzed. As in the previous study, no correlations were found between the mean of skin temperature and CK activity, pain level, as well as fatigue, and recovery perception ($r < 0.2$, $p > 0.05$).

6.4 Hypothesis of an Inverted U Relationship Between Skin Temperature and Training Load

Recently, the hypothesis of an inverted U relationship between skin temperature and internal training loads was proposed to explain the variability of results observed in the literature [14]. For this hypothesis, physical exercise with high training loads but without leading to exercise-induced muscle damage or DOMS would increase skin temperature due to an increased regional blood flow. However, when muscle damage and DOMS are induced, vasoconstriction happens and an opposite behavior could be observed, with a reduction or no alteration in skin temperature. As mentioned in the previous section, the literature on skin blood flow and DOMS is still limited, especially when skin temperature is included in the measurements. Therefore, this hypothesis has not yet been confirmed.

Several studies have sought to verify this theory and determine the optimal time window for measuring skin temperature. In this regard, it has been observed that 24 h after running 10 km, individuals accustomed to this type of exercise exhibit higher skin temperatures in the lower limbs compared to a control group that did not exercise [41]. Additionally, another study evaluated three groups: a control group with no exercise, a group of militaries who performed a simulated combat protocol, and a third group that completed the same combat simulation along with a muscle damage induction protocol consisting of 100 drop jumps [42]. The key finding was that while the group that only performed the simulated combat exhibited increased skin temperature 24 h post exercise, neither the control group nor the muscle damage group showed significant changes, with their skin temperature following a similar pattern [42]. Notably, this study found no differences in DOMS at 24 h between groups, but higher DOMS for the muscle damage group at 48 h post exercise, suggesting that skin temperature (and possibly peripheral blood flow) is more influenced by muscle damage-related processes than by pain perception during the first 24 h. Figure 6.4 illustrates the evolution of skin temperature in both studies [41, 42]. Furthermore, both studies highlight the effect of the circadian cycle, demonstrating that skin temperature increases throughout the day regardless of the type of physical exercise performed, with a peak observed between 6 and 9 h post exercise [41, 42].

The inverted U hypothesis could also explain why some studies report increased skin temperature at 24 or 48 h post exercise, assuming that muscle damage has not reached significant levels. This may account for the findings obtained during a triathlon training camp when skin temperature increased over the course of 24 and 48 h following the start of training and the accumulation of physical exercise [43]. Similarly, studies involving maximal knee contractions on a dynamometer have reported comparable results [44].

Moreover, this hypothesis is based on the influence of muscle damage on skin blood flow. In this sense, a study, that used a cold-stress test to evaluate the skin blood flow capacity of marathon runners, observed higher decreases of skin temperature during the cold-stress test 24 h after the marathon suggesting a higher skin vasoconstriction state of the runners [14]. These results agree with the idea that higher skin vasoconstriction is present after muscle damage and encourage further investigation assessing DOMS and muscle damage by the combination of infrared thermography and cold-stress test (see Chap. 10 for more details about these tests). Moreover, a notable exception that nonetheless aligns with this hypothesis is a study conducted during a marathon in Costa Rica [45]. This event took place under hot and humid conditions (~28 °C and 81% of relative humidity), and increases in skin temperature were observed 24 h after the marathon race (Fig. 6.5) [45]. A plausible explanation is that the elevated ambient temperature led to sustained peripheral vasodilation, preventing the skin vasoconstriction typically induced by muscle damage. As a result, heat dissipation from the muscle to the skin was facilitated, maintaining higher skin temperatures despite the presence of muscle damage. This

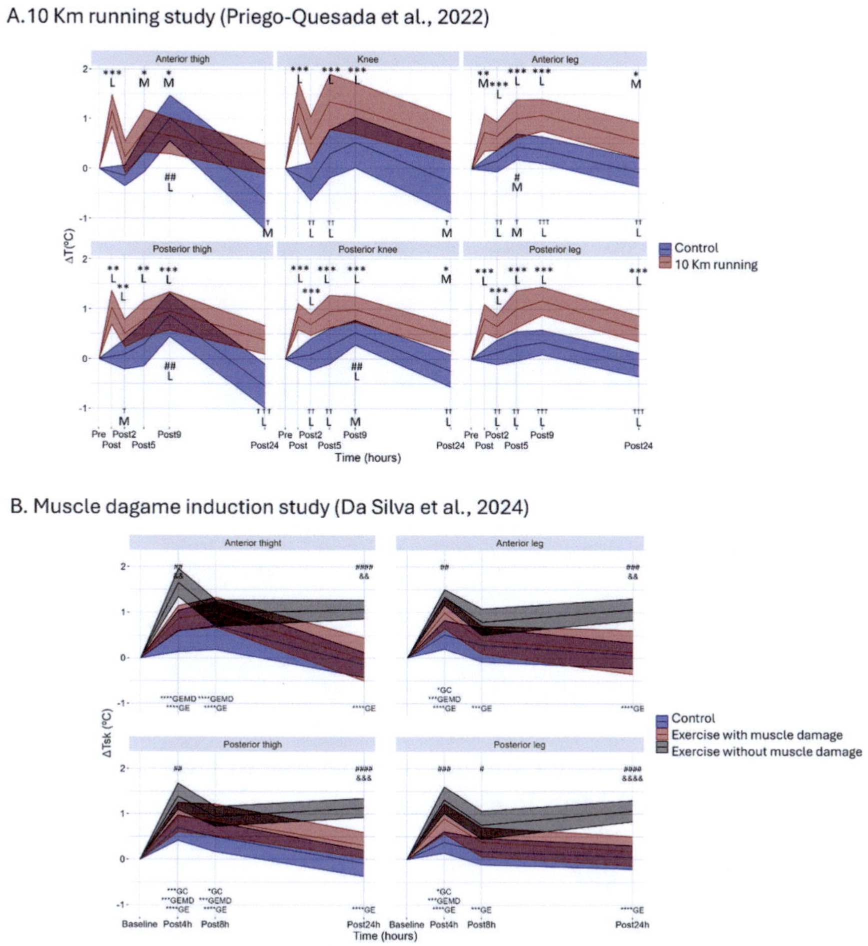

Fig. 6.4 Skin temperature variations (ΔT) over the following 24 h at different exercise intensities. (Figure reproduced from Da Silva et al. [42] and Priego-Quesada et al. [41])

explanation could also be attributed to the study of de Andrade Fernandes et al. [31] when environmental conditions were ~28 °C and 62% of relative humidity.

Despite all these results, the hypothesis of an inverted U relationship between skin temperature and internal training loads needs to be validated by further studies with longitudinal data that include the response to multiple exercise loads. Currently, the results of many of the studies are in agreement with the idea that physical exercise without muscle damage results in an increase in baseline skin temperature [41–43], and that further accumulating exercise without this damage results in greater increases [31, 43]. Also, when DOMS or muscle damage occurs, this response is generally inhibited [32, 35, 42]. Figure 6.6 shows how the results of these studies could be located in this hypothesis. However, further research is still

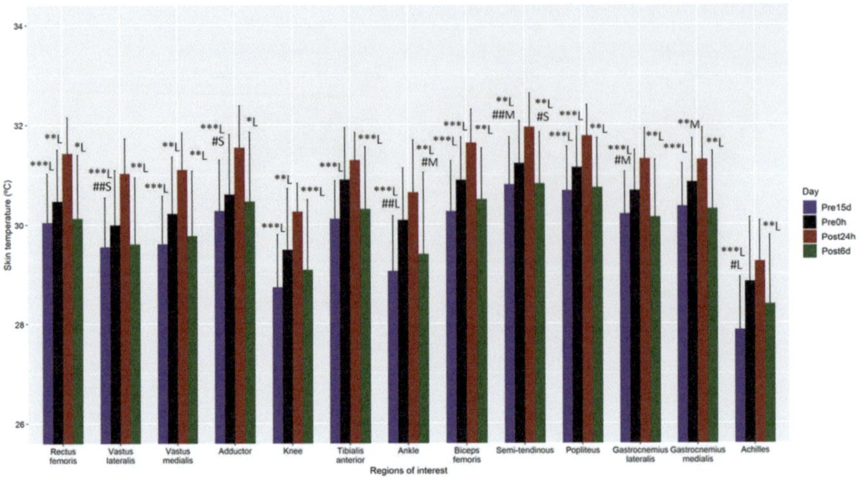

Fig. 6.5 Skin temperature before and after a marathon performed in a hot environmental condition. (Figure reproduced from Rojas-Valverde et al. [45])

needed to know if this relationship is an inverted U-shape or really has another type of shape.

6.5 Skin Temperature Changes in Response to Exercise Intensity and Muscle Fatigue

6.5.1 Muscle Fatigue

During intense exercise, it is common to experience muscle fatigue. Muscle fatigue is characterized by a reduction or loss of the ability to produce force and power, leading to performance decline or task abandonment [46]. Thus, muscle fatigue shares similarities with DOMS in a way that both involve subjective and objective factors for its detection. Subjective factors include the perception of exhaustion and the sensation of strength loss, while objective factors encompass measurements such as muscle strength, contractility, and power output, among others [47]. In biomechanics, surface electromyography (EMG) is one of the most common techniques used for detecting muscle fatigue [48, 49]. Exploratory studies have compared EMG results with infrared thermography and demonstrated that infrared thermography holds potential as a strategy for monitoring muscle fatigue in protocols targeting a single exercised muscle, by analyzing comparisons between baseline and post-exercise values [50, 51]. On the other hand, the relationship between muscle fatigue detected through EMG and skin temperature variations remains a topic of debate in the literature, highlighting the need for robust methodological criteria to

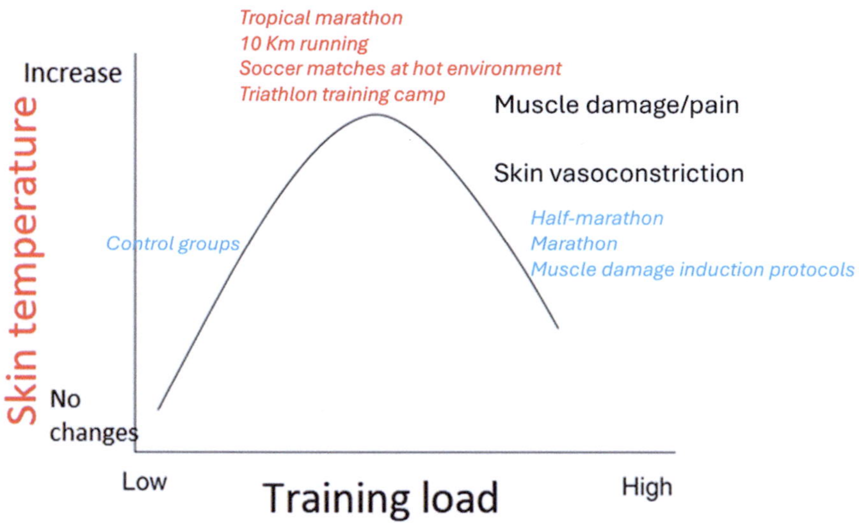

Fig. 6.6 Location of the different studies results in the hypothesis of an inverted U relationship between skin temperature and internal training loads. Studies with increases or no responses 24 and/or 48 h post exercise are shown in red and blue letters, respectively

establish this technique as a valid alternative [52]. In this section, we will discuss different methodological approaches to clarify the most appropriate choices for various experimental designs in the study of skin temperature responses to exercise and muscle fatigue.

The basic approach to the study of skin temperature responses to the exercise involves the measurement of a basal condition before exercise that serves as a comparator. This initial measurement can be compared to a postprotocol measurement. By comparing the skin temperature recorded before and after an exercise leading to fatigue, it becomes possible to determine whether a significant variation has occurred, providing a potential parameter for assessing physiological responses to exercise. Additionally, a pre- and post-exercise analysis can also include comparisons between limbs, specifically assessing asymmetries, provided the induced fatigue is unilateral [28]. The analysis of asymmetry not only reveals the warming of the exercised limb but may also indicate cooling of the control limb due to systemic effects of the exercise [53], thus creating a more pronounced contrast between the exercised and nonexercised conditions. However, simply comparing a region's temperature before and after exercise may underestimate the effects of muscle fatigue on skin temperature [44]. Changes in skin temperature are also related to the overall physical condition of the participant and thermoregulation. In this regard, the association between the changes in skin temperature of different ROIs seemed to rely on the physical fitness of the participant. It was observed that participants achieving larger overall muscle activation and reduced low frequency component for vastus lateralis activation presented a better adaptive response of their thermoregulatory system by showing fewer changes in skin temperature after incremental cycling test [54].

In addition, it is possible to assess skin temperature throughout an exercise session to understand its behavior over time. Since infrared thermography is an imaging technique, it allows for the possibility of capturing sequential images throughout the protocol, creating a timeline report of temperature values with the expectation of observing temperature changes during the task that may not be captured in a simple pre- and post-exercise comparison [55–57]. Korman et al. [58] described the changes in the skin temperature of the lower limbs throughout a running protocol to exhaustion and during a recovery period, recording data every 3 min, for a total protocol duration of 30 min. In this study, it was observed that when starting to exercise until exhaustion, the endurance athlete cools the skin significantly, generating a recovery in temperature only after the task is completed. In an effort to gather more information, a study investigated different exercise intensities captured thermographic images of the hands at a sampling rate of 0.1 Hz over a period exceeding 10 min, and it was possible to observe for the first time that the thermographic response is task-dependent [59].

Like other physiological outcomes, exercise intensity may also influence skin temperature responses. It is possible to use infrared thermography to compare the physiological responses to exercises performed at different intensities. During running at maximal exercise, the temperature of the chest region drops to a greater extent and does not show increases in the final moments, in contrast to what occurs during sub-maximal exercise [60]. On the other hand, in a study that monitored the temperature behavior of the backs of the hands—an area of smaller size with less contribution to heat dissipation and distal in relation to the muscles exercised—exercise intensity variation did not produce different patterns in temperature behavior [59]. It is important to note that experimental design choices can significantly impact the ability to observe exercise intensity effects on skin temperature using infrared thermography. For example, after a prolonged warm-up (40 min), high-performance athletes exhibited a reduction in skin temperature that remained unchanged when they began a high-intensity exercise [61]. This is because thermoregulation is highly sensitive to changes from a resting state. Moreover, during a pyramidal intensity cycling test, it was observed that skin temperature of the limbs (calf and forearm) was oppositely related to workload and heart rate, presenting higher decreases of skin temperature at the higher workloads [56]. However, facial regions did not present this relationship with the workload [56]. These studies are some examples that suggest that skin temperature measurements using infrared thermography result in variation during exercise that appears to be intensity dependent.

In addition to being intensity dependent, the study of fatigue using infrared thermography also appears to be population dependent. Even among trained individuals, the infrared thermography responses in the lower limbs during the same running protocol to exhaustion differ depending on whether the individual is trained in speed sports or endurance sports [58]. This study suggested a limitation in the composition of experimental groups in infrared thermography fatigue studies: groups should be matched based on their usual physical activities. In a similar context, a submaximal exercise for the gastrocnemius in trained and untrained women also revealed differences in infrared thermography patterns [34], indicating that trained individuals are more specialized in heat dissipation mechanisms, particularly through sweating. Therefore, physical fitness of the participants is an important characteristic to consider in the study of skin temperature changes due to fatigue.

It is important to note that there is an important sex-related component in the ability to produce sweat in response to exercise [62], which may reflect in different thermographic patterns. Women differ compared to men in heat dissipation through the skin for having a larger ratio of body surface to body mass [63]. The recovery pattern of normal temperature after exercise differs between sexes [64]. Additionally, skin temperature is affected by the menstrual cycle [65], underscoring the need to consider sex as a factor when composing studies using thermography in the context of muscle fatigue.

Infrared thermography can also be explored in relation to subjective measures of muscle fatigue. One study followed triathletes subjected to cumulative training loads, primarily targeting the lower limbs, and found a correlation between perceived exertion and the average temperature in the muscles acting on the knee joint [43]. On the other hand, when monitoring football players over a season, no relationship was found between skin temperature in the lower limbs and subjective markers of fatigue [40]. This striking difference between studies with similar objectives seems to be related to the extent of control over variables that may affect skin temperature, such as sleep, nutrition, and, most importantly, the characteristics of the exercises (with or without contact/impacts).

6.5.2 Cumulative Fatigue

The monitoring of changes in skin temperature in response to exercise workloads, intensity and its cumulative effects also highlights questions about the influence of recovery periods on the associations between skin temperature and exercise outcomes. Exercise-induced fatigue does not lead to permanent losses in efficiency. In order to restore biochemical homeostasis and habitual performance, the exercised muscle requires a period of rest, which varies depending on the population, muscle fiber type, exercise, and other factors, but generally requires more than 1 day of rest. Occasionally, the reduction of fatigue can be optimized through active recovery strategies; however, their application depends on the guidance of highly trained professionals [66]. In certain sports and rehabilitation protocols, there is a need to resume physical activity the day after the initial fatigue induction, without providing rest or active recovery, leading to a cumulative effect of fatigue. Cumulative fatigue is described here as the disruption of muscle regeneration and the recovery of muscle energy stores, or alternatively, it can be defined as the induction of fatigue over 2 or more consecutive days. Exercise performed in place of a rest day leads to an intensification of muscle damage resulting from oxidative stress triggered by the release of free radicals and the inflammatory process, which contributes to the loss of contractility and strength [67].

It is necessary to differentiate cumulative fatigue from the typical notion of muscle fatigue, even though their final outcomes may appear similar. Cumulative fatigue occurs due to the addition of factors that amplify the loss of performance in contractile units, such as the free radicals previously mentioned, which impair the coupling of contractile units [68], as well as the intensification of muscle damage promoted by oxidative stress [69]. To date, there is no consensus on the concept of cumulative fatigue [70], and as such, it may also be described as a series of exercises that accumulate stimuli until fatigue, in contrast to muscle fatigue induced by a single exercise bout [71]. The relationship between muscle fatigue and biochemical markers is well-established. By inducing cumulative fatigue in the quadriceps of highly trained cyclists through concentric and eccentric contractions on a leg extension machine, a loss of muscle contractility was observed during post-exercise cycling, accompanied by an intensification of muscle damage and oxidative stress [72]. This allows for speculation that an increased heat production will be experienced by the participants, which could be diagnosed using infrared thermography.

The hypothesis is that an intensified muscle damage could lead to a more pronounced inflammatory process, and consequently, greater heating, which can be transmitted to the skin more intensely. Being able to make this diagnosis could be useful for optimizing training protocols, especially in sports, by allowing for quick identification of athletes who have recovered or not, enabling immediate adjustment of training load [73]. In an assessment of women subjected to unilateral cumulative fatigue of the biceps brachii with 2 days of exhaustive exercise, it was observed that the exercised arm exhibited higher maximum temperatures compared to the control arm. Furthermore, it was observed that the minimum and average temperatures of the observed region were correlated with the loss of maximal isometric strength in the fatigued limb. This suggests that infrared thermography may be capable of indicating the presence of cumulative fatigue in a muscle group [74]. One of the negative effects of inducing additional muscle damage to a previously fatigued muscle is the increase in oxidative stress, which, at levels beyond the tolerable threshold of the individual's antioxidant defenses, can lead to biochemical damage to muscle tissues [67]. Therefore, the correlation between increased local skin temperature and strength loss opens up the possibility of estimating the contractile condition of the muscle without subjecting it to a maximal strength production protocol, which could be uncomfortable and further delay tissue regeneration [74]. Further studies are also necessary to understand the effects of cumulative fatigue on skin temperature.

The infrared thermography has the advantage of enabling broad analyses, covering the entire body or multiple segments at once, without incurring significant financial cost, although it requires more data processing time. This allows for observations such as the combination of cycling training and core strengthening at high intensity over 2 consecutive days, leading to an increase in skin temperature in almost all regions of the legs, but with less significant results in the upper limbs [43]. This suggests that infrared thermography describes cumulative fatigue at the local level with specificity to the muscle groups that were most engaged during exercise, in contrast to techniques like joint stiffness analysis in jumping tasks, which indicate fatigue in limbs but are unable to track the muscles most affected [71].

6.6 Future Directions

Considering the current literature on the use of infrared thermography to detect DOMS, muscle damage, and muscle fatigue, we propose the following general research directions:

- Increase the number of studies combining infrared thermography with techniques that measure skin blood flow. Most studies interpret their results based on skin blood flow, yet its direct measurement remains limited. Advancing basic knowledge on the interrelationship between these variables is essential.
- Conduct long-term follow-up studies on athletes to better understand responses to different internal loads. The majority of studies involve short-term interventions, and more long-term data are needed. Additionally, publishing these datasets in open repositories is encouraged to facilitate the combination of results from different studies and explore the application of artificial intelligence for improved data interpretation.
- Enhance gender representation in studies. Most research samples consist predominantly of males, despite clear evidence that thermophysiological responses are significantly influenced by sex-related factors.
- Expand research on the application of dynamic thermography (e.g., cold-stress tests; see Chap. 10) to obtain more insights into athletes' vascular state and capacity.

6.7 Conclusions

Infrared thermography offers significant advantages over other techniques, combining noninvasiveness with the ability to obtain results quickly and conduct statistical analyses with relative ease. Therefore, its implementation in the routine of sports clubs or rehabilitation clinics and hospitals can optimize assessment time without compromising the quality of the information. Although many expectations regarding the applicability of IRT to determine skin temperature responses to DOMS and fatigue have yet to be fully confirmed by the literature, this chapter has demonstrated its potential for assessing muscle condition in various contexts.

Infrared thermography presents difficulties in predicting muscle damage and DOMS from changes in skin temperature, which was observed in both controlled laboratory conditions and a practical field. The inverted U hypothesis seems promising to explain the difficulty in establishing this relationship. This hypothesis debates that DOMS induces peripherical vasoconstriction, limiting heat transfer to the skin. Future studies should be performed to confirm this hypothesis and establish the potential of IRT in assessing DOMS.

Similarly, skin temperature variation appears to correlate with muscle fatigue. However, several precautions must be considered. First and foremost, the initial thermal response of the skin to intense exercise is a reduction in heat due to sweating and venous conduction. However, the effectiveness of this response depends on the chosen exercise, the target population, and the monitored region. Consequently, the ability of IRT to assess muscle fatigue status depends on these same factors. Although the mechanisms underlying cumulative fatigue differ from those of immediate muscle fatigue, similar precautions should be taken when designing data collection setups. Cumulative fatigue generates a feedback loop of muscle damage, causing inflammation-induced heat production in the muscle to persist and intensify. In this regard, IRT shows promising potential for identifying muscles under cumulative fatigue, potentially leading to time and cost savings in sports settings. However, studies on cumulative fatigue remain preliminary and require further investigation to establish IRT as a reliable diagnostic tool.

References

1. Armstrong RB (1984) Mechanisms of exercise-induced delayed onset muscular soreness: a brief review. Med Sci Sports Exerc 16(6):529–538
2. Cheung K, Hume P, Maxwell L (2003) Delayed onset muscle soreness: treatment strategies and performance factors. Sports Med Auckl NZ 33(2):145–164
3. Sadacharan CM, Seo S (2021) Effect of large versus small range of motion in the various intensities of eccentric exercise-induced muscle pain and strength. Int J Exerc Sci 14(7):1–18
4. Hotfiel T, Freiwald J, Hoppe M, Lutter C, Forst R et al (2018) Advances in Delayed-Onset Muscle Soreness (DOMS): part I: pathogenesis and diagnostics. Sportverletz Sportschaden 32(04):243–250
5. Smith LL (1991) Acute inflammation: the underlying mechanism in delayed onset muscle soreness? Med Sci Sports Exerc 23(5):542–551
6. Dos Santos MA, Lemos AL, Machado MS, Lazaro LDOC, Paz MM et al (2024) Effects of triceps surae exercise-induced delayed onset muscle soreness on control of body stability in different postures. J Electromyogr Kinesiol 76:102882
7. Lemos AL, Stoelben KJV, Pappas E, Carpes FP (2024) Differences in landing biomechanics in the presence of delayed onset muscle soreness following or not active recovery. J Electromyogr Kinesiol 79:102948
8. Lau WY, Muthalib M, Nosaka K (2013) Visual analog scale and pressure pain threshold for delayed onset muscle soreness assessment. J Musculoskelet Pain 21(4):320–326
9. McCarney L, Lythgo N, Fazalbhoy A, Moreland A (2025) Objective measures of stiffness and ratings of pain and stiffness in the gastrocnemii following delayed-onset muscle soreness. J Bodyw Mov Ther 41:187–193
10. Borg GA (1982) Psychophysical bases of perceived exertion. Med Sci Sports Exerc 14(5):377–381
11. Clarkson PM, Hubal MJ (2002) Exercise-induced muscle damage in humans. Am J Phys Med Rehabil 81(11 Suppl):S52–S69

12. Brancaccio P, Lippi G, Maffulli N (2010) Biochemical markers of muscular damage. Clin Chem Lab Med 48(6):757–767
13. Pérez-Guarner A, Priego-Quesada JI, Oficial-Casado F, Cibrián Ortiz De Anda RM, Carpes FP, Palmer RS (2019) Association between physiological stress and skin temperature response after a half marathon. Physiol Meas 40(3):034009
14. Priego-Quesada JI, Pérez-Guarner A, Gandia-Soriano A, Oficial-Casado F, Galindo C et al (2020b) Effect of a marathon on skin temperature response after a cold-stress test and its relationship with perceptive, performance, and oxidative-stress biomarkers. Int J Sports Physiol Perform 15(10):1467–1475
15. MacIntyre DL, Reid WD, McKenzie DC (1995) Delayed muscle soreness: the inflammatory response to muscle injury and its clinical implications. Sports Med 20(1):24–40
16. Lewis PB, Ruby D, Bush-Joseph CA (2012) Muscle soreness and delayed-onset muscle soreness. Clin Sports Med 31(2):255–262
17. Safran MR, Seaber AV, Garrett WE (1989) Warm-up and muscular injury prevention. An update. Sports Med Auckl NZ 8(4):239–249
18. Gulick DT, Kimura IF (1996) Delayed onset muscle soreness: what is it and how do we treat it? J Sport Rehabil 5(3):234–243
19. Connolly DAJ, Sayers SP, McHugh MP (2003) Treatment and prevention of delayed onset muscle soreness. J Strength Cond Res 17(1):197–208
20. Fridén J, Lieber RL (1992) Structural and mechanical basis of exercise-induced muscle injury. Med Sci Sports Exerc 24(5):521–530
21. Hertel J (1997) The role of nonsteroidal anti-inflammatory drugs in the treatment of acute soft tissue injuries. J Athl Train 32(4):350–358
22. Vane J, Botting R (1987) Inflammation and the mechanism of action of anti-inflammatory drugs. FASEB J 1(2):89–96
23. Markus I, Constantini K, Hoffman JR, Bartolomei S, Gepner Y (2021) Exercise-induced muscle damage: mechanism, assessment and nutritional factors to accelerate recovery. Eur J Appl Physiol 121(4):969–992
24. Hody S, Croisier J-L, Bury T, Rogister B, Leprince P (2019) Eccentric muscle contractions: risks and benefits. Front Physiol 10:536
25. Paulsen G, Egner IM, Drange M, Langberg H, Benestad HB et al (2010) A COX-2 inhibitor reduces muscle soreness, but does not influence recovery and adaptation after eccentric exercise. Scand J Med Sci Sports 20(1):e195–e207
26. Peake JM, Neubauer O, Della Gatta PA, Nosaka K (2017) Muscle damage and inflammation during recovery from exercise. J Appl Physiol (Bethesda Md 1985) 122(3):559–570
27. Campbell I (2008) Body temperature and its regulation. Anaesth Intensive Care Med 9(6):259–263
28. Hildebrandt C, Raschner C, Ammer K (2010) An overview of recent application of medical infrared thermography in sports medicine in Austria. Sensors 10(5):4700–4715
29. Zambolin F, Giuriato G, Laginestra FG, Ottaviani MM, Favaretto T et al (2022) Effects of nociceptive and mechanosensitive afferents sensitization on central and peripheral hemodynamics following exercise-induced muscle damage. J Appl Physiol 133(4):945–958
30. Maeda H, Saito M, Ishida K, Akima H (2020) Relationships between resting blood flow and the indices of muscle damage after eccentric contractions. Eur J Appl Physiol 120(10):2183–2192
31. de Andrade Fernandes A, Pimenta EM, Moreira DG, Sillero-Quintana M, Marins JCB et al (2017) Skin temperature changes of under-20 soccer players after two consecutive matches. Sport Sci Health 13(3):635–643
32. da Silva W, Machado ÁS, Souza MA, Kunzler MR, Priego-Quesada JI, Carpes FP (2018) Can exercise-induced muscle damage be related to changes in skin temperature? Physiol Meas 39(10):104007
33. Thedon T, Mandrick K, Foissac M, Mottet D, Perrey S (2011) Degraded postural performance after muscle fatigue can be compensated by skin stimulation. Gait Posture 33(4):686–689

34. Formenti D, Ludwig N, Gargano M, Gondola M, Dellerma N et al (2013) Thermal imaging of exercise-associated skin temperature changes in trained and untrained female subjects. Ann Biomed Eng 41(4):863–871
35. da Silva W, Machado ÁS, Lemos AL, de Andrade CF, Priego-Quesada JI, Carpes FP (2021) Relationship between exercise-induced muscle soreness, pain thresholds, and skin temperature in men and women. J Therm Biol 100:103051
36. Bartelink ML, Wollersheim H, Theeuwes A, Van Duren D, Thien T (1990) Changes in skin blood flow during the menstrual cycle: the influence of the menstrual cycle on the peripheral circulation in healthy female volunteers. Clin Sci 78(Pt 5):527–532
37. Jimenez-Perez I, Gil-Calvo M, Priego-Quesada JI, Aparicio I, Pérez-Soriano P, Ortiz de Anda RMC (2020) Effect of prefabricated thermoformable foot orthoses on plantar surface temperature after running: a gender comparison. J Therm Biol 91:102612
38. Marins JCB, Formenti D, Costa CMA, de Andrade FA, Sillero-Quintana M (2015) Circadian and gender differences in skin temperature in militaries by thermography. Infrared Phys Technol 71:322–328
39. Barboza JAM, Ferreira JJDA, Cerqueira MS, Maciel DG, de Barros ACM et al (2022) Can skin temperature be altered after different magnitudes of eccentric exercise-induced muscle damage? Res Q Exerc Sport 93(4):702–709
40. de Carvalho G, Girasol CE, Gonçalves LGC, Guirro ECO, Guirro RRDJ (2021) Correlation between skin temperature in the lower limbs and biochemical marker, performance data, and clinical recovery scales. PLoS One 16(3):e0248653
41. Priego-Quesada JI, Catalá-Vilaplana I, Bermejo-Ruiz JL, Gandia-Soriano A, Pellicer-Chenoll MT et al (2022) Effect of 10 km run on lower limb skin temperature and thermal response after a cold-stress test over the following 24 h. J Therm Biol 105:103225
42. Da Silva W, Godoy-López JR, Machado ÁS, Lemos AL, Sendra-Pérez C et al (2024) Effect of different volumes of exercise on skin temperature responses over the following 24 hours. J Therm Biol 123:103923
43. Priego-Quesada JI, Oficial-Casado F, Gandia-Soriano A, Carpes FP (2019) A preliminary investigation about the observation of regional skin temperatures following cumulative training loads in triathletes during training camp. J Therm Biol 84:431–438
44. Stewart IB, Moghadam P, Borg DN, Kung T, Sikka P, Minett GM (2020) Thermal infrared imaging can differentiate skin temperature changes associated with intense single leg exercise, but not with delayed onset of muscle soreness. J Sports Sci Med 19(3):469–477
45. Rojas-Valverde D, Gutiérrez-Vargas R, Sánchez-Ureña B, Gutiérrez-Vargas JC, Priego-Quesada JI (2021) Relationship between skin temperature variation and muscle damage markers after a marathon performed in a hot environmental condition. Life 11(8):725
46. Enoka RM, Duchateau J (2008) Muscle fatigue: what, why and how it influences muscle function. J Physiol 586(1):11–23
47. Enoka RM, Duchateau J (2016) Translating fatigue to human performance. Med Sci Sports Exerc 48(11):2228–2238
48. Moritani T, Nagata A, Muro M (1982) Electromyographic manifestations of muscular fatigue. Med Sci Sports Exerc 14(3)
49. Rodrigues SB, De Faria LP, Monteiro AM, Lima JL, Barbosa TM, Duarte JA (2022) EMG signal processing for the study of localized muscle fatigue—pilot study to explore the applicability of a novel method. Int J Environ Res Public Health 19(20):13270
50. Bartuzi P, Roman-Liu D, Wiśniewski T (2012) The influence of fatigue on muscle temperature. Int J Occup Saf Ergon 18(2):233–243
51. Shakhih MFM, Ridzuan N, Wahab AA, Zainuddin NF, Delestri LFU et al (2021) Non-obstructive monitoring of muscle fatigue for low intensity dynamic exercise with infrared thermography technique. Med Biol Eng Comput 59(7–8):1447–1459
52. Priego-Quesada JI, De La Fuente C, Kunzler MR, Perez-Soriano P, Hervás-Marín D, Carpes FP (2020a) Relationship between skin temperature, electrical manifestations of muscle fatigue,

and exercise-induced delayed onset muscle soreness for dynamic contractions: a preliminary study. Int J Environ Res Public Health 17(18):6817
53. Kenny GP, McGinn R (2017) Restoration of thermoregulation after exercise. J Appl Physiol 122(4):933–944
54. Priego Quesada JI, Carpes FP, Bini RR, Salvador Palmer R, Pérez-Soriano P, Ortiz C, de Anda RM (2015) Relationship between skin temperature and muscle activation during incremental cycle exercise. J Therm Biol 48:28–35
55. Hadžić V, Širok B, Malneršič A, Čoh M (2019) Can infrared thermography be used to monitor fatigue during exercise? A case study. J Sport Health Sci 8(1):89–92
56. Hillen B, Andrés López D, Marzano-Felisatti JM, Sanchez-Jimenez JL, Cibrián Ortiz de Anda RM et al (2023) Acute physiological responses to a pyramidal exercise protocol and the associations with skin temperature variation in different body areas. J Therm Biol 115:103605
57. Kuniszyk-Jóźkowiak W, Jaszczuk J, Czaplicki A (2018) Changes in electromyographic signals and skin temperature during standardised effort in volleyball players. Acta Bioeng Biomech. 032018 ISSN 1509-409X
58. Korman P, Kusy K, Straburzyńska-Lupa A, Kantanista A, Quintana MS, Zieliński J (2024) Response of skin temperature, blood ammonia and lactate during incremental exercise until exhaustion in elite athletes. Sci Rep 14(1):2237
59. Zontak A, Sideman S, Verbitsky O, Beyar R (1998) Dynamic thermography: analysis of hand temperature during exercise. Ann Biomed Eng 26(6):988–993
60. Balci GA, Basaran T, Colakoglu M (2016) Analysing visual pattern of skin temperature during submaximal and maximal exercises. Infrared Phys Technol 74:57–62
61. Korman P, Straburzyńska-Lupa A, Kusy K, Kantanista A, Zieliński J (2016) Changes in body surface temperature during speed endurance work-out in highly-trained male sprinters. Infrared Phys Technol 78:209–213
62. Buono MJ, Sjoholm NT (1988) Effect of physical training on peripheral sweat production. J Appl Physiol 65(2):811–814
63. Kaciuba-Uscilko H, Grucza R (2001) Gender differences in thermoregulation. Curr Opin Clin Nutr Metab Care 4(6):533–536
64. Vargas NT, Chapman CL, Sackett JR, Johnson BD, Gathercole R, Schlader ZJ (2019) Thermal behavior differs between males and females during exercise and recovery. Med Sci Sports Exerc 51(1):141–152
65. Baker FC, Siboza F, Fuller A (2020) Temperature regulation in women: effects of the menstrual cycle. Temperature (Austin Tex) 7(3):226–262
66. Adamczyk JG (2023) Support Your Recovery Needs (SYRN) – a systemic approach to improve sport performance. Biomed Hum Kinet 15(1):269–279
67. Agrawal S, Chakole S, Shetty N, Prasad R, Lohakare T, Wanjari M (2023) Exploring the role of oxidative stress in skeletal muscle atrophy: mechanisms and implications. Cureus 15(7):e42178
68. Prochniewicz E, Spakowicz D, Thomas DD (2008) Changes in actin structural transitions associated with oxidative inhibition of muscle contraction. Biochemistry 47(45):11811–11817
69. Lian D, Chen M-M, Wu H, Deng S, Hu X (2022) The role of oxidative stress in skeletal muscle myogenesis and muscle disease. Antioxid Basel Switz 11(4):755
70. Kataoka R, Vasenina E, Hammert WB, Ibrahim AH, Dankel SJ, Buckner SL (2022) Is there evidence for the suggestion that fatigue accumulates following resistance exercise? Sports Med 52(1):25–36
71. Dindorf C, Bartaguiz E, Dully J, Sprenger M, Becker S et al (2023) In vivo monitoring of acute and intermittent fatigue in sport climbing using near-infrared spectroscopy wearable biosensors. Sports 11(2):37
72. Machado AS, da Silva W, Souza MA, Carpes FP (2018) Green tea extract preserves neuromuscular activation and muscle damage markers in athletes under cumulative fatigue. Front Physiol 9(1):9

73. Gómez-Carmona P, Fernández-Cuevas I, Sillero-Quintana M, Arnaiz-Lastras J, Navandar A (2020) Infrared thermography protocol on reducing the incidence of soccer injuries. J Sport Rehabil 29(8):1222–1227
74. Machado ÁS, da Silva W, Priego-Quesada JI, Carpes FP (2023) Can infrared thermography serve as an alternative to assess cumulative fatigue in women? J Therm Biol 115:103612

Chapter 7
The Use of Infrared Thermography in the Study of Sport and Exercise Physiology

Damiano Formenti, David Perpetuini, Arcangelo Merla, and Jose Ignacio Priego-Quesada

Abstract Infrared thermography (IRT) is considered a well-established methodology in the field of exercise physiology. Skin temperature distribution derives from muscular activity, skin blood flow as well as perspiration patterns in specific body parts. This chapter aims to provide a general overview on the literature about the study of the skin temperature response to exercise assessed by IRT and its relationship with other thermoregulatory variables, exercise characteristics, and performance factors.

7.1 Introduction: Why Is It Important to Monitor Skin Temperature in Sport and Exercise Physiology?

At rest, human presents a thermal homeostasis by the regulation of core temperature in values about 36.8 °C [1]. This regulation is managed mainly by the hypothalamus, which receives sensory inputs from central and peripheral sites, and it activates or inhibits the parallel efferent thermoregulatory pathways [2].

However, physical exercise and repetitive effort are a challenge to thermal homeostasis [1, 3, 4]. It is well established that physical activity induces complex thermoregulatory processes where part of heat in excess is dissipated through the

---

D. Formenti (✉)
Department of Biotechnology and Life Sciences (DBSV), University of Insubria, Varese, Italy
e-mail: damiano.formenti@uninsubria.it

D. Perpetuini · A. Merla
Department of Engineering and Geology (INGEO), University "G. d'Annunzio" of Chieti-Pescara, Pescara, Italy
e-mail: david.perpetuini@unich.it; arcangelo.merla@unich.it

J. I. Priego-Quesada
Research Group in Sport Biomechanics (GIBD), Department of Physical Education and Sports, University of Valencia, Valencia, Spain
e-mail: j.ignacio.priego@uv.es

J. I. Priego-Quesada (ed.), *Application of Infrared Thermography in Sports Science*,
https://doi.org/10.1007/978-3-031-93311-0_7

skin to the external environment [3]. During exercise, the thermoregulatory control of skin blood flow has the fundamental role to maintain normal core temperature and leads to changes in hemodynamics affecting thermal signals [5, 6]. Skin temperature is the result of the heat transferred by the skin blood flow, thermoregulatory responses of heat dissipation such as conduction, convection, radiation and sweat evaporation, environmental conditions, and different biophysical factors such as body surface area or body composition [7, 8]. Therefore, measuring skin temperature provides useful information about the complex thermal control systems.

Research attention has been devoted especially to understand mechanisms limiting performance during exercise in hot and warm environment [4, 9–12]. Among the mechanisms limiting performance during exercise at submaximal intensities in the heat, cardiovascular factors play an important role [3, 9]. Research works have put emphasis on the limitations related to the central nervous system, and the role of high core temperature [9, 13–15]. It has been proposed that high core temperature (~40 °C) would be the primary factor affecting CNS, thus limiting aerobic performance at submaximal intensity [9, 10, 13, 15].

While the increase in core temperature occurs proportionately to the exercise intensity, skin temperature is mainly related to environmental conditions, the heat loss capacity [11, 16, 17], but also the region measured [18]. An increase in core temperature during exercise also affects skin temperature, as heat in excess is transferred from the inner to the superficial parts of the body via cutaneous vasodilation, where it is dissipated through the skin [19]. Therefore, an isolated increment in skin blood flow accompanies an increment in skin temperature [20]. However, this increment in skin blood flow is commonly accompanied by other processes, such as sweat evaporation. As in many situations, skin temperature can result in a decrease instead of an increase. However, studies supporting the hypothesis of the critical core temperature were conducted inducing simultaneously high core and skin temperatures [9, 10, 13]. During a lot of time, it was assumed that core temperature was the main factor responsible for fatigue during aerobic exercise in the heat. However, in the last years, this theory has been questioned showing that skin temperature also plays an important role, and therefore, the emphasis in research has shifted toward the role of skin temperature [11, 12]. It has been recently demonstrated that the relationship between skin temperature and core temperature, and not only core temperature, has a key role in limiting exercise performance [11, 12, 21]. In particular, critical reduction in the core to skin temperature gradient is the "primary" factor responsible for impaired aerobic capacity, rather than a critical core temperature [11, 12, 21]. For example, a field-based study at the 2019 Doha World Athletics Championships revealed that skin temperature, its rate of decrease, and the core-to-skin temperature gradient were crucial indicators of heat tolerance. These findings provide further evidence on the importance of monitoring skin temperature not focusing solely on core temperature for assessing heat stress responses [12].

Based on this evidence, in the last years, research attention has been devoted to studying the thermal balance between the human body and the environment during exercise through the assessment of skin temperature. During the initial phases of the exercise, skin temperature tends to decrease due to the cutaneous vasoconstrictor response to exercise [22, 23]. After that, skin temperature primarily increases when blood flow shifts from internal tissues to the skin, thus dissipating heat in excess related to an increased metabolic activity [23–25]. Nevertheless, especially in moderate environments, skin temperature may decrease due to heat dissipation by convection, radiation and evaporation [26, 27].

All in all, skin temperature in specific regions of the body is the result of a thermal balance between muscle activity, vasodilation and sweat evaporation [3, 27].

This chapter aims to provide a general overview of the literature about the study of the skin temperature response to exercise assessed by means of infrared thermography (IRT). More specifically, the chapter is made in order to provide the reader with an understanding of the effect of exercise on skin temperature. Specifically, some important factors that affect the behavior of skin temperature during exercise and sport will be presented and discussed, as well as some of the most significant studies on infrared thermography to date. In order to fulfill all the above points, the chapter is structured as follows:

1. First, the chapter will discuss some methodological issues that could affect the skin temperature behavior during exercise, such as the instrument used (e.g., thermal contact sensors or IRT) as well as the characteristics of the exercise.
2. Second, the chapter will show the relationship between the skin temperature and other thermoregulatory variables (i.e., blood blow, core temperature and sweat rate).
3. Third, the chapter will review the effect of the exercise on skin temperature in relation with different experimental and exercise conditions.
4. Finally, the chapter will show the relationship between skin temperature and performance variables (VO_{2max}, heart rate, neuromuscular activation and body composition).

7.2 Measuring Skin Temperature in Exercise Physiology: Methodological Aspects

7.2.1 Measurement Instrument: Contact Thermal Sensors Versus Infrared Thermography

Typically, skin temperature is measured using contact thermal sensors, such as thermocouples or wired thermistors, usually considered as gold standard methods (despite its limitations, as this section will show). A thermocouple is a temperature-measuring device consisting of two dissimilar conductors that contact each other at one or more joint locations. Their system is based on the Seebeck effect, which occurs when a difference in electric potential is created between two conductor materials at different temperatures [28, 29]. Thermistors are resistors in which resistance varies with temperature, allowing stored calibration data within the circuit to convert this to a temperature value [30].

In the last decade, thermocouples have also been developed in the wireless system, providing a solution to the limitations of hard-wired methods. In fact, wireless sensors provide subjects with great mobility as they do not interfere with their movements, such in case of physical exercise during high dynamic situations, both in laboratory and in the field [29, 31]. These sensors also allow measuring below or in-between clothing layers [32].

Although contact thermal sensors present some advantages, there are some limitations in their use. Determining temperature in just one single point can limit the understanding of human thermal response, as the evaluated body part could not be properly represented [33]. Furthermore, thermal interactions between contact sensors and environment can affect the reliability of the measurements [29, 30]. Contact sensors are usually attached to skin using different types of clinical tapes. The attaching method of contact sensors to the skin has been shown to affect the local heat transfer [34, 35], thus influencing the local thermal regulation and skin temperature [30, 31, 36].

An alternative device to measure skin temperature is IRT. IRT has become popular in the last years in sport and exercise physiology research due to its noncontact and noninvasive character. Furthermore, IRT obtains a thermal image, which can measure the temperature at all points, thus allowing to measure large areas instead of single points (such as contact sensors) [31, 33]. As can be seen, IRT can solve some of the limitations of contact sensors. For this reason, apart from studies applying IRT in the measurement of skin temperature modifications related to exercise, research attention has also been devoted to investigate the validity of IRT against thermal contact sensors during both resting and exercising conditions [31, 33, 37, 38].

De Andrade Fernandes et al. [31] published a study aimed to compare the mean skin temperature measured using thermocouples to those measured using IRT and to check the agreement between the two methods before, during, and after exercise performed in a thermo-neutral environment. Two identical experimental protocols were performed in two different days: one with the thermocouple measures, and one with the IRT measures. In each of the 2 days, the subjects completed an interval test on a treadmill consisting of 12 blocks of 5 min, each one separated by an interval of 1 min, during which temperature was measured. The exercise intensity was individually fixed to 60% of the VO_{2max} speed obtained in the presession experiment. Low agreement between the mean skin temperature values measured using thermocouples and IRT was found in each of the three phases considered (before, during, and after exercise), demonstrating the discrepancies between the two methods. Thermocouples showed higher mean skin temperature values during exercise than IRT, while IRT showed higher values before and after exercise. The results indicated that the interpretation will vary depending on the method employed (Fig. 7.1). The authors suggested that the low agreement between

thermocouples and IRT could be mainly due to the contact character of thermocouples. In fact, several factors could contribute to explain the observed differences: the pressure exerted by fixation methods and the use of substances for the fixation [36]. Furthermore, increased heat loss by convection and evaporation in those

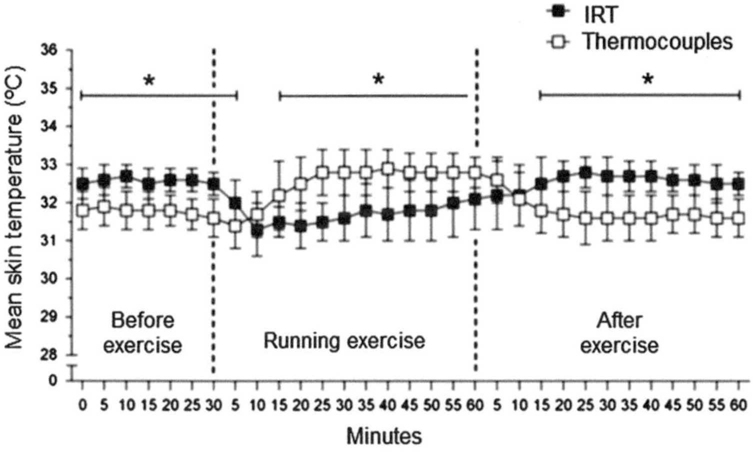

Fig. 7.1 Mean skin temperature values pre-, during and post-exercise obtained from thermocouples and IRT. Significant differences between methods are shown using *. (Figure modified from de Andrade Fernandes et al. [31])

regions where thermocouples were fixed could increase the mean skin temperature measured by thermocouples [34, 36].

Another study investigated the agreement between hard-wired thermistors, telemetry thermistors (i.e., wireless sensors), and IRT during exercise in a hot and humid environment [37]. Subjects performed an incremental running exercise test in a climatic chamber with an environmental temperature of 31.9 ± 1 °C and relative humidity of 61 ± 8.9%. Skin temperature was measured at four different sites (i.e., chest, arm, thigh, calf), and mean skin temperature was computed with the weighted average: chest 30%, arm 30%, thigh 20%, and calf 20% [39]. Exercise started with speed between 8 and 10 km h^{-1}. Each subject completed five stages of three minutes, with speed increasing by 1 km h^{-1}. At the end of each stage, subjects straddled the treadmill and the thermal camera recorded skin temperature at each site. The authors concluded that IRT tended to underestimate the temperature values measured by hard-wired thermistors. Authors suggested that the differences between methods could be due to the contact and fixation of the thermistors as in the previous study, but mainly by a measurement error of IRT due to the alteration of the skin emissivity by the sweat. Due to this interpretation, the authors suggested that IRT may be a useful tool for measuring skin temperature in static and controlled environments, but its use is not recommended for live monitoring during dynamic conditions, such as during exercise.

Another study compared IRT and thermal contact sensors for measuring skin temperature in various parts of the body in a moderate cycling scenario and subsequent cooling down phase [33]. Furthermore, the novelty of this study is that the authors compared the results of human scenario with a more controlled scenario: surface temperature of a hot plate system during dry and wet conditions. This instrumented test was conducted to exclude human thermoregulatory processes when

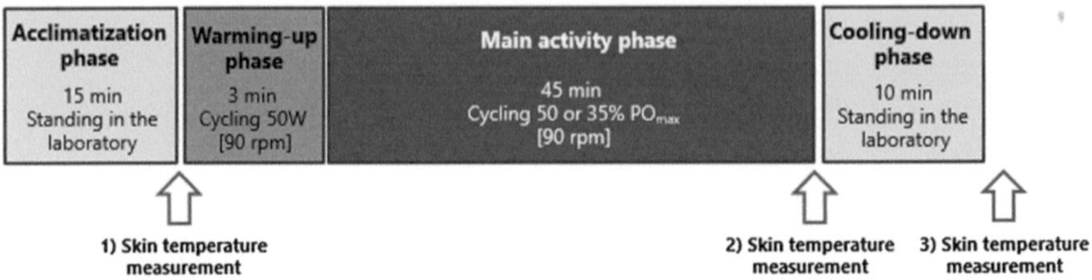

Fig. 7.2 Experimental protocol of the study (PO_{max} corresponds to the peak power output if pedaling at 90 rpm determined in pretest). (Figure adapted from Priego Quesada et al. [33])

simulating heat exchange in dry and wet conditions. The exercise test protocol was structured in the following phases and is resumed in Fig. 7.2:

1. Thermal adaptation phase (15 min): standing in the laboratory. The investigator attached the skin temperature sensors within the first minutes of this phase. Thermal images were recorded at the end of this phase (precycling).
2. Cycling phase (48 min): After warming up for 3 min at 50 W and 90 rpm, participants cycled at 50% of their maximal power output at 90 ± 3 rpm. Thermal images were recorded immediately at the end of the exercise (post cycling).
3. Cooling-down phase (10 min): standing in the laboratory. Thermal images were recorded at the end of this phase (post cooling).

Immediately after cycling, IRT provided lower temperature values with respect to thermal contact sensors, whereas it presented higher temperatures after the cooling-down phase [33]. In the exercise trial, the correlation coefficient between the two methods was high when measured before cycling ($r = 0.92$) whereas it was reduced immediately after cycling ($r = 0.82$) and after the cooling-down phase ($r = 0.59$). However, the discrepancies in temperature between the two methods remained within the accuracy of the thermal camera (±2 °C). Comparable results were observed between dry and wet states when comparing exercise test and instrumented tests, in which no water coating was present, suggesting that in such an exercising scenario, sweat delivery was not high enough to produce a continuous water layer on the skin. Furthermore, they suggested that the comparable differences between the instrumented test and the human scenario show that differences between both methods are mainly related to the effect of the contact sensors on the reduction of heat loss (via radiation and sweat evaporation). The authors stressed that these findings demonstrate the possibility of using IRT in the field of sports and exercise physiology for assessing skin temperature.

A preliminary study investigated the effect of sweat accumulation on skin temperature measurements after moderate-intensity running [38]. Their findings indicated that thermal contact sensors placed on the skin throughout the protocol (fixed sensors) recorded higher skin temperatures after running (33.9 ± 1.3 °C) compared to infrared thermography (33.2 ± 1.4 °C). This difference was attributed to reduced convective and evaporative heat loss at the sensor site. Furthermore, fixed sensors exhibited significantly higher relative humidity (101 ± 2%) than nonfixed sensors

(83 ± 16.2%; attached only 5 minutes before measurement) after exercise, suggesting sweat saturation caused by sensor attachment.

Based on the findings of these methods-comparison studies, IRT seems to be appropriate for measuring skin temperature modifications associated to exercise. IRT allows to measure skin temperature distribution with high sensitivity (0.05 °C), thus having the main advantage to be capable of analyzing large and thus, more representative areas on the body surface. Due to its noninvasiveness, heat transfer process at skin level is not affected by attachment of sensors. Nevertheless, thermal contact sensors have some advantages with respect to IRT, especially when measuring absolute temperatures in particular setting conditions. For example, contact sensors have the possibility to measure temperature when subjects are clothed, or when body posture and space do not allow having optimal conditions to record infrared thermal images [33]. Furthermore, sweat on the skin produced during and after exercise may act as a filter for infrared radiation and that could lead to an error in the estimation of the skin temperature using IRT [40]. However, this issue needs future studies to explore the real physical effect of the sweat on the calculation of skin temperature using IRT in the different exercise scenarios.

To conclude this section, it is important to take into account that advantages and disadvantages of both methods should be considered before designing a study assessing skin temperature changes associated to exercise, and cautions should be taken to control possible sources of error. Furthermore, it is important to consider the measurement method during the interpretation of the result of one study.

7.2.2 Exercise Characteristics and Experimental Approach

In the last decades, many authors have described the skin temperature response to exercise in different body areas under different conditions and different experimental approaches [17, 41–47]. Although drawing a certain conclusion on the skin temperature response to exercise seems to be difficult due to huge discrepancies between studies in aim and methodology, it is generally accepted that skin temperature modification associated to exercise—and its physiological meaning—varies in relation to several factors. Among them, the most relevant are characteristics of the exercise and different experimental approaches.

In general, exercise can be classified on the basis of the discipline (i.e., the type of discipline studied, such as running, cycling, or specific resistance exercise). For example, Merla et al. [45] investigated the skin temperature response to running exercise, as well as Priego Quesada et al. [46] did for cycling exercise. However, in the case of resistance exercise, it is important to distinguish between single-joint resistance exercise and multijoint resistance exercise. Single-joint exercises usually isolate a specific muscle or muscle group (such as leg extension), whereas multijoint exercises usually recruit one or more large muscle groups as agonist muscles (such as squat), and other muscles as coacting muscles (for example, for stabilization). Most sport activity and daily movements consist of multijoint movements.

The more similar the training activity to the actual sport and daily movements, the greater positive transfer of resistance exercise to these movements (i.e., the specificity concept). In general, from both scientists and practitioners point of views, multijoint exercises are considered more suitable for improving sport and daily performance with respect to single-joint exercises [48]. The fact of recruiting single muscle group (single-joint exercise) rather than large muscle groups (multijoint exercise) is likely to differently influence thermoregulatory processes (especially vasoregulation), thus in turn to influence also skin temperature response to exercise in different way. As regards to the study of the skin temperature response to single-joint and multijoint exercises, both are present in the literature. For example, Formenti et al. [44] studied the skin temperature response of anterior thigh to a typical multijoint exercise, such as squat; on the other hand, Ferreira et al. [42] investigated the skin temperature response on posterior thigh to a single joint exercise (knee flexion).Furthermore, it has been shown that skin temperature response varies according to the intensity and the duration of the exercise. For example, Balci et al. [41] examined the difference in the skin temperature dynamics during a submaximal constant load cycling exercise and a maximal incremental cycling exercise. It was found that the skin temperature time course during the two trials exhibited a different dynamic. On the other hand, Priego Quesada et al. [17] assessed the effect of different cycling workloads on core and skin temperature. They observed that cycling workload did not have any effect on the skin temperature in the almost totality of the body regions due to the higher heat loss of the thermoregulatory system, and only Regions of Interest (ROIs) that are mostly constituted by connective, bone and fat tissues were affected (Fig. 7.3). These ROIs presented higher reductions of the skin temperature at the higher workload (50%). The authors suggested that these regions, that are not located on top of active muscles, are affected by a temperature decrease due to a higher overall sweat rate according to a higher intensity, rather than temperature increases through rising workloads. On the other hand, core temperature increased continuously throughout the exercise, as expected.

Investigations have been conducted with the aim to study localized effect of the exercise, or systemic effect of the exercise. These are different experimental approaches, which are both present in the literature. For instance, Zontak et al. [47] measured the skin temperature response of hands before, during, and after two bouts of cycling exercise. In this study, while the cycling exercise involved only lower limbs, the skin temperature was measured on the hands. Thanks to this approach, the authors were interested in studying the systemic effect, since the skin temperature was measured over parts of the body not involved in the exercise directly. Another experimental approach was used by Formenti et al. [43], which compared the skin temperature response to standing calf rise exercise in a sample of trained and untrained subjects. In this paper, the authors aimed to study the localized effect of the exercise, since skin temperature was measured on the calves area of the subjects, i.e., the muscle tendon unit directly involved in the exercise movement. A simultaneous evaluation of systemic and localized effect is also possible since both measurements can be combined in the same study. For example, Priego Quesada et al. [17] investigated the influence of cycling workload on core and skin

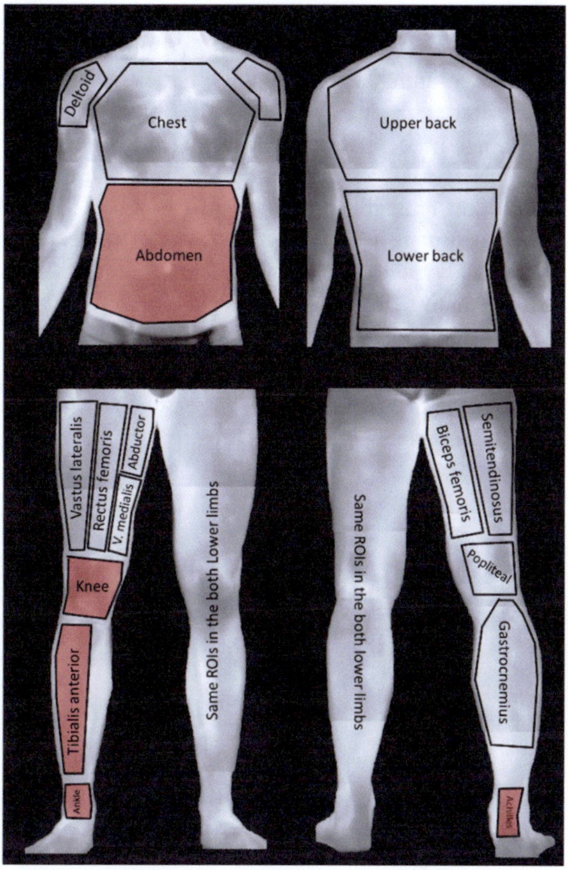

Fig. 7.3 ROI representation of the differences in the variation of skin temperature between 50 and 35% PO_{max}, using a red background in the ROI, in the study of Priego Quesada et al. [17]

temperature measured in 17 different regions of interests on lower limbs (i.e., over the muscles involved in exercise) and on the trunk.

Moreover, the experimental approach is different when researchers are interested in studying the dynamic of the skin temperature time course during exercise, or when researchers are interested in a pre–post exercise comparison of skin temperature. These are different methodological approaches to the problem, with different physiological meanings. Merla et al. [45] measured skin temperature of various parts of the body continuously during running exercise using a sampling rate of 50 Hz. The authors were clearly interested in studying the dynamic of skin temperature changes occurring throughout the duration of the exercise. Similarly, more recent investigations focused on the analysis of different body regions during cycling and running incremental exercises, providing valuable information on the dynamics of the skin temperature distribution modifications occurring throughout the exercises [18, 49]. This was possible thanks to advancements on both instrumentation and data analysis, especially exploring artificial intelligence-based algorithms for thermal image analysis and automatic ROI selections and tracking [50].

Conversely, the pre–post comparison approach utilizes the exercise as a stimulus, recording skin temperature by IRT before and after the cessation of the exercise. This is a common approach in the studies investigating the exercise-associated skin temperature changes by IRT. For example, a study by Priego Quesada et al. [46] investigated the relationship between neuromuscular activation (measured using surface electromyography) and skin temperature (measured using infrared thermography) during cycling exercise. In this study, three thermographic measurements were performed in different moments of the experiment: before the cycling test, immediately after the cessation of the cycling test, and 10 min after the cessation of the cycling test. This approach is usually followed by a type of analysis considering the skin temperature variation, as follows [17, 46]:

- ΔT: difference between temperature immediately after the cycling test and before, expressed in °C.
- ΔT_{10}: difference between temperature 10 min after the cycling test and before, expressed in °C.
- ΔT_{after}: difference between temperature 10 min after the cycling test and immediately after, expressed in °C.

All in all, among the variables influencing skin temperature modifications associated to exercise, exercise characteristics and experimental approach are probably the most relevant, and those that researchers should be aware when designing experiments.

7.3 Relationship of Other Thermoregulatory Variables with Skin Temperature During Exercise

Skin layer constitutes the interface between the human body and the external environment. It plays a fundamental role in thermoregulation processes, thus regulating heat transfer produced by metabolism between the core to the external thermal conditions of environment [51]. Modifications of skin temperature are primarily modulated by skin perfusion, which is a function of microvascular anatomy and vasoactive control of the autonomic nervous system [2, 5]. Under conditions of hyperthermic stress, skin blood flow can be regulated, thus increasing in the skin surface for transfers of heat from the body, and the sweat rate increases to dissipate the heat from the skin to the environment [2]. Conversely, under cold environments, cutaneous vasoconstriction is produced and the skin plays the role of insulating the body from external environment [2].

Infrared imaging provides a visual map of the skin temperature of the body surface. However, it is worth noticing that an infrared image cannot quantify measurements of skin blood flow. The infrared image provides a map of temperatures of the skin, which in turn is a function of blood perfusion to the surface area [52]. Nevertheless, thermal changes on the skin may also be influenced by external thermal stressors. Therefore, skin temperature is the result of a complex interaction between skin blood perfusion, environmental temperature, heat loss processes and biophysical characteristics [7, 8].

The process of repeated and/or sustained contraction of the muscles during an exercise bout produces heat, leading to an increase in core temperature [24, 53]. Therefore, during exercise, thermoregulatory reflexes are activated with the aim to reduce body temperature, and to maintain thermal homeostasis [2]. Cardiovascular system during periods of exercise aims to provide sufficient amounts of oxygen and nutrients to the active muscle [3]. To this purpose, as exercise begins, muscles involved in exercise rapidly require increased oxygen delivery. This need is mediated by the sympathetic stimulation of vessels in those areas where blood flow can be recruited (e.g., splanchnic and renal circulations), thus increasing blood flow in the exercising muscles. Meanwhile, thermoregulatory reflexes, aiming to reduce core temperature, lead to an increase in skin blood flow [2]. Therefore, thermoregulatory reflexes induced by the thermogenic effect of the exercise are in conflict with nonthermoregulatory cardiovascular responses to exercise [22]. For instance, during intense exercise and/or during exercise in warm environment, there is a competition between thermoregulatory and nonthermoregulatory reflexes. The first for increasing blood flow to the skin, the second for increasing blood flow to the active muscles. In general, these competing demands are resolved thanks to the increase in cardiac output [3, 23]. However, increasing cardiac output may not be enough to resolve deficits resulting from combined intense exercise and heat stress (also when the dehydration is present), thus possibly leading to cardiovascular collapse [3].

The sympathetic control of skin blood flow is unique as there are sympathetic vasoconstrictor fibers (similar to those present in skeletal muscles) and sympathetic active vasodilator fibers in the skin surface. During exercise, as body core temperature rises, there is an initial vasoconstriction response caused by the increased demand of blood flow to the active muscles [23]. However, once a specific body core temperature threshold is reached, skin blood flow begins to increase by the activation of the sympathetic active vasodilator response. Such an increase in skin blood flow during exercise aims to promote heat loss, because blood flow is the first mechanism of heat transfer from the inner part of the body to the skin. Therefore, an increase in skin blood flow is likely to accompany an increase in skin temperature when sweat evaporation is not present.

Acute thermoregulation adaptations to exercise have been extensively studied by monitoring skin blood flow [6, 54, 55]. However, exercise is associated with large hemodynamic responses involving a series of thermoregulatory processes. Since exercise causes heat production within the body and invokes cutaneous thermoregulatory processes, modifications of skin temperature associated to exercise can be monitored by infrared thermography. Recent studies have investigated the interaction of skin temperature measured by thermography with other variables influencing thermoregulation in response to exercise. Some studies will be presented and discussed briefly.

Priego Quesada and colleagues [17] investigated the relationship between core and skin temperature measured in different sites of the body. Fourteen cyclists performed 2 45-min cycling tests at 35 and 50% of peak power output on different days, with a constant cadence (i.e., 90 rpm). Core temperature was measured using an ingestible pill continuously throughout the test. Local skin temperature was recorded using infrared thermography before, immediately after, and 10 min after the end of the cycling tests. In general, from pre to post exercise skin temperature decreased in trunk regions, whereas increased in lower limb regions. Core and local skin temperatures showed moderate negative correlations for regions presenting the highest sweat rates over the body (such as trunk) whereas some positive correlations were observed in regions in which sweat production was low (lower limbs). These findings highlight the difficulty of finding a relationship between skin temperature and core temperature, probably due to the thermoregulatory system efficiency in the increase of the thermal gradient (i.e., the difference between core and skin temperature) [11, 21], together with the multifactorial nature of the skin temperature [8, 33].

In another paper, Formenti and colleagues [44] studied the skin temperature response to two types of resistance exercises modulating the amount of skin blood flow. The rationale was that low intensity resistance training with slow movement and tonic force generation has been shown to create blood flow restriction within muscles (and therefore an important muscle deoxygenation), thus maximizing the hypertrophic response even with low intensity. This condition of restricted muscle blood flow may affect skin blood flow, and in turn thermoregulation through the skin. Therefore, the authors investigated the effect of two speeds of exercise execution on skin temperature dynamics using infrared thermography. Thirteen active males performed randomly two sessions of squat exercise (normal speed, 1 s eccentric/1 s concentric phase, 1 s; slow speed, 5 s eccentric/5 s concentric phase, 5 s), using *50% of 1 maximal repetition. Thermal images of skin temperature of the muscles quadriceps were recorded before the exercise (to determine basal skin temperature) and for 480 s following the initiation of the exercise (to determine the nonsteady-state time course of skin temperature). In summary, this study showed that squat exercises performed at two different speeds (1 s vs. 5 s) until exhaustion determined different profiles of skin temperature of the muscles quadriceps. Although the two exercises determined comparable maximal excursions of skin temperature, the rate of change of skin temperature was strongly reduced during the 5 s exercise as compared to the 1 s exercise (Fig. 7.4). Since the 5 s exercise mimics a condition of blood flow restriction within the muscle, these data provided a detailed portrait of the skin temperature modifications associated with this condition, thus laying the basis for similar investigations using IRT coupled with Doppler flowmetry.

In the study by Merla et al. [45], the authors investigated the whole body anterior skin temperature modifications in well-trained subjects during an incremental running test until exhaustion. Fifteen males volunteers performed an incremental treadmill test until reaching their individual maximal heart rate. Skin temperature of total body was continuously monitored by infrared thermography. Skin temperature decreased as the subjects began the exercise, whereas a further skin temperature decrement occurred with the progress of the exercise. This was attributed to the continuous vasoconstrictor response, attributable to an increase in catecholamine

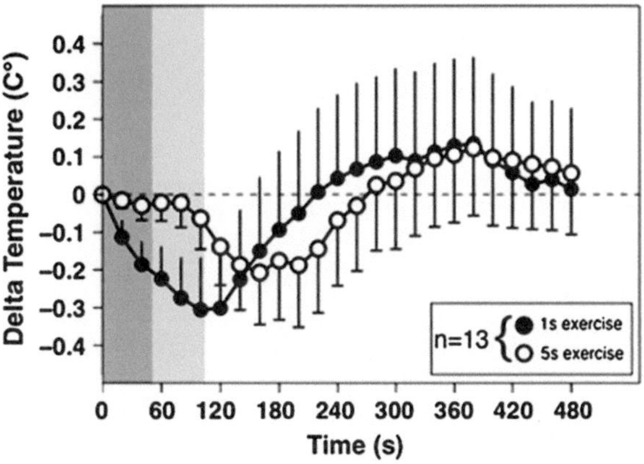

Fig. 7.4 Skin temperature variation from baseline recorded during 1 s exercise and 5 s exercise speed of a squat exercise. *Dark grey* and *light grey bar* represent the mean duration of 1 s exercise (55 s) and 5 s exercise (100 s), respectively. (Figure obtained from Formenti et al. [44])

and other vasoconstrictor hormones released as the exercise intensity increases [56, 57]. Thighs and forearms exhibited the earliest response. At the exercise interruption (i.e., at exhaustion), skin temperature values were 3–5 °C lower with respect to the baseline, and began to increase during recovery. Thighs and forearms exhibited the faster increase, followed by the total body skin temperature increase. Thermal images documented the presence of hyperthermal spots due to the presence of muscle perforator vessels during recovery (Fig. 7.5). The diffusion of heat from the hyperthermal spots to the surrounding cutaneous tissue suggests a possible hemodynamic and thermoregulatory role for the perforator vessels during/after exercise [58]. The findings of this study indicate that infrared thermography permits the quantitative evaluation of acute cutaneous whole body thermal adaptations occurring in response to an incremental exercise until exhaustion. It is such an incremental character of the exercise that caused a continuous vasoconstriction response, thus requiring blood flow to the active muscles from the other parts of the body. However, the continuous decrease in skin temperature observed during exercise should be attributed not only to the decrease in skin blood flow, but may also be related to a small amount of sweat delivery, even though it seems to play a secondary role [33].

All in all, the interpretation of the dynamic process of skin thermal regulation obtained from thermal imaging requires a basic understanding of physiological mechanisms of skin blood flow and factors that influence thermoregulation processes. Researchers have combined the use of infrared thermography with techniques providing more direct measurements of blood flow, to better understand the complexity of thermoregulation processes. For this purpose, a study by Merla et al. [59] proposed a noninvasive method to calculate blood flow by means of thermal infrared imaging and bio-heat transfer modeling. The method, able to provide high time-resolution series of cutaneous blood flow images, was tested against a standard laser Doppler imaging system, which is considered the gold standard for noninvasive assessment of skin blood flow, on both healthy subjects and patients suffering from systemic sclerosis [59]. However, to the best of our knowledge, no studies have combined the measure of skin temperature by infrared thermography with the

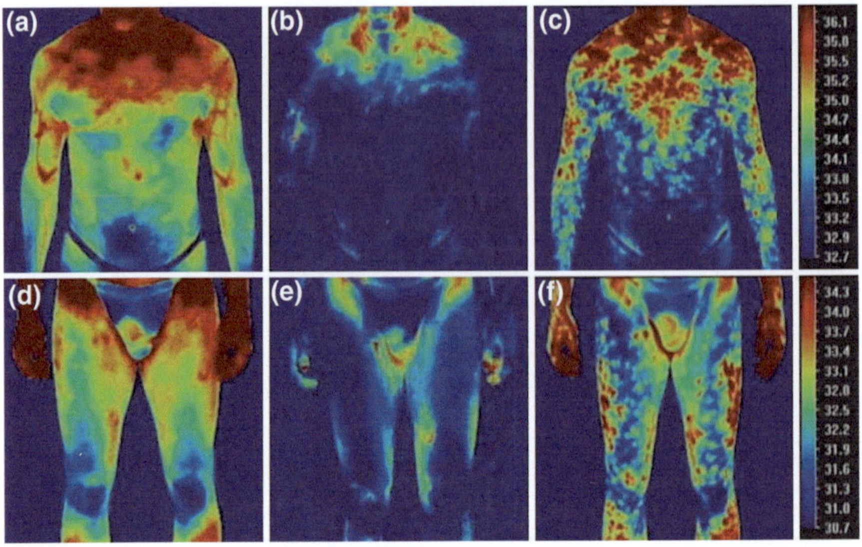

Fig. 7.5 Anterior view of the upper and lower body of a representative participant taken before running (**a**, **d**), immediately after reaching the age-predicted maximal heart rate value during incremental running test (**b**, **e**); and during recovery from exercise (**c**, **f**). (Figure obtained from Merla et al. [45])

measure of skin blood flow by laser Doppler during exercise. This should be the further step for advancing the knowledge about the thermoregulatory processes occurring at the skin level during different types of exercise and/or conditions. It should be noticed that the association of skin temperature with other thermoregulatory variables during a pyramidal load cycling exercise has been recently studied using a combined approach that assessed skin temperature of various body regions, regional sweat rate, and core temperature [18]. To allow for the simultaneous monitoring of the skin temperature changes in different body regions (e.g., locomotor muscles and facial regions), a multi-infrared cameras system was used. Each ROI exhibited a distinct pattern of skin temperature change: the nose tip temperature increased continuously, the forehead temperature decreased continuously, and the calves temperature followed exercise intensity. As core temperature and local sweat rate increased throughout the exercise, the relationship between these thermoregulatory parameters was strictly dependent on the specific ROI [18]. This study confirms that skin temperature change during exercise has its own behaviors depending on the ROI, highlighting the complexity of thermoregulation and the need for further investigation into its interactions with the other thermoregulatory and physiological variables [18].

7.4 Effect of the Exercise Characteristics on Skin Temperature Response

As it was mentioned in the previous sections of this chapter, skin temperature could present increases or decreases depending on the characteristics of the study and the ROI investigated. In order to understand all the factors that could affect the skin temperature behavior during exercise, first it is important to check the heat balance equation, i.e., an equation that is often used in the mathematical calculation of thermoregulation:

$$\text{Heat production} = \text{Heat loss, and more specifically :}$$
$$\mathrm{M} - \mathrm{W} = (\mathrm{K} + \mathrm{C} + \mathrm{R} + \mathrm{ESK}) + \mathrm{S}$$

where M is the rate of metabolic heat production, W is the rate of mechanical work, K is the rate of conductive heat loss, C is the rate of convective heat loss from the skin, R is the rate of radiative heat loss from the skin, ESK is the rate of evaporative heat loss from the skin, and S is the rate of body heat storage.

Higher values on one side of the equation could explain greater increases or reductions in skin temperature. Thus, high increases of skin temperature would be explained by high heat production and low capacity of heat loss, and high reductions by low heat production and high capacity of heat loss. Below is a summary of what are the factors that affect heat production and heat loss:

- Heat production (metabolic rate—mechanical work): metabolic cost of the exercise, duration and intensity.
- Heat loss: heat dissipation by conduction, convection, radiation, and evaporation, and external work. At the same time, these processes depend mainly on the following factors: environmental conditions (temperature, humidity, wind speed and radiation), clothing, body surface and sweat rate.

After sum up all the factors that can influence skin temperature, it is easy to understand why skin temperature does not always follow the same pattern. Other factors that could affect skin temperature (e.g., circadian rhythm) are explained in Chap. 4 of this book.

To develop this section, it was decided to conduct a review of the effect of exercise on skin temperature in the different studies, taking into account some of the above factors. The overview of the studies is shown in Table 7.1. In particular, the table aims to summarize the studies assessing both the magnitude and the direction of skin temperature changes in response to exercise with respect to basal pre-exercise condition. Ten studies were selected using the following criteria in order to present the last and most representative researches: (1) published after 2005, (2) more than ten participants, (3) ROIs excluding foot and hands for its particular physiology, (4) published in international peer-reviewed journals and written in English, and (5) avoid repetitive characteristics. The last criterion means that if there are some studies with different objectives but with similar experimental

Table 7.1 Overview of some of the most representative IRT studies in sport and exercise physiology

No.	Study	Participants characteristics	Environmental conditions	Exercise characteristics	Result ΔSkin temperature
Stable workload					
1	de Andrade Fernandes et al. [31]	12 males physically active	25 ± 1 °C, 62 ± 6%	Running. 12 blocks of 5 min each at 60% of the VO_{2max} separated by an interval of 1 min	Mean skin temperature: ~↓0.5 ° C Arm: Not changes Thigh: ~↑0.5 °C
2	Priego Quesada et al. [60]	29 males runners and 15 females runners	24 ± 1 °C, 49 ± 11%	Running, 10 min warm up and 20 min at 75% Maximal aerobic speed (aerobic high intensity)	Knee extensors: ~↑1 °C Knee: ~↑2 °C Anterior ankle: ~↑1.5 °C Achilles: ~↑3 °C Rest of lower limb ROIs: Not changes
3	Priego Quesada et al. [61]	19 males trained cyclists	24 ± 1 °C, 45 ± 12%	Cycling, 45 min at 50% Peak power	Knee extensors: ~↑1 °C Rest of body ROIs (trunk and lower limbs): Not changes
Incremental workload					
4.	Merla et al. [45]	15 males trained runners	23–24 °C, 50 ± 5%	Running, incremental test until exhaustion	Trunk: ~↓2 °C Forearm: ~↓4 °C Thigh: ~↓5 °C
5	Abate et al. [62]	20 trained runners and 20 sedentary males	23 ± 1 °C, 55 ± 10%	Cycling. Three steps: 5 min at 100 W, 5 min at 130 W and 5 min at 160 W	Trunk: ↓ ~ 0.9 °C trained Not changes untrained Upper limbs: ↓ ~ 0.8 °C trained Not changes untrained
6	Arfaoui et al. [63]	11 male cyclists	20 ± 1 °C	Cycling, incremental test during 18 min until 250 W	Gastrocnemius: ↓ ~ 1 °C
7	Priego Quesada et al. [46]	10 males physically active	20 ± 1 °C, 63 ± 3%	Cycling, incremental test until exhaustion	Knee extensors: ~↑1.5 °C Biceps femoris/ gastrocnemius medialis: Not changes

(continued)

Table 7.1 (continued)

No.	Study	Participants characteristics	Environmental conditions	Exercise characteristics	Result ΔSkin temperature
8	Hillen et al. [18]	17 healthy physically active	23–24 °C, 45 ± 4%	Pyramidal cycling load: Five stages at light, moderate, hard, moderate and light perceived exertion.	Forehead: Continuous decrease ~↓1.5 °C Eye: No changes Nose tip: Small increase until hard intensity, and higher increase after reduction of the intensity ~↑3.5 °C Calves: ~↓2.0 °C during intensity increase; ~↑2.0 °C during intensity decrease Forearm: ~↓1.5 °C during intensity increase; ~↑1.0 °C during intensity decrease
Training game session					
9	Chudecka and Lubkowska [26]	16 professional males handball players	20 °C training and 25 °C thermography room, 55%	Training session 90 min: Endurance training and handball game	Upper extremities: ~↓2.4 °C
	Shida et al. [64]	20 athletes with spinal cord injury: 10 with cervical spinal cord Injuries (CSCI); 10 with thoracolumbar spinal cord injuries (TSCI)	28–29 °C 60–70%	2 hours of wheelchair rugby (CSCI) or wheelchair basketball (TSCI)	Forehead: ~↑0.6–0.9 °C Upper limbs(CSCI): ~↑1.2 °C Upper limbs (TSCI): ~↑0.7 °C Lower limbs(CSCI): ~↑2.3 °C Lower limbs (TSCI): ~↑0.9 °C
Resistance exercise					
10	Ferreira et al. [42]	14 elderly and 15 young	22–24 °C, <50%	Resistance exercise. One limb 3 min knee extension and flexion with a 1 kg weight.	Posterior thigh exercised limb: Not changes Posterior thigh contralateral rest limb: ~↓0.5 °C
11	Formenti et al. [43]	7 trained and 7 untrained females	22–23 °C, 60 ± 5%	Resistance exercise with body weight, 2 min standing calf raise.	Calves: ~↑1.0 °C trained ~↑0.4 °C untrained

conditions, only one study was selected (i.e., the study with the higher number of participants). An example of this criterion is the studies performed by Priego Quesada and colleagues [17, 33, 61, 65] where the objectives are different but in all of them the characteristics of the exercise (45 min of moderate cycling) and environmental conditions are the same.After reviewing Table 7.1, it is possible to extract some ideas:

- The environmental temperature ranges from ~20 to ~25 °C, with relative humidity from 45% to 63%.
- Skin temperature response to exercise has been investigated following cycling or running exercises, sports game activities and resistance exercises. In cycling and running exercise, it is worth distinguishing between constant load sub-maximal exercise [17], and incremental maximal exercise until exhaustion [45, 46].
- The different studies suggested that there are differences in the skin temperature response between the ROIs analyzed. As previously mentioned, skin temperature in specific regions of the body is the result of a thermal balance between muscle activity, vasodilation and sweat evaporation [3, 27, 61]. Athletes who have altered physiological function (e.g., spinal cord injury population) could present different thermal responses [64].
- The studies that assessed stable load submaximal exercise [31, 60, 61] showed a skin temperature increment of active muscle. Although it is difficult to affirm that skin temperature of active muscles increases in constant exercise, it is possible to suggest that in this kind of exercise the heat production of this area is higher than no active muscles, showing larger increases or lower decreases of skin temperature during exercise.
- There are discrepancies in skin temperature response between the different studies. For example, in the study by Merla and colleagues [33], an incremental maximal exercise [45] induced a large decrease in skin temperature of active muscles during running, whereas incremental maximal cycling exercise tested by Priego-Quesada et al. [46] induced an increase. Differences between the two exercises could be related with the kind of exercise (running vs. cycling). However, it is important to remind that other factors could affect these differences: duration of the exercise, environmental differences (i.e. differences in relative humidity), differences between participants in fitness level, age, body composition, etc.
- Resistance exercise with body weight studied by Formenti et al. [43] induced an increase in skin temperature of the calf area. Conversely, resistance exercise with very low load (i.e., 1 kg) studied by Ferreira et al. [42] did not produce any modifications on active muscles. Resistance exercises may have a lower sweat production (in some cases nonexistent). This could result in a direct association between intensity of exercise and skin temperature response, contrary to what happens to other exercises (e.g., cycling and running) that result in a whole-body sweating. However, further studies are necessary to research this idea.

All in all, there is a large heterogeneity in the methodology used, that also depends on the purpose of the study. Differences can be found in the sample characteristics, exercise characteristics, as well as ROI considered. Due to this heterogeneity, it is generally difficult to find a common point among the studies.

7.5 Relationship Between Skin Temperature and Performance Variables

Most of the studies that have investigated the skin temperature modifications associated to exercise by IRT have focused on identifying possible relationships between skin temperature values and performance parameters. Among them, the performance variables most studied are undoubtedly cardiorespiratory parameters, such as maximal oxygen uptake (VO_{2max}) and heart rate.

The relationship of skin temperature changes with cardiorespiratory parameters associated to exercise has been studied both during team sport training [26], endurance cycling exercise [63], and resistance exercise [66]. Chudecka and Lubkowska [26] published a study aimed to investigate the temperature changes in arms and forearms in response to a 90-min physical exercise session. Furthermore, the authors investigated also the impact of physiological factors (such as VO_{2max} and heart rate) on the dynamics of temperature changes. Sixteen professional handball players performed a training session lasting 90 min, which contained elements of the actual game, such as sport-specific plays. The session was performed in a sports hall with air temperature of 20 °C and air humidity of 55%, i.e., thermal neutral environment. The skin temperatures of arms and forearms were registered before, immediately after and 10 min after the training session. They obtained a regression model with a significant correlation when the participants with a higher fitness level (by VO_{2max}), higher percentage of the maximum heart rate during the training season, and lower skin-fat fold on the arm, presented a higher decrease in skin temperatures after exercise. The reduction in the skin temperature after the training session was primarily attributed to the efficacy of sweat evaporation mechanism. The authors concluded stressing that players with higher fitness level (VO_{2max} and lower fat body composition) may be beneficial by better thermoregulatory response. Furthermore, a possible explanation regarding the higher decrease in skin temperature in players with a higher percentage of the maximum heart rate was that these players exercised with greater intensity, thus probably sweating more. This in turn implicates a more efficient heat elimination, and of consequence larger decrease in skin temperature after the end of the training session [26].

If Chudecka and Lubkowska [26] investigated the relationship between performance variables and skin temperature after a handball training session, another research group published two studies investigating such relationship in cycling exercise [63, 67]. In the more recent study, Arfaoui et al. [63] investigated the relationship of skin temperature of gastrocnemius with heart rate during an incremental cycling exercise. Eleven subjects performed an incremental cycling exercise on a cycle ergometer. The exercise began at a power output of 100 W (10 min), and then the intensity increased in increments of 50 W every 3 min, each up to 200 W, whereas the last load was performed at 250 W for 2 min. Skin temperature (measured by infrared thermography) and heart rate were recorded continuously during the experiment. Skin temperature values at each stage of the exercise were reported as variations with respect to basal values. Dynamics of skin temperature and heart rate during exercise are reported in Fig. 7.6. Immediately at the onset of the exercise, skin temperature of the calves drastically decreased. This rapid fall in skin temperature during initial stage of the exercise was not attributed to increased evaporative cooling but to vasoconstriction related to the muscular blood recruitment [6]. This phase was followed by a steady state period, thus meaning the presence of a thermal equilibrium between heat production and heat dissipation reflected on the skin. Prolonging exercise increased the metabolic heat production thus invoking thermoregulatory processes. In fact, an important decrease in the skin temperature occurred as exercise intensity increased (i.e., from 150 W at each load). Heart rate showed a quasi-perfect reverse dynamic with respect to skin temperature, thus proving the presence of a strong relationship between the two variables. This was also confirmed by the significant correlations found by the authors between heart rate and the skin temperature variations during the phase of 100 W (r = 0.9), and the incremental phase until 250 W (r = 0.99).

The relationship between skin temperature and heart rate was also investigated and assessed during resistance exercise. Neves et al. [66] studied the association between the skin temperature change and the heart rate response during two

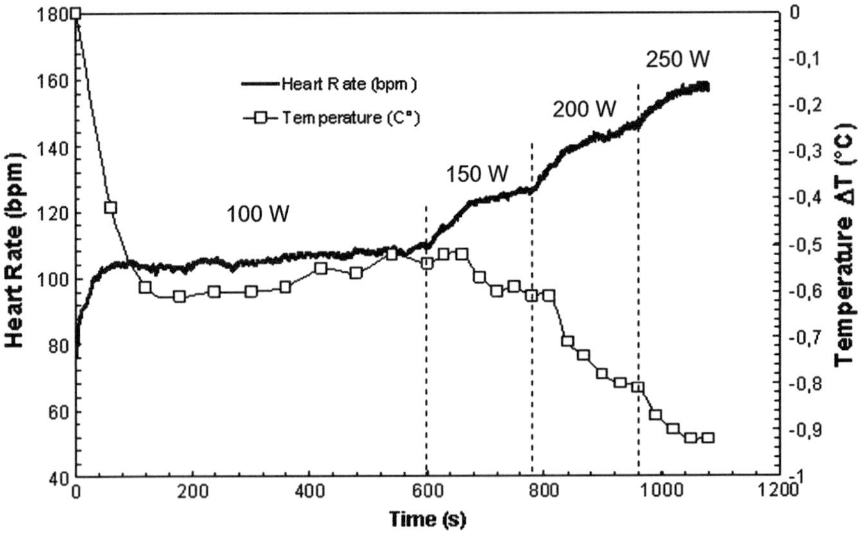

Fig. 7.6 Heart rate and skin average temperature gradient during incremental cycling test. (Figure modified from Arfaoui et al. [63])

sessions of resistance exercises at different intensities. A sample of 31 female subjects were divided into two groups: 15 subjects performing unilateral biceps curl, and 16 subjects back half squat at 90°. The exercise protocol of both biceps curl and squat was repeated in two randomized sessions (70% of 10 RM and 80% of 10 RM). The exercise protocol consisted in four sets with ten repetitions in each set, and with 30 s to rest between them. The heart rate and skin temperature in the two groups were recorded continuously before, during, and after exercise. Skin temperature on the active muscles decreased during exercise and during the first minutes of the recovery period, returning to basal values within1 hour of recovery after exercise. There were no statistical differences in thermal response to exercises in 70% and 85% of 10 RM in both biceps curl and squat group. Significant negative correlations between heart rate and skin temperature during exercise were found in 70% of 10 RM sessions only in the squat group, whereas in 85% of 10 RM session both in the biceps curl and squat group. Moreover, significant negative correlations were also found in the recovery in 70% and 85% of 10 RM, both in the biceps curl and squat group. The authors concluded stressing that skin temperature on the active muscles decreased during exercise and during the first minutes of the recovery, returning to basal values slowly. These negative associations between skin temperature and heart rate were also observed in the calve ROI during a pyramidal cycling load test [18].

Infrared thermography has also been used to assess the difference in thermoregulation between subjects with different level of physical fitness [43, 62, 68]. Based on the fact that higher fitness level is associated with higher skin blood flow during exercise [69], Merla et al. [68] conducted an experiment aimed to compare the skin temperature response to an incremental cycling exercise between trained and untrained subjects. After an initial trend that was similar in both groups, trained subjects exhibited a higher decreasing temperature rate with respect to the untrained ones. These findings seem to support the hypothesis that the muscular capability to recruit blood to supply its needs during an exercise is better in trained subjects than in untrained ones [68].

Further evidence in this direction has been more recently provided by the studies by Abate et al. [62] and by Formenti et al. [43]. In these studies, the authors investigated the influence of physical fitness on exercise-associated skin temperature changes. It was hypothesizing that exercise can induce differences in the trends of skin temperature between trained and untrained subjects. As regards to the study by Formenti and colleagues [43], seven trained and seven untrained female subjects performed standing calf rise exercise lasting 2 min, with a constant pace consisting in 1 s eccentric and 1 s concentric phase, with the aim of a metronome. Skin temperature of the calves was measured before, during, and for 7 min after the exercise. By comparing the parameters describing the skin temperature time course in two groups, it was found that trained subjects, after a slight decrease due to vasoconstriction recruiting blood flow to the active muscles, increased their skin temperature differently with respect to untrained subjects. Specifically, trained subjects responded to exercise more quickly than untrained controls.

Also, the study by Abate et al. [62] supports the findings by Formenti et al. [43]. In this study [62], the authors investigated the differences in the cutaneous temperature among trained and untrained subjects. A total of 20 trained and 20 untrained male subjects performed a standard cycling warm up, divided in three steps: (1) 0–5 min at 100 W; (2) 5–10 min at 130 W; and (3) 10–15 min at 160 W. Thermal images of thorax and upper limbs were recorded during the exercise. With respect to baseline, trained subjects exhibited a significant temperature decrement in the third step in both trunk and upper limbs, while no difference was observed in untrained subjects. In the comparison between groups, a statistically significant difference was observed in both regions of interest (i.e., trunk and upper limbs), in the second and in the third step. In conclusion, the findings of this study support the notion that the level of physical fitness improves the ability to rapidly activate thermoregulatory processes in response to exercise [43, 62].

Korman et al. [70] explored the relationship between skin temperature, lactate and ammonia during incremental running test until exhaustion. Initially, skin temperature decreases due to sympathetic vasoconstriction, which redistributes blood flow from the skin to active muscles to optimize oxygen and substrate delivery. However, as exercise progresses and metabolic heat production increases, skin temperature stabilizes and eventually rises during recovery due to vasodilation and enhanced peripheral blood flow for thermoregulation. Meanwhile, lactate and ammonia concentrations steadily increase with exercise intensity, with lactate reflecting the shift toward anaerobic metabolism and ammonia indicating accelerated purine nucleotide degradation. Interestingly, during recovery, a positive correlation emerges, as higher skin temperature coincides with enhanced blood perfusion and lactate clearance. Endurance athletes tend to exhibit lower skin temperature at exhaustion and a more gradual lactate accumulation compared to sprinters, likely due to their superior oxidative capacity and thermoregulatory efficiency. These findings highlight the complex interplay between thermoregulation and metabolic byproducts, emphasizing the role of body composition and aerobic fitness in modulating these responses.

Since surface electromyography permits to estimate the magnitude of electrical neuromuscular activity during exercise, the existence of a relationship between muscle activation during exercise and associated skin temperature modifications was also hypothesized [46, 71]. The study by Bartuzi was probably the first assessing this relationship [71]. An inverse relationship between biceps brachii activation and skin temperature during fatigue isometric contractions was observed. Based on these findings on isometric exercise, another group of research investigated the relationship between neuromuscular activation (measured using surface electromyography) and skin temperature of different body area (measured using infrared thermography) during an incremental exercise on a cycle ergometer [46]. Subjects performed an incremental cycling test until exhaustion on a cycle ergometer. The incremental cycling test started with initial workload of 50 W lasting 3 min, and it was followed by increments of 25 W/min until reaching the complete exhaustion. Pedaling cadence was controlled at 90 ± 3 rpm. Peak power output was identified as the workload of the last stage. Skin temperature of front and back surface of subjects' thighs was recorded in three different times during the experiment. Neuromuscular activation was monitored using surface EMG from the right and left rectus femoris, vastus lateralis, biceps femoris and gastrocnemius medialis throughout the duration of the test. No significant correlations between skin temperature and neuromuscular activation were found in rectus femoris, biceps femoris and gastrocnemius medialis. In fact, the relationship observed was in vastus lateralis, and it was more related with fitness level and not with the heat generated by the muscle. Participants with higher overall neuromuscular activation and lower frequency content in activation for vastus lateralis presented lower increments of skin temperature, or what is the same, better thermoregulatory response. This profile of neuromuscular activations was associated with better fitness level. More recently, the association between muscular activity, fatigue and skin temperature was also investigated by using machine learning approaches. Specifically, electromyography metrics were successfully estimated from thermal features, demonstrating that thermal imaging can deliver metrics of muscle activity and fatigue, indicating its potential for sports and clinical applications [72]. The association between skin temperature changes and fatigue during resistance training sessions was also studied monitoring facial regions, thus considering the central system contribution to the perception of fatigue. The study confirmed significant correlations between frequency-domain thermal features and fatigue perception, indicating the ability of infrared thermography in assessing exercise-induced fatigue. Importantly, this finding corroborated the hypothesis that facial temperature variations may reflect the emotional state related to fatigue, suggesting a link between temperature and fatigue perception [73]. The associations between skin temperature, muscle fatigue, and delayed onset muscle soreness (DOMS) were also investigated by Priego-Quesada and colleagues in response to concentric–eccentric bicep curls, revealing that skin temperature—despite an association with fatigue—is not a reliable predictor of DOMS [74].

Finally, it is important to remark the association between fat composition and skin temperature. Priego-Quesada et al. [46] observed that participants with larger thigh skinfolds presented lower increments of skin temperature in ROIs of the thigh ($p < 0.01$ and $r > -0.7$). These authors discussed two points about this association. Firstly, it is clear that body fat tissue had an insulation capacity resulting in impairment of the heat dissipation between the core and the skin [23, 26, 75]. Moreover, it is worth considering that fat tissue is related to the physical fitness [18]. Body composition is one of the factors that change with the training and this variable can be one of the factors that explain the differences between people with different fitness capacity. On the other hand, Chudecka and Lubkowska [26] selected this ROI in their study in relation with the fat tissue composition. They suggested that upper limbs are appropriate ROIs to analyze the effect of a training session for its lower adipose tissue and then less effective thermal insulation. These inverse associations between skin temperature and adipose tissue during exercise were also observed by a later study [76].

7.6 Conclusion

In this chapter, we presented a general overview on the literature about the study of the skin temperature response to exercise assessed by means of IRT and its relationship with other thermoregulatory variables, exercise characteristics and performance factors. Some conclusions could be extracted from the chapter:

- IRT presents some differences in the skin temperature calculation in comparison with the thermal contact sensors, and also some advantages and limitations. Generally, IRT presents lower values of skin temperature during exercise than thermal contact sensors. Furthermore, IRT could provide the measurement of large areas without interfere in the heat exchange of the subject. However, it is not possible to measure when the subjects are clothed and the sweat could interfere in the temperature calculation. These advantages and limitations are important to take into account before and after a study.
- The experimental approach (e.g., measure during exercise or before–after exercise) and the characteristics of the exercise will influence the skin temperature, and researchers should be aware in the design of the experiment.
- Skin temperature is the result of different thermoregulatory variables such as blood flow, sweat rate, and core temperature. It is important to know the role of all these variables in order to interpret correctly the skin temperature changes.
- Differences in skin temperature response to exercise between studies in the literature are due to heterogeneity in exercise characteristics, as well as region of interests considered.
- People with better physical fitness present a more effective thermoregulatory system or, what is the same, a greater capacity of heat dissipation. Because of this, different studies have observed relationships between some performance parameters and skin temperature.

References

1. Lim CL, Byrne C, Lee JK (2008) Human thermoregulation and measurement of body temperature in exercise and clinical settings. Ann Acad Med Singap 37:347–353
2. Charkoudian N (2016) Human thermoregulation from the autonomic perspective. Auton Neurosci Basic Clin 196:1–2. https://doi.org/10.1016/j.autneu.2016.02.007
3. González-Alonso J (2012) Human thermoregulation and the cardiovascular system. Exp Physiol 97:340–346. https://doi.org/10.1113/expphysiol.2011.058701
4. Nybo L (2010) Cycling in the heat: performance perspectives and cerebral challenges. Scand J Med Sci Sports 20(Suppl 3):71–79. https://doi.org/10.1111/j.1600-0838.2010.01211.x
5. Charkoudian N (2010) Mechanisms and modifiers of reflex induced cutaneous vasodilation and vasoconstriction in humans. J Appl Physiol Bethesda Md 1985(109):1221–1228. https://doi.org/10.1152/japplphysiol.00298.2010
6. Kenney WL, Johnson JM (1992) Control of skin blood flow during exercise. Med Sci Sports Exerc 24:303–312

7. Cramer MN, Jay O (2016) Biophysical aspects of human thermoregulation during heat stress. Auton Neurosci Basic Clin 196:3–13. https://doi.org/10.1016/j.autneu.2016.03.001
8. Fernández-Cuevas I, Bouzas Marins JC, Arnáiz Lastras J et al (2015) Classification of factors influencing the use of infrared thermography in humans: a review. Infrared Phys Technol 71:28–55. https://doi.org/10.1016/j.infrared.2015.02.007
9. González-Alonso J, Teller C, Andersen SL et al (1985) (1999) influence of body temperature on the development of fatigue during prolonged exercise in the heat. J Appl Physiol Bethesda Md 86:1032–1039
10. Nielsen B, Savard G, Richter EA et al (1990) Muscle blood flow and muscle metabolism during exercise and heat stress. J Appl Physiol 69:1040–1046
11. Sawka MN, Cheuvront SN, Kenefick RW (2012) High skin temperature and hypohydration impair aerobic performance. Exp Physiol 97:327–332. https://doi.org/10.1113/expphysiol.2011.061002
12. Racinais S, Havenith G, Aylwin P, Ihsan M, Taylor L, Adami PE, Adamuz M-C, Alhammoud M, Alonso JM, Bouscaren N, Buitrago S, Cardinale M, van Dyk N, Esh CJ, Gomez-Ezeiza J, Garrandes F, Holtzhausen L, Labidi M, Lange G et al (2022) Association between thermal responses, medical events, performance, heat acclimation and health status in male and female elite athletes during the 2019 Doha World Athletics Championships. Br J Sports Med 56(8):439–445. https://doi.org/10.1136/bjsports-2021-104569
13. Nielsen B, Hales JR, Strange S et al (1993) Human circulatory and thermoregulatory adaptations with heat acclimation and exercise in a hot, dry environment. J Physiol 460:467–485
14. Nybo L, Møller K, Volianitis S et al (2002) Effects of hyperthermia on cerebral blood flow and metabolism during prolonged exercise in humans. J Appl Physiol 93:58–64Nybo L, Nielsen B (2001) hyperthermia and central fatigue during prolonged exercise in humans. J Appl Physiol 91:1055–1060
15. Galloway SD, Maughan RJ (1997) Effects of ambient temperature on the capacity to perform prolonged cycle exercise in man. Med Sci Sports Exerc 29:1240–1249
16. Priego Quesada JI, Martínez N, Salvador Palmer R et al (2016a) Effects of the cycling workload on core and local skin temperatures. Exp Thermal Fluid Sci 77:91–99. https://doi.org/10.1016/j.expthermflusci.2016.04.008
17. Fujii N, Honda Y, Komura K et al (2014) Effect of voluntary hypocapnic hyperventilation on the relationship between core temperature and heat loss responses in exercising humans. J Appl Physiol 117:1317–1324. https://doi.org/10.1152/japplphysiol.00334.2014
18. Hillen B, Andrés López D, Marzano-Felisatti JM, Sanchez-Jimenez JL, Cibrián Ortiz de Anda RM, Nägele M, Salvador-Palmer MR, Pérez-Soriano P, Schömer E, Simon P, Priego-Quesada JI (2023a) Acute physiological responses to a pyramidal exercise protocol and the associations with skin temperature variation in different body areas. J Therm Biol 115:103605. https://doi.org/10.1016/j.jtherbio.2023.103605
19. Cheuvront SN, Kenefick RW, Montain SJ, Sawka MN (2010) Mechanisms of aerobic performance impairment with heat stress and dehydration. J Appl Physiol 109:1989–1995. https://doi.org/10.1152/japplphysiol.00367.2010
20. Cuddy JS, Hailes WS, Ruby BC (2014) A reduced core to skin temperature gradient, not a critical core temperature, affects aerobic capacity in the heat. J Therm Biol 43:7–12. https://doi.org/10.1016/j.jtherbio.2014.04.002
21. Torii M, Yamasaki M, Sasaki T, Nakayama H (1992) Fall in skin temperature of exercising man. Br J Sports Med 26:29–32
22. Kenny GP, Reardon FD, Zaleski W et al (2003) Muscle temperature transients before, during, and after exercise measured using an intramuscular multisensor probe. J Appl Physiol 94:2350–2357. https://doi.org/10.1152/japplphysiol.01107.2002
23. Chou T-H, Coyle EF (2023) Cardiovascular responses to hot skin at rest and during exercise. Temperature 10(3):326–357. https://doi.org/10.1080/23328940.2022.2109931
24. Saltin B, Gagge AP, Stolwijk JA (1970) Body temperatures and sweating during thermal transients caused by exercise. J Appl Physiol 28:318–327

25. Chudecka M, Lubkowska A (2010) Temperature changes of selected body's surfaces of handball players in the course of training estimated by thermovision, and the study of the impact of physiological and morphological factors on the skin temperature. J Therm Biol 35:379–385
26. Smith CJ, Havenith G (2011) Body mapping of sweating patterns in male athletes in mild exercise-induced hyperthermia. Eur J Appl Physiol 111:1391–1404. https://doi.org/10.1007/s00421-010-1744-8
27. Childs PRN (2001) Practical temperature measurement. Butterworth-Heinemann
28. Smith ADH, Crabtree DR, Bilzon JLJ, Walsh NP (2010) The validity of wireless iButtons and thermistors for human skin temperature measurement. Physiol Meas 31:95–114. https://doi.org/10.1088/0967-3334/31/1/007
29. James CA, Richardson AJ, Watt PW, Maxwell NS (2014) Reliability and validity of skin temperature measurement by telemetry thermistors and a thermal camera during exercise in the heat. J Therm Biol 45:141–149. https://doi.org/10.1016/j.jtherbio.2014.08.010
30. MacRae BA, Annaheim S, Spengler CM, Rossi RM (2018) Skin temperature measurement using contact thermometry: a systematic review of setup variables and their effects on measured values. Front Physiol 9. https://doi.org/10.3389/fphys.2018.00029
31. Niedermann R, Wyss E, Annaheim S et al (2013) Prediction of human core body temperature using non-invasive measurement methods. Int J Biometeorol 58:7–15. https://doi.org/10.1007/s00484-013-0687-2
32. Priego Quesada JI, Martínez Guillamón N, Cibrián O, de Anda RM et al (2015a) Effect of perspiration on skin temperature measurements by infrared thermography and contact thermometry during aerobic cycling. Infrared Phys Technol 72:68–76. https://doi.org/10.1016/j.infrared.2015.07.008
33. Buono MJ, Jechort A, Marques R et al (2007) Comparison of infrared versus contact thermometry for measuring skin temperature during exercise in the heat. Physiol Meas 28:855–859. https://doi.org/10.1088/0967-3334/28/8/008
34. Psikuta A, Niedermann R, Rossi RM (2013) Effect of ambient temperature and attachment method on surface temperature measurements. Int J Biometeorol:1–9
35. Tyler CJ (2011) The effect of skin thermistor fixation method on weighted mean skin temperature. Physiol Meas 32:1541–1547. doi:10.1088/0967-3334/32/10/003. Ramanathan NL (1964) A new weighting system for mean surface temperature of the human body. J Appl Physiol 19:531–533
36. Ammer K (2009) Does neuromuscular thermography record nothing else but an infrared sympathetic skin response? Thermol Int 19:107–108
37. de Andrade FA, dos Santos Amorim PR, Brito CJ et al (2014) Measuring skin temperature before, during and after exercise: a comparison of thermocouples and infrared thermography. Physiol Meas 35:189
38. Priego-Quesada JI, Machado AS, Gil-Calvo M, Jimenez-Perez I, Cibrian Ortiz de Anda RM, Salvador Palmer R, Perez-Soriano P (2020b) A methodology to assess the effect of sweat on infrared thermography data after running: preliminary study. Infrared Phys Technol 109:103382. https://doi.org/10.1016/j.infrared.2020.103382
39. Balci GA, Basaran T, Colakoglu M (2016) Analysing visual pattern of skin temperature during submaximal and maximal exercises. Infrared Phys Technol 74:57–62. https://doi.org/10.1016/j.infrared.2015.12.002
40. Ferreira JJA, Mendonça LCS, Nunes LAO et al (2008) Exercise-associated thermographic changes in young and elderly subjects. Ann Biomed Eng 36:1420–1427. https://doi.org/10.1007/s10439-008-9512-1
41. Formenti D, Ludwig N, Gargano M et al (2013) Thermal imaging of exercise-associated skin temperature changes in trained and untrained female subjects. Ann Biomed Eng 41:863–871. https://doi.org/10.1007/s10439-012-0718-x
42. Formenti D, Ludwig N, Trecroci A et al (2016) Dynamics of thermographic skin temperature response during squat exercise at two different speeds. J Therm Biol 59:58–63. https://doi.org/10.1016/j.jtherbio.2016.04.013

43. Merla A, Mattei PA, Di Donato L, Romani GL (2010) Thermal imaging of cutaneous temperature modifications in runners during graded exercise. Ann Biomed Eng 38:158–163. https://doi.org/10.1007/s10439-009-9809-8
44. Priego Quesada JI, Carpes FP, Bini RR et al (2015b) Relationship between skin temperature and muscle activation during incremental cycle exercise. J Therm Biol 48:28–35. https://doi.org/10.1016/j.jtherbio.2014.12.005
45. Zontak A, Sideman S, Verbitsky O, Beyar R (1998) Dynamic thermography: analysis of hand temperature during exercise. Ann Biomed Eng 26:988–993
46. Kraemer WJ, Ratamess NA (2004) Fundamentals of resistance training: progression and exercise prescription. Med Sci Sports Exerc 36:674–688. https://doi.org/10.1249/01.MSS.0000121945.36635.61
47. Kenney WL, Wilmore JH, Costill DL (2012) Physiology of sport and exercise. Human Kinetics
48. Schlager O, Gschwandtner ME, Herberg K et al (2010) Correlation of infrared thermography and skin perfusion in Raynaud patients and in healthy controls. Microvasc Res 80:54–57. https://doi.org/10.1016/j.mvr.2010.01.010
49. Hillen B, Andrés López D, Pfirrmann D, Neuberger EW, Mertinat K, Nägele M, Schömer E, Simon P (2023b) An exploratory, intra- and interindividual comparison of the deep neural network automatically measured calf surface radiation temperature during cardiopulmonary running and cycling exercise testing: a preliminary study. J Therm Biol 113:103498. https://doi.org/10.1016/j.jtherbio.2023.103498
50. Perpetuini D, Formenti D, Cardone D, Filippini C, Merla A (2021) Regions of interest selection and thermal imaging data analysis in sports and exercise science: a narrative review. Physiol Meas 42(8):08TR01. https://doi.org/10.1088/1361-6579/ac0fbd
51. González-Alonso J, Calbet JA, Nielsen B (1999) Metabolic and thermodynamic responses to dehydration-induced reductions in muscle blood flow in exercising humans. J Physiol 520(Pt 2):577–589
52. Johnson JM (1985) Kellogg DL (2010) local thermal control of the human cutaneous circulation. J Appl Physiol Bethesda Md 109:1229–1238. https://doi.org/10.1152/japplphysiol.00407.2010
53. Simmons GH, Wong BJ, Holowatz LA, Kenney WL (2011) Changes in the control of skin blood flow with exercise training: where do cutaneous vascular adaptations fit in? Exp Physiol 96:822–828. https://doi.org/10.1113/expphysiol.2010.056176
54. Brengelmann GL, Johnson JM, Hermansen L, Rowell LB (1977) Altered control of skin blood flow during exercise at high internal temperatures. J Appl Physiol 43:790–794
55. Vainer BG (2005) FPA-based infrared thermography as applied to the study of cutaneous perspiration and stimulated vascular response in humans. Phys Med Biol 50:R63. https://doi.org/10.1088/0031-9155/50/23/R01
56. Taylor GI, Gianoutsos MP, Morris SF (1994) The neurovascular territories of the skin and muscles: anatomic study and clinical implications. Plast Reconstr Surg 94:1–36
57. Merla A, Di Donato L, Romani GL et al (2008) Comparison of thermal infrared and laser doppler imaging in the assessment of cutaneous tissue perfusion in scleroderma patients and healthy controls. Int J Immunopathol Pharmacol 21:679–686
58. Priego Quesada JI, Carpes FP, Salvador Palmer R et al (2016b) Effect of saddle height on skin temperature measured in different days of cycling. Springerplus 5:205–214. https://doi.org/10.1186/s40064-016-1843-z
59. Priego Quesada JI, Lucas-Cuevas AG, Salvador Palmer R et al (2016c) Definition of the thermographic regions of interest in cycling by using a factor analysis. Infrared Phys Technol 75:180–186. https://doi.org/10.1016/j.infrared.2016.01.014
60. Arfaoui A, Bertucci WM, Letellier T, Polidori G (2014) Thermoregulation during incremental exercise in masters cycling. J Sci Cycl 3:33–41
61. Abate M, Di Carlo L, Di Donato L et al (2013) Comparison of cutaneous termic response to a standardised warm up in trained and untrained individuals. J Sports Med Phys Fitness 53:209–215

62. Neves EB, Cunha RM, Rosa C et al (2016) Correlation between skin temperature and heart rate during exercise and recovery, and the influence of body position in these variables in untrained women. Infrared Phys Technol 75:70–76. https://doi.org/10.1016/j.infrared.2015.12.018
63. Bertucci W, Arfaoui A, Janson L, Polidori G (2013) Relationship between the gross efficiency and muscular skin temperature of lower limb in cycling: a preliminary study. Comput Methods Biomech Biomed Engin 16(Suppl 1):114–115. https://doi.org/10.1080/10255842.2013.815902
64. Shida N, Furukawa Y, Nitta O (2018) Skin temperature responses in a hot environment among wheelchair Rugby and Basketball players with spinal cord injury. Int J Phys Med Rehab 06(04). https://doi.org/10.4172/2329-9096.1000478
65. Priego Quesada JI, Lucas-Cuevas AG, Gil-Calvo M et al (2015c) Effects of graduated compression stockings on skin temperature after running. J Therm Biol 52:130–136. https://doi.org/10.1016/j.jtherbio.2015.06.005
66. Merla A, Iodice P, Tangherlini A et al (2005) Monitoring skin temperature in trained and untrained subjects throughout thermal video. Conf Proc Annu Int Conf IEEE Eng Med Biol Soc IEEE Eng Med Biol Soc Annu Conf 2:1684–1686. https://doi.org/10.1109/IEMBS.2005.1616767
67. Fritzsche RG, Coyle EF (2000) Cutaneous blood flow during exercise is higher in endurance-trained humans. J Appl Physiol 88:738–744
68. Bartuzi P, Roman-Liu D, Wiśniewski T (2012) The influence of fatigue on muscle temperature. Int J Occup Saf Ergon JOSE 18:233–243
69. Chudecka M, Lubkowska A, Kempińska-Podhorodecka A (2014) Body surface temperature distribution in relation to body composition in obese women. J Therm Biol 43:1–6. https://doi.org/10.1016/j.jtherbio.2014.03.001
70. Korman P, Kusy K, Straburzyńska-Lupa A, Kantanista A, Quintana MS, Zieliński J (2024a) Response of skin temperature, blood ammonia and lactate during incremental exercise until exhaustion in elite athletes. Sci Rep 14(1):2237. https://doi.org/10.1038/s41598-024-52374-z
71. Savastano DM, Gorbach AM, Eden HS et al (2009) Adiposity and human regional body temperature. Am J Clin Nutr 90:1124–1131. https://doi.org/10.3945/ajcn.2009.27567
72. Perpetuini D, Formenti D, Cardone D, Trecroci A, Rossi A, Di Credico A, Merati G, Alberti G, Di Baldassarre A, Merla A (2023) Can data-driven supervised machine learning approaches applied to infrared thermal imaging data estimate muscular activity and fatigue? Sensors 23(2):2. https://doi.org/10.3390/s23020832
73. Perpetuini D, Formenti D, Iodice P, Cardone D, Filippini C, Chiarelli AM, Michielon G, Trecroci A, Alberti G, Merla A (2022) Central and peripheral thermal signatures of brain-derived fatigue during unilateral resistance exercise: a preliminary study. Biology 11(2):2. https://doi.org/10.3390/biology11020322
74. Priego-Quesada JI, De la Fuente C, Kunzler MR, Perez-Soriano P, Hervás-Marín D, Carpes FP (2020a) Relationship between skin temperature, electrical manifestations of muscle fatigue, and exercise-induced delayed onset muscle soreness for dynamic contractions: a preliminary study. Int J Environ Res Public Health 17(18). https://doi.org/10.3390/ijerph17186817
75. Johnson W, de Ruiter I, Kyvik KO et al (2014) Genetic and environmental transactions underlying the association between physical fitness/physical exercise and body composition. Behav Genet. https://doi.org/10.1007/s10519-014-9690-6
76. Korman P, Kusy K, Straburzyńska-Lupa A, Sillero-Quintana M, Zarębska E, Zieliński J (2024b) Exploring the correlation of skin temperature and body composition in athletes undergoing exhaustive physical exercise. J Therm Biol 123:103918

Chapter 8
Effects of Sports Disciplines on Skin Temperature Asymmetries

Joaquín Martín Marzano-Felisatti, Alberto Hermo Argibay, Roberto Sanchis-Sanchis, and Alberto Encarnación-Martínez

Abstract This chapter analyses thermoregulatory responses in relation to different sports disciplines, focusing on cyclic sports (such as running, cycling, and rowing) and non-cyclic sports (such as racket sports and handball). The main objective is to determine whether thermoregulatory asymmetries exist between body segments and to discuss methodological limitations found in current studies. Findings suggest that in cyclic sports, there are no evident thermoregulatory asymmetries. The thermal response varies throughout the performance, initially involving vasoconstriction, followed by vasodilation as exercise progresses. Additionally, better-trained athletes (such as elite runners and cyclists) exhibit more efficient thermoregulatory strategies compared to less-trained individuals. However, several factors, such as body fat percentage, may influence thermographic analysis and limit the accuracy of these studies. Therefore, further research is needed to assess long-term responses and the potential existence of asymmetries. In contrast, non-cyclic sports show an asymmetric thermoregulatory pattern, even in non-injured athletes. This asymmetry appears to be sport-specific, as the segments most involved in the activity require greater thermoregulation.

J. M. Marzano-Felisatti · R. Sanchis-Sanchis · A. Encarnación-Martínez
Research Group in Sports Biomechanics (GIBD), Department of Physical Education and Sports, University of Valencia, Valencia, Spain
e-mail: joaquin.marzano@uv.es; roberto.sanchis@uv.es; alberto.encarnacion@uv.es

A. H. Argibay (✉)
Foundation for the Promotion of Health and Biomedical Research in the Valencian Region (FISABIO), Department of Endocrinology and Nutrition, University Hospital Doctor Peset, Valencia, Spain
e-mail: alberto.hermo@fisabio.es

J. I. Priego-Quesada (ed.), *Application of Infrared Thermography in Sports Science*,
https://doi.org/10.1007/978-3-031-93311-0_8

8.1 Introduction: Why Is It Interesting to Analyse Skin Temperature Asymmetries Related to Sports Techniques?

Skin temperature (Tsk) provides valuable information about the physiological processes taking place in the athlete during sports practice [1]. Physical activity induces complex thermoregulatory mechanisms to dissipate excess heat generated by movement, primarily through peripheral circulation, heat transfer (conduction, convection and radiation) and sweat evaporation [2, 3]. Moreover, environmental conditions and individual factors such as body mass composition and the characteristics of the region of interest (ROI) influence Tsk behaviour [2, 3]. Consequently, Tsk reflects a balance between muscle activity, vasodilation, and sweat evaporation, offering a non-invasive way to monitor the body's thermal responses and hemodynamic changes during and after sports practice [2–4].

Healthy humans are thermally symmetrical, meaning that under normal healthy conditions, no contralateral Tsk asymmetries should be present [5]. Deviations from this symmetry may indicate localised physiological processes, such as inflammation, muscle fatigue, or adaptation to mechanical load [6]. This makes the analysis of contralateral asymmetries highly relevant for both clinical and sports science. In this sense, there are sport disciplines where the technique is predominantly asymmetrical (handball, tennis, archery, etc.) and thermal asymmetries can be expected and other sports where laterality is not so evident and therefore, Tsk asymmetries cannot be related with the technique execution.

Traditionally, the study of Tsk asymmetries in sports science has focused on injury prevention [7]. Specifically, baseline asymmetries exceeding 0.5 °C in healthy athletes have been associated with an increased injury risk [5]. However, recent research suggests that thermal asymmetries above this threshold may also reflect acute or chronic adaptations to specific sports' technical and physical demands [8–10].

In this context, thermal asymmetries exceeding 0.5 °C have been observed before and after sports practice [9–11]. Findings suggest that these asymmetries are not necessarily pathological but may be sport-specific thermal patterns related to movement efficiency, load distribution, and repetitive use of specific muscle groups as shown in tennis, padel, Valencian handball and archery [9–12]. Such findings highlight the potential of infrared thermography (IRT) application beyond injury prevention, offering valuable insights into acute and chronic adaptations to sports techniques and constant pattern repetition.

In recent years, numerous studies have investigated this topic by applying IRT before, during, and after sports practice in different disciplines. However, findings in this field remain heterogeneous [8]. While some sports exhibited Tsk asymmetries related to technical execution [8–12], others did not reflect this trend [8, 13, 14]. This results variability underscores the idea that each sport's specific technical, physical, and contextual demands shape the behaviour of Tsk asymmetries in unique ways. Consequently, in our aim to deepen the literature that has addressed Tsk asymmetries concerning sports technique, the classification of cyclic and non-cyclic sports was followed in connection to the technical pattern execution required in each sport.

Therefore, studying contralateral Tsk asymmetries associated with sports techniques offers crucial insights into variations in muscle activation patterns, peripheral blood flow, and metabolic activity between limbs or body regions [8]. Understanding these variations is essential for interpreting athletes' chronic and acute adaptations in response to their training and competitive contexts [8, 9, 12]. Furthermore, this knowledge has the potential to inform strategies for optimising sports techniques and enhancing performance [8].

For all the above reasons, the primary aim of this chapter is to explore the scientific evidence regarding contralateral Tsk asymmetries associated with sports techniques. To do so, firstly, the scientific literature will be summarised, explaining how Tsk behaves in sports practice and what influences this behaviour. Secondly, results obtained from cyclic and non-cyclic sports research will be addressed in two separate sections focusing on Tsk asymmetries. And finally, practical applications and future research lines will be indicated to conclude with the main conclusions.

8.2 Background: Effects of Sports on Skin Temperature Responses

As discussed in the introduction and previous chapters, the application of IRT for the analysis of Tsk has gained significant interest in sports science. The diversity of published studies with different approaches and research proposals highlights the complexity of Tsk behaviour in sports. In this sense, the effects of sport on Tsk should be approached as a complex phenomenon where physiological (thermoregulation, peripheral circulation, sweating, etc.), environmental (temperature, humidity, wind, etc.), individual (body composition, age, sex, fitness level, clothing, etc.) and contextual (exercise intensity, type of effort, movement characteristics, type of contraction, sports discipline, etc.) factors interact [2, 15, 16]. Therefore, in this chapter section, the effects of sport discipline on Tsk's behaviour will be addressed, assuming its complexity and inter-factors dependency.

To understand the physiological aspects of thermoregulation during sports activities, it is important to consider how the body maintains its internal temperature. The human body naturally strives to maintain internal thermal homeostasis, while Tsk fluctuates in response to physical demands and environmental conditions [1]. During sports practice, athletes' movement requires energy production, which leads to the generation of a significant amount of heat, increasing body temperature [2, 3, 16]. This increase in temperature is detected by the central nervous system, specifically the hypothalamus, which activates thermoregulatory mechanisms to maintain internal thermal homeostasis [4, 16, 17]. These mechanisms include peripheral vasoconstriction or vasodilation, along with the activation of sweating according to the athlete's needs, allowing the body to manage heat dissipation through conduction, convection, and radiation affecting Tsk [2, 4].

Going deeper into these concepts, unifying the Tsk behaviour in relation to sports practice is complex. Considering the type of physical effort, in endurance sports, Tsk tends to decrease at the beginning of the activity due to the energy and cardiorespiratory demands, producing peripheral vasoconstriction [18–21]. If the effort is continuous, the internal temperature tends to rise after reaching a stabilisation point or plateau, where it produces the opposite effect [19, 21, 22]. This causes peripheral vasodilatation to favour heat dissipation and sweating increasing [2, 4, 16]. Furthermore, if the exercise is performed incrementally until exhaustion, Tsk has been shown to decrease constantly throughout the whole activity, likely due to metabolic and circulatory adjustments [19, 23, 24]. What is clear is that once the activity is over, either in constant or incremental endurance effort, Tsk return to basal values, reflecting thermal recovery processes [19, 21, 23, 25].

Studies results can vary significantly in localised strength exercises or in sports involving contralateral asymmetrical technical action. While some research showed an increase in Tsk in the exercised ROI [9, 12, 26–28], other studies reported no differences or Tsk decreases [19, 29–31]. These discrepancies among study results can be attributed to various factors, including differences in the research protocols, the muscle groups involved, the type of contraction, the actions performed, and individual, environmental, and contextual variables [16, 19, 32]. These factors make it difficult to generalise Tsk's responses across studies.

Another important aspect to be considered when discussing Tsk responses is ROI location and the measurement moments [16, 32]. Distal ROIs tend to have more significant Tsk alterations, while proximal ones show more Tsk stability [1, 6]. Studies have reported higher Tsk in the thighs than distal areas, justifying this difference by their proximity to the core area, where the main organs are located and most heat is produced [1, 6, 33].

In relation to the moments of measurement, the proposals vary depending on the study aim, finding continuous or specific moments analysis [16, 19]. These moments can be before (basal), during (between series) or after the sports practice (POST, POST10 min, POST15 min, etc.), finding differences in Tsk responses [16]. A tendency towards thermal equilibrium and contralateral symmetry in pre-exercise measurements is shown, except in the presence of injury, overload or chronic adaptations to sports practice [5, 7]. In post-exercise, the tendency is to seek thermal homeostasis by increasing or decreasing Tsk, depending on the final versus initial Tsk comparison [1, 19–21].

Athletes' characteristics also influence Tsk [32]. Key factors such as fitness level, anthropometric measurements, and body composition play a crucial role in how Tsk behaves during exercise [19, 32]. Well-trained athletes exhibit a lower increase in internal temperature compared to non-trained athletes at the same exercise intensity due to more efficient thermoregulatory responses [19, 25, 28, 34, 35]. Furthermore, variations in baseline Tsk among athletes are often linked to differences in anthropometric characteristics, such as body mass index and body fat distribution [2, 32]. In this sense, adipose tissue acts as a thermal insulator, which reduces heat transfer from the body's core to the skin surface, affecting Tsk responses [2, 32, 36, 37]. Moreover, clothes and materials used by athletes significantly impact Tsk, as they can either facilitate or hinder heat dissipation through convection and evaporation mechanisms [38].

Finally, environmental factors, such as temperature, humidity, and wind, are critical in shaping Tsk responses during sports practice [2, 3, 24]. At high temperatures, the body increases sweat rate to enhance evaporative cooling, although Tsk during exercise remains relatively consistent [1, 24]. Under hyperthermic stress, increased skin blood flow facilitates heat dissipation by enhancing heat transfer from the body core to the external environment [2, 4, 16]. Conversely, in cold environments, cutaneous vasoconstriction occurs to preserve core temperature by reducing peripheral heat loss, and the skin acts as an insulator against external thermal stress [4]. The dynamic behaviour of Tsk throughout the duration of exercise reflects the body's adaptive mechanisms to balance internal heat production and external heat exchange [2, 4, 16]. This interaction underscores the importance of considering environmental conditions when analysing Tsk responses, as they directly influence thermoregulatory processes [2, 32]. Such insights are essential for optimising performance and ensuring safety, particularly in extreme environmental conditions where thermoregulation is challenged [2, 16].

In conclusion, Tsk's behaviour during sports practice is highly complex and influenced by multiple factors. The interaction between physiological thermoregulatory mechanisms, individual athlete characteristics, and environmental conditions, along with intrinsic aspects of the sports practice, such as exercise intensity, energy demands, and work-rest ratios, determines Tsk's response. While general patterns can be identified, individual and contextual variations make it challenging to generalise findings, highlighting the need for tailored approaches in interpreting Tsk data. The following sections will examine the behaviour of contralateral Tsk asymmetries in different sports disciplines, focusing on both cyclic and non-cyclic sports to assess the body's adaptation to symmetric and asymmetric movement patterns. However, before these sections, it is necessary to take into account some methodological considerations.

8.3 Methodological Considerations

Before delving into Tsk asymmetries related to athletes' technique in cyclic and non-cycling sports, it is important to highlight some specific methodological considerations that will aid in understanding the following sections. When analysing Tsk asymmetries related to sports techniques, it is essential to determine when the IRT images must be taken, which body ROIs will be analysed with their contralateral classification, and which final variable will be calculated for asymmetry determination.

Concerning the IRT moments, images are commonly taken at basal situations before training (PRE), at the end of training (POST), and ten minutes after the end of training (POST10) [1, 8, 10, 12]. Knowing and comparing the Tsk at these stages can determine whether the asymmetries are due to acute adaptations caused by the sport technique, excluding possible thermal asymmetry before training or their maintenance after it. Some studies have extended POST IRT images to 15 minutes, 1 hour or 24 hours after sports practice, but with a different aim than measuring acute response to athletes' technique [39, 40].

To continue with the ROIs selection and classification, ROIs are usually determined considering the study aims and the sports' characteristics [32]. The whole body [8, 9, 12, 41], the upper limbs and the trunk [1, 10, 26], the lower limbs [42] or a specific ROI highly involved in a sport modality, as the hand in Valencian Handball can be analysed [11]. For contralateral comparison, ROI classification must be determined. Usually, dominant and non-dominant classification is the widely used and most recommended option [9, 11, 12], but right and left [26], lead and rear [41], or bow and drawing [10], among others, can be found in the literature. This hinders an overall comparison between studies but, at the same time, highlights the need for term adaptation to the sports' technical characteristics.

Finally, the difference and the Tsk variation, considering the moment when the images were taken (PRE, POST, POST10) and the ROIs classification (e.g., dominant vs non-dominant), are the main variables calculated for contralateral asymmetry comparison [8, 10–12]. Differences (e.g., dominant vs non-dominant) are calculated at the POST moment to determine which side has more temperature and if this difference is greater than the normal symmetrical variance of 0.5 °C [5]. Tsk variation is calculated based on the moment when the images were captured (e.g. POST-PRE), and the contralateral comparison allows us to know which side (e.g. dominant vs non-dominant) presented more significant variation (magnitude and direction of the variation) after the spot practice [9, 11, 12].

8.4 Cyclic Sports

Cyclic sports are characterised by the constant repetition of a technical gesture, where performance is based on maintaining continuous movement through a repeated motor pattern, such as strides in running or strokes in rowing. Sports events are defined by distance, and the goal is to cover it in the shortest possible time [43]. Due to these characteristics, the influence of external factors or decision-making is minimal during the competition, unlike in acyclic sports. Therefore, physiological variables are crucial for athletic performance. Athletes who can produce and maintain a higher amount of energy output over the competition distance will result in a higher performance [43].

To achieve maximum performance in a cyclic sport, the athlete must produce a propulsive force (force on the pedals, in the stroke, etc.), which is the result of energy expenditure by the muscles involved in the technical gesture. However, of all the energy expenditure the body must make to produce this muscular effort, approximately 10 to 25% of the total energy produced is used for movement, while the remaining 75–90% is heat that must be dissipated [44, 45]. This point is particularly relevant, as gross efficiency, measured as the amount of energy used by the body that is converted into mechanical energy for movement, plays a fundamental role in athletic performance [46]. According to Jeukendrup et al. [47], a 1% improvement in gross efficiency would translate to a 46-second improvement in a 40 km time trial for an elite cyclist.

Muscles generate heat when they contract due to a process called ATP hydrolysis. During muscle contraction, ATP (adenosine triphosphate) breaks down into ADP (adenosine diphosphate) and inorganic phosphate, releasing energy in the form of heat [48] (Fig. 8.1). This process is essential for muscle contraction, as the energy released allows the actin and myosin filaments within the muscle to slide over each other, causing contraction. This heat is crucial for maintaining core temperature and is an important part of the body's homeostasis, but at the same time, an excess of temperature is detrimental to both sports performance and the stability of physiological systems [49]. The heat generated during mechanical work is an inevitable byproduct of energy metabolism and contributes to the increase in core temperature (when the heat generated is uncompensable). The more intense the exercise, the higher the metabolic rate and, therefore, the greater the heat production [45].

Exercise causes an increase in blood flow to the active muscles to supply them with the oxygen and nutrients needed for muscle contraction. This increase in blood

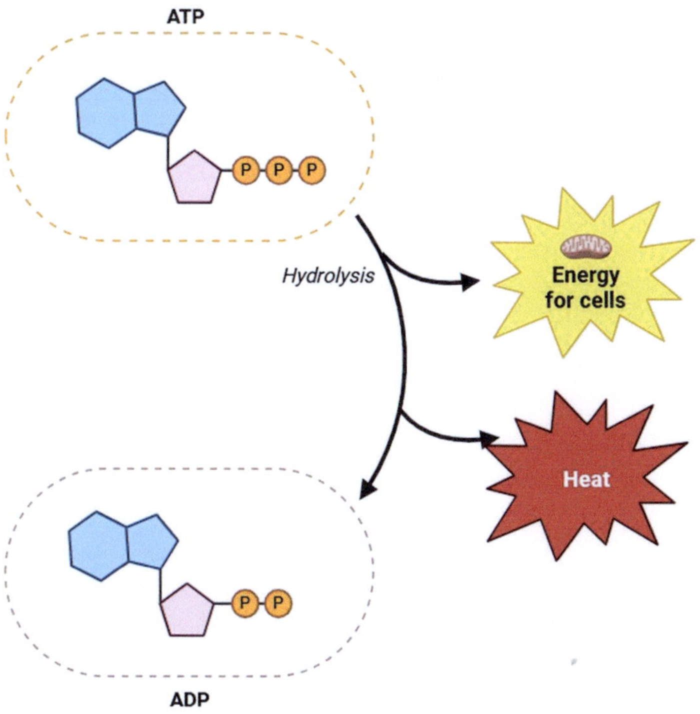

Fig. 8.1 The process of ATP hydrolysis and heat production. Abbreviations ATP adenosine triphosphate; ADP adenosine diphosphate

flow also transports the heat generated in the muscles to other parts of the body. Blood acts as a means of transporting heat, distributing it throughout the body and raising the overall core temperature [50]. Additionally, the increase in blood flow to the skin dissipates heat through the body's surface [51].

Cyclic sports are generally practised in open environments and exposed to environmental factors, which play a significant role in thermoregulation (Fig. 8.2). Environmental temperature and relative humidity significantly influence the body's ability to dissipate heat [52]. In hot and humid environments, sweat evaporation is less efficient due to high humidity, reducing the body's ability to cool down. This can reduce the efficiency of heat dissipation and lead to an additional increase in core temperature [53]. Conversely, in dry environments, sweat evaporation is more efficient, facilitating heat dissipation [53]. However, in both cases, hot environment can hinder heat loss, increasing the risk of overheating [51].

Temperature plays a very significant role in cyclic sports performance. Elevating core temperature imposes great physiological stress on the body [54]. However, measuring core temperature is very invasive [55]. Thermography allows for non-invasive measurement of the Tsk variations, providing insight into the athlete's thermoregulation process [56] and its impact on sports performance [57]. In this sense, recent research showed that a decrease in the core-skin gradient results in a higher

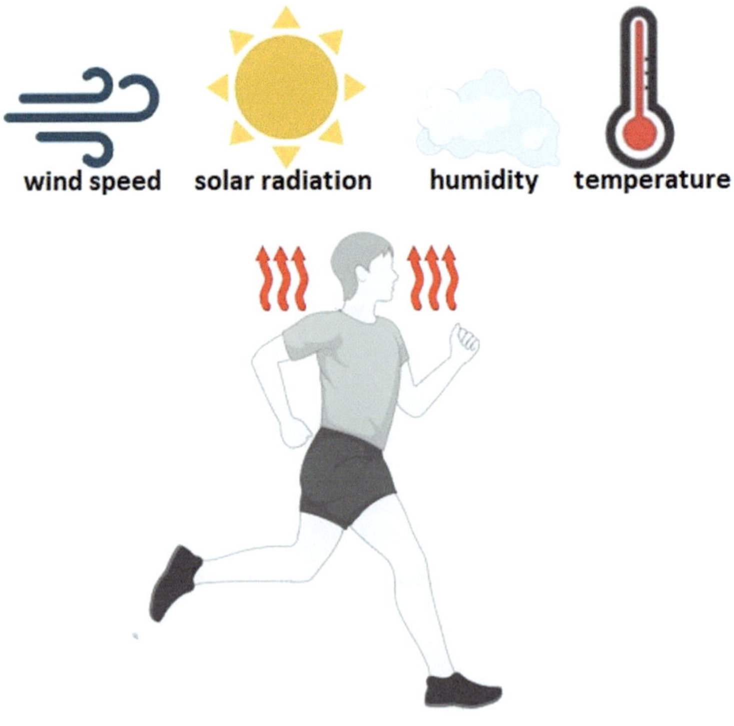

Fig. 8.2 Environmental factors affecting thermoregulatory capacity during physical exercise

risk of heat medical problems during a sport competition than just a high core temperature [58]. This has led to an increasing interest in using IRT cyclic sports [19].

Upon reviewing the scientific literature pertaining to cyclic sports, several common patterns can be discerned. In running, despite the significant methodological variability, as running protocols of very different duration and intensity are carried out, as well as analysing different ROI, studies indicate that no thermal asymmetries exist between lateral body regions, neither in training protocols [13, 59–61] nor when comparing the evolution during a training period [62]. Therefore, during a continuous and cyclic gesture, the response to Tsk behaves the same on both sides of the body due to the redistribution of blood flow for heat dissipation.

Regarding the behaviour of Tsk, the Tsk of runners generally decreases at the beginning of physical exercise, which is more noticeable in the peripheral regions. In contrast, at the end of physical exercise, the temperature increases during recovery [33]. This response would be mainly due to the thermoregulatory mechanisms of the body (sweating and conduction, mainly) as well as vasoconstriction [63]. This occurs due to muscle activity, as the demands for oxygen and substrates by the muscle increase considerably, so the body must constrict certain pathways to favour blood supply to the target tissue, in this case, the executing muscle [64].

We also observe that technical modifications, such as plantar eversion (a relevant aspect of running technique), do not seem to affect temperature behaviour [65]. However, running forcibly in an asymmetric manner (with a load) would generate higher temperatures without showing differences between sides [13]. This could indicate that the technical aspect of a cyclic sport discipline is not as relevant to the impact on skin temperature. Nevertheless, the effect of forcing a movement could make it less efficient, thereby generating more heat, but in a symmetrical manner.

Regarding equipment, the use of different orthoses (insoles) did not modify plantar temperature [66, 67]. However, compression garments influence Tsk by increasing it [68], as they could be altering blood circulation and heat transport. Hence, differences are only observed with the use of compression garments.

From a performance perspective, Racinais et al. [58] found that a lower Tsk before a long-distance competition was associated with better performance, while those who did not finish the race exhibited higher Tsk. This could be explained from a physical point of view, as lower Tsk generates a greater temperature gradient, meaning the difference between internal and external temperature is greater and therefore improving the cooling, dissipation, and heat transport capacity, consequently leading to an improvement in physical capacity [69, 70]. We observed a similar response in studies with cyclists. When compared to untrained subjects, they presented lower temperatures at the same internal load (VT1 or VT2), suggesting that they are more efficient at dissipating heat at the same load [25]. However, at maximum power output, those who were able to generate more power had higher Tsk, which could be due to a higher metabolic rate [25].

In studies focusing on cycling sport, similar results in running were observed. Studies reported a decrease in Tsk in the early moments of physical activity [14, 23, 34, 65, 71] due to cutaneous vasoconstriction, allowing blood flow to be redirected to active muscles. This thermal response is a common phenomenon due to vascular adjustments observed in different types of exercise [72], such as kayaking or rowing [73].

Mechanical efficiency plays a crucial role in cyclic sports, determining how effectively muscular force is converted into movement. Mechanical efficiency is influenced by biomechanical factors, such as the force transmission lever [74], and technical factors specific to each sport modality [75]. In order to improve this efficiency, athletes make biomechanical modifications, such as adjusting saddle height in cycling, yet studies indicate that these modifications do not significantly alter Tsk responses [76]. This finding is consistent with running studies in healthy runners, where changes in technique did not influence thermal behaviour [13, 65].

Other factors that play a very important role, and which align with the idea that athletes with better physical condition present lower temperatures, is the role that subcutaneous adipose tissue plays, as it could influence heat transfer due to its insulating capacity [77]. Likewise, sweating and vascularisation are adaptations that improve with athlete training [51]. More trained athletes, and therefore with better physical condition, have a better sweating capacity, as well as an increase in the number of capillaries both in muscle tissue and on the skin surface [78] (Fig. 8.3). This would indicate that these athletes have a greater capacity to transport heat and cool down during physical activity.

Regarding temperature asymmetries between opposite limbs, this phenomenon does not occur due to the characteristics of the effort and the continuous chaining of motor patterns [8]. This makes the response much more generalised, and there are no variations between sides, unlike in acyclic sports, as seen in the study of Chudecka

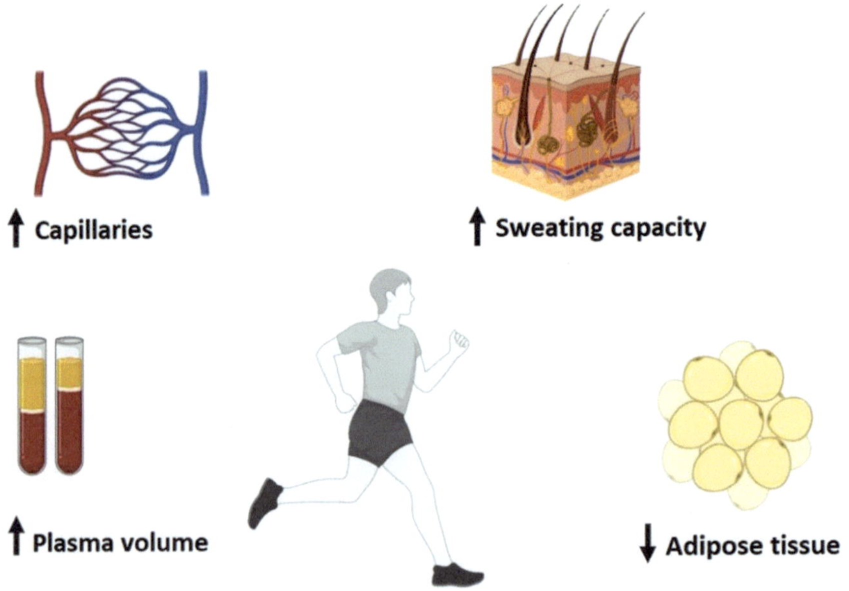

Fig. 8.3 Main adaptations in improving thermoregulation in trained subjects

et al. [8], where a group of rowers (cyclic sport) is directly compared to a handball team (acyclic sport). Here, thermal asymmetries between cyclic and acyclic sports are directly evidenced. Both groups, after their competitive effort (indoor rowing simulation or handball match), exhibited distinct thermal responses. The rowers, following the simulated competition, showed a symmetrical distribution of Tsk. In contrast, the handball players displayed significant temperature differences between both sides of the upper limbs.

8.5 Non-cyclic Sports

Unlike cyclic sports, such as running or cycling, non-cyclic sports are characterised by movements and actions that do not follow repetitive patterns [79]. These sports involve dynamic and variable actions influenced by a specific aim in a particular context, generating a constant interaction between athletes' perceptions and actions [80]. These actions determine the different sport's techniques, mainly adaptative asymmetric movements with high contralateral dominance [81–83].

Athletes' asymmetries in non-cyclic sports have been a hot topic in the sports science literature. Mainly focused on understanding contralateral asymmetries for injury prevention, new trends suggest that these asymmetries are an inherent feature of sports performance rather than a direct injury risk factor [81, 84, 85]. In this sense, the latest IRT studies in non-cyclic sports such as tennis, padel, archery or handball, among others, have found contralateral differences in Tsk associated with acute and chronic adaptations to specific technical demands [8–10, 12]. Therefore, this section aims to deepen the effects of athletes' technique on Tsk in non-cyclic sports to explain the thermal behaviour under these asymmetries.

8.5.1 Skin Temperature Asymmetries in Non-cyclic Sports

As outlined in Sect. 8.4, no asymmetric thermal behaviours have been observed in cyclic sports; however, scientific evidence has confirmed significant differences in contralateral Tsk comparisons in non-cyclic sports [9, 10, 12]. These differences represent acute adaptations linked to athletes' technical executions and exhibit different characteristics depending on the sports discipline, providing valuable data to researchers and coaches for athletes' technical analysis [8, 9, 26]. This section will deepen the physiological response behind this asymmetrical thermal behaviour by discussing reference literature and highlighting results observed across different non-cyclic sports.

In asymmetric sports, the Tsk is influenced by the specific demands of the body segments involved and the physiological mechanisms that regulate the heat generated by athletes' movement [8, 9, 12]. As mentioned in the previous section, during sports practice, movement demands activate metabolic processes (ATP hydrolysis) to produce the energy required for muscle contractions, generating a significant amount of heat [2, 27]. This heat increases body temperature and is distributed via blood flow to other body parts, including the skin, where it can be dissipated through sweating and thermal conduction [1, 2, 4, 27]. However, in non-cyclic sports, differences in neuromuscular activation and metabolic demands between the body segments can result in Tsk asymmetries [8–10]. This can occurs because certain areas require more energy and increased blood flow, while others experience vasoconstriction to prioritise the delivery of oxygen and nutrients to the active muscles [1, 4, 27].

Moreover, Tsk also depends on the thermal properties of the tissues underlying the skin, such as the musculature and the adipose tissue, as well as on peripheral vascularisation [2, 4, 27]. In this sense, athletes trained in specific disciplines develop chronic adaptations in these tissues, altering their structure, composition, and capillarisation, increasing sweating and improving peripheral vascularisation, which enhances their ability to dissipate heat [86, 87]. These adaptations are evident in Tsk asymmetries during non-cyclic sports activities, resulting in acute Tsk contralateral asymmetries after practice [8–10, 12, 26].

To continue, it is difficult to unify and generalise athletes' chronic and acute adaptations after performing different non-cyclic sports, so it is interesting to know the Tsk behaviours related to the different sports' asymmetric techniques before and after practice. Therefore, we will deepen the main findings of interventions in non-cyclic sports that analysed acute thermal asymmetries resulting from sports techniques, categorising them into Team Sports, Racket Sports, Archery, and Valencian handball.

8.5.1.1 Team Sports

Although much reference literature has applied IRT to measure thermal asymmetries in team sports such as football, volleyball or handball, among others, most have not delved into thermal asymmetries related to sports techniques [88–90]. Within these interventions, most have focused on injury prevention, performing only pre-training baseline measurements. Authors have also analysed the training load assimilation throughout the season, the performance in specific physical tests, the influence of environmental variables or the application of precooling on performance, making it difficult to analyse acute response to athletes' spots technique [88–90]. Therefore, a gap in the literature addressing this issue in team sports has been identified.

Focusing on the research that has considered the influence of sports technique on Tsk, we find studies carried out in handball that, by employing PRE, POST and POST10 measurements, were able to analyse contralateral asymmetry [1, 8] (Fig. 8.4). These studies have found significant contralateral differences (right vs. left) in the arm and forearm on both the anterior and posterior sides [8]. With greater temperature variations (POST-PRE) on the dominant side, these differences did not exceed the 0.5 °C threshold presented in the literature [8]. The authors concluded

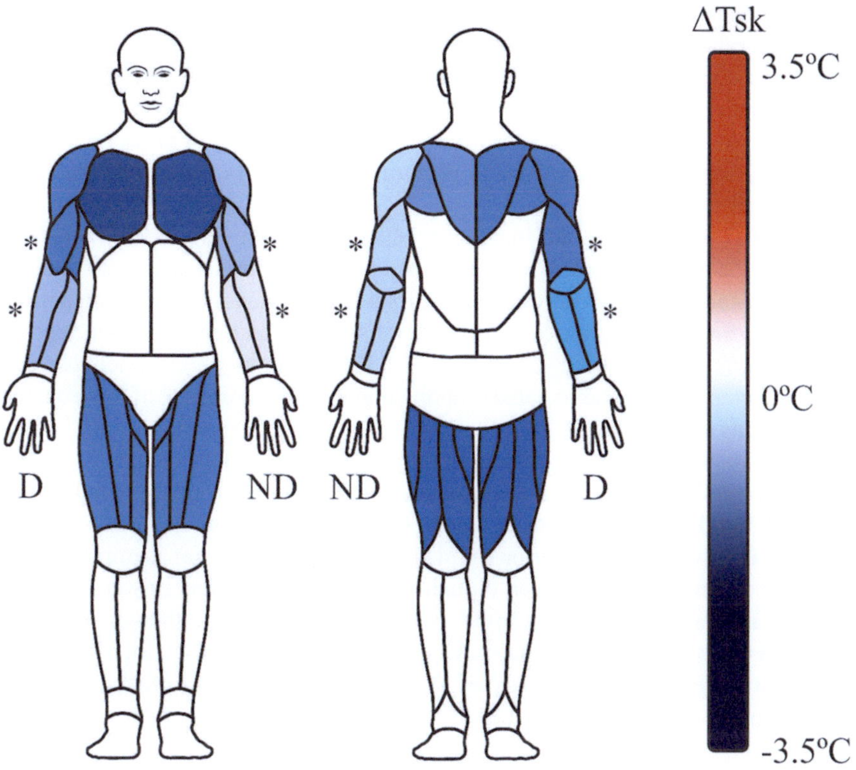

Fig. 8.4 Skin temperature behaviour in handball based on [8]. ΔTsk shows the variation of Tsk in each ROI (POST-PRE), (D) indicates the dominant side, (ND) indicates the non-dominant side, and (*) points out the significant differences in contralateral Tsk comparisons.

that these differences in thermal behaviour were due to greater activity in the dominant arm, caused by handball's asymmetrical technical requirements, where the dominant side's involvement in ball handling, throwing, and passing is greater [8]. Therefore, it was concluded that IRT can be used to improve and monitor athletes' sports technique and motor coordination [8].

8.5.1.2 Racket Sports

Concerning racket sports, studies have monitored the Tsk behaviour for contralateral asymmetry with PRE and POST measurements in tennis (full body or upper limbs), padel (full body) and badminton (lower limbs) [9, 12, 26, 42].

In tennis, higher temperatures were found on the dominant side in the posterior forearm and biceps after 40 serving actions, with differences between 0.3 to 1.8 °C [26]. Similar results were found monitoring the whole body Tsk before and after 2.5 hours of standardised training sessions [9]. The anterior and posterior forearms were the only areas presenting significant differences, with 1.1 ± 0.5 °C higher mean contralateral Tsk values on the dominant side [9] (Fig. 8.5). These post-exercise contralateral asymmetries were associated with acute responses to the sports technique, specifically provoked by the constant racket grip (isometric force required to hold the racket) and the different stroke movements of the dominant side (forearm prone-supination actions to perform serve, volleys, forehand, backhand, etc) [9, 26].

A similar trend was found in padel tennis intervention (Fig. 8.6), where after analysing the whole body Tsk before and after a standardised 3 ± 1 hours training, the contralateral post-exercise comparison found higher values on the dominant side in the anterior and posterior forearm, anterior shoulder and posterior arm regions, with the difference between 0.7 and 1.7 °C [12]. As in the previous tennis study, these asymmetries were associated with the upper limb dominance of the technical requirements of the sport [9, 12].

Finally, badminton research focused only on the lower limbs (Fig. 8.7), finding an increase in POST exercise Tsk in the anterior knee and anterior leg of both dominant and non-dominant sides and higher Tsk values only in the non-dominant side of the anterior thigh and posterior leg [42]. These demonstrate an asymmetric contralateral thermal behaviour even though no significant differences were found in the contralateral post-exercise comparison [42]. In this sense, although significant contralateral Tsk asymmetries may not appear on the POST measurements, the thermal variation (POST-PRE) of analysed ROIs can give valuable information about its implication on the sports technique.

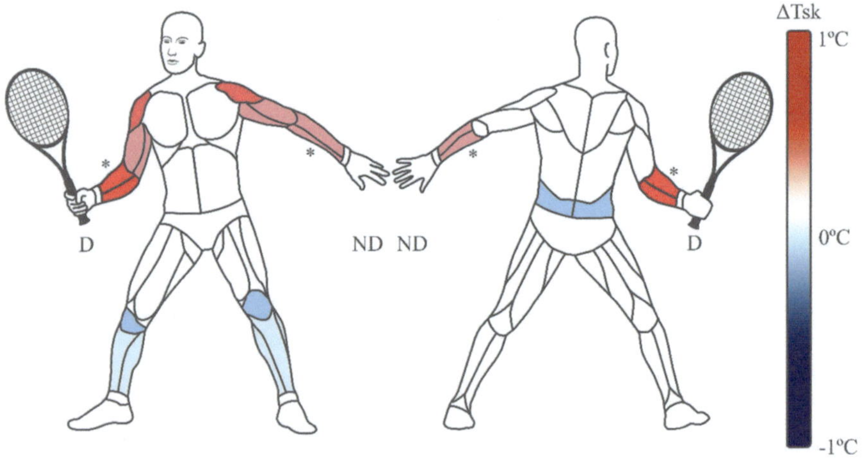

Fig. 8.5 Skin temperature behaviour in tennis based on [9]. ΔTsk shows the variation of Tsk in each ROI (POST-PRE), (D) indicates the dominant side, (ND) indicates the non-dominant side, and (*) points out the significant differences in contralateral Tsk comparisons.

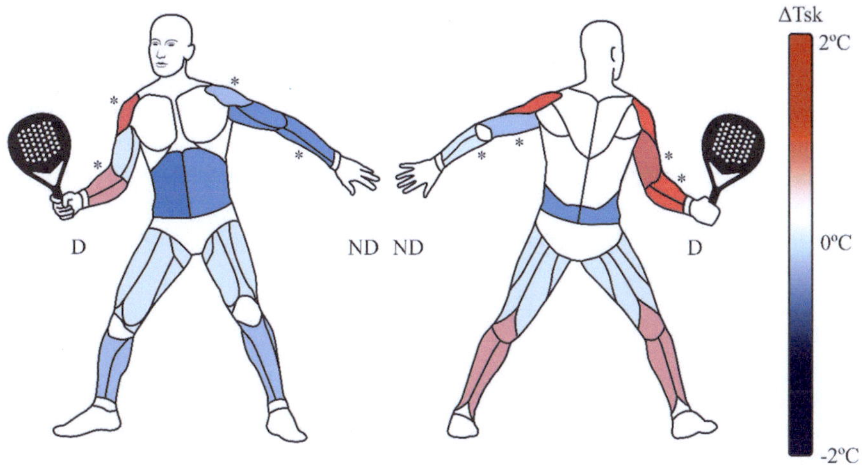

Fig. 8.6 Skin temperature behaviour in padel based on [12]. ΔTsk shows the variation of Tsk in each ROI (POST-PRE), (D) indicates the dominant side, (ND) indicates the non-dominant side, and (*) points out the significant differences in contralateral Tsk comparisons.

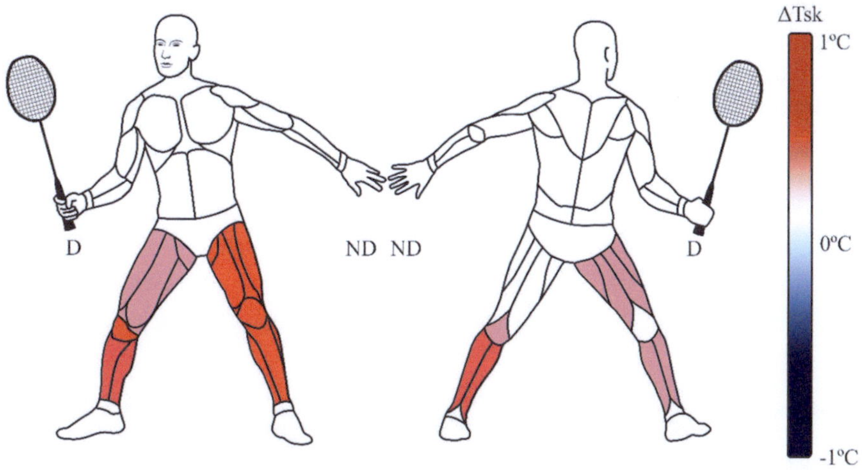

Fig. 8.7 Skin temperature behaviour in badminton based on [42]. ΔTsk shows the variation of Tsk in each ROI (POST-PRE), (D) indicates the dominant side, (ND) indicates the non-dominant side, and (*) points out the significant differences in contralateral Tsk comparisons.

8.5.1.3 Archery

Contralateral Tsk asymmetries were also studied in archery (Fig. 8.8), considering its high asymmetric technique [10]. To establish this comparison, the side holding the bow was classified as the *"bow side"*, while the side that pulls the string was named the *"drawing side"* [10]. After PRE, POST and POST10 Tsk measurements during a simulated competition, a contralateral thermal asymmetry was found in the

POST measurement, with higher Tsk values on the “bow side” in the upper back, medial forearm, posterior shoulder, lateral arm and lateral elbow [10]. The temperature differences, ranging from 0.3 to 0.6 °C, were linked to the asymmetrical technique used in archery [10]. This technique involved isometric muscle activation in the shoulder, elbow, and wrist and synchronised activation of the forearm, shoulder girdle, and triceps brachii [10, 91, 92]. These muscles stabilise the bow arm, helping athletes maintain balance and precision when pulling the bowstring, as the force from the bow tends to flex the arm joint [10, 93, 94].

8.5.1.4 Valencian Handball

Another asymmetrical Tsk analysis was carried out in Valencian handball (Fig. 8.9), in the Raspall discipline (invasion type), where the dominant hand normally hits the ball and the ground during the contact action [11]. In this intervention with PRE, POST and POST10 measurements, athletes presented thermal asymmetries PRE match with lower Tsk values in the dominant hand in five ROIs (thumb, middle finger, ring finger, thenar eminence and palm) of the eight analysed (index, little finger and wrist) [11]. These asymmetries were due to the chronic adaptations produced by the repeated hits given with the dominant hand to the ball and the ground, generating alterations in the skin, subcutaneous tissues and superficial vascularisation, reducing the peripheral blood flow together with the chronic effects of common previous injuries [11, 95–97]. Regarding POST contralateral Tsk measurements,

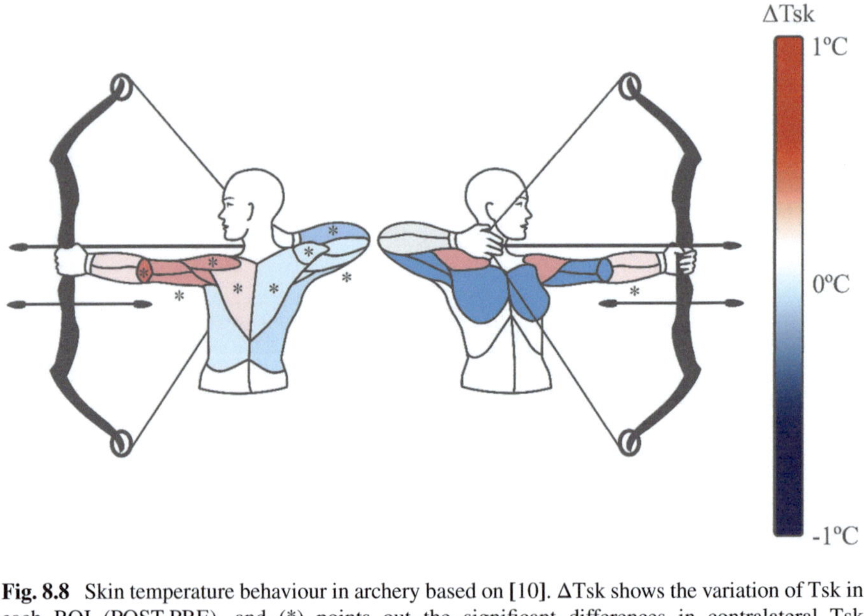

Fig. 8.8 Skin temperature behaviour in archery based on [10]. ΔTsk shows the variation of Tsk in each ROI (POST-PRE), and (*) points out the significant differences in contralateral Tsk comparisons.

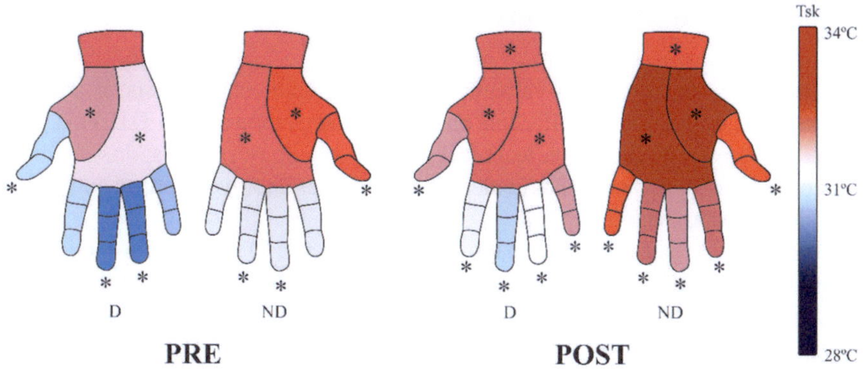

Fig. 8.9 Skin temperature in Valencian handball based on [11]. Tsk in each ROI in PRE and POST moments, (D) indicates the dominant side, (ND) indicates the non-dominant side, and (*) points out the significant differences in contralateral Tsk comparisons.

significant asymmetries were found in all ROIs with lower Tsk values of the dominant hand, although Tsk of both hands and all ROIs increased [11]. This reflected an acute post-match response associated with the athletes' technical executions during the match but with a different tendency from those in racket sports.

8.5.2 *Final Considerations*

Tsk contralateral asymmetries have been identified in non-cyclic sports and were associated with the athletes' functional movements and technical executions. Post-exercise Tsk measurements have provided information about the ROI implication during sports practice. The magnitude and direction of contralateral asymmetry can differ based on the athlete's acute and chronic adaptations to the demands of the sport. Racket sports have shown contralateral asymmetries, with higher Tsk in the dominant arm with differences that can exceed 1.0 °C [9, 12, 26]. In contrast, higher Tsk values were found in the non-dominant hand in the Valencian handball [11]. Furthermore, Tsk variation can be considered through PRE and POST Tsk comparison. In handball, greater variations were found in the dominant arm. Meanwhile, in badminton, different behaviours in the ROIs of the lower limbs were shown, with the anterior thigh and the posterior leg of the non-dominant side showing higher Tsk variations after practice. For all the above, it can be concluded that IRT can be used for Tsk contralateral comparison in non-cycling sports to determine the implications of ROIs in the technical execution.

8.6 Practical Applications

Thermography emerges as a valuable tool for the technical analysis of athletes, allowing the evaluation of both acute and chronic adaptations in the Tsk. This technique provides crucial data on technical execution in non-cyclical sports, facilitating coaches and sports scientists in identifying areas for improvement and optimising technical performance [8, 9, 26].

Additionally, thermography plays an important role in injury prevention. The identification of contralateral thermal asymmetries can reveal imbalances in neuromuscular activation and metabolic demands, which could indicate a higher risk of injury. This is particularly useful in team and racket sports, where technical demands are often asymmetrical [12, 98].

Regarding performance, as discussed in this chapter, thermography allows us to observe Tsk's behaviour during physical exercise [23] or under different environmental conditions in our athletes. This provides important insights into the stress the body is experiencing to thermoregulate, which can limit athletic performance [63]. Furthermore, we can assess the adaptation of our athletes by observing how their thermoregulatory capacity improves with training, resulting in lower or less elevated Tsk [32]. Pre- or post-cooling protocols can be evaluated based on their impact on Tsk, thereby creating a greater temperature gradient that facilitates pre-competition thermoregulation [99] or enhances post-competition recovery [100].

According to the review by Masur et al. [101], there has been increasing interest in recent years in correlating infrared thermography data with internal load markers, such as creatine kinase or lactate dehydrogenase. Establishing this correlation would be highly useful to avoid the need for invasive techniques to measure these values. However, the results are inconsistent, and no consistent correlations can be drawn. One of the key points highlighted by the authors is the methodological variability within the analysed studies (different measurement times and methods for calculating IRT parameters). Therefore, unifying protocols could yield more robust results [101].

In summary, thermography offers multiple practical applications in the sports field, from technical analysis and injury prevention to training load monitoring, performance optimisation, and the evaluation of physiological adaptations.

8.7 Future Research

Based on all the information presented throughout this chapter, we outline below the main directions that future research should take to improve the applicability of IRT in sports.

Firstly, methodological standardisation is essential. This would allow for the unification of results, obtaining firm conclusions, and improving the reproducibility of the thermal data obtained from athletes, allowing for more reliable comparisons across studies. This point could also generate reference values for Tsk under certain situations or stimuli, as well as tables for the correct interpretation of thermal responses to different efforts or exercises, thus distinguishing functional patterns from injurious or pathological ones, whether due to asymmetry or skin hyper/hypothermia. Additionally, conducting longitudinal studies, that evaluate the long-term thermoregulatory response in cyclical and non-cyclical disciplines, would provide more robust data, allowing the identification of key values to establish thresholds from which an athlete may be in suboptimal conditions for physical performance or at risk of injury.

Another important point for improving the use of this technique would be to enhance the technology to enable reliable, valid, and precise measurements in specific sports settings, as it is currently still a technology focused on the laboratory. Given the strong influence of environmental factors on thermoregulation [102, 103], optimising IRT for field use will ensure more ecologically valid assessments that better reflect athletes' real physiological responses.

Finally, the study of thermography in a more diverse athletic population should be increased to see how certain physiological factors can affect the measurement of Tsk, such as the menstrual cycle, the age of the subjects, body composition, or even the use of different supplements or antipyretic drugs used to enhance sports performance [104].

8.8 Conclusions

The analysis of Tsk asymmetries in sports has evolved beyond its traditional focus on injury prevention, providing valuable insights into athletes' physiological adaptations to sport-specific technical demands. Findings from cyclic and non-cyclic sports confirm that Tsk distribution is closely linked to movement patterns, muscle activation, and thermoregulatory mechanisms.

In cyclic sports, where motor patterns are repetitive and symmetrical, Tsk remains evenly distributed between contralateral limbs. Studies in running, cycling, and rowing have consistently shown that no significant thermal asymmetries occur, as blood flow and heat dissipation mechanisms are balanced across both sides of the body. This highlights the role of vascular regulation in maintaining thermal equilibrium, ensuring efficient heat dissipation without localised temperature imbalances.

In contrast, non-cyclic sports exhibit consistent contralateral Tsk asymmetries, reflecting the asymmetric demands of technical execution. Research in racket sports, archery, handball, and Valencian handball has identified higher Tsk in the dominant limb due to greater muscle activation and metabolic heat production. However, specific adaptations, such as chronic tissue remodelling in Valencian handball, can result in inverse asymmetry patterns, where the non-dominant limb shows higher post-exercise temperatures. These findings emphasise the importance of sport-specific thermal analysis, as Tsk asymmetries cannot be generalised across all disciplines.

The methodological considerations discussed in this chapter reinforce the need for standardised protocols in IRT assessments. The selection of measurement moments, ROI classification, and asymmetry calculation methods significantly influence data interpretation. Future research should focus on refining these protocols to ensure greater consistency and comparability between studies.

From an applied perspective, IRT presents great potential for sports science, offering a non-invasive tool for monitoring technical efficiency, physiological adaptations, and recovery processes. Coaches and practitioners can leverage Tsk analysis to optimise training programmes, ensuring that athletes develop efficient thermoregulatory responses while minimising excessive thermal asymmetries that may indicate overload or improper technique.

In conclusion, Tsk asymmetries should not be viewed solely as indicators of injury risk but as valuable markers of sport-specific adaptations. The integration of IRT in performance assessment and training monitoring represents an important step forward in understanding how athletes' bodies respond to the technical and physical demands of their discipline. Future investigations should further explore the interplay between thermoregulation, biomechanics, and performance, contributing to a more comprehensive approach to sports science and athlete development.

References

1. Chudecka M, Lubkowska A (2010) Temperature changes of selected body's surfaces of handball players in the course of training estimated by thermovision, and the study of the impact of physiological and morphological factors on the skin temperature. J Therm Biol 35(8):379–385. https://doi.org/10.1016/j.jtherbio.2010.08.001
2. Cramer MN, Jay O (2016) Biophysical aspects of human thermoregulation during heat stress. Auton Neurosci 196:3–13. https://doi.org/10.1016/j.autneu.2016.03.001
3. Gisolfi CV, Wenger CB (1984) Temperature regulation during exercise: old concepts, new ideas. Exerc Sport Sci Rev 12(1):399
4. Charkoudian N (2003) Skin blood flow in adult human thermoregulation: how it works, when it does not, and why. Mayo Clin Proc 78(5):603–612. https://doi.org/10.4065/78.5.603
5. Vardasca R, Ring EFJ, Plassmann P, Jones C (2012) Termal symmetry of the upper and lower extremities in healthy subjects. Thermol Int 22:53–60
6. Bouzas Marins JC, de Andrade Fernandes A, Gomes Moreira D, Souza Silva F, Costa CMA, Pimenta EM, Sillero-Quintana M (2014) Thermographic profile of soccer players' lower limbs. Revista Andaluza de Medicina Del Deporte 7(1):1–6. https://doi.org/10.1016/S1888-7546(14)70053-X
7. Fernández-Cuevas I, Arnáiz Lastras J, Escamilla Galindo V, Gómez Carmona P (2017) Infrared thermography for the detection of injury in sports medicine. In: Priego Quesada JI (ed) Application of infrared thermography in sports science. Springer International Publishing, pp 81–109. https://doi.org/10.1007/978-3-319-47410-6_4
8. Chudecka M, Lubkowska A, Leźnicka K, Krupecki K (2015) The use of thermal imaging in the evaluation of the symmetry of muscle activity in various types of exercises (symmetrical and asymmetrical). J Hum Kinet 49:141–147. https://doi.org/10.1515/hukin-2015-0116
9. Marzano-Felisatti JM, Martinez-Amaya A, Priego-Quesada JI (2023) Preliminary analysis of skin temperature asymmetries in elite young tennis players. Appl Sci 13(1):628. https://doi.org/10.3390/app13010628

10. Sanchis-Sanchis R, Priego-Quesada JI, Ribas-Garcia V, Carpes FP, Encarnacion-Martinez A, Perez-Soriano P (2020) Effects of asymmetrical exercise demands on the symmetry of skin temperature in archers. Physiol Meas 41(11):114002. https://doi.org/10.1088/1361-6579/abc020
11. Sánchez-Jiménez JL, Tejero-Pastor R, de Calzadillas-Valles MC, Jimenez-Perez I, de Cibrián Ortiz Anda RM, Salvador-Palmer R, Priego-Quesada JI (2022) Chronic and acute effects on skin temperature from a sport consisting of repetitive impacts from hitting a ball with the hands. Sensors 22(21):21. https://doi.org/10.3390/s22218572
12. De León-Muñoz A, Priego-Quesada JI, Marzano-Felisatti JM, Sanchez-Jimenez JL, Sendra-Pérez C, Aparicio-Aparicio I (2024) Preliminary application of infrared thermography to monitoring of skin temperature asymmetries in professional padel players. Sensors 24(14):14. https://doi.org/10.3390/s24144534
13. Gil-Calvo M, Herrero-Marco J, de González-Peña RJ, Perez-Soriano P, Priego-Quesada JI (2020) Acute effect of induced asymmetrical running technique on foot skin temperature. J Therm Biol 91:102613. https://doi.org/10.1016/j.jtherbio.2020.102613
14. Trecroci A, Formenti D, Ludwig N, Gargano M, Bosio A, Rampinini E, Alberti G (2018) Bilateral asymmetry of skin temperature is not related to bilateral asymmetry of crank torque during an incremental cycling exercise to exhaustion. PeerJ 6:e4438. https://doi.org/10.7717/peerj.4438
15. Charkoudian N (2016) Human thermoregulation from the autonomic perspective. Auton Neurosci 196:1–2. https://doi.org/10.1016/j.autneu.2016.02.007
16. Formenti D, Merla A, Priego Quesada JI (2017) The use of infrared thermography in the study of sport and exercise physiology. In: Priego Quesada JI (ed) Application of infrared thermography in sports science. Springer International Publishing, pp 111–136. https://doi.org/10.1007/978-3-319-47410-6_5
17. Boulant JA (1996) Hypothalamic neurons regulating body temperature. In: Comprehensive physiology, supplement 14: handbook of physiology, environmental physiology, pp 105–126. https://doi.org/10.1002/cphy.cp040106
18. de Fernandes AA, Amorim PRDS, Brito CJ, Sillero-Quintana M, Bouzas Marins JC (2016) Regional skin temperature response to moderate aerobic exercise measured by infrared thermography. Asian J Sports Med 7(1):e29243. https://doi.org/10.5812/asjsm.29243
19. Hillen B, Pfirrmann D, Nägele M, Simon P (2020) Infrared thermography in exercise physiology: the dawning of exercise Radiomics. Sports Med 50(2):263–282. https://doi.org/10.1007/s40279-019-01210-w
20. Tanda G (2015) The use of infrared thermography to detect the skin temperature response to physical activity. J Phys Conf Ser 655(1):012062. https://doi.org/10.1088/1742-6596/655/1/012062
21. Tanda G (2018) Total body skin temperature of runners during treadmill exercise. J Therm Anal Calorim 131(2):1967–1977. https://doi.org/10.1007/s10973-017-6634-4
22. Korman P, Straburzyńska-Lupa A, Kusy K, Kantanista A, Zieliński J (2016) Changes in body surface temperature during speed endurance work-out in highly-trained male sprinters. Infrared Phys Technol 78:209–213. https://doi.org/10.1016/j.infrared.2016.08.003
23. Ludwig N, Trecroci A, Gargano M, Formenti D, Bosio A, Rampinini E, Alberti G (2016) Thermography for skin temperature evaluation during dynamic exercise: A case study on an incremental maximal test in elite male cyclists. Appl Opt 55(34):D126–D130. https://doi.org/10.1364/AO.55.00D126
24. Torii M, Yamasaki M, Sasaki T, Nakayama H (1992) Fall in skin temperature of exercising man. Br J Sports Med 26(1):29–32
25. Priego-Quesada JI, Sampaio LT, Bini RR, Rossato M, Cavalcanti V (2017b) Multifactorial cycling performance of Cyclists and Non-Cyclists and their effect on skin temperature. J Therm Anal Calorim 127(2):1479–1489. https://doi.org/10.1007/s10973-016-5971-z
26. Amaro AM, Paulino MF, Neto MA, Roseiro L (2019) Hand-arm vibration assessment and changes in the thermal map of the skin in tennis athletes during the service. Int J Environ Res Public Health 16(24):E5117. https://doi.org/10.3390/ijerph16245117

27. Cooper T, Randall WC, Hertzman AB (1959) Vascular convection of heat from active muscle to overlying skin. J Appl Physiol 14(2):207–211. https://doi.org/10.1152/jappl.1959.14.2.207
28. Escamilla-Galindo VL, Estal-Martínez A, Adamczyk JG, Brito CJ, Arnaiz-Lastras J, Sillero-Quintana M (2017) Skin temperature response to unilateral training measured with infrared thermography. J Exerc Rehabil 13(5):526–534. https://doi.org/10.12965/jer.1735046.523
29. Adamczyk JG, Boguszewski D, Siewierski M (2014) Thermographic evaluation of lactate level in capillary blood during post-exercise recovery. Kinesiology 46(3):186–193
30. Ferreira JJA, Mendonça LCS, Nunes LAO, Andrade Filho ACC, Rebelatto JR, Salvini TF (2008) Exercise-associated thermographic changes in young and elderly subjects. Ann Biomed Eng 36(8):1420–1427. https://doi.org/10.1007/s10439-008-9512-1
31. Neves EB, Cunha RM, Rosa C, Antunes NS, Felisberto IMV, Vilaça-Alves J, Reis VM (2016) Correlation between skin temperature and heart rate during exercise and recovery, and the influence of body position in these variables in untrained women. Infrared Phys Technol 75:70–76. https://doi.org/10.1016/j.infrared.2015.12.018
32. Priego Quesada JI, Kunzler MR, Carpes FP (2017) Methodological aspects of infrared thermography in human assessment. In: Priego Quesada JI (ed) Application of infrared thermography in sports science. Springer International Publishing, pp 49–79. https://doi.org/10.1007/978-3-319-47410-6_3
33. Merla A, Mattei PA, Di Donato L, Romani GL (2010) Thermal imaging of cutaneous temperature modifications in runners during graded exercise. Ann Biomed Eng 38(1):158–163. https://doi.org/10.1007/s10439-009-9809-8
34. Akimov EB, Son'kin VD (2011) Skin temperature and lactate threshold during muscle work in athletes. Hum Physiol 37(5):621–628. https://doi.org/10.1134/S0362119711050033
35. Formenti D, Ludwig N, Gargano M, Gondola M, Dellerma N, Caumo A, Alberti G (2013) Thermal imaging of exercise-associated skin temperature changes in trained and untrained female subjects. Ann Biomed Eng 41(4):863–871. https://doi.org/10.1007/s10439-012-0718-x
36. Salamunes ACC, Stadnik AMW, Neves EB (2017) The effect of body fat percentage and body fat distribution on skin surface temperature with infrared thermography. J Therm Biol 66:1–9. https://doi.org/10.1016/j.jtherbio.2017.03.006
37. Savastano DM, Gorbach AM, Eden HS, Brady SM, Reynolds JC, Yanovski JA (2009) Adiposity and human regional body temperature. Am J Clin Nutr 90(5):1124–1131. https://doi.org/10.3945/ajcn.2009.27567
38. Luong MP (2017) Assessment of equipment using infrared thermography in sports. In: Priego Quesada JI (ed) Application of infrared thermography in sports science. Springer International Publishing, pp 185–209. https://doi.org/10.1007/978-3-319-47410-6_8
39. Brito CJ, Moreira DG, Ferreira JJ, Díaz-de-Durana AL, Miarka B, Marins JCB, Sillero-Quintana M (2020) Immune response related with skin thermal pattern in judokas: a new application for infrared thermography? J Strength Cond Res 34(10):2886. https://doi.org/10.1519/JSC.0000000000002672
40. Gomes Moreira D, José Brito C, de Almeida Ferreira JJ, Bouzas Marins JC, López Díaz de Durana A, Couceiro Canalejo J, Butragueño Revenga J, Sillero-Quintana M (2021) Lactate concentration is related to skin temperature variation after a specific incremental judo test. J Strength Cond Res 35(8):2213. https://doi.org/10.1519/JSC.0000000000003095
41. del Estal A, Brito C-J, Galindo V-E, Diaz L, de Durana A, Franchini E, Sillero-Quintana M (2017) Thermal asymmetries in striking combat sports athletes measured by infrared thermography. Sci Sports 32(2):e61–e67. https://doi.org/10.1016/j.scispo.2016.09.005
42. Bravo-Sánchez A, Abián-Vicén J, Montalbán AT, Abián-Vicén P (2018) Acute effects of badminton practice on the surface temperature of lower limbs introduction. Archivos de Medicina Del Deporte 35(4):239–244. Scopus
43. Bompa OT, Buzzichelli C (2019) Periodization: theory and methodology of training (Sixth edition). Human Kinetics. https://us.humankinetics.com/products/periodization-6th-edition-pdf

44. Bertucci W, Arfaoui A, Janson L, Polidori G (2013) Relationship between the gross efficiency and muscular skin temperature of lower limb in cycling: a preliminary study. Comput Methods Biomech Biomed Engin 16(Suppl 1):114–115. https://doi.org/10.1080/10255842.2013.815902
45. Gaesser GA, Brooks GA (1975) Muscular efficiency during steady-rate exercise: effects of speed and work rate. J Appl Physiol 38(6):1132–1139. https://doi.org/10.1152/jappl.1975.38.6.1132
46. Hopker J, Coleman D, Passfield L (2009) Changes in cycling efficiency during a competitive season. Med Sci Sports Exerc 41(4):912–919. https://doi.org/10.1249/MSS.0b013e31818f2ab2
47. Jeukendrup AE, Craig NP, Hawley JA (2000) The bioenergetics of world class cycling. J Sci Med Sport 3(4):414–433. https://doi.org/10.1016/S1440-2440(00)80008-0
48. De Meis L (2001) Role of the sarcoplasmic reticulum Ca2+-ATPase on heat production and thermogenesis. Biosci Rep 21(2):113–137. https://doi.org/10.1023/a:1013640006611
49. Gauer R, Meyers BK (2019) Heat-related illnesses. Am Fam Physician 99(8):482–489
50. González-Alonso J (2012) Human thermoregulation and the cardiovascular system. Exp Physiol 97(3):340–346. https://doi.org/10.1113/expphysiol.2011.058701
51. Périard JD, Racinais S, Sawka MN (2015) Adaptations and mechanisms of human heat acclimation: applications for competitive athletes and sports. Scand J Med Sci Sports 25(Suppl 1):20–38. https://doi.org/10.1111/sms.12408
52. Beigtan M, Gonçalves M, Weon BM (2024) Heat transfer by sweat droplet evaporation. Environ Sci Technol 58(15):6532–6539. https://doi.org/10.1021/acs.est.4c00850
53. Pilch W, Szygula Z, Palka T, Pilch P, Cison T, Wiecha S, Tota Ł (2014) Comparison of physiological reactions and physiological strain in healthy men under heat stress in dry and steam heat saunas. Biol Sport 31(2):145–149. https://doi.org/10.5604/20831862.1099045
54. Kenny GP, Wilson TE, Flouris AD, Fujii N (2018) Heat exhaustion. Handb Clin Neurol 157:505–529. https://doi.org/10.1016/B978-0-444-64074-1.00031-8
55. Hymczak H, Gołąb A, Mendrala K, Plicner D, Darocha T, Podsiadło P, Hudziak D, Gocoł R, Kosiński S (2021) Core temperature measurement-principles of correct measurement, problems, and complications. Int J Environ Res Public Health 18(20):10606. https://doi.org/10.3390/ijerph182010606
56. Aylwin P, Havenith G, Cardinale M, Lloyd A, Ihsan M, Taylor L, Adami PE, Alhammoud M, Alonso J-M, Bouscaren N, Buitrago S, Esh C, Gomez-Ezeiza J, Garrandes F, Labidi M, Lange G, Moussay S, Mtibaa K, Townsend N et al (2023) Thermoregulatory responses during road races in hot-humid conditions at the 2019 Athletics World Championships. J Appl Physiol 134(5):1300–1311. https://doi.org/10.1152/japplphysiol.00348.2022
57. Sawka MN, Cheuvront SN, Kenefick RW (2012) High skin temperature and hypohydration impair aerobic performance. Exp Physiol 97(3):327–332. https://doi.org/10.1113/expphysiol.2011.061002
58. Racinais S, Ihsan M, Taylor L, Cardinale M, Adami PE, Alonso JM, Bouscaren N, Buitrago S, Esh CJ, Gomez-Ezeiza J, Garrandes F, Havenith G, Labidi M, Lange G, Lloyd A, Moussay S, Mtibaa K, Townsend N, Wilson MG, Bermon S (2021) Hydration and cooling in elite athletes: relationship with performance, body mass loss and body temperatures during the Doha 2019 IAAF World Athletics Championships. Br J Sports Med 55(23):1335–1341. https://doi.org/10.1136/bjsports-2020-103613
59. Barboza JAM, Souza LIS, Cerqueira MS, de Andrade PR, dos Santos HH, de Ferreira JJA (2020) Skin temperature of middle distance runners after a maximum effort test. Acta Sci Health Sci 42. https://www.redalyc.org/journal/3072/307264461016/html/:e48114
60. Fournet D, Ross L, Voelcker T, Redortier B, Havenith G (2013) Body mapping of thermoregulatory and perceptual responses of males and females running in the cold. J Therm Biol 38(6):339–344. https://doi.org/10.1016/j.jtherbio.2013.04.005

61. Robles Dorado V (2016) Variaciones termométricas en la planta del pie y piernas valorada en corredores antes y después de correr 30 km. Revista Internacional de Ciencias Podológicas 10(1):31–40
62. Tumilty S, Adhia DB, Smoliga JM, Gisselman AS (2019) Thermal profiles over the Achilles tendon in a cohort of non-injured collegiate athletes over the course of a cross country season. Phys Ther Sport 36:110–115. https://doi.org/10.1016/j.ptsp.2019.01.009
63. Périard JD, Eijsvogels TMH, Daanen HAM (2021) Exercise under heat stress: thermoregulation, hydration, performance implications, and mitigation strategies. Physiol Rev 101(4):1873–1979. https://doi.org/10.1152/physrev.00038.2020
64. Just TP, Cooper IR, DeLorey DS (2016) Sympathetic vasoconstriction in skeletal muscle: adaptations to exercise training. Exerc Sport Sci Rev 44(4):137–143. https://doi.org/10.1249/JES.0000000000000085
65. Priego-Quesada JI, Gil-Calvo M, Jimenez-Perez I, Lucas-Cuevas ÁG, Pérez-Soriano P (2017a) Relationship between foot eversion and thermographic foot skin temperature after running. Appl Opt 56(19):5559–5565. https://doi.org/10.1364/AO.56.005559
66. Gil-Calvo M, Priego-Quesada JI, Jimenez-Perez I, Lucas-Cuevas A, Pérez-Soriano P (2019) Effects of prefabricated and custom-made foot orthoses on skin temperature of the foot soles after running. Physiol Meas 40(5):054004. https://doi.org/10.1088/1361-6579/ab1c8c
67. Jimenez-Perez I, Gil-Calvo M, Priego-Quesada JI, Aparicio I, Pérez-Soriano P, Ortiz de Anda RMC (2020) Effect of prefabricated thermoformable foot orthoses on plantar surface temperature after running: A gender comparison. J Therm Biol 91:102612. https://doi.org/10.1016/j.jtherbio.2020.102612
68. Priego-Quesada JI, Lucas-Cuevas AG, Gil-Calvo M, Giménez JV, Aparicio I, Cibrián Ortiz de Anda RM, Salvador Palmer R, Llana-Belloch S, Pérez-Soriano P (2015) Effects of graduated compression stockings on skin temperature after running. J Therm Biol 52:130–136. https://doi.org/10.1016/j.jtherbio.2015.06.005
69. Chou T-H, Allen JR, Hahn D, Leary BK, Coyle EF (2018) Cardiovascular responses to exercise when increasing skin temperature with narrowing of the core-to-skin temperature gradient. J Appl Physiol 125(3):697–705. https://doi.org/10.1152/japplphysiol.00965.2017
70. Cuddy JS, Hailes WS, Ruby BC (2014) A reduced core to skin temperature gradient, not a critical core temperature, affects aerobic capacity in the heat. J Therm Biol 43:7–12. https://doi.org/10.1016/j.jtherbio.2014.04.002
71. Cholewka A, Kasprzyk T, Stanek A, Sieroń-Stołtny K, Drzazga Z (2016) May thermal imaging be useful in cyclist endurance tests? J Therm Anal Calorim 123(3):1973–1979. https://doi.org/10.1007/s10973-015-4662-5
72. Romão W, Mello D, Neves EB, Dias T, dos Santos AOB, Alkmim R, Vale R (2021) The use of infrared thermography in endurance athletes: A systematic review. Motricidade 17(2):2. https://doi.org/10.6063/motricidade.21116
73. Rynkiewicz M, Korman P, Zurek P, Rynkiewicz T (2015) Application of thermovisual body image analysis in the evaluation of paddling effects on a kayak ergometer. Med Sport 1(68):31–42. https://www.minervamedica.it/en/journals/medicina-dello-sport/article.php?cod=R26Y2015N01A0031
74. Martin JC, Spirduso WW (2001) Determinants of maximal cycling power: crank length, pedaling rate and pedal speed. Eur J Appl Physiol 84(5):413–418. https://doi.org/10.1007/s004210100400
75. Korff T, Romer LM, Mayhew I, Martin JC (2007) Effect of pedaling technique on mechanical effectiveness and efficiency in cyclists. Med Sci Sports Exerc 39(6):991–995. https://doi.org/10.1249/mss.0b013e318043a235
76. Priego Quesada JI, Carpes FP, Salvador Palmer R, Pérez-Soriano P, Cibrián Ortiz de Anda RM (2016) Effect of saddle height on skin temperature measured in different days of cycling. Springerplus 5:205. https://doi.org/10.1186/s40064-016-1843-z
77. Korman P, Kusy K, Straburzyńska-Lupa A, Sillero-Quintana M, Zarębska E, Zieliński J (2024) Exploring the correlation of skin temperature and body composition in athletes under-

going exhaustive physical exercise. J Therm Biol 123:103918. https://doi.org/10.1016/j.jtherbio.2024.103918

78. Ross M, Kargl CK, Ferguson R, Gavin TP, Hellsten Y (2023) Exercise-induced skeletal muscle angiogenesis: impact of age, sex, angiocrines and cellular mediators. Eur J Appl Physiol 123(7):1415–1432. https://doi.org/10.1007/s00421-022-05128-6
79. Casas A (2008) Physiology and methodology of intermittent resistance training for acyclic sports. J Human Sport Exerc 3(1):23–52
80. Correia V, Araújo D, Vilar L, Davids K (2013) From recording discrete actions to studying continuous goal-directed behaviours in team sports. J Sports Sci. https://www.tandfonline.com/doi/abs/10.1080/02640414.2012.738926
81. Afonso J, Peña J, Sá M, Virgile A, García-de-Alcaraz A, Bishop C (2022) Why sports should embrace bilateral asymmetry: a narrative review. Symmetry 14(10):1993. https://doi.org/10.3390/sym14101993
82. Bartlett R, Wheat J, Robins M (2007) Is movement variability important for sports biomechanists? Sports Biomech 6(2):224–243. https://doi.org/10.1080/14763140701322994
83. Grouios G (2004) Motoric dominance and sporting excellence: training versus heredity. Percept Mot Skills 98(1):53–66. https://doi.org/10.2466/pms.98.1.53-66
84. Bishop C, Turner A, Read P (2018) Effects of inter-limb asymmetries on physical and sports performance: a systematic review. J Sports Sci 36(10):1135–1144. https://doi.org/10.1080/02640414.2017.1361894
85. Maloney SJ (2019) The relationship between asymmetry and athletic performance: a critical review. J Strength Cond Res 33(9):2579–2593. https://doi.org/10.1519/JSC.0000000000002608
86. Johnson JM (1998) Physical training and the control of skin blood flow. Med Sci Sports Exerc 30(3):382
87. Lorenzo S, Minson CT (2010) Heat acclimation improves cutaneous vascular function and sweating in trained cyclists. J Appl Physiol 109(6):1736–1743. https://doi.org/10.1152/japplphysiol.00725.2010
88. Duarte TS, Ferreira-Júnior JB, Oliveira JCG, Costa CMA, Coelho DB, Marins JCB, Bara-Filho MG (2025) Is skin temperature associated with training load during preparatory period of a professional volleyball team? J Therm Biol:104038. https://doi.org/10.1016/j.jtherbio.2024.104038
89. Lino-Samaniego Á, de la Rubia A, Sillero-Quintana M (2022) Acute effect of auxotonic and isometric contraction evaluated by infrared thermography in handball players. J Therm Biol 109:103318. https://doi.org/10.1016/j.jtherbio.2022.103318
90. Majano C, Garcia-Unanue J, Fernández-Cuevas I, Escamilla-Galindo V, Alonso-Callejo A, Sanchez-Sanchez J, Gallardo L, Felipe JL (2023) Association between physical demands, skin temperature and wellbeing status in elite football players. Sci Rep 13(1):13780. https://doi.org/10.1038/s41598-023-40396-y
91. Hennessy MP, Parker AW (1990) Electromyography of arrow release in archery. Electromyogr Clin Neurophysiol 30(1):7–17
92. Nishizono H, Shibayama H, Izuta T, Saito K (1987) Analysis of archery shooting techniques by means of electromyography. In: ISBS - Conference Proceedings Archive. https://ojs.ub.uni-konstanz.de/cpa/article/view/2332
93. Kim RN, Lee J-H, Hong SH, Jeon JH, Jeong WK (2018) The characteristics of shoulder muscles in archery athletes. Clin Should Elbow 21(3):145–150. https://doi.org/10.5397/cise.2018.21.3.145
94. Kolayiş İE, Ertan H (2016) Differences in activation patterns of shoulder girdle muscles in recurve archers. Pamukkale J Sport Sci 7(1):1
95. Li WW, Carter MJ, Mashiach E, Guthrie SD (2016) Vascular assessment of wound healing: a clinical review. Int Wound J 14(3):460–469. https://doi.org/10.1111/iwj.12622

96. Montaner Sesmero AM, Llana Belloch S, Gámez Payá J, Montaner Sesmero C (2013) Estudio epidemiológico en pelota valenciana. Revista Internacional de Medicina y Ciencias de la Actividad Fisica y del Deporte 13(50):235–255
97. Pritchard MH, Pugh N, Wright I, Brownlee M (1999) A vascular basis for repetitive strain injury. Rheumatology 38(7):636–639. https://doi.org/10.1093/rheumatology/38.7.636
98. Hildebrandt C, Raschner C, Ammer K (2010) An overview of recent application of medical infrared thermography in sports medicine in Austria. Sensors (Basel, Switzerland) 10(5):4700–4715. https://doi.org/10.3390/s100504700
99. Bongers CCWG, Thijssen DHJ, Veltmeijer MTW, Hopman MTE, Eijsvogels TMH (2015) Precooling and percooling (cooling during exercise) both improve performance in the heat: a meta-analytical review. Br J Sports Med 49(6):377–384. https://doi.org/10.1136/bjsports-2013-092928
100. Costello JT, McInerney CD, Bleakley CM, Selfe J, Donnelly AE (2012) The use of thermal imaging in assessing skin temperature following cryotherapy: a review. J Therm Biol 37(2):103–110. https://doi.org/10.1016/j.jtherbio.2011.11.008
101. Masur L, Brand F, Düking P (2024) Response of infrared thermography related parameters to (non-)sport specific exercise and relationship with internal load parameters in individual and team sport athletes—a systematic review. Front Sports Active Living 6. https://doi.org/10.3389/fspor.2024.1479608
102. Folk GE, Riedesel ML, Thrift DL (1998) Principles of integrative environmental physiology. Austin & Winfield
103. Lim CL, Byrne C, Lee JK (2008) Human thermoregulation and measurement of body temperature in exercise and clinical settings. Ann Acad Med Singap 37(4):347–353
104. Grgic J (2022) What is the effect of paracetamol (acetaminophen) ingestion on exercise performance? Current findings and future research directions. Sports Med 52(3):431–439. https://doi.org/10.1007/s40279-021-01633-4

Chapter 9
Infrared Thermography in Water Sports

Ricardo Vardasca, Adérito Seixas, Joaquim Gabriel, and João Vilas-Boas

Abstract Like any other sport, water sports can also be objectively assessed through a physiological method such as thermal imaging. These sports present an extra parameter, which is the particular environment where it happens. This chapter outlines the underlying concepts, challenges and methods, along with the related published research to date in the field with practical examples and future lines of research are proposed.

9.1 Introduction

Exercise training causes adaptations in the skin blood flow response to exercise, including changes in cutaneous microvascular reactivity. Increases in skin blood flow at a given exercise core temperature appear to be related to the expansion of

R. Vardasca (✉)
LABIOMEP, INEGI, LAETA, University of Porto, Porto, Portugal

ISLA Santarém – Instituto Politécnico, Santarém, Portugal

LABIOMEP, CIFI2D, Faculty of Sport, University of Porto, Porto, Portugal
e-mail: ricardo.vardasca@islasantarem.pt

A. Seixas
LABIOMEP, INEGI, LAETA, University of Porto, Porto, Portugal

Faculdade de Ciências da Saúde, Universidade Fernando Pessoa, Porto, Portugal
e-mail: aderito@ufp.edu.pt

J. Gabriel
LABIOMEP, INEGI, LAETA, University of Porto, Porto, Portugal

Faculdade de Engenharia, Universidade do Porto, Porto, Portugal
e-mail: jgabriel@fe.up.pt

J. Vilas-Boas
LABIOMEP, CIFI2D, Faculty of Sport, University of Porto, Porto, Portugal
e-mail: jpvb@fade.up.pt

J. I. Priego-Quesada (ed.), *Application of Infrared Thermography in Sports Science*,
https://doi.org/10.1007/978-3-031-93311-0_9

blood volume and increased cardiac output that characterize the trained state. In contrast, adaptations in the cutaneous microvasculature are mediated by changes in the biological availability or activity of endothelium-derived vasoactive compounds [1]. Changes in skin temperature are known to be affected by blood flow [2], and skin temperature are passive of being passively captured and posteriorly analyzed through thermal imaging, a non-contact, non-invasive and non-ionizing imaging technique that registers the surface emitted thermal radiation into radiometric temperature maps [3].

Water sports, swimming being one of the major examples, induce changes in blood flow through exercise, where heat is produced by the muscles, and the contact with water highly influences skin thermoregulation. These two aspects can be object of monitoring using thermal imaging to investigate skin temperature changes.

One of the basic functions of the human body is to keep core body temperature constant between 36 and 37.5 °C, keeping its vital functioning. This is maintained by the hypothalamus, which receives constant information from the autonomous nervous system, taking advantage of the different body thermal sensors placed close to the vital organs internally and at the periphery at the skin dermis. The skin is the major organ of the human body and plays a major role in the interface with the surrounding environment. The neutral zone for the skin thermal body sensors is within an ambient temperature between 18 and 24 °C at a relative humidity of less than 50%. Below this value the temperature recovery mechanisms such as vasoconstriction, muscle shivering and increasing of the metabolic rate are activated. On the other hand, above 24 °C the temperature release mechanisms are activated, consisting in vasodilatation and sweating [4, 5].

Water sports and in particular swimming, like any other sport, challenge the thermal balance of the human body, increasing the skin temperature [6] through a complex thermoregulation process that intends to preserve core temperature around 37.5 °C [7]. Despite the study of Wade and Veghte [8] and the increasing number of papers in sport research reporting the use of thermal imaging, its use in swimming research is limited and few researchers have informed the Use of this Technology in Swimming Activities [6, 9, 10].

9.2 Relationship Between Skin Temperature, Exercise and Water Exposure

Water is one of the simplest examples of matter that can easily change its state with the internal molecules' activity. It is known that at temperature below 0 °C it solidifies, become solid, and at temperatures above 100 °C it boils, becoming a gas. Within these intervals the water stays in its liquid state, when in contact with it the human body starts to exchange thermal energy through forced convection. The physical properties of the water change according to its state, at a liquid state its emissivity value is of 0.97 [11], very close to the human skin value (0.98). However,

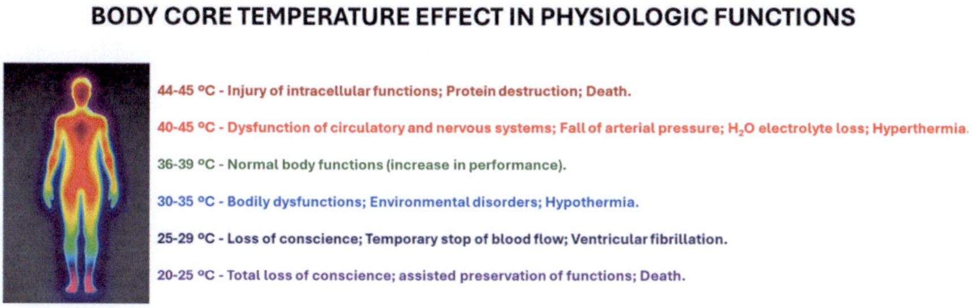

Fig. 9.1 Effect that body core temperature has in the physiological functions, describing its undesired risks

the presence of water in the skin causes an undesirable effect of blocking the spontaneous skin ability of radiating thermal energy. It worth also worth mentioning that water is a much better heat conductor than air, about 25 times more effective [5]. The first major challenge of this type of research is to measure the thermal effect of any exercise, minimizing the water-associated issues related to thermoregulation. Good understanding of how water conditions and exposure can affect thermoregulation and the effect on human body core temperature is required to prevent undesired loss of physiological functions, as presented in Fig. 9.1 [12].

In water temperatures of 25 °C and below, oxygen consumption (VO_2) increases at a faster rate than air at temperatures of 25 °C, and cardiac frequency is reduced [13, 14]. Oxygen consumption increases because the muscle tremor generates thermogenesis and possibly a decrease in mechanical ability during exercise. The cardiac output increases linearly with an increase in peripheral vasoconstriction in cold water. When water temperatures are at 30 °C and above, metabolic rate and cardiac frequency are the same in both air and water environments.

If the increase in pulse volume is because of hydrostatic pressure from the increasing peripheral vasoconstriction, then cardiac frequency could be depressed in warm water. Heat dissipates faster in cold water, and swimming in submaximal temperatures reduces the human body's core temperature, even if the metabolic rate increases [15]. The increase in metabolic rate is inversely proportionate to the increase in water temperature. The levels of lactic acid are increased during submaximal swimming distances in cold water, while muscle temperatures could alter physical and chemical functions in the muscle's interior during swimming, ultimately resulting in the reduction of mechanical ability in cold water [12]. Immersion of the skin in water results in a characteristic whitening and wrinkled appearance associated with the uptake of significant quantities of water. The rate of evaporation from hydrated skin surfaces is considerably more than that from non-hydrated areas. Although, prolonged immersion suppresses sweating on the palm of the hands and feet sole [16], not being these areas recommended for skin temperature assessment with thermal imaging.

After stabilization of body temperature at rest in the coldest possible water (12 °C), exercise reduced internal insulation only in muscular parts of the limbs.

Exercise also increased heat loss elsewhere by exposing skin of protected regions such as flexural surfaces of joints. During exercise, total heat production increased rather more than heat loss in unreactive subjects, but less than loss in subjects whose heat production had already risen to a high level when they were at rest in cold water [17].

In warm water (37 °C), tissue insulations were lower and much more uniform between subjects and between different body regions than in the cold. Even in the warm, however, insulation remained rather higher in fat than thin subjects, higher at rest than during exercise, and usually higher in the limbs than the upper trunk [17]. Repeated cold water immersion produces acclimation to cold air in humans and the cold acclimation is primarily of the insulative type in that skin temperatures are lower, probably due to greater sympathetic nervous system activation mediating stronger cutaneous vasoconstriction. The degree of cold acclimatization achieved is probably related to the intensity (frequency and duration of reduction in core temperature) of the acclimatization procedures employed [18].

Water and electrolyte balances are critical for a person's ability to thermoregulate and perform exercise in the heat. During exercise in the heat, water and electrolyte losses primarily occur from sweating. A person's sweating rate is dependent upon the climatic conditions, clothing worn and exercise intensity. Daily fluid requirements range (for sedentary to very active persons) from 2 to 4 l day in a cool climate up to 8–16 l/day in very hot climates. Over a 24 h period, fluid and electrolyte losses will be replaced if people consume their normal diet. During exercise, sweat output often exceeds water intake, producing a water deficit or hypohydration. The water deficit lowers both intracellular and extracellular fluid volumes [19].

It also results in plasma hypertonicity and plasma hypovolemia, both of which adversely affect heat loss responses. Aerobic exercise tasks are likely to be adversely affected by hypohydration, with the potential effect being greater in warm environments. Hypohydration increases heat storage by reducing sweating rate and skin blood flow responses for a given core temperature. Hyperhydration provides no advantages over euhydration regarding thermoregulation and exercise performance in the heat [19].

A cold beverage taken in ecological conditions (i.e., small volumes taken regularly during training sessions) had a significant effect on body core temperature, thermal sensation, and heart rate during both training and high-intensity training in high-level swimmers in a tropical climate. The effects were optimal in the evening, certainly in relation to a more stressful environment and/or sensation. Because the aim of using cold water is to decrease the impact of stress to improve performance, studies exploring the effect of cold-water ingestion on performance during real competitive events are needed, in both acclimated and unacclimatized high-level subjects [20]. The Fig. 9.2 shows the effects that water temperature has on the human body functions [12].

Water temperature immediately affects bodily functions upon entry into the water, throughout immersion and during swimming, gradually provoking changes, which affect the condition of the organism, as well as its overall performance. The increase or decrease of the water temperature beyond the normal range may act

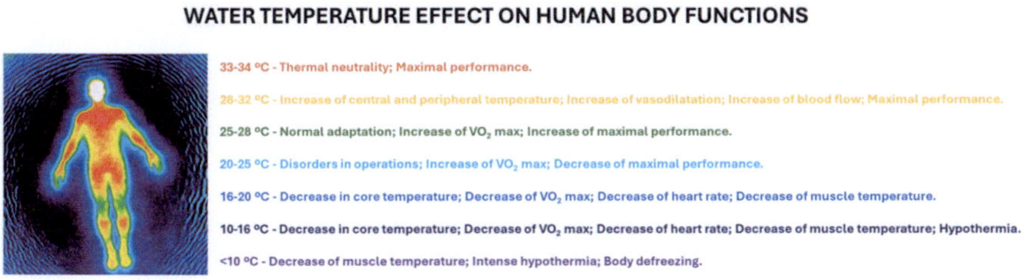

Fig. 9.2 Effect of water temperature in human body functions

therapeutically, entertain, or to enhance the body's physical fitness, but it does not contribute to the improvement of performance [12].

When the human body is immersed in water, body temperature changes. These changes are due to and influenced by several factors, such as: water temperature, air temperature, air currents, environmental moisture, the composition and percentage rate of tissues in the body, the percentage rate of fat tissue, and the percentage of immersion of the body in water. During exercise in water temperatures between 26 and 28 °C, in healthy adult individuals, a low and pleasant body temperature is observed. This temperature is present in most swimming pools and they have a strong effect on the human body, such as: only a few minutes after immersion of the body in water, skin temperature is equalized with water temperature (with a difference of approximately 1 °C), since heat conductivity is 25 times greater in water than in air [12].

In a state of rest-immobility, in water temperatures of 33–34 °C, the human body acquires thermal neutrality; this allows for mild swimming activities, with entertaining and therapeutic properties. These high temperatures are appropriate for mild activities for infants and children, the elderly, and fragile or injured individuals [12].

At water temperatures below 30 °C, hypothermia progressively sets in. In competitive swimming, any increase or decrease of the water temperatures beyond the normal [21], chain reactions and effects between organs, systems, mechanisms and substances are observed, on the surface and inside the body; these result in changes in their functions and/or activities, and strongly affect the outcome of the swimming effort [12].

The Table 9.1 presents the survival times of an average human subject immersed in cold water without protective clothing, characterizing the loss of dexterity, time of exhaustion or unconsciousness and expected time of survival [22].

When the water temperature is in the range from 28 to 32 °C, there is a tendency for performance to improve, primarily in speed events. However, at temperatures higher than that, performance decreases and other goals are pursued. In temperatures much higher than this range, staying in the water becomes impossible and pronounced phenomena of hypothermia occur [12].

At water temperatures ranging from 25 to 27 °C, there is a tendency for performance to improve, primarily in endurance events. However, at lower temperatures

Table 9.1 The survival times of an average human subject immersed in cold water without protective clothing

Water temperature (°C)	Loss of dexterity with no protective clothing	Exhaustion or unconsciousness	Expected time of survival
>26.5	2–12 h	Indefinite	Indefinite
21–26.5	1–2 h	2–12 h	3 h to indefinite
15.5–21	30–40 min	2–7 h	2–40 h
10–15.5	10–15 min	1–2 h	1–6 h
4.5–10	Under 5 min	30–60 min	1–3 h
0.3–4.5	Under 3 min	15–30 min	30–90 min
<0.3	Under 2 min	Under 15 min	Under 15–45 min

performance drops, and perhaps there is a need to investigate the appropriate temperature for marathon and hyper-marathon swimming [12].

Within the interval between 25 and 20 °C in water temperature, malfunctions appear in the organism, the swimming effort and performance is reduced and remaining in the water becomes difficult, with phenomena of hypothermia [12].

If the water temperature ranges between 20 and 12 °C, pronounced malfunctions appear in the organism, the swimming effort and performance is reduced and remaining in the water becomes difficult, with gradual phenomena of hypothermia. Only practiced, fit individuals can swim in these conditions [12].

When water temperatures drop below 12 °C, pronounced and immediate malfunctions appear in the organism. There is a complete inability to swim or perform, inability to remain in the water for longer than 1 h, and pronounced phenomena of hypothermia and cooling of the body [12].

An investigation in the effects of regular winter swimming on brown adipose tissue (BAT) activity and overall thermoregulation was performed in Denmark [23]. It involved young, healthy male participants, including experienced winter swimmers who engaged in cold-water immersion followed by hot sauna sessions 2-3 times per week, and a control group with no such exposure. The researchers measured core body temperature, BAT activity, cold-induced thermogenesis, and supraclavicular skin temperature (a region associated with BAT presence) in both groups.

It was found from that study [23] that:

- Winter swimmers exhibited a lower core body temperature and reported reduced thermal comfort compared to the control group.
- Upon cold exposure, winter swimmers demonstrated a greater increase in energy expenditure, indicating enhanced cold-induced thermogenesis.
- Winter swimmers showed a more significant rise in supraclavicular skin temperature during cold exposure, suggesting heightened BAT activity.
- All participants experienced a nocturnal decrease in supraclavicular skin temperature; however, winter swimmers exhibited a distinct peak between 4:30–5:30 a.m., indicating altered BAT thermoregulation.

It suggests that habitual winter swimming may lead to enhanced BAT activity and cold-induced thermogenesis, contributing to improved thermoregulatory responses. However, further research is required to address the limitations and fully understand the implications of these findings.

Another study aiming to examine how regular exposure to cold water affects body surface temperature in habitual winter swimmers was conducted in Poland [24]. It involved 20 male subjects, which underwent approximately four-minute immersions in ice-cold water during the winter swimming season, totaling 14 immersions. Body surface temperatures were measured before and after immersion using a thermographic camera (A20M) connected to the FLIR ThermaCam Researcher 2.8 program. Temperatures were recorded for both the anterior and posterior body surfaces.

Results showed a significant decrease in temperature was observed in 12 body areas after immersion, particularly in the lower limbs and torso, accounting for 60% of the observed changes. The anterior (frontal) body areas, which are generally less muscular, exhibited a more substantial temperature drop compared to the posterior areas. A correlation was found between body surface temperatures before and after immersion, influenced by the cooling environment (water and air). The strongest associations were noted in the lower limbs during water immersion.

This study [24] provided insights into how regular winter swimmers' bodies adapt to cold water immersion, particularly in terms of body surface temperature changes. However, further research is necessary to explore individual variability, long-term adaptations, and the broader health implications of this practice.

9.3 Assessment of Skin Temperature in Water Sports

This part is divided into five subsections: acclimatization, skin drying, swimming, injury prevention and skin temperature and performance.

9.3.1 Acclimatization

Acclimatization is important to achieve a pre-imaging thermal equilibrium between the body skin and surrounding environment [25], but the time required is not yet consensual and previous published research recommends time periods of 10 min [26], 15 min [27] and 20 min [28]. These authors suggest that acclimatization should be undertaken at rest in an examination room with controlled thermal environment to obtain baseline thermograms when using dynamic thermal imaging. Swimming exercise is performed in the water and if baseline thermograms are obtained in such conditions the analysis would be wrong and meaningless.

9.3.2 *Drying*

Water is an isotropic substance and opaque to skin radiation [29]. Therefore, swimmers must be dry before thermal imaging acquisition. In this process, skin friction must be avoided since it can influence local microcirculation and cause unwanted increases in the skin temperature values measured with the thermal images.

These issues were researched by Zaidi et al. [6]. However, their work was limited, further research is recommended.

9.3.3 *Swimming*

The first known application of thermal imaging in water sports was performed by Wade and Veghte [8] on investigating the regional differences in skin temper- ature in four competitive swimmers of varying body compositions, prior to and after 5 min of immersion in water at 23.5 °C, and after a 500-m freestyle swim at a training pace. Decreases in skin temperature that correlated with skinfold thickness were seen post-immersion. After swimming, skin temperatures were highest in regions overlying active muscle masses (deltoids, trapezius, triceps and biceps brachi, and pectorals) and were independent of skinfold thickness. It was also found that thin swimmers were found to dissipate faster the heat than fat subjects.

Another study a few years later attempted to determine the thermal and metabolic effects of wearing a rubberized wet suit when compared with a competitive swimming suit while swimming for 30 min in 20.1, 22.7, and 25.6 °C water, with a 15 min resting period between them. Nine subjects were investigated, and skin temperature was obtained not with thermal imaging but with skin temperature sensors (TX-4 Columbus instruments). The rubberized wet suit presents a significant higher body temperature than the competitive swimming suit, that difference decreased when the water temperature becomes warmer [30].

Hue et al. demonstrated that training in a tropical climate significantly enhances performance in swimmers (400 m freestyle), 30 days after return to neutral climate [31]. A sample of 16 swimmers was used and the climates compared were the tropical and the neutral, there were three moments used, the baseline, 10 days after and 30 days after. No relation was obtained between the 10- and 30-days post training. A preliminary experimental study was undertaken for studying the feasibility of using thermal imaging in the discipline of swimming and for quantifying the influence of the swimming technique (within the framework of a well defined protocol) on the distributions of skin temperatures [6]. They have used an uncooled FLIR ThermaCAM FLIR SC1000 (sensor array size of 256 × 256, NETD of <70 mK at 30 °C and measurement repeatability of ±2% of the overall reading) and a cooled CEDIP TITANIUM HD 560 M (sensor array size of 640 × 256, NETD of <40 mK at 30 °C and measurement repeatability of ±2% of the overall reading) thermal cameras. Significant variations were found in the cutaneous temperature according to the swimming styles. From the examination of infrared thermography, the whole-body surface has respectively increased by 2.16 °C for the butterfly, 2.56 °C for the backstroke, 1.78 °C for the breaststroke and 2.00 °C for the freestyle, after the single participant had performed the 400 m test in each technique on different days. The swimmer had spent 10 min of acclimatization period in the pool with water at 27 °C. The thermal images were taken from swimmers after the period of acclimatization and after 100 m of each swimming technique (butterfly, backstroke, breaststroke and free style). Between each swimming technique performance there was a 10-min rest period to achieve thermal balance and the thermal imaging measurements were performed after this period were the baseline for the swimming technique performed after. The authors presented changes in skin temperature after each swimming technique and concluded that each technique changed the skin temperature differently. Although this study presented a new methodology to assess skin temperature after swimming exercise, some questions are still open, only one swimmer was assessed and the fact that all swimming techniques were performed on the same day since, with a 10-min resting period to achieve thermal balance between different exercises, is it enough? Does fatigue play any role in the observed temperature changes? Nothing was mentioned describing the velocity and intensity of each swimming technique, which may bias the presented results.

Another study compared the physical activities of swimming and practicing yoga in pregnant women [32]. It was found that swimming reduced the skin temperature in the breast and belly regions whereas yoga as presented an increase in most skin temperature areas, this may be due to the cooling effect of water. A FLIR T335 (sensor array size of 320 × 240, NETD of <50 mK at 30 °C and measurement repeatability of ±2% of the overall reading) uncooled thermal camera was used, the participants had engaged in a 10 min acclimatization period in a controlled environment room, the images after practice were taken 10 min after the exercise at the same conditions and in the swimming group after the participants dried the skin with towels without rubbing [32]. The acclimatization period and the drying method and subsequent 10 min waiting period may have biased the obtained results.

Five of twenty muscle regions have notable tendencies of increased temperatures, including a front part of deltoideus at right side which is very active by spreading arm forward and beginning of swimming stroke. In nine areas of main agonist swimmers' movement forward included—triceps brachii, it was found no significant lowering temperatures. That can be due to the cooling effect of the swimmer in the water. A Fluke TiR (sensor array size of 160 x 120, NETD of <90 mK at 30 °C and measurement repeatability of ±2% of the overall reading) uncooled thermal camera was used. The water temperature was 27 °C, it was no acclimatization in the water and the post-exercise images were taken 15 min after coming out of the swimming pool and being dried with a towel avoiding rubbing [33]. The use of an acclimatization period with different conditions from the exercise and a non-recommend low specifications thermal camera for human skin screening [34, 35] may have biased the obtained results.

A study investigated the possible relationship between energy expenditure and skin temperature during 2 × 300 m protocols (front crawl and backstroke), performed at the same intensity and velocity, by 2 male experimented swimmers [9]. Both swimming techniques increased skin temperature but with different thermal patterns. Higher values of energy expenditure during the 300 m backstroke test were found comparing with the values obtained for the same distance performed in front crawl and higher difference between skin temperatures values were associated with lower energy expenditure values. The swimmers have refrained an acclimatization period of 10 min immersed still in water at the pool till the head at a temperature of 27 °C. A FLIR A325sc (sensor array size of 320 × 240, NETD of <70 mK at 30 °C and measurement repeatability of ±2% of the overall reading) uncooled thermal camera was used.

A research conducted in Czech Republic has found that there is a significant increase (normalized units) 15 min after swimming in triceps brachii (on the right prior to swimming 0.950 and after swimming 0.994; on the left prior to swimming 0.947 and after 0.990), and in side, rear and front parts of the deltoid muscles [36]. On the contrary, there was a significant relative decrease in temperature in pectoralis, rhombic and lower trapezius, erector spinae lumbalis and latissimus dorsi. It can be concluded that swimming 1000 m breaststroke affected significant increase in the temperature of regions of interest, i.e., corresponding to agonists and synergists of upper extremities for the swimmer's forward motion. A relative decrease in temperature occurred rather in body muscles. The problem of biased results due to water-cooling was solved by using thermograms taken only in the 15th minute after getting out of water and calculating relative temperatures with normalized units. A Fluke TiR (sensor array size of 160 x 120, NETD of <90 mK at 30 °C and measurement repeatability of ±2% of the overall reading) uncooled thermal camera was used, which does not meet the minimum specifications for human skin screening [34, 35].

A study attempted to develop a methodology to investigate swimmers and verify whether there is a considerable change in skin temperature during different swimming techniques (7 × 200m protocol in front crawl and backstroke) and, if confirmed, to identify the most affected regions [10]. A FLIR A325sc (sensor array size of 320 × 240, NETD of <70 mK at 30 °C and measurement repeatability of ±2% of the overall reading) uncooled thermal camera was used. A total of 10 experienced swimmers had been investigated in two consecutive days at same time, performing on the first day the crawl technique and on the second day the back- stroke technique.

The regions used were: For anterior view (Fig. 9.3), anterior neck right; anterior neck left; anterior torax right; anterior torax left; anterior arm right; anterior arm left; anterior forearm right; anterior forearm left; anterior abdomen right; anterior abdomen left; anterior thigh right; anterior thigh left; anterior leg right; and anterior leg Left. For dorsal view (Fig. 9.4), dorsal neck left; dorsal neck right; dorsal upper back left; dorsal upper back right; dorsal arm left; dorsal arm right; dorsal middle back left; dorsal middle back right; dorsal forearm left; dorsal forearm right; dorsal lower back left; dorsal lower back right; dorsal thigh left; dorsal thigh right; dorsal leg left; and dorsal leg right (DLR).

There were collected half-body view images and full-body view images, the first being collected at 2 m and the second at 4 m. In both types of views, the ROIs were

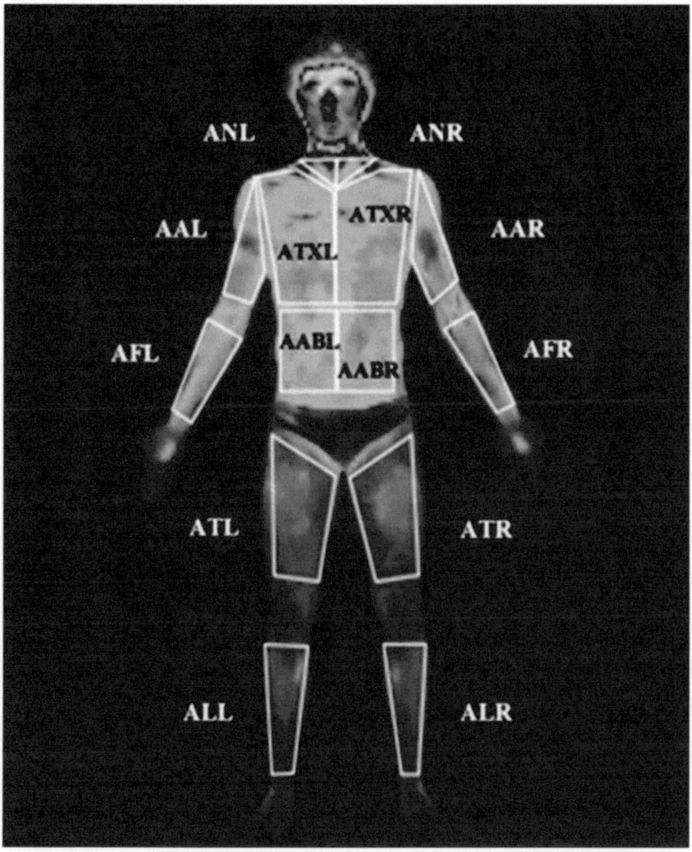

Fig. 9.3 Anterior regions of interest used in the study: Anterior Neck Right (ANR); Anterior Neck Left (ANL); Anterior Torax Right (ATXR); Anterior Torax Left (ATXL); Anterior Arm Right (AAR); Anterior Arm Left (AAL); Anterior Forearm Right (AFR); Anterior Forearm Left (AFL); Anterior Abdomen Right (AABR); Anterior Abdomen Left (AABL); Anterior Thigh Right (ATR); Anterior Thigh Left (ATL); Anterior Leg Right (ALR); Anterior Leg Left (ALL)

analyzed in terms of mean temperature and between bilateral ROIs, the thermal symmetry [37] was analyzed.

Thermal symmetry was defined as the "degree of similarity" between two ROIs, mirrored across the human body's longitudinal main axes which are identical in shape, identical in size and as near identical in position as possible. The degree of similarity is measured in terms of the respective mean temperatures obtained in the ROIs [37].

The capture protocol consisted of the swimmer being subjected to a 10 min acclimatization period in the pool, maintaining a static position and immersing all parts of the body, except the head. To avoid the negative influence of water in thermograms, the swimmer was quickly dried with microfiber towels, avoiding any friction (Fig. 9.5).

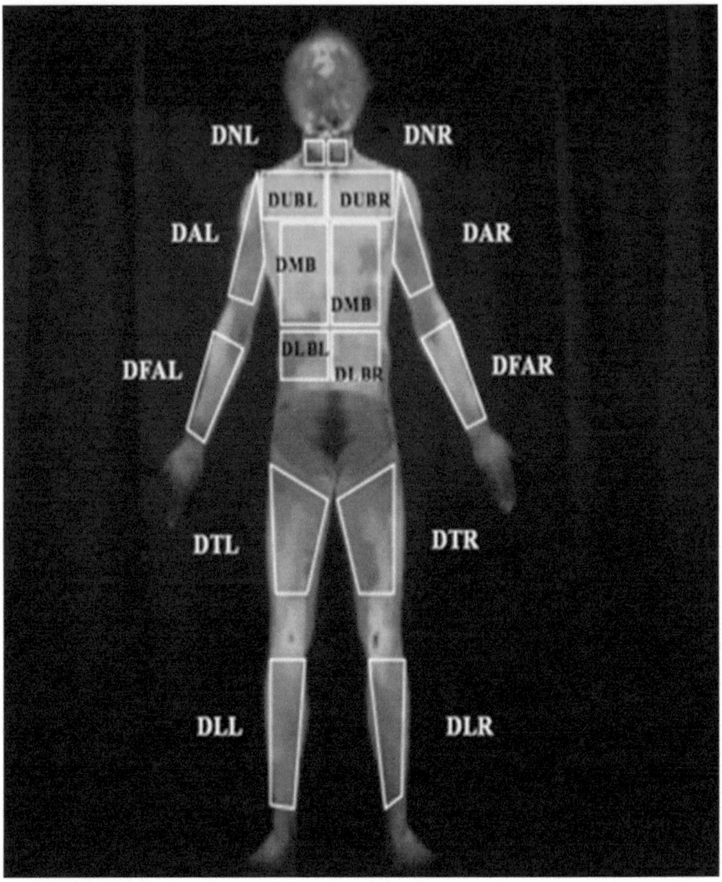

Fig. 9.4 Dorsal regions of interest used in the study: dorsal neck left (DNL); dorsal neck right (DNR); dorsal upper back left (DUBL); dorsal upper back Right (DUBR); dorsal arm left (DAL); dorsal arm right (DAR); dorsal middle back left (DMBL); dorsal middle back right (DMBR); dorsal forearm left (DFAL); dorsal forearm right (DFAR); dorsal lower back left (DLBL); dorsal lower back right (DLBR); dorsal thigh left (DTL); dorsal thigh right (DTR); dorsal leg left (DLL); dorsal leg right (DLR)

Regarding the differences in temperature distribution associated with exercise. In all studied regions, the temperature variation was greater than zero, which indicates an increase in skin temperature, as can be seen in Fig. 9.6 for anterior ROIs of half-body views in both swimming techniques and in Fig. 9.7 for posterior ROIs. The regions with higher temperature increases for both techniques are the upper limb (arm and forearm) and upper trunk region (thoracic, upper back and middle back). Although the abdomen and lower back are located in the trunk, they do not indicate significant differences in temperature, as well as, in the neck and leg regions. Example full-body view images, including pre and post-swimming technique exercise) are presented from Figs. 9.8, 9.9, 9.10, 9.11, 9.12, and 9.13.

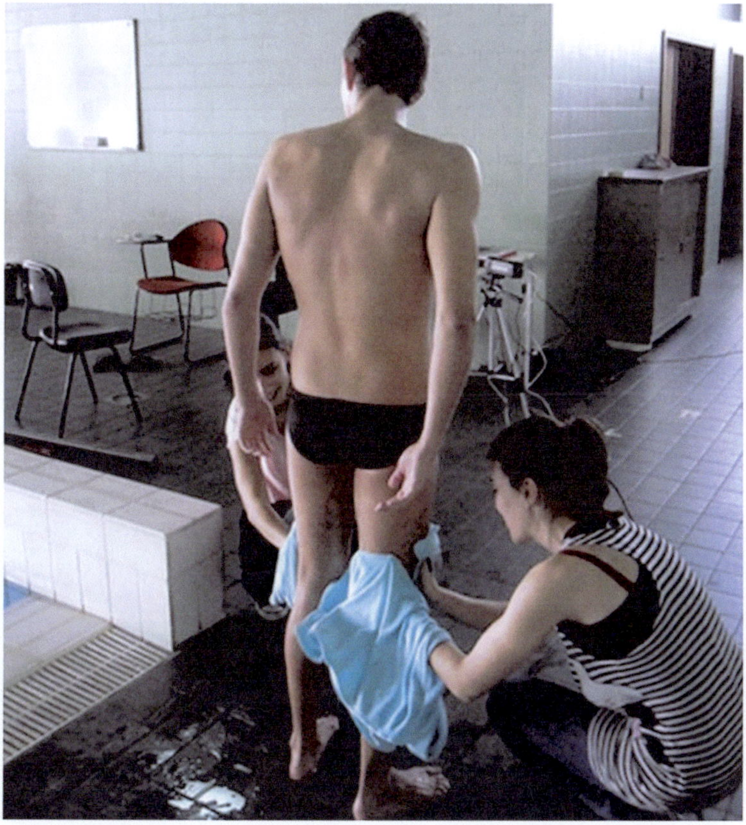

Fig. 9.5 Example of participants drying without any friction using microfiber towels

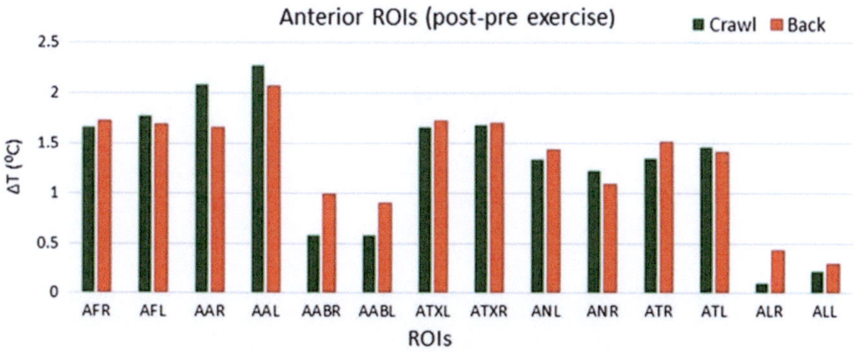

Fig. 9.6 Anterior half-body views regions of interest variation post-pre exercise for both swimming techniques. Anterior forearm right (AFR); anterior forearm left (AFL); anterior arm right (AAR); anterior arm left (AAL); anterior abdomen right (AABR); anterior abdomen left (AABL); anterior thorax right (ATXR); anterior thorax left (ATXL); anterior neck right (ANR); anterior neck left (ANL); anterior thigh right (ATR); anterior thigh left (ATL); anterior leg right (ALR); anterior leg left (ALL)

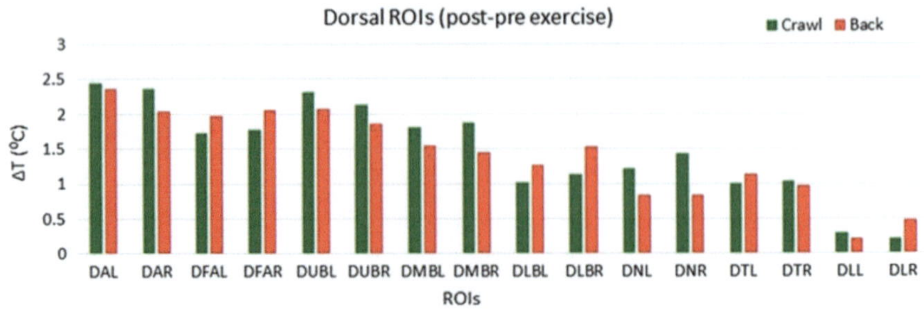

Fig. 9.7 Dorsal half-body views regions of interest variation post-pre exercise for both swimming techniques. Dorsal arm left (DAL); dorsal arm right (DAR); dorsal forearm left (DFAL); dorsal forearm right (DFAR); dorsal upper back left (DUBL); dorsal upper back right (DUBR); dorsal middle back left (DMBL); dorsal middle back right (DMBR); dorsal lower back left (DLBL); dorsal lower back right (DLBR); dorsal neck left (DNL); dorsal neck right (DNR); dorsal thigh left (DTL); dorsal thigh right (DTR); dorsal leg left (DLL); dorsal leg right (DLR)

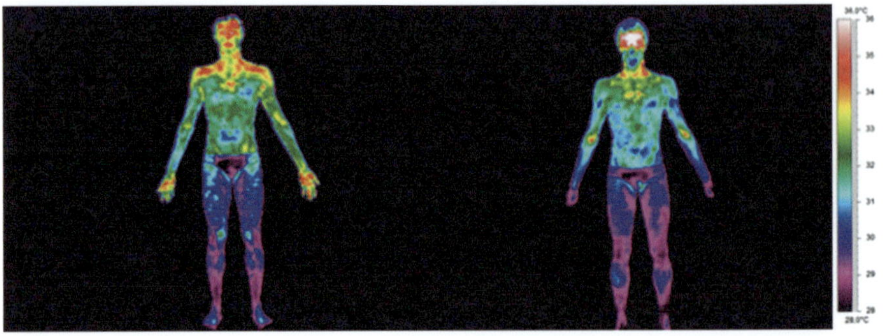

Fig. 9.8 Example of thermal images of pre (*left*) and post (*right*) exercise in full-body anterior view

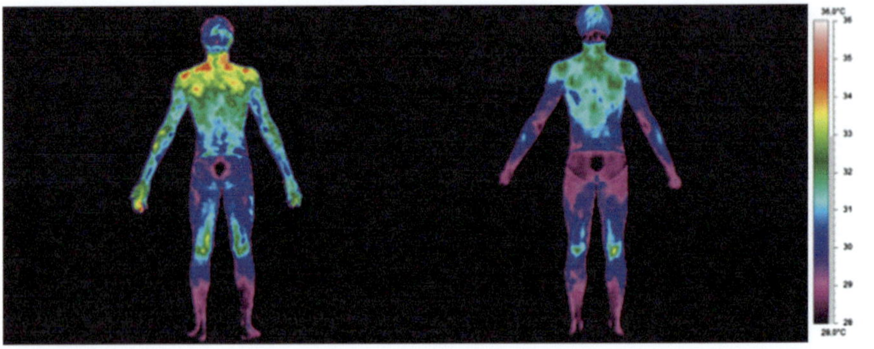

Fig. 9.9 Example of thermal images of pre (*left*) and post (*right*) exercise in full-body dorsal view

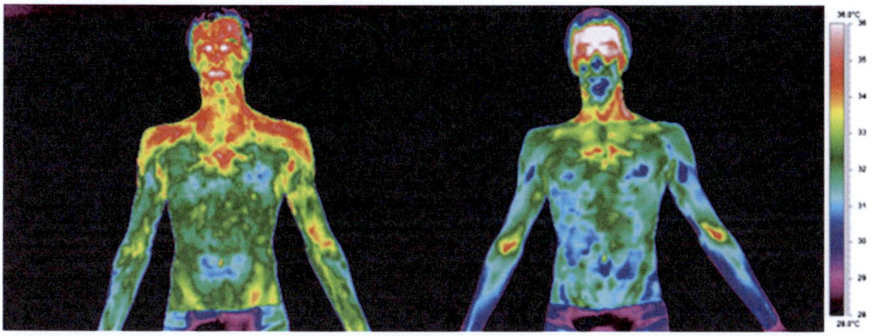

Fig. 9.10 Example of thermal images of pre (*left*) and post (*right*) exercise in half-body anterior-top view

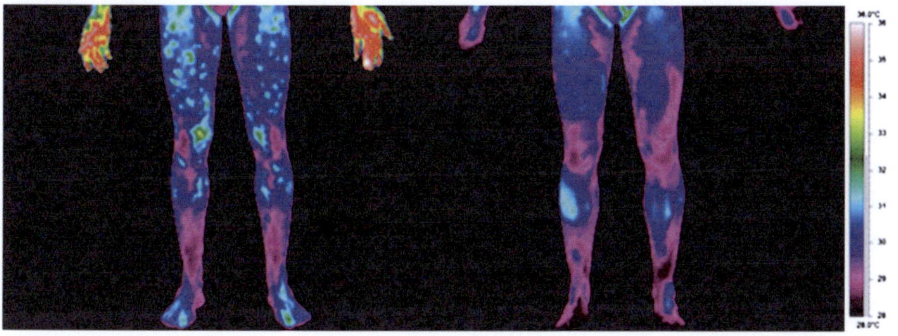

Fig. 9.11 Example of thermal images of pre (*left*) and post (*right*) exercise in half-body anterior-bottom view

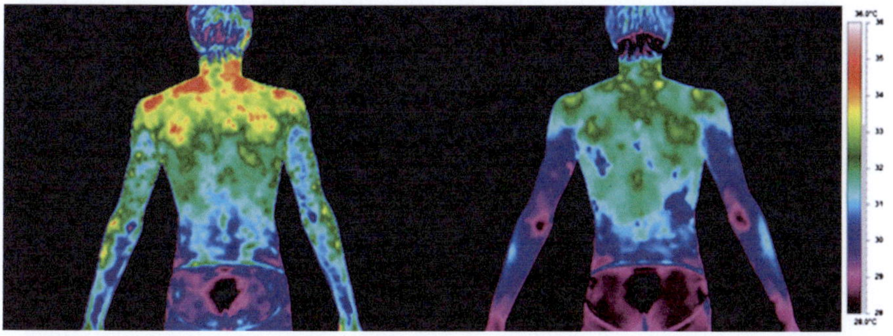

Fig. 9.12 Example of thermal images of pre (*left*) and post (*right*) exercise in half-body dorsal-top view

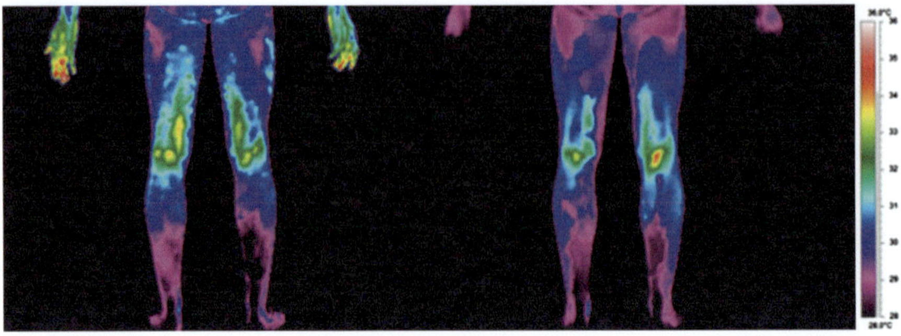

Fig. 9.13 Example of thermal images of pre (*left*) and post (*right*) exercise in half-body dorsal-bottom view

The temperature difference between left and right regions did not exceed more than 0.5 °C for all ROIs. These differences were expected due to the health condition of athletes and also because the swimming techniques performed were symmetrical, only small noticeable changes were found in the arms (for crawl swimming technique) and thighs (for backstroke swimming technique) regions (Figs. 9.14, 9.15, 9.16, and 9.17). There was no significant variation between full and half-body view, although the usage of half-body views is recommended because it allows bigger regions of interest and future better discrimination in the case of any pathology.

One weakness of this research was not having the participants again inside water during the time of swimming practice for better estimation of the difference between effect of exercise and water immersion.

Both swimming techniques significantly increased skin temperature in every ROI ($p < 0.05$); however, technique discrimination was not possible since no significant differences were found. From the analysis through the related-samples Wilcoxon signed rank test, it is possible to observe that there is statistical evidence ($p < 0.05$) for every ROI between pre and post-exercise in half-body views. Differentiating swimming techniques through the ROIs and their differences in mean temperatures (post-pre exercise) was not possible due to the absence of statistical evidence in avoiding the null hypothesis ($p > 0.05$) with the independent-samples Kruskal–Wallis test. No significant statistical evidence of differences was found between the two studied swimming techniques [10].

Another research which aimed to assess the symmetrical and dynamics of skin temperature changes in Polish swimmers after intensive exercise using infrared thermography (IRT) [38, 39]. A total of 12 male swimmers with 11-16 years of swimming experience participated in the study; each participant performed a 2-minute front crawl exercise on a VASA Swim Ergometer set to a load of 5. Skin surface temperatures of the back and upper limbs were measured using a FLIR E60 infrared camera at periods such as before exercise, immediately after, and during a 12-minute recovery period.

A significant decrease in skin surface temperature was observed in all assessed areas immediately after exercise, particularly in the back and posterior arms.

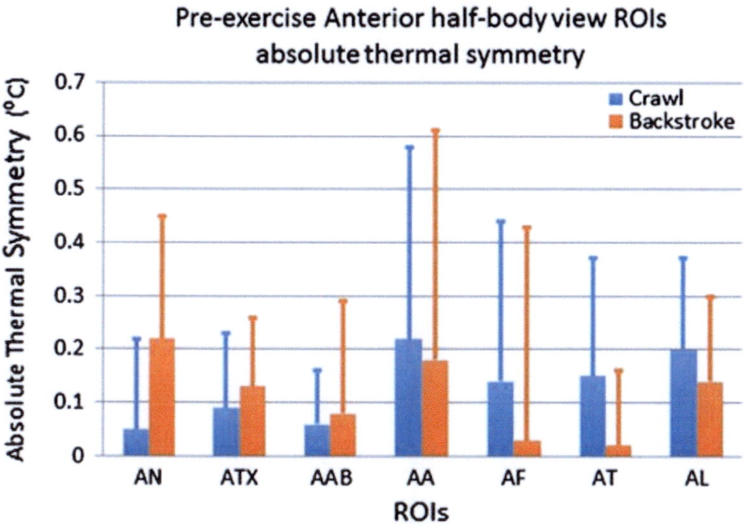

Fig. 9.14 The thermal symmetry absolute values (*left–right* ROI) for both swimming techniques pre-exercise in anterior half-body view. Anterior neck (AN); Anterior Torax (ATX); Anterior Abdomen (AAB); Anterior Arm (AA); Anterior Forearm (AF); Anterior Thigh (AT); Anterior Leg (AL)

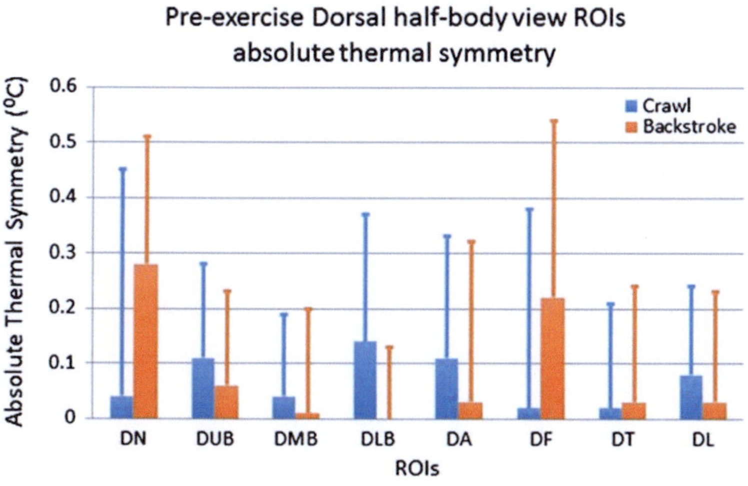

Fig. 9.15 The thermal symmetry absolute values (*left–right* ROI) for both swimming techniques pre-exercise in dorsal half-body view. Dorsal neck (DN); dorsal upper back (DUB); dorsal middle back (DMB); dorsal lower back (DLB); dorsal arm (DA); dorsal forearm (DF); dorsal thigh (DT); dorsal leg (DL)

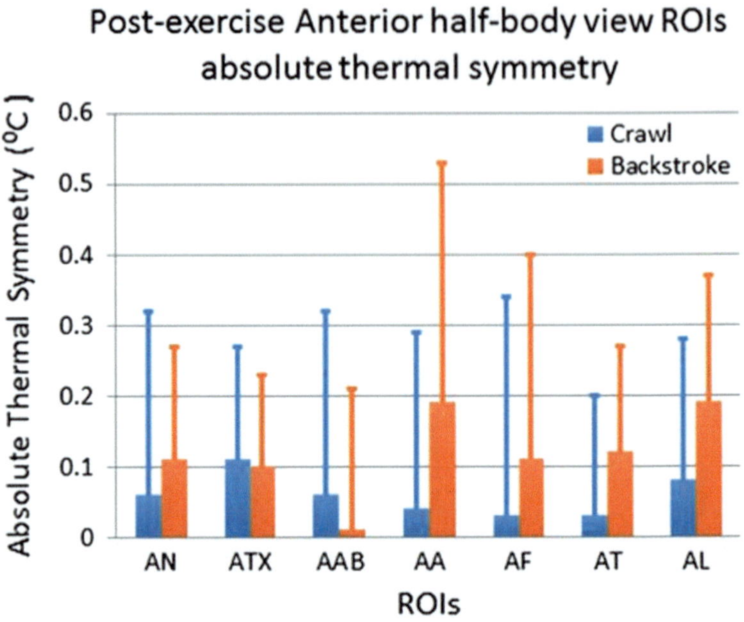

Fig. 9.16 The thermal symmetry absolute values (*left–right* ROI) for both swimming techniques post-exercise in anterior half-body view. Anterior Neck (AN); anterior torax (ATX); anterior abdomen (AAB); anterior arm (AA); anterior forearm (AF); anterior thigh (AT); anterior leg (AL)

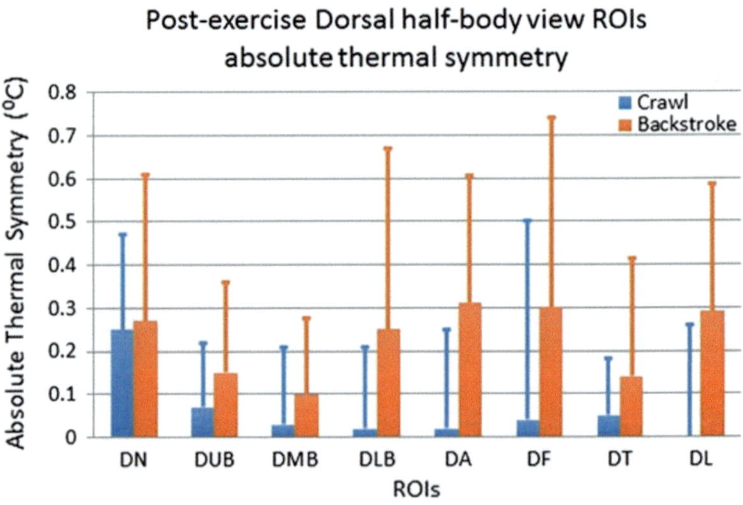

Fig. 9.17 The thermal symmetry absolute values (*left–right* ROI) for both swimming techniques post-exercise in dorsal half-body view. Dorsal neck (DN); dorsal upper back (DUB); dorsal middle back (DMB); dorsal lower back (DLB); dorsal arm (DA); dorsal forearm (DF); dorsal thigh (DT); dorsal leg (DL)

Temperatures returned to baseline within 12 minutes of post-exercise. No significant temperature asymmetry was found between the right and left upper limbs, indicating symmetrical muscle involvement during exercise [38, 39].

9.3.4 Injury Prevention

A study on five male athletes from the Brazilian Olympic Swimming Team was made aiming to explore the application of medical IRT in monitoring elite swimmers during high-altitude training. The athletes underwent thermographic evaluations to monitor skin temperature distributions, and to identify potential areas of concern that could lead to injuries or impact performance [40].

IRT enabled the identification of thermal asymmetries and anomalies in skin temperature, which were indicative of underlying musculoskeletal issues. Early detection allowed timely interventions, preventing the progression of injuries. By monitoring thermal patterns, coaches and medical staff could tailor training programs to address specific physiological responses, thereby enhancing overall performance. However, addressing the challenges of standardization, integration, and individual variability is essential for its broader application in sports medicine [40].

9.3.5 Skin Temperature and Performance

An investigation on how different durations between warm-up and performance affect skin temperature and subsequent swimming performance was performed in 9 adolescent male swimmers in Porto, Portugal [40]. After a standardized warm-up, each swimmer performed three maximal 100-meter front crawl trials. These trials were conducted after transition periods of 10, 20, and 45 minutes, respectively. The skin temperature was assessed along with performance metrics (time, stroke frequency, stroke length, stroke index, and propelling efficiency), heart rate, blood lactate levels, and perceived effort.

This study results [41] showed that skin temperature increased logarithmically over time during the transition phases, stabilizing after approximately 15 minutes. No significant differences in performance metrics or physiological responses were observed across the different transition durations. There was no meaningful relationship between pre-exercise skin temperature and 100-meter front crawl performance times.

From this study it can be understood that within the tested parameters, pre-exercise skin temperature does not significantly influence 100-meter front crawl performance. Addressing the identified challenges in further research could provide a deeper understanding of the relationship between thermal physiology and athletic performance [41].

Another more recent research [42] aiming to study the impact of inspiratory muscle pre-activation on various performance metrics in young swimmers was conducted in 14 young Brazilian swimmers. Swimmers performed inspiratory muscle pre-activation (IN_{PA}) at 40% of their maximal inspiratory pressure, consisting of 2 sets of 15 repetitions. Each participant completed a 30-second all-out tethered swimming test under two conditions: with and without prior IN_{PA}. Mechanical parameters such as force, impulse, and fatigue index, physiological responses such as skin temperature and blood lactate levels, and subjective ratings of perceived exertion and dyspnea were assessed.

It was found that the IN_{PA} protocol did not lead to significant changes in swimming force, skin temperature, blood lactate concentration, or subjective perceptions of exertion and dyspnea during high-intensity tethered swimming. Regardless of IN_{PA} application, there were positive correlations between mean force and blood lactate levels, as well as between impulse and blood lactate. This research demonstrated that the applied IN_{PA} protocol did not significantly influence the measured performance or physiological parameters in young swimmers. It was recommended that optimizing the IN_{PA} protocol and exploring individual responses to enhance swimming performance should be object of further investigation [42].

9.4 Proposed Methodology for Measuring Skin Temperature in Water Sports

Based on the previously described existing literature, it is known that skin temperature is influenced by several factors. Assessing skin temperature in water sports upon on the internationally accepted human body thermal imaging capture protocols ([25, 43, 44], which specify subject preparation previously and during examination, recording equipment preparation, and examination room preparation, there are other intrinsic aspects related with water immersion related exercising.

The athletes should refrain from having a heavy meal, smoking, taking alcohol or drugs, having tea or coffee in the 2 h before examination, having oil or ointments in the skin, using any jewelry. The recording equipment must be switched on at a safe interval to prevent the startup drift [45], and at the local examination, actions should be taken to prevent any unwanted thermal reflection from any possible source.

Instead of having the acclimatization period at room constant air environment, since the activity takes place in the water, its temperature value has to be constant and depends on the exercise type and research question, but it can be assumed the for most exercises the recommended value would be 27 °C, the subject must stay sill during that period, which is in line with the proposed knowledge [12]. The challenge after acclimatizing in water and after performing the water activity is to dry the skin surface with minimal effect of the whole process, which can be achieved through having minimal contact with highly absorbent microfiber towels, but it must be done in a minimal time (typically 1 min) and using the same direction and

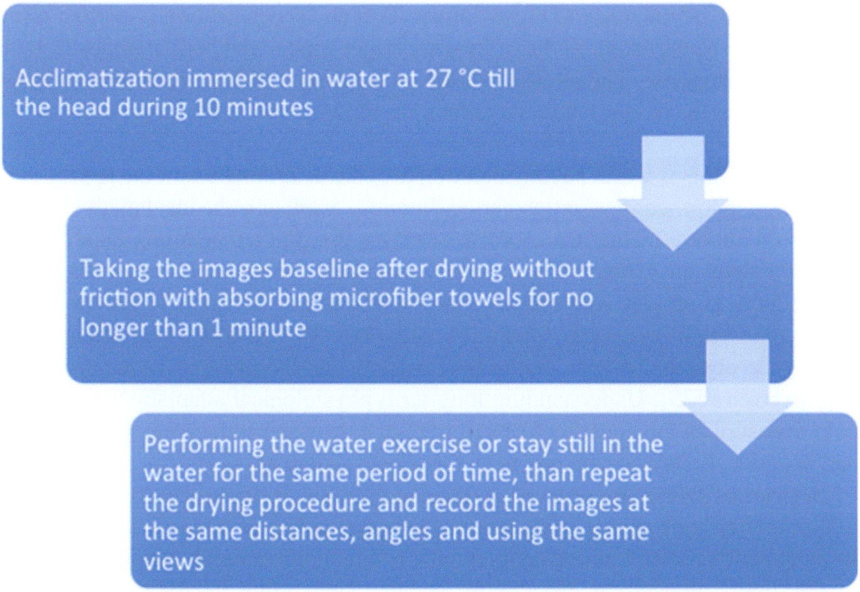

Fig. 9.18 The main points of the proposed methodology for water sports screening with thermal imaging

speed to reduce the influencing variables. This action with limit the cooling and evaporation effect that water exposure has on skin surface. For a better understanding of the exercise skin temperature influence, it is recommended that the subjects perform a test of staying still at the same condition of the exercise test and during the same time. It is also recommended to use half-body views and keep about 2 m from the target and use a 90 angle for images capture. The images should be assessed using regions of interest over the muscular areas and measuring the mean, minimum and maximum temperatures and standard deviations along with the histograms. Based on the literature, distal extremities such as hand and feet should not be considered because of the higher effect that water contact has on them. Figure 9.18 resumes the main points of the proposed methodology for water sports screening with thermal imaging.

9.5 Future Research Directions in Water Sports

Future research is needed and suggested for any water activity involving body immersion for a period higher than 5 min. In the particular case of swimming, it is required to know if it has any influence in total skin temperature or in the temperature of individual ROIs; the existing studies [6, 8–10, 30–33, 36, 46] give that indication but had a limited number of participants to take wider and accurate conclusions. The influence of the swimmer level and specialization is another issue that needs to be explored, as the influence of age, gender and body composition on the thermal response. The intensity of the swimming technique, the different types of swimming techniques and its effect on skin temperature also needs to be addressed.

An aspect that is worth paying attention to is the determination of the appropriate water immersion acclimatization period duration. Another is the ideal method to dry the skin without affecting the exercise effect on the skin temperature. Establishing causality between winter swimming and enhanced thermogenic responses, long-term effects of regular cold exposure on metabolic health and BAT function with longitudinal studies to assess potential benefits or risks, and underlying mechanisms driving the observed adaptations in BAT activity and thermogenesis due to regular cold exposure [23]. Understanding long-term physiological adaptations to regular cold exposure and the health effects of regular cold-water immersion, both positive and negative, to inform safe practices for winter swimmers [24].

In terms of swimming research, exploring different exercise intensities and durations could provide a more comprehensive understanding of thermal responses, and investigate long-term training adaptations on skin temperature dynamics [38, 39].

For injury prevention, developing standardized protocols for thermographic assessments in sports settings is necessary to ensure consistency and reliability across different teams and conditions. Combining thermography with other physiological monitoring tools could provide a more comprehensive understanding of an athlete's condition, but effective integration methods need to be established. Accounting for individual differences in thermal responses remains a challenge, necessitating personalized baseline assessments for accurate interpretation [40].

Toward performance monitoring, assess environmental conditions that can influence skin temperature and performance and explore the effects of longer transition periods between warm-up and practice to a more comprehensive understanding of skin temperature dynamics and their potential impact on performance [41]. Explore alternative protocols to IN_{PA}, possibly with increased volume or intensity, to determine effective strategies for performance improvement [42].

There are a lot of opportunities in research in this field, most till now were based on swimming but there are other water activities that worth attention and interest such as hydrotherapy, sauna, Turkish bath, diving, water polo, water jogging and any other emerging underwater activity.

9.6 Conclusion

In this chapter, the reasons for measuring skin temperature changes during water sports and underlying physiological aspects were outlined. The aspects of the relationship between water contact, exercise and skin temperature were addressed. Examples of applications of skin temperature assessment with thermal imaging involve the effect of different body masses in water immersion [8], using different types of swimming suits [30], training in different climates [31], assessing different swimming techniques with water acclimatization [6, 10, 46], relating that with energy expenditure [9], comparing yoga and swimming in pregnant women [32], fifteen minutes in muscle areas after breaststroke swimming [33, 37], injury prevention [40] and water sports performance monitoring [41, 42]. All these research studies involved small samples but presented interesting and promising results.

A new investigation methodology for thermal imaging screening in water sports was proposed based on the previous studies findings. A future challenge can be repeating some of the applications described with larger samples for confirming the initial results and conclusions.

One thing is certain, using standardized medical thermal imaging has proven to be a useful tool to evaluate temperature differences after sports activity, and only by implementing it, it is possible to contribute for its development, growing usage and credibility, and attracting new users.

References

1. Simmons GH, Wong BJ, Holowatz LA, Kenney WL (2011) Changes in the control of skin blood flow with exercise training: where do cutaneous vascular adaptations fit in? Exp Physiol 96:822–828. https://doi.org/10.1113/expphysiol.2010.056176
2. Edholm OG, Fox RH, Macpherson RK (1956) The effect of body heating on the circulation in skin and muscle. J Physiol 134:612–619
3. Ring EFJ, Ammer K (2012) Infrared thermal imaging in medicine. Physiol Meas 33:R33–R46. https://doi.org/10.1088/0967-3334/33/3/R33
4. Tortora G, Grabowski (2003) Principles of anatomy and physiology, 10th edn. Wiley, New York
5. Houdas Y, Ring EFJ (2013) Human body temperature: its measurement and regulation. Springer Science & Business Media
6. Zaidi H, Fohanno S, Polidori G, Taiar R (2007) The influence of swimming type on the skin-temperature maps of a competitive swimmer from infrared thermography. Acta Bioeng Biomech 9:47
7. Bertucci W, Arfaoui A, Janson L, Polidori G (2013) Relationship between the gross efficiency and muscular skin temperature of lower limb in cycling: a preliminary study. Comput Methods Biomech Biomed Engin 16(Suppl 1):114–115. https://doi.org/10.1080/10255842.2013.815902
8. Wade CE, Veghte JH (1977) Thermographic evaluation of the relative heat loss by area in man after swimming. Aviat Space Environ Med 48:16–18
9. Seixas A, Gonjo T, Vardasca R et al (2014) A preliminary study on the relationship between energy expenditure and skin temperature in swimming. Bordeaux, France, pp 90–97
10. Vardasca R, Domingues AS, Barbosa F et al (2015) Thermographic evaluation of swimming techniques. Thermol Int 25:125–126
11. Robinson PJ, Davies JA (1972) Laboratory determinations of water surface emissivity. J Appl Meteorol 11:1391–1393
12. Alexiou S (2014) The effect of water temperature on the human body and the swimming effort. Biol Exerc 10:9–23
13. Craig AB (1969) Comparison of exercise in air and in water of different temperatures. Med Sci Sports Exerc 1:124–130
14. McArdle WD, Magel JR, Lesmes GR, Pechar GS (1976) Metabolic and cardiovascular adjustment to work in air and water at 18, 25, and 33 °C. J Appl Physiol 40:85–90

15. Holmér I, Bergh U (1974) Metabolic and thermal response to swimming in water at varying temperatures. J Appl Physiol 37:702–705
16. Randall WC, Peiss CN (1957) The relationship between skin hydration and the suppression of sweating. J Invest Dermatol 28:435–441
17. Hayward MG, Keatinge WR (1981) Roles of subcutaneous fat and thermoregulatory reflexes in determining ability to stabilize body temperature in water. J Physiol 320:229–251
18. Young AJ, Muza SR, Sawka MN et al (1986) Human thermoregulatory responses to cold air are altered by repeated cold water immersion. J Appl Physiol Bethesda Md 1985 60:1542–1548
19. Sawka MN, Montain SJ, Latzka WA (2001) Hydration effects on thermoregulation and performance in the heat. Comp Biochem Physiol A Mol Integr Physiol 128:679–690
20. Hue O, Monjo R, Lazzaro M et al (2013) The effect of time of day on cold water ingestion by high-level swimmers in a tropical climate. Int J Sports Physiol Perform 8:442–451
21. Wales RE (1994) Fina handbook, constitution rules. Wales
22. Hayward JS, Eckerson JD, Collis ML (1975) Thermal balance and survival time prediction of man in cold water. Can J Physiol Pharmacol 53:21–32
23. Søberg S, Löfgren J, Philipsen FE et al (2021) Altered brown fat thermoregulation and enhanced cold-induced thermogenesis in young, healthy, winter-swimming men. Cell Reports Med 2(10):100408. https://doi.org/10.1016/j.xcrm.2021.100408
24. Gruszka K, Jedrzejewski G, Sobiech KA, Chwalczynska A (2018) Body surface temperature adaptations after ice-cold water immersion in regular winter swimmers. Biol Exerc 14(1):87–102. https://doi.org/10.4127/jbe.2018.0133
25. Ring EFJ, Ammer K (2000) The technique of infrared imaging in medicine. Thermol Int 10:7–14
26. Ferreira JJA, Mendonça LCS, Nunes LAO et al (2008) Exercise-associated thermographic changes in young and elderly subjects. Ann Biomed Eng 36:1420–1427. https://doi.org/10.1007/s10439-008-9512-1
27. Zontak A, Sideman S, Verbitsky O, Beyar R (1998) Dynamic thermography: analysis of hand temperature during exercise. Ann Biomed Eng 26:988–993
28. Merla A, Mattei PA, Di Donato L, Romani GL (2010) Thermal imaging of cutaneous temperature modifications in runners during graded exercise. Ann Biomed Eng 38:158–163. https://doi.org/10.1007/s10439-009-9809-8
29. Viskanta R, Toor JS (1972) Radiant energy transfer in waters. Water Resour Res 8:595–608
30. Trappe TA, Starling RD, Jozsi AC et al (1995) Thermal responses to swimming in three water temperatures: influence of a wet suit. Med Sci Sports Exerc 27:1014–1021
31. Hue O, Antoine-Jonville S, Sara F (2007) The effect of 8 days of training in tropical environment on performance in neutral climate in swimmers. Int J Sports Med 28:48–52. https://doi.org/10.1055/s-2006-923958
32. Sillero Quintana M, Conde Pascual E, Gómez Carmona PM et al (2012) Effect of yoga and swimming on body temperature of pregnant women. Thermol Int 22:143–149
33. Rybářová S, Novotný J (2014) Skin temperature changes of muscle regions in training swimmers. Int Christmas Sport Sci Conf, Proc
34. ISO (2017) Particular requirements for the basic safety and essential performance of screening thermographs for human febrile temperature screening. TC121/SC3-IEC SC62D
35. Howell KJ, Smith RE (2009) Guidelines for specifying and testing a thermal camera for medical applications. Thermol Int 19:3–14
36. Novotny J, Rybarova S, Zacha D et al (2015) The influence of breaststroke swimming on the muscle activity of young men in thermographic imaging. Acta Bioeng Biomech 17:121
37. Vardasca R, Ring F, Plassmann P, Jones C (2012) Thermal symmetry of the upper and lower extremities in healthy subjects. Thermol Int 22:53–60
38. Novotny J, Rybarova S, Zacha D, Novotny JJ, Bernacikova M, Ramadan W (2017) Thermographic evaluation of muscle activity after front crawl swimming in young men. Acta Bioeng Biomech 19(4):109–116. https://doi.org/10.5277/ABB-00924-2017-03

39. Knyszyńska A, Radecka A, Lubkowska A (2021) Thermal imaging of exercise-associated skin temperature changes in swimmers subjected to 2-min intensive exercise on a VASA swim bench ergometer. Int J Environ Res Public Health 18(12):6493. https://doi.org/10.3390/ijerph18126493
40. Côrte ACR, Lopes GR, Moraes M, De Oliveira RM, Brioschi ML, Hernandez AJ (2018) The importance of thermography for injury prevention and performance improvement in olympic swimmers: a series of case study. Int Phys Med Rehab J 3:137–141. https://doi.org/10.15406/ipmrj.2018.03.00089
41. Jimenez-Perez I, Gil-Calvo M, Vardasca R, Fernandes RJ, Vilas-Boas JP (2021) Pre-exercise skin temperature evolution is not related with 100 m front crawl performance. J Therm Biol 98:102926. https://doi.org/10.1016/j.jtherbio.2021.102926
42. Soares de Araujo L, Marostegan AB, Menezes Scariot PP et al (2024) Inspiratory muscles pre-activation in young swimmers submitted to a tethered swimming test: effects on mechanical, physiological, and skin temperature parameters. Sci Rep 14(1):5975. https://doi.org/10.1038/s41598-024-52312-z
43. Schwartz RG (2006) Guidelines for neuromusculoskeletal thermography. Thermol Int 16:5–9
44. Ammer K (2008) The Glamorgan protocol for recording and evaluation of thermal images of the human body. Thermol Int 18:125–144
45. Ring EFJ, Ammer K, Wiecek B et al (2007) Quality assurance for thermal imaging systems in medicine. Thermol Int 17:103–106
46. Arfaoui A, Polidori G, Taiar R, Popa C (2012) Infrared thermography in sports activity. Infrared Thermogr 141–168

Chapter 10
Dynamic Infrared Thermography for Assessing Human Physiological Responses

Clara Carrión-González, Carlos Sendra-Pérez, Inmaculada Aparicio-Aparicio, and Jose Ignacio Priego-Quesada

Abstract Dynamic thermography involves applying thermal stress (cold or heat) to the skin, with the recorded thermal response reflecting the tissue's reaction to restore baseline conditions. This technique has significantly advanced the understanding of the complex mechanisms of thermoregulation, allowing to provide more detailed physiological information about peripheral vascularization capacity and heat transfer from deeper tissues. Consequently, its applications have proven particularly relevant in medicine (for the diagnosis, prevention, and treatment of pathological conditions) and sports science (for performance enhancement, injury prevention and recovery optimization).

10.1 Introduction: Dynamic Thermography and Thermoregulatory Foundations

Infrared thermography (IRT) is a non-contact and non-invasive imaging technique that uses infrared cameras to measure and visualize skin temperature (Tsk), whether in localized regions or large areas, in a quick, easy, and cost-effective manner [1]. The IRT can be applied in several fields, including material characterization and medical diagnostics [2].

C. Carrión-González · I. Aparicio-Aparicio · J. I. Priego-Quesada (✉)
Research Group in Sports Biomechanics (GIBD), Department of Physical Education and Sports, Universitat de València, Valencia, Spain
e-mail: clacagon@alumni.uv.es; inmaculada.aparicio@uv.es; j.ignacio.priego@uv.es

C. Sendra-Pérez (✉)
Research Group in Sports Biomechanics (GIBD), Department of Physical Education and Sports, Universitat de València, Valencia, Spain

Department of Education and Specific Didactics, Jaume I University, Castellon, Spain
e-mail: carlos.sendra@uv.es

J. I. Priego-Quesada (ed.), *Application of Infrared Thermography in Sports Science*,
https://doi.org/10.1007/978-3-031-93311-0_10

One of the physiological keys of IRT is in the study of human thermoregulation, the body's ability to maintain its core temperature within a narrow range. Thermoregulation is defined as the capacity of the body to regulate the core temperature approximately between 36 and 37.5 °C to maintain normal functions [3]. Outside this range, certain physiological functional may be compromised [4]. In the human body, Tsk varies according to multiple factors such as sex, age, individual anatomy (e.g. body surface), circadian rhythm or emotions experienced (e.g. stress) [5]. For this reason, humans have a control system that allows them to respond to changes in core temperature [3]. To manage core temperature variations, the hypothalamus is responsible for receiving all the inputs (thermal and non-thermal information) and emitting the signals that result in physiological mechanisms that cause heat loss or gain [6]. Thermoreceptors are distributed throughout the body; internally, they take the temperature of the organs and peripherally, the temperature below the epidermis [6]. Therefore, whether there is some alteration in the core temperature, the nerve impulse are sent to the preoptic zone, which in turn activates the heat-promoting center through the production of hormones that activate heat-gain mechanisms such as peripheral vasoconstriction, shivering, and a slow increase in the metabolism rate [3, 7]. If an increase in the core temperature is perceived, thermoreceptors send nerve signals to the preoptic area, which generates hormones that inhibit the heat-promoting center and activates the heat loss center [3, 7], which in turn activates core temperature-lowering mechanisms such as peripheral vasodilation and stimulation of sweat glands that activate perspiration through the sympathetic nervous system [8].

There are different types of methodologies of IRT (i.e., static and dynamic thermography). Static IRT measures steady-state conditions (i.e., baseline conditions), and the experiments are commonly conducted at controlled environments considering environmental temperature (22–23 °C) and relative humidity (45–55%) [9]. One of the most relevant applications of static IRT in medicine and sports is detecting side-to-side asymmetries in Tsk, which can help identify abnormalities, prevent injuries, and monitor their healing processes [10–12]. However, it is important to be cautious of false negatives, as static IRT captures only the instantaneous temperature distribution. The lesion may be too deep to influence Tsk, and the pathology might not be detectable under baseline conditions, or individual factors—such as the amount of subcutaneous fat covering the observed tissue—could interfere with detection.

An alternative that would solve this problem would be the use of dynamic IRT because this technique allows observing the variation of the temperature distribution by constructing a time series of thermal images [13]. With dynamic IRT, it is possible to apply mechanical or thermal stresses to induce changes in the skin with the aim of observing the thermal response after this stress. As we will show in the following sections, this thermal response could be affected by pathologies such as diabetes or Raynaud pathology [14, 15]. Therefore, dynamic IRT was suggested as a methodology to assess vascular capacity and reduce the risk of false negatives. In this chapter, we will focus on the measurement of temperature after the application of thermal stressors (i.e., cool or heat), which are applied for a certain time and alter the temperature of the subsurface structures (vasculature) [16].

The thermal response in dynamic IRT is recorded as tissue reacts to stress in an effort to restore baseline conditions. In the case of cold stress, this involves cooling the skin to assess its vascularization capacity. During this process, increased sympathetic activity induces peripheral vasoconstriction, leading to a decrease in Tsk. Once cooling ceases, vasodilation occurs, causing the Tsk to rise as blood flow returns [17]. The rate of rewarming and the time required to reach pre-stress temperature provide valuable functional insights into tissue state and vascular capacity [16, 18]. Conversely, heat stress triggers peripheral vasodilation, resulting in an increase in muscle temperature, which is reflected on the skin surface [19]. This rise in Tsk occurs due to vasodilation and enhanced blood flow to the surface, facilitating heat dissipation. Similar to cold stress, analyzing how the body regulates temperature and returns to baseline can provide valuable insights into tissue function and thermoregulation capacity.

Although Tsk rewarming primarily depends on skin blood flow, it can also occur passively due to heat conduction from deeper tissues and environmental factors, with the duration of the cooling phase further influencing the process [17]. In general, Tsk is determined by multiple factors, including heat conduction from central areas, blood perfusion in superficial and deep skin layers, and heat exchange with the surrounding environment [20]. Among these, blood perfusion and epidermis thickness are considered primary determinants of Tsk variation [21].

Dynamic IRT provides insight into the peripheral nervous system, enables physiological monitoring of the measurement area, and offers an indirect evaluation of skin blood flow [1]. Since abnormal thermal responses can be linked to pathologies such as diabetes or breast cancer [15, 22], dynamic IRT enhances the ability to detect physiological abnormalities. For instance, as tumors tend to exhibit atypical thermal responses, analyzing rewarming dynamics may provide valuable diagnostic information [18, 23].

Compared to static methods, dynamic IRT enhances diagnostic capabilities by incorporating thermoregulatory mechanisms into the analysis. Additionally, it has improved the IRT reliability, expanding its potential applications in both clinical and research settings, including disease prevention [24, 25] and treatment monitoring [26]. Finally, dynamic IRT can be implemented through various methodologies and analytical techniques, which will be explored in the following sections.

10.2 Methodology

IRT can provide additional insights through the application of thermal stress (i.e., cold or heat). The following sections will explore the different methods of applying thermal stress in dynamic IRT and their methodological aspects. Additionally, various analytical approaches used to process data extracted from dynamic IRT will be presented, such as regression coefficient graphs and similar metrics.

10.2.1 Cold Application

Several studies that applied cold to dynamic IRT are presented in Table 10.1. The researchers have used cold thermal stress in various regions (e.g., the hand, lower limbs, knee, and foot). However, each study applied a different methodology, varying in aspects such as the duration of the test and the temperature of the thermal stress. Moreover, sport studies that use cold thermal stress are scarce compared with its use in clinical setting, when the number is high and some representative studies were selected for this table. Nevertheless, all studies shared a comparable overall approach.

Firstly, regarding the overall cooling methodology, most studies have employed cold or ice water immersion, ensuring that the skin remains dry to prevent interference with thermographic measurements. For instance, in hand immersion protocols, gloves are typically used to keep the skin dry and are removed after cooling to capture thermographic images. However, alternative cooling methods that do not require such precautions have also been explored. These include cold gel packs [40], electronic cryotherapy systems [43, 44], and direct hand contact with an aluminum plate [41].

The majority of studies applied cold thermal stress for durations ranging between 1 and 5 min [27–29, 31–33, 35, 36, 38, 39, 41–43]. However, a few studies have used shorter durations (i.e., <1 min) [14, 30, 34, 40] while only one study applied cold thermal stress for 30 min [37]. Regarding cold stress temperatures, findings vary considerably. While some studies used lower temperatures (0–9 °C) [27, 30, 34, 42], others employed moderate water temperatures (around 15 °C) [28, 29, 31, 32]. Interestingly, no differences were observed in temperature selection based on the anatomical region where the cold was applied.

Furthermore, rewarming time exhibits a wide range, as some studies recorded short rewarming periods (90–400 s) [30, 34–36, 41–43], while others monitored Tsk responses for up to 30 min [14, 29, 33, 35]. It is important to consider that post-cold stress responses follow a logarithmic pattern, where the rate of temperature change is fastest within the first 30 s, after which the rewarming velocity gradually decreases [42, 43]. In this regard, Ramírez-Arbeláez et al. [45] found that after applying cold thermal stress to the hand and capturing images for 10 min post-stimulus, healthy subjects exhibited nearly complete thermal restoration to baseline values.

Finally, the variables most used in the studies comments are the mean Tsk, Tsk gradient of the different regions assessed and several indices of rewarming: thermal recovery time and the distal distal-dorsal difference (DDD) for studies that cooled the hands [31, 35]. More information about these parameters is shown in the Sect. 10.2.3 of this chapter.

Table 10.1 Studies that have applied cold stress temperature to perform dynamic thermography to obtain more physiological information about human condition

Clinical					
Study	**Region**	**Participants**	**Temperature**	**Methods**	**Measures**
Coughlin et al. [27]	Palm of the hand	2 groups (10 controls and 21 HAVS patients)	5 °C during 1-min and 10-min of rewarming	Cold water immersion with thin, closely fitting, latex gloves Images were taken every 30-s for 10 min	Tsk mean of the fingertip and finger base Tsk gradient
Poole et al. [28]	Palm of the hand	2 groups (21 controls and 33 HAVS patients)	15 °C during 5-min and 10-min of rewarming	Cold water immersion, up to the wrist crease, with thin plastic gloves loosely taped at the wrist Images were taken in basal and during 10-min of recuperation post thermic stress	Absolute fingertip Tsk and Tsk of the middle of the finger Gradient between fingertip and base Tsk Gradient between fingertip and middle Tsk Gradient between the middle of the finger and the base
Foerster et al. [29]	Finger	3 groups (53 controls; 40 RP patients and 46 patients with scleroderma)	12 °C during 90-s and 30-min of rewarming	Cold water immersion (250 ml plastic cup) Dried with three brief swaps using a paper towel Images were taken in basal, thermic stress, and 30-min later or when the temperature has recovered to 63%	Fingertip temperature Cold response index Thermal recovery time
Stefańczyk et al. [30]	Dorsum of the hand	2 groups (16 controls and 24 RP patients)	0 °C during 12-s and 12-min of rewarming	Ice-cold water immersion to the wrist level Dried with a paper towel Images were taken in basal, immediately after of stress and at 1-, 2-, 3-, 6-, 12-min after stress	Tsk mean at the nail plate level of the middle finger Tsk mean at the wrist level

(continued)

Table 10.1 (continued)

Anderson et al. [31]	Dorsum of the hand	152 patients divided into five groups	15 °C during 1-min and 15-min of rewarming	Cold water immersion with thin latex gloves Images were taken immediately following 1-min and during all rewarming (videotaped during 15-min)	Tsk mean of the dorsum of the hand of each hand DDD Maximum DDD across all fingers of hands Number of fingers with a DDD >1 °C The rewarming curve after cold challenge Lag time during 15-min of rewarming Maximum Tsk during rewarming The percentage of recovery
Horikoshi et al. [14]	Dorsum of the hand	2 groups (40 controls and 60 RP patients)	10 °C during 10-s and 30-min of rewarming	Cold water immersion Images were taken immediately following 3-,5-,10-,15-,20- and 30-min	Index of nail fold recovery of each fact Disparity of the nail fold temperature of the five fingers. Distal-dorsal difference
Mason et al. [32]	Palm of the hand	1 group (1051 miners)	15 °C during 5-min and 10-min of rewarming	Cold water immersion with appropriately sized, long-wristed, thin plastic gloves, taped at the wrist Images were taken during the 10-min of rewarming	Tsk of the right hand. Tsk of the left hand. The Stockholm scale to analyze vascular status. Griffin test—to analyze the paleness of the fingers.

Schuhfried et al. [33]	Dorsum of the hand	86 RP patients	16 °C during 1-min and 20-min of rewarming	Cold water immersion with thin, disposable latex gloves Images were taken immediately, each min the firsts 7-min, and at 15-min and 20-min	Absolute Tsk of fingertips before 10–20-min after stress in degrees Gradient of Tsk between the fingertips and the radial carpometacarpal region before, 10- and 20-min after stress Average area under the curve of thermal recovery of buds during 20-min after stress Percentage of thermal recovery Thermal recovery index of yolks at 20-min of thermal stress
Suzuki et al. [34]	Dorsum of the hand and 2nd fingertip	1 group (38 SMID patients)	0 °C during 10-s and 6-min of rewarming	Cold water immersion to the wrist level Wiped and dried with a dry towel Images were taken immediately after cold stress and 6-min	The recovery rate of Tsk in both regions at 6-min: $RR = (T_6 - T_0)/(T_{base} - T_0) * 100(\%)$ DDD
Sternbersky et al. [35]	Dorsum of the hand	2 groups (20 controls and 150 RP patients)	9 °C during 5-min and 30-min of rewarming	Cold water immersion up to the wrist level Images were taken immediately after cold stress, and 5-, 10-, 20-, and 30-min	DDD Tsk mean of the distal phalanges of the fingers on both hands Tsk mean of the fingers of both hands Tsk mean of the dorsal part of the palm Distal difference (temperature difference of fingers before and after thermal stress)
Lahiri et al. [36]	Lower limbs	1 group (33 healthy subjects)	12 °C during ~180 s and 400-s of rewarming	Ice surface (temperature increased by ~0.4 ± 0.1 °C, after ~180 s: formation a thin water layer over the ice surface) Images were taken at 0-, 50-, 100-, 150-, 200-, 250-, 300-, 350- and 400-sec	Gradient of Tsk (i.e., differences between stimulated and non-stimulated) Gradient of Tsk (Function of. Time) (i.e., d(ΔTst)/dt)

(continued)

Table 10.1 (continued)

Norrbrand et al. [37]	Hand and foot	1 group (70 healthy subjects)	Heat stress: 35 °C during 5-min and 1-min before cold stress Cold stress: 8 °C during 30-min and 15-min of rewarming	1. Heat water immersion 2. Cold water immersion 3. Dried with a towel if necessary Both immersions covered with a thin plastic bag (thickness of 0.025 mm), sealed with air-permeable tape to the skin (~10 cm above the wrist and ankle, respectively) Hand immersion: up to the ulnar and radial styloids Foot immersion: up to the tibial and fibular styloids Images were taken immediately after thermal stress and during 1-min, 5-min, 10-min and 15-min	Tsk mean of the volar and the dorsal side of the distal phalanx of each finger. Tsk mean of the total palm. Tsk mean of the dorsal metacarpal area. Tsk mean of the he volar and the dorsal side of the distal phalanx of each toe. Tsk mean of the total sole. Tsk mean of the dorsal metatarsal area.
Selfe et al. [38]	Knee	1 group (58 AKP patients)	NA °C during 3-min and 20-min of rewarming	Cryo/Cuff knee sleeve (Aircast, New Jersey, USA) with a flask elevated 40 cm above the knee to fill with iced water for 1-min Images were taken immediately after thermal stress and at minute intervals for 20-min	Tsk mean The area under the curve was calculated between 0- and 20-min after knee sleeve removal
Balbinot et al. [39]	Plantar	2 group (10 diabetics and 10 non-diabetics)	15 °C during 1-min and 10-min of rewarming	Cold water immersion covered by a thin, waterproof plastic bag Images were taken one pre-colds stress test and other 10-min after the cold stress	Tsk mean Gradient of Tsk Rewarming index (*The relative mean temperature post-cold stress test was calculated in relation to the mean temperature pre-cold stress test.*)
Carbonell et al. [40]	Plantar	2 group (15 diabetics and 7 non-diabetics)	0 °C during 10-s and 10-min of rewarming	Cold gel application under the skin Images were taken immediately after mechanical stress and after thermal stress during 10-min	Recovery time of the basal foot temperature of each ROI Tsk mean of each ROI

Sports					
Study	**ROI**	**Participants**	**Temperature**	**Methods**	**Measures**
Sánchez-Jiménez et al. [41]	Palm of the hand	2 groups (10 controls and 13 athletes)	NA °C during 2-min and 5-min of rewarming	Pressuring hands on aluminum plate Images were taken 2-min after the match and every minute during 5-min after the cold stress test	Gradient of Tsk of each ROI (non-dominant—dominant) Gradient of Tsk of each ROI (baseline—post-match) Gradient of cold stress temperature (immediately after and each minute after cold stress temperature)
Priego-Quesada et al. [42]	Lower limbs	2 groups (14 experimental and 14 controls)	0–3 °C during 3-min and 180-s of rewarming	Electronic cryotherapy system (Game Ready with moderate pneumatic compression Circumferential wrap used: full leg boot for cooling the entire lower limb Images were taken at 30-, 60-, 120-, and 180-s	Gradient of tsk of anterior thigh, posterior thigh, knee, posterior knee, anterior leg, and posterior leg Logarithmic equation for each ROI constant (β0) and slope (β1) *Gradient Tsk = β0 + β1 * ln (time after the cold stress test (s))*
Muñoz-Alcamí et al. [43]	Lower limbs	1 group (10 experimental)	0–3 °C during 3-min and 180-s of rewarming	Electronic cryotherapy system (Game Ready with moderate pneumatic compression using an assembled thigh band Images were taken at 15-, 30-, 45-, 60-, 90-, 120-, 150- and 180-s	Logarithmic equation for each ROI constant (β0) and slope (β1) *Gradient Tsk = β0 + β1 * ln (time after the cold stress test (s))*

Note: *Tsk* skin temperature, *HAVS* Hand-arm vibration syndrome, *RP* Raynaud's phenomenon, *VWF* Vibration-induced white finger syndrome, *DDD* distal-dorsal difference, *SMID* Severe motor and intellectual disabilities, *AKP* Anterior knee pain

10.2.2 Heat Application

Although most clinical studies have used cold stress to assess peripheral vascularization by analyzing Tsk responses, some have instead applied heat stress for this purpose. Table 10.2 presents various protocols, detailing differences in duration, anatomical regions, and applied temperatures. Our review identified only two studies that applied heat stress in different anatomical regions (the knee and hands) [46, 47]. Ziegler et al. [47] applied heat stress following local exposure to 39 °C, while Trejo-Chavez et al. [46] divided participants into subgroups: one exposed to cold thermal stress (4.6 °C for 10 min using a hot pack) and another to heat thermal stress (49.6 °C for 15 min using a hot pack). Additionally, the variables extracted from the thermal stress experiments included Tsk of the target regions, temperature gradient, and mean temperature.

Table 10.2 Studies that have applied heat thermal stress to perform dynamic thermography to obtain more physiological information about human condition

Clinical					
Study	Region	Participants	Temperature	Methods	Measures
Trejo-Chavez et al. [46]	Knee	2 groups (24 controls and 24 experimental)	Heat stress: 49.5 °C during 15-min	Heat stress: hot pack wrapped in a large towel, 10 layers were placed underneath for skin contact and 2 layers on top	Tsk at baseline Tsk asymmetries (right knee—left knee) Gradient of Tsk
			Cold stress: 4.6 °C during 10-min	Cold stress: cold compress wrapped in a pullman towel was used, covering it with one layer above and one layer below Images were taken for 15-min with intervals of 3-min for each image	
Ziegler et al. [47]	Dorsum of the hand	2 groups (13 controls and 38 RP and VWF patients)	Heat stress: 39 °C Cold stress: 20 °C and 20-min of rewarming	1. Heat stress 2. Cold stress Local exposure Images were taken in 5 moments (basal values, during heat stress, during cold stress, 10-min and 20-min after cold stress)	Tsk of thumbs Tsk means of digits II-V

Note: *Tsk* skin temperature, *RP* Raynaud's phenomenon, *VWF* Vibration-induced white finger syndrome

10.2.3 Analysis

There are different analyses for assessing the thermal response after thermal stress. The most common approach is to assess the mean Tsk or the Tsk gradient of a specific region before and after thermal stress [28, 40, 41]. The asymmetry value has been another parameter calculated from the mean Tsk differences between both sides (e.g., right knee minus left knee) with the application of thermal stress [46]. The percentage of temperature recovery has also been used for analysis, calculated as [(temperature increase/initial temperature decrease) × 100%] at different time points during the rewarming process [31, 33].

Furthermore, some studies have employed a rewarming index. For instance, Balbinot et al. [39] used an equation that considers only the mean Tsk before and after the test (Eq. 10.1), while Suzuki et al. [34] used a similar approach but incorporated the Tsk gradient as well (Eq. 10.2):

$$\text{Rewarming Index} = \frac{\text{Average Temperature PostCold Stress Test}}{\text{Average Temperature PreCold Stress Test}} \times 100 \tag{10.1}$$

Equation 10.1 Rewarming index from Balbinot et al. [39] used in plantar region.

$$\text{Recovery rate} = \frac{T_6 - T_0}{T_{\text{base}} - T_0} \times 100 \tag{10.2}$$

Equation 10.2 Rewarming rate from Suzuki et al. [34] used in dorsum of the hand and second fingertip. T_{base}: baseline skin temperature measured, T_0: skin temperature immediately after cold water stress, and T_6: skin temperature 6 min after cold water stress.

Other studies on the cold stress thermal during rewarming used a logarithmic equation for each region and constant (β_0) and slope (β_1) coefficients were obtained [42–44]. To obtain the coefficients of the logarithmic equations, Tsk variations were calculated as the difference between each moment and the pre—measurement moment (Eq. 10.3 and Fig. 10.1). Other studies proposed the use of the TAR index (tissue activity ratio) with active dynamic IRT after cooling stress as a parameter to identify carotid artery stenosis or predict the risk of post-operative necrosis following breast reconstruction surgery [48, 49]. The TAR is computed for each pixel and represents a dimensionless ratio comparing the rewarming rate to the cooling rate [49, 50]. Finally, Tsk can be represented as shown in Fig. 10.2, where the rewarming time can be observed:

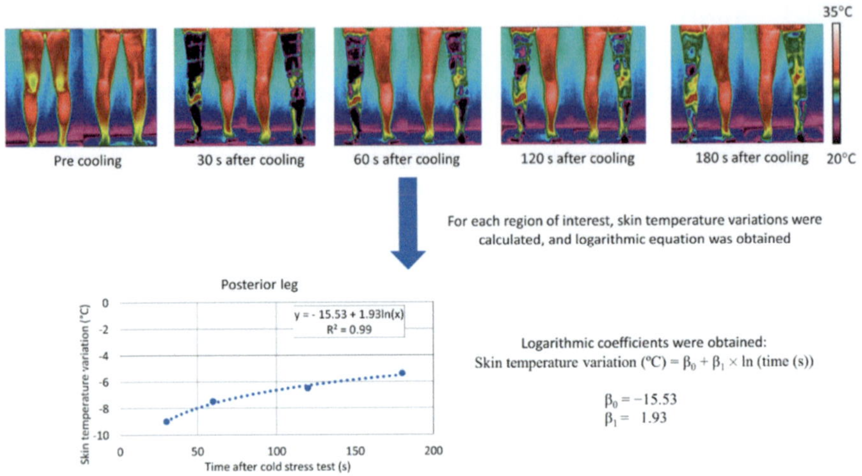

Fig. 10.1 Representation of the process to obtain the logarithmic coefficients. (The figure has been extracted from Priego-Quesada et al. [17])

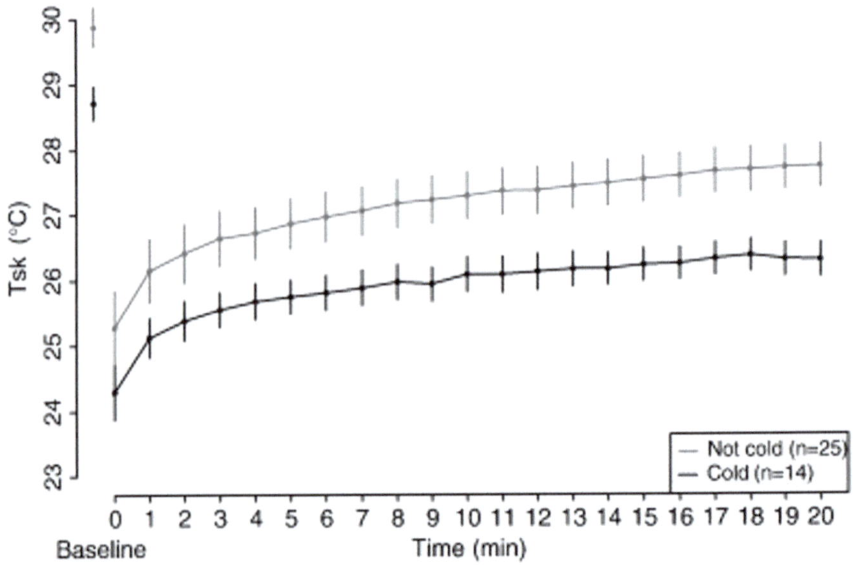

Fig. 10.2 Skin temperature after cold thermal stress and during rewarming curves. (The figure has been extracted from Selfe et al. [38])

$$\text{Skin temperature variation} = \beta 0 + \beta 1 \times \ln\left(\text{time after the cold stress test}\left(\text{s}\right)\right) \tag{10.3}$$

Equation 10.3 Logarithmic equation for each region from Priego-Quesada et al. [44].

10.3 Applications

As we commented in Sect. 10.2.1, the use of dynamic IRT in sports science remains limited. Therefore, this section will first present studies conducted in medical research to provide readers with new perspectives on its potential applications in sports science. Following this, the scarce existing literature on the use of dynamic thermography in sports science will be reviewed.

10.3.1 Human Medicine

The consolidation of IRT in medicine in the 1960s marked the beginning of its practical application [13], with remarkable findings such as its use during that decade in the United Kingdom for the early assessment of burns, which improved the outcomes of skin grafts through non-invasive imaging [51]. In its early stages, the use of IRT in the medical sector was controversial. However, as discussed in Chap. 1, its recognition as a viable diagnostic tool by the American Medical Council in 1987 strengthened its position in the medical field. Since then, specialized institutions in IRT have played a crucial role in developing clinical standards and protocols to ensure the accuracy and reliability of this technology in medical applications [12, 52].

The advent of next-generation infrared detectors has significantly enhanced the accuracy of infrared thermal imaging [13], and the latest advancements in the development of low-cost IRT cameras have already been tested in various medical applications [53, 54]. As research in this area continues to evolve, the outlook for dynamic IRT becomes increasingly promising. The potential of thermographic studies using thermal stress in the medical field lies primarily in facilitating the differential diagnosis of pathological conditions [14, 39, 40, 45, 55–57]. However, dynamic IRT is not only contributing to diagnostic advancements but is also expanding our understanding of human physiology, positioning itself as a high-potential tool for a broader range of medical applications. The following sections discuss in greater detail the examples of medical research where this technique plays a significant role. Although these studies are not directly related to sports science, they provide valuable insights into the potential applications of this technique in the sports field.

10.3.1.1 Vascular Alterations

It was demonstrated that the temperature dynamics of human skin recorded by dynamic IRT are shown to depend mostly on peripheral blood flow variations [58]. As a result, dynamic IRT has gained an important role in the investigation of conditions where vascularization is compromised. This tool has been applied with cold thermal stress in diabetic patients for foot assessment [39, 40]. It is known that cold exposure induces vasoconstriction in peripheral arteries, resulting in a decrease in

Tsk. In these patients, the alteration appears to involve a greater decrease in foot temperature and a prolonged rewarming time after the cooling stimulus compared to healthy subjects [40]. The repeatability of rewarming index in plantar region after cold thermal stress in diabetic patients was good (i.e., <5%) [39]. Therefore, dynamic IRT is a tool that can reflect microvascular dysfunction and would serve as a valid technique for the early diagnosis of diabetic foot conditions.

Cold thermal stress dynamic IRT has also been employed for the evaluation of vascular disorders in hand-arm vibration syndrome, with white finger as its most typical manifestation, classified as a type of secondary Raynaud's phenomenon [27, 28, 59]. Dynamic IRT has been applied as a diagnostic aid for this condition in a similar way to diabetic foot, as prolonged finger rewarming times indicate abnormalities in blood flow associated with the syndrome [39, 59].

In addition, IRT has been used to analyze various therapies that rely on temperature-induced physiological responses in the target region, such as changes in vascularization (e.g., vasoconstriction due to cold). This approach enables an empirical evaluation of their effectiveness by analyzing the thermal variations they induce. Among these therapies are (i) cryotherapy, (ii) laser therapy, (iii) electrotherapy, (iv) diathermy, and (v) massage [26]. Advances in this area suggest that therapies can be prescribed more appropriately based on the magnitude of the thermal effects produced by each one. They have also been studied as thermal stressors themselves (e.g., cryotherapy acting as a cold stressor in dynamic IRT) to support the validation of this method across various disciplines, such as physiotherapy and sports science, by improving our understanding of thermodynamic responses [60].

10.3.1.2 Breast Cancer

Multiple studies have evaluated Tsk distribution in breast cancer [61–65] (Fig. 10.3) since it was first suggested by Lawson in 1956 [66]. This is a medical application of IRT that has been prioritized due to its potential relevance as a screening technique. Currently, the use of dynamic IRT for breast cancer detection is supported because of the limited efficacy of mammography in women with dense breast tissue or implants despite being the standard screening method [56]. An example of dynamic IRT application in breast cancer can be found in the review by Kandlikar et al. [56], which describes a method involving the capture of thermograms after exposing the breasts to cold thermal stress using air cooling (20–5 °C) for a short duration (less than 5 min).

The goal of the current studies in this context is to establish IRT as a standard for detection, achieving a more accurate and validated diagnosis of the tumor. Thus, a recent software has been developed for the automatic detection of this type of cancer [67] using an algorithm that extracts characteristic features of the breast to detect signs of the disease through thermal image processing. This software was able to classify the thermal images of 140 individuals as normal or suspicious, achieving an accuracy rate of 98.95%. Although, Ekici and Jawzal [67] did not consider the use

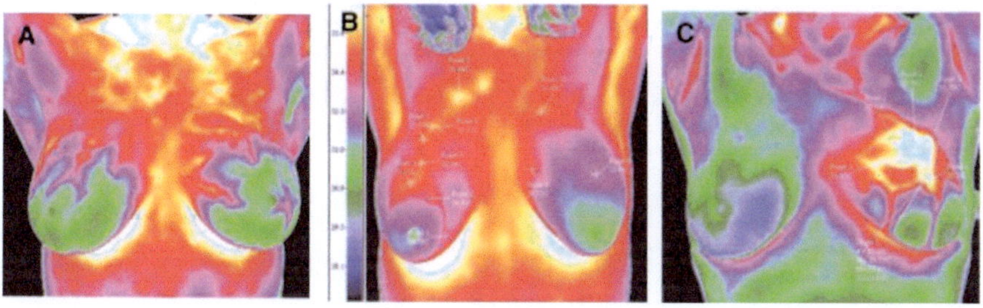

Fig. 10.3 Examples of thermographic breast images: (**a**) normal; (**b**) showing early stage of cancer in the right breast; (**c**) showing advanced cancer in the left breast. (Figure obtained from [64])

of dynamic IRT, they considered this as a promising research area to determine whether its application could enhance the detection of breast cancer.

In this context, dynamic IRT with thermal stress becomes particularly relevant, as baseline Tsk asymmetry differences may be minimal, and by applying thermal stress, dynamic IRT enhances the contrast between a breast with a malignant tumor and the contralateral healthy breast, making tumor-related heat patterns more distinguishable and improving tumor visibility in thermal imaging. [56]. Dynamic thermal imaging has been shown to significantly increase accuracy (from 54% to 82%) when compared with static IRT [68]. Thus, current research proposes the development of predictive algorithms by combining dynamic IRT with deep learning techniques, with the aim of standardizing this diagnostic tool [57]. Dynamic thermographic analysis gains particular value in this context because it is a non-invasive, fast, and safe procedure.

10.3.1.3 Cold Injuries

Exposure to hypothermic conditions can lead to vascular and nervous system disorders, as well as injuries such as frostnip and chilblains, which disrupt human thermoregulation [69]. The risk of these conditions increases in individuals with pre-existing medical conditions, such as Raynaud's phenomenon [69]. A recent study applied cold thermal stress to the hands of military personnel by immersing them in 20 °C water for 1 min, analyzing different time points: before exposure, immediately after cold exposure, and at 1-min intervals during the 4-min rewarming period [25]. As illustrated in Fig. 10.4, three groups of healthy individuals with distinct rewarming capacities were identified: fast rewarmers (72% of the sample), intermediate rewarmers (18%), and slow rewarmers (10%). By analyzing variability in reacclimatization capacity, researchers can identify individuals predisposed to hypothermia-related pathologies, as slow rewarmers have been shown to be at greater risk of cold-induced injuries [24]. Consequently, dynamic IRT has become an important tool in the prevention of cold-related conditions.

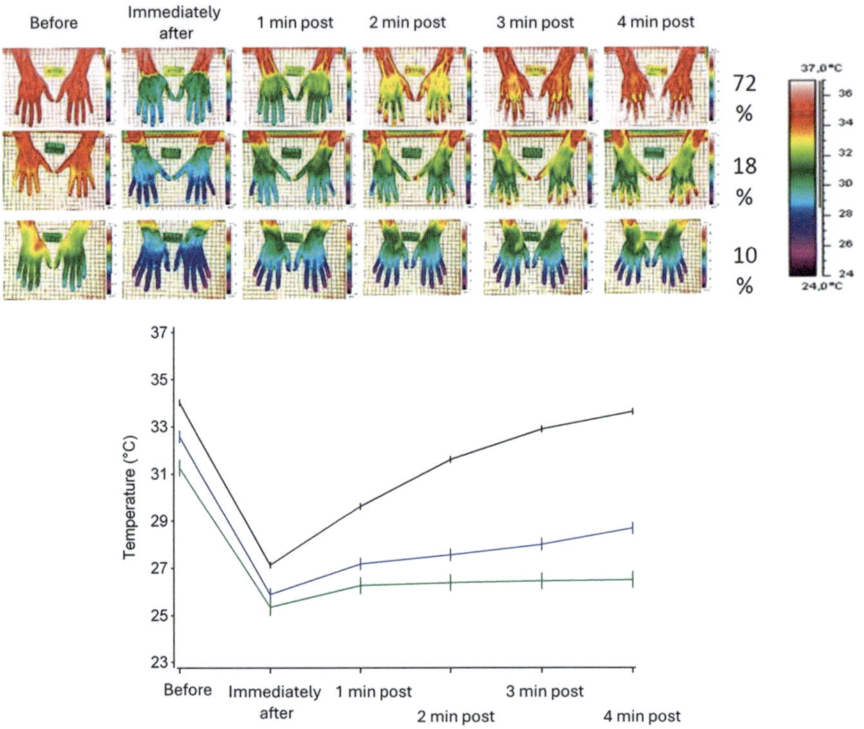

Fig. 10.4 Example of dynamic IRT in three different individuals (a fast rewarmer—top row; an intermediate rewarmer—middle row; and a slow rewarmer—bottom row). (The figure has been extracted and modified from Norheim et al. [25])

In addition, for the specific conditions of Raynaud's phenomenon and cold intolerance, a protocol for image acquisition using dynamic IRT after thermal stress was validated in 2015 with the aim of enabling its reproduction as a contribution to the medical diagnostic field [45]. In this sense, the authors established a cold thermal stress consisting of immersing the hand in 15 °C water for 30 s while maintaining an immobile position, with image capture performed for 10 min after the stimulus. In this way, five regions of interest corresponding to the distal phalanges of the fingers were selected, and temperature curves were constructed considering different conditions: (i) thermal restitution to baseline values in healthy subjects, (ii) incomplete recovery in Raynaud's with thermal alterations focused on the fingertips, and (iii) slow Tsk recovery in patients with cold intolerance, with a wide and diffuse distribution of cold areas throughout the hand [45].

10.3.1.4 Burn Injuries

A 2024 review on IRT and burn diagnosis provides a fairly up-to-date perspective on the topic [55]. While IRT is not yet a routine tool in clinical practice for wound evaluation and diagnosis, the authors emphasize its risk-free nature and its advantage of enabling dynamic monitoring. Although the use of static vs. dynamic IRT in this context remains a subject of debate due to the lack of consensus and standardized protocols, dynamic IRT has demonstrated greater accuracy [55, 70, 71], reinforcing its applicability in cases of nonspecific intermediate burns. Its potential benefits include screening capabilities and counteracting evaporative heat loss, which can otherwise lead to false positives in the diagnosis of deep wounds.

10.3.2 Sports Science

There is limited evidence on the use of dynamic IRT in sports science. However, some medical studies have referred to walking as a form of mechanical stress [40], suggesting that measuring Tsk after physical exercise could be considered a type of dynamic IRT. This chapter, however, focuses on the application of dynamic thermography through thermal stress (cold or heat), whether applied to athletes or before/after physical exercise, without considering the mechanical stress induced by exercise itself. Additionally, certain sports contexts can be perceived as a form of thermal stress. For example, in swimming, water temperature directly impacts bodily functions post-immersion, leading to physiological changes that influence performance outcomes [72, 73]. Another example is performing exercise in hot or cold environments, in which the extreme environment is an added factor to the exercise that allows understanding specific thermophysiological responses, with important applications in Tsk monitoring [19, 74], sports performance [75, 76], thermoregulatory sex differences [77], or clothing design [78]. However, we reiterate that although these interpretations are completely valid, in this chapter, we have focused on the direct application of cold or heat, knowing that this entails a limitation in the number of studies discussed.

10.3.2.1 Effect of Muscle Damage and Exercise on Skin Temperature Response After Cold Stress Test

In the field of performance, studies evaluating changes in Tsk in response to the application of cold or heat thermal stress following a fatigue or muscle damage protocol are particularly notable [44]. One study assessed dynamic IRT in marathon runners 48 and 24 h before a marathon and 24 and 48 h after running the marathon competition [44]. When cold thermal stress was applied, the posterior legs exhibited lower constant parameters and higher slope parameters in the algorithmic equations fitted for skin temperature recovery 24 h after a marathon. This occurred despite baseline temperature remaining unchanged at both 24 and 48 h post-exercise [44]. The authors suggested that this could be due to increased peripheral vasoconstriction caused by muscle damage. This study highlighted the potential application of dynamic IRT for muscle damage assessment, as well as its use for the indirect evaluation of skin blood flow and the stress state of athletes. Moreover, rewarming after

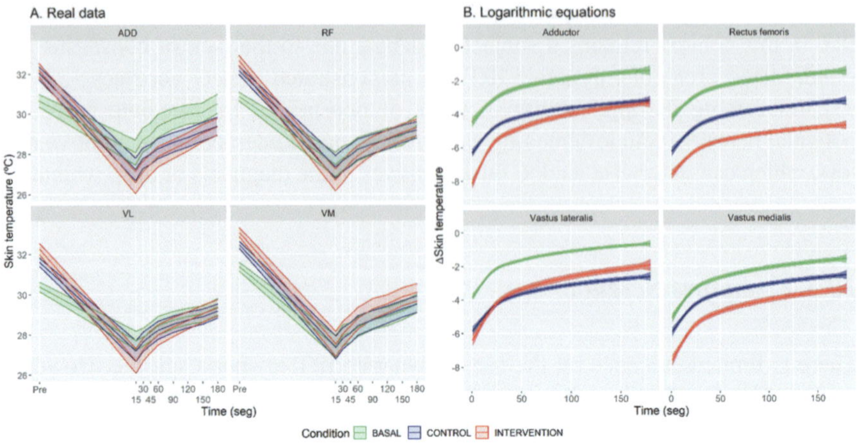

Fig. 10.5 Real data ((**a**): mean and standard deviation) and representation of the logarithmic coefficients obtained ((**b**), see more in Sect. 10.2.3) for baseline condition, control (lower limb that did not perform strength exercise after the protocol) and intervention lower limb (lower limb that performed the strength exercise protocol). (The figure has been extracted from Muñoz-Alcamí et al. [43])

cold thermal stress was analyzed over 24 h following a 10 km run with low or no muscle damage [42]. The authors concluded that muscle damage is a key determinant of the Tsk response, as baseline Tsk increased throughout the day, peaking between 5 and 9 h post-exercise, yet without significantly affecting the thermal response to a cold stress test.

However, the mechanisms by which exercise modifies Tsk after cold thermal stress were not clearly understood. More recently, a study analyzing Tsk responses following cold application after a fatiguing strength exercise protocol has revealed a temperature reduction and a faster rewarming post-exercise, emphasizing the influence of exercise on thermal recovery dynamics [43] (Fig. 10.5). Moreover, it was suggested that the effect of cold thermal stress on Tsk may vary depending on the type of exercise, highlighting the need for further research on this topic.

10.3.2.2 Therapeutic Applications

Thermal treatments are increasingly being used by physiotherapists due to their stronger scientific backing, which also becomes relevant in the sports field in terms of rehabilitation and both physiological and pathological conditions [26]. As we have seen, in other disciplines such as medicine, the term “cryotherapy” is used to define thermal stress applied in situations requiring cold thermal treatment. Thus, cryotherapy is currently employed with the aim of providing analgesia, contributing to the field of athletic recovery and injury prevention. In that sense, the most popular cold thermal stress in sports has been evaluated using IRT, with findings that whole-body cryotherapy (−110 °C) decreases both muscle and core temperatures to a greater extent compared to cold water immersion (8 °C), although both treatments produced significant reductions [60].

More specifically, in the context of injuries, cold sensitivity is one of the chronic symptoms of non-freezing cold injury (NFCI), caused by prolonged exposure to cold and/or wet environments [79]. Eglin et al. [79] have demonstrated the effectiveness of exercise prior to a cold sensitivity test, using IRT, as a means to improve the accuracy of NFCI prognosis. Thus, exercise is closely linked to the medical field, particularly when aiming to improve the diagnosis and prognosis of an injury or disorder. Additionally, this evaluation provides valuable information for prevention in the study of sports environments under extreme thermal conditions, such as winter sports. Therefore, the study with dynamic IRT involves a series of physiological responses that are specifically used to analyze certain situations or conditions sensitive to cold or heat thermal stress.

10.4 Future Research

The applications described above are not exclusive to the medical or sports field but rather form a framework that helps us understand that dynamic IRT is a multidimensional research tool that should continue to be explored and holds great potential in the field of health. Currently, there are numerous studies on thermal sensitivity to cold or heat [80–82], but there is a need for more studies employing more objective research techniques, such as dynamic IRT, in order to clarify the thermoregulatory mechanisms of the human body on a solid foundation.

Additionally, both sexes should be considered, as women appear to experience thermal sensations of greater magnitude than men [80], being more sensitive to cold than to heat [81], with important thermophysiological differences (e.g. body fat, skin blood flow, and Tsk) [83, 84], and also with the effect of menstrual cycle [85]. The heat thermal stress in this regard becomes particularly relevant due to the different responses in cutaneous blood flow and sweating threshold depending on the underlying hormonal variations of the different phases of the menstrual cycle [86]. Giersch et al. [77] concluded that, although no significant differences in mean Tsk and sweating rate were observed in the studies analyzed, women appear to have increased core temperature in the luteal phase compared to the follicular phase, both pre-exercise and post-exercise. This thermographic approach provides an objective basis for incorporating a gender perspective into thermal analysis, as data on women in sports remain scarce and are often excluded from physiological research [87].

Age is another important factor that affects Tsk responses [88, 89]. Therefore, exploring how thermal stress responses vary across different age groups is of particular interest. Moreover, since physical exercise is the most effective intervention against age-related health issues, and aging is closely linked to various cardiovascular pathologies [90], the application of dynamic IRT for monitoring these populations could offer valuable insights for both preventive and clinical purposes.

The application of dynamic IRT in the field of sports science must be well-founded, always considering the specific thermal pattern of the sport and the athlete's individual temperature profile, as incorrect conclusions or criteria may be drawn by focusing on thermal imbalances that are within the normal range. Thus, analyzing the specific responses of each sports discipline would be interesting to optimally address training (see more in Chap. 8). The impact of clothing, equipment, or the environment can be studied more in-depth for sports events where it is relevant, such as sports that are increasingly experiencing higher temperatures and relative humidity as consequences of climate change.

Another approach would be to continue using this technique to understand the hemodynamics of the human body in sports performance and recovery, with the aim of assessing and improving the evolution of muscle damage. On one hand, training would gain a perspective of more effective and efficient work, even reducing the risk of injury, which is closely related to athlete fatigue, by detecting early signs through extreme temperatures. On the other hand, it would be possible to optimize the treatment of muscle damage itself, allowing for the design of more tailored recovery programs through the study of thermal variations and localized muscle fatigue.

From a perspective more closely related to sports medicine, dynamic IRT could be used to detect early alterations in athletes with a history of or any predisposition to inflammatory, metabolic, or circulatory diseases, contributing to early diagnosis studies. In a similar manner, it could be studied in individuals with pathological conditions to analyze different types of training and determine their impact on thermal regulation and their medical implications.

The growing research is allowing the development of predictive algorithms for the creation of software that facilitates the analysis of thermal images [67]. In the specific field of sports science, machine learning software has been validated showing a time-saving in analysis for research in this area (reduction of 86% of the time) with excellent reliability compared to manual procedures [91]. Consequently, creating predictive models that analyze thermal patterns following the application of cold and heat to predict physiological responses or risks would be the next step.

10.5 Conclusions

Dynamic IRT is a methodology that provides more information than static IRT for studying human physiology. Its ability to assess tissue responses to thermal stress (cold or heat) has significantly advanced our understanding of thermoregulation. This technology offers several key advantages: it is non-invasive, safe, and objective, delivering rapid results with the aid of advanced software.

In medicine, dynamic thermal stress IRT plays a crucial role in the diagnosis, prevention, and treatment of various vascular and neurological disorders, as well as breast cancer. In the field of sports science, while this technique has the potential to enhance performance, prevent injuries, and optimize recovery, research in this area remains limited. We firmly believe that dynamic IRT holds great promise, not only in medicine but also for healthcare practitioners and sports scientists. Further research is essential to deepen our understanding of human thermoregulation, and the application of thermal stress in IRT may be the key to unlocking its full potential.

References

1. Vardasca R, Magalhaes C, Silva P, Abreu P, Mendes J, Restivo MT (2019) Biomedical musculoskeletal applications of infrared thermal imaging on arm and forearm: a systematic review. J Therm Biol 82:164–177. https://doi.org/10.1016/j.jtherbio.2019.04.008
2. Rösner H, Netzelmann U, Hoffmann J, Karpen W, Kramb V, Meyendorf N (2004) Thermographic materials characterization. In: Meyendorf NGH, Nagy PB, Rokhlin SI (eds) Nondestructive materials characterization: with applications to aerospace materials. Springer, pp 246–285. https://doi.org/10.1007/978-3-662-08988-0_9
3. Vardasca R, Simoes R (2013) Current issues in medical thermography. In: Tavares JMRS, Natal Jorge RM (eds) Topics in medical image processing and computational vision. Springer, pp 223–237. https://doi.org/10.1007/978-94-007-0726-9_12
4. Schieber AMP, Ayres JS (2016) Thermoregulation as a disease tolerance defense strategy. Pathog Dis 74(9):ftw106. https://doi.org/10.1093/femspd/ftw106
5. Fernández-Cuevas I, Bouzas Marins JC, Arnáiz Lastras J, Gómez Carmona PM, Piñonosa Cano S, García-Concepción MÁ, Sillero-Quintana M (2015) Classification of factors influencing the use of infrared thermography in humans: a review. Infrared Phys Technol 71:28–55. https://doi.org/10.1016/j.infrared.2015.02.007
6. Charkoudian N (2016) Human thermoregulation from the autonomic perspective. Auton Neurosci Basic Clin 196:1–2. https://doi.org/10.1016/j.autneu.2016.02.007
7. González-Alonso J (2012) Human thermoregulation and the cardiovascular system. Exp Physiol 97(3):340–346. https://doi.org/10.1113/expphysiol.2011.058701
8. Shibasaki M, Crandall CG (2010) Mechanisms and controllers of eccrine sweating in humans. Front Biosci (Schol Ed) 2:685–696
9. Renkielska A, Nowakowski A, Kaczmarek M, Dobke MK, Grudziński J, Karmolinski A, Stojek W (2005) Static thermography revisited—an adjunct method for determining the depth of the burn injury. Burns 31(6):768–775. https://doi.org/10.1016/j.burns.2005.04.006
10. Escamilla-Galindo V-L, Felipe JL, Alonso-Callejo A, Van-der-Horst R, Torre-Combarros A d l, Minafra P, Fernández-Muñoz D, Fernández-Cuevas I (2024) Return-to-play criteria based on infrared thermography during anterior cruciate ligament rehabilitation in football players. Biol Sport 42(2):161–167. https://doi.org/10.5114/biolsport.2025.144295
11. Gómez-Carmona P, Fernández-Cuevas I, Sillero-Quintana M, Arnaiz-Lastras J, Navandar A (2020) Infrared thermography protocol on reducing the incidence of soccer injuries. J Sport Rehabil 1-6:1222. https://doi.org/10.1123/jsr.2019-0056
12. Hildebrandt C, Raschner C, Ammer K (2010) An overview of recent application of medical infrared thermography in sports medicine in Austria. Sensors (Basel, Switzerland) 10(5):4700–4715. https://doi.org/10.3390/s100504700
13. Lahiri BB, Bagavathiappan S, Jayakumar T, Philip J (2012) Medical applications of infrared thermography: a review. Infrared Phys Technol 55(4):221–235. https://doi.org/10.1016/j.infrared.2012.03.007
14. Horikoshi M, Inokuma S, Kijima Y, Kobuna M, Miura Y, Okada R, Kobayashi S (2016) Thermal disparity between fingers after cold-water immersion of hands: a useful indicator of disturbed peripheral circulation in Raynaud phenomenon patients. Intern Med 55(5):461–466. https://doi.org/10.2169/internalmedicine.55.5218

15. Zeng S, Chen Q, Wang X-W, Hong K, Li J-X, Li P, Cheng X-S, Su H (2016) Longer rewarming time in finger cooling test in association with HbA1c level in diabetics. Microvasc Res 107:72–75. https://doi.org/10.1016/j.mvr.2016.05.003
16. Pirtini Çetingül M, Herman C (2011) The assessment of melanoma risk using the dynamic infrared imaging technique. J Therm Sci Eng Appl 3(031006). https://doi.org/10.1115/1.4004424
17. Priego-Quesada JI, Gandia-Soriano A, Pellicer-Chenoll MT, Catalá-Vilaplana I, Bermejo-Ruiz JL, Encarnación-Martínez A, Salvador-Palmer R, Cibrián Ortiz de Anda R (2021) Reproducibility of skin temperature response after cold stress test using the game ready system: preliminary study. Int J Environ Res Public Health 18(16):8295. https://doi.org/10.3390/ijerph18168295
18. Burkes SA, Patel M, Adams DM, Hammill AM, Eaton KP, Wickett RR, Visscher MO (2016) Infantile hemangioma status by dynamic infrared thermography – a preliminary study. Int J Dermatol 55(10):e522–e532. https://doi.org/10.1111/ijd.13298
19. Rojas-Valverde D, Gutiérrez-Vargas R, Sánchez-Ureña B, Gutiérrez-Vargas JC, Priego-Quesada JI (2021) Relationship between skin temperature variation and muscle damage markers after a Marathon performed in a hot environmental condition. Life 11(8):Article 8. https://doi.org/10.3390/life11080725
20. Zontak A, Sideman S, Verbitsky O, Beyar R (1998) Dynamic thermography: analysis of hand temperature during exercise. Ann Biomed Eng 26(6):988–993. https://doi.org/10.1114/1.33
21. Saiko G (2022) Skin temperature: the impact of perfusion, epidermis thickness, and skin wetness. Appl Sci 12(14):Article 14. https://doi.org/10.3390/app12147106
22. Gonzalez-Hernandez J-L, Recinella AN, Kandlikar SG, Dabydeen D, Medeiros L, Phatak P (2019) Technology, application and potential of dynamic breast thermography for the detection of breast cancer. Int J Heat Mass Transf 131:558–573. https://doi.org/10.1016/j.ijheatmasstransfer.2018.11.089
23. Wu Z, Liu HH, Lebanowski L, Liu Z, Hor PH (2007) A basic step toward understanding skin surface temperature distributions caused by internal heat sources. Phys Med Biol 52(17):5379–5392. https://doi.org/10.1088/0031-9155/52/17/019
24. Brändström H, Grip H, Hallberg P, Grönlund C, Angquist K-A, Giesbrecht GG (2008) Hand cold recovery responses before and after 15 months of military training in a cold climate. Aviat Space Environ Med 79(9):904–908. https://doi.org/10.3357/asem.1886.2008
25. Norheim AJ, Borud E, Wilsgaard T, De Weerd L, Mercer JB (2018) Variability in peripheral rewarming after cold stress among 255 healthy Norwegian army conscripts assessed by dynamic infrared thermography. Int J Circumpolar Health 77(1):1536250. https://doi.org/10.1080/22423982.2018.1536250
26. Lubkowska A, Pluta W (2022) Infrared thermography as a non-invasive tool in musculoskeletal disease rehabilitation—the control variables in applicability—a systematic review. Appl Sci 12(9):Article 9. https://doi.org/10.3390/app12094302
27. Coughlin PA, Chetter IC, Kent PJ, Kester RC (2001) The analysis of sensitivity, specificity, positive predictive value and negative predictive value of cold provocation thermography in the objective diagnosis of the hand-arm vibration syndrome. Occup Med (Oxford, England) 51(2):75–80. https://doi.org/10.1093/occmed/51.2.075
28. Poole K, Elms J, Mason H (2006) Cold-provocation testing for the vascular component of hand-arm vibration syndrome in health surveillance. Ind Health 44(4):577–583. https://doi.org/10.2486/indhealth.44.577
29. Foerster J, Kuerth A, Niederstrasser E, Krautwald E, Pauli R, Paulat R, Eweleit M, Riemekasten G, Worm M (2007) A cold-response index for the assessment of Raynaud's phenomenon. J Dermatol Sci 45(2):113–120. https://doi.org/10.1016/j.jdermsci.2006.11.006
30. Stefańczyk L, Woźniakowski B, Pietrzak P, Majos A, Grzelak P (2007) Comparison of thermography and Doppler sonography in the evaluation of the cold immersion test in women with excessive vasospastic reaction. Med Sci Monit 13(1):121–128

31. Anderson ME, Moore TL, Lunt M, Herrick AL (2007) The 'distal–dorsal difference': a thermographic parameter by which to differentiate between primary and secondary Raynaud's phenomenon. Rheumatology 46(3):533–538. https://doi.org/10.1093/rheumatology/kel330
32. Mason HJ, Poole K, Saxton J (2003) A critique of a UK standardized test of finger rewarming after cold provocation in the diagnosis and staging of hand–arm vibration syndrome. Occup Med 53(5):325–330. https://doi.org/10.1093/occmed/kqg096
33. Schuhfried O, Vacariu G, Lang T, Korpan M, Kiener HP, Fialka-Moser V (2000) Thermographic parameters in the diagnosis of secondary Raynaud's phenomenon. Arch Phys Med Rehabil 81(4):495–499. https://doi.org/10.1053/mr.2000.4870
34. Suzuki Y, Kobayashi M, Kuwabara K, Kawabe M, Kikuchi C, Fukuda M (2013) Skin temperature responses to cold stress in patients with severe motor and intellectual disabilities. Brain Dev 35(3):265–269. https://doi.org/10.1016/j.braindev.2012.04.003
35. Sternbersky J, Tichy M, Zapletalova J (2021) Infrared thermography and capillaroscopy in the diagnosis of Raynaud's phenomenon. Biomed Papers 165(1):90–98. https://doi.org/10.5507/bp.2020.031
36. Lahiri BB, Bagavathiappan S, Philip J (2020) Infrared thermal imaging based study of localized cold stress induced thermoregulation in lower limbs: the role of age on the inversion time. J Therm Biol 94:102781. https://doi.org/10.1016/j.jtherbio.2020.102781
37. Norrbrand L, Kölegård R, Keramidas ME, Mekjavic IB, Eiken O (2017) No association between hand and foot temperature responses during local cold stress and rewarming. Eur J Appl Physiol 117(6):1141–1153. https://doi.org/10.1007/s00421-017-3601-5
38. Selfe J, Sutton C, Hardaker NJ, Greenhalgh S, Karki A, Dey P (2010) Anterior knee pain and cold knees: a possible association in women. Knee 17(5):319–323. https://doi.org/10.1016/j.knee.2009.10.005
39. Balbinot LF, Robinson CC, Achaval M, Zaro MA, Brioschi ML (2013) Repeatability of infrared plantar thermography in diabetes patients: a pilot study. J Diabetes Sci Technol 7(5):1130–1137
40. Carbonell L, Priego Quesada JI, Retorta P, Benimeli M, Cibrián Ortiz De Anda RM, Salvador Palmer R, González Peña RJ, Galindo C, Pino Almero L, Blasco MC, Mínguez MF, Macián-Romero C (2019) Thermographic quantitative variables for diabetic foot assessment: preliminary results. Comput Methods Biomech Biomed Eng Imaging Vis 7(5–6):660–666. https://doi.org/10.1080/21681163.2018.1542349
41. Sánchez-Jiménez JL, Tejero-Pastor R, Calzadillas-Valles M d C, Jimenez-Perez I, Cibrián Ortiz de Anda RM, Salvador-Palmer R, Priego-Quesada JI (2022) Chronic and acute effects on skin temperature from a sport consisting of repetitive impacts from hitting a ball with the hands. Sensors 22(21):Article 21. https://doi.org/10.3390/s22218572
42. Priego-Quesada JI, Catalá-Vilaplana I, Bermejo-Ruiz JL, Gandia-Soriano A, Pellicer-Chenoll MT, Encarnación-Martínez A, Cibrián Ortiz de Anda R, Salvador-Palmer R (2022) Effect of 10 km run on lower limb skin temperature and thermal response after a cold-stress test over the following 24 h. J Therm Biol 105:103225. https://doi.org/10.1016/j.jtherbio.2022.103225
43. Muñoz-Alcamí M, Priego-Quesada JI, Gimeno Raga M, Durán Lozano Á, Gil-Calvo M (2021) Effect of fatigue strength exercise on anterior thigh skin temperature rewarming after cold stress test. J Therm Biol 101:103098. https://doi.org/10.1016/j.jtherbio.2021.103098
44. Priego-Quesada JI, Pérez-Guarner A, Gandia-Soriano A, Oficial-Casado F, Galindo C, Anda RMCO d, Piñeiro-Ramos JD, Sánchez-Illana Á, Kuligowski J, Barbosa MAG, Vento M, Palmer RS (2020) Effect of a Marathon on skin temperature response after a cold-stress test and its relationship with perceptive, performance, and oxidative-stress biomarkers. Int J Sports Physiol Perform 15(10):1467–1475. https://doi.org/10.1123/ijspp.2019-0963
45. Ramírez-Arbeláez LM, Jiménez-Díaz KT, Correa-Castañeda AC, Giraldo-Restrepo JA, Fandiño-Toro HA (2015) Protocolo de adquisición de imágenes diagnósticas por termografía infrarroja. Med Lab 21:161–178. https://medicinaylaboratorio.com/index.php/myl/article/view/118/102

46. Trejo-Chavez O, Priego-Quesada JI, Gonzalez-Hernandez MP, Morales-Hernandez LA, Cruz-Albarran IA (2023) Knee skin temperature response of patients with bilateral patellofemoral syndrome before and after heat and cold stress. J Therm Biol 115:103601. https://doi.org/10.1016/j.jtherbio.2023.103601
47. Ziegler S, Zöch C, Gschwandtner M, Eckhardt G, Windberger U, Minar E, Rüdiger H, Osterode W (2005) Thermoregulation and rheological properties of blood in primary Raynaud's phenomenon and the vibration-induced white-finger syndrome. Int Arch Occup Environ Health 78(3):218–222. https://doi.org/10.1007/s00420-004-0601-7
48. Saxena A, Ng EYK, Raman V, Syarifuddin Bin Mohamed Hamli M, Moderhak M, Kolacz S, Jankau J (2019a) Infrared (IR) thermography-based quantitative parameters to predict the risk of post-operative cancerous breast resection flap necrosis. Infrared Phys Technol 103:103063. https://doi.org/10.1016/j.infrared.2019.103063
49. Saxena A, Ng EYK, Lim ST (2020) Active dynamic thermography to detect the presence of stenosis in the carotid artery. Comput Biol Med 120:103718. https://doi.org/10.1016/j.compbiomed.2020.103718
50. Saxena A, Raman V, Ng EYK (2019b) Study on methods to extract high contrast image in active dynamic thermography. Quant InfraRed Thermogr J 16(3–4):243–259. https://doi.org/10.1080/17686733.2019.1586376
51. Ring EFJ, Ammer K (2012) Infrared thermal imaging in medicine. Physiol Meas 33(3):R33–R46. https://doi.org/10.1088/0967-3334/33/3/R33
52. Moreira DG, Costello JT, Brito CJ, Adamczyk JG, Ammer K, Bach AJE, Costa CMA, Eglin C, Fernandes AA, Fernández-Cuevas I, Ferreira JJA, Formenti D, Fournet D, Havenith G, Howell K, Jung A, Kenny GP, Kolosovas-Machuca ES, Maley MJ et al (2017) Thermographic imaging in sports and exercise medicine: a Delphi study and consensus statement on the measurement of human skin temperature. J Therm Biol 69:155–162. https://doi.org/10.1016/j.jtherbio.2017.07.006
53. Passos M d D, Rocha AF d (2022) Evaluation of infrared thermography with a portable camera as a diagnostic tool for peripheral arterial disease of the lower limbs compared with color Doppler ultrasonography. Arch Med Sci Atheroscler Dis 7:e66. https://doi.org/10.5114/amsad/150716
54. Qu Y, Meng Y, Fan H, Xu RX (2022) Low-cost thermal imaging with machine learning for non-invasive diagnosis and therapeutic monitoring of pneumonia. Infrared Phys Technol 123:104201. https://doi.org/10.1016/j.infrared.2022.104201
55. Avilés-García Á, López-Casanova P, Segura-Jordá G, Verdú-Soriano J, Berenguer-Pérez M, Avilés-García Á, López-Casanova P, Segura-Jordá G, Verdú-Soriano J, Berenguer-Pérez M (2024) Termografía infrarroja como método de diagnóstico en quemaduras: Una revisión exploratoria de la literatura. Gerokomos 35(1):62–66. https://scielo.isciii.es/scielo.php?script=sci_abstract&pid=S1134-928X2024000100011&lng=es&nrm=iso&tlng=es
56. Kandlikar SG, Perez-Raya I, Raghupathi PA, Gonzalez-Hernandez J-L, Dabydeen D, Medeiros L, Phatak P (2017) Infrared imaging technology for breast cancer detection – current status, protocols and new directions. Int J Heat Mass Transf 108:2303–2320. https://doi.org/10.1016/j.ijheatmasstransfer.2017.01.086
57. Timadius ED, Wongso R, Baihaqi T, Gunawan AAS, Setiawan KE (2024) Breast cancer image classification obtained through dynamic thermography using deep learning. In: 2024 4th international conference of science and information technology in smart administration (ICSINTESA), pp 207–212. https://doi.org/10.1109/ICSINTESA62455.2024.10748049
58. Gulyaev YV, Markov AG, Koreneva LG, Zakharov PV (1995) Dynamical infrared thermography in humans. IEEE Eng Med Biol Mag 14(6):766–771. Scopus. https://doi.org/10.1109/51.473272
59. Jankovic S, Stankovic S, Borjanovic S, Tenjovic L, Bogdanovic M (2008) Cold stress dynamic thermography for evaluation of vascular disorders in hand-arm vibration syndrome. J Occup Health 50(5):423–425. https://doi.org/10.1539/joh.L7140

60. Costello JT, Culligan K, Selfe J, Donnelly AE (2012) Muscle, skin and core temperature after −110 °C cold air and 8 °C water treatment. PLoS One 7(11):e48190. https://doi.org/10.1371/journal.pone.0048190
61. Etehadtavakol M, Ng EYK (2013) Breast thermography as a potential non-contact method in the early detection of cancer: A review. J Mech Med Biol 13(02):1330001. https://doi.org/10.1142/S0219519413300019
62. Gautherie M (1983) Thermobiological assessment of benign and malignant breast diseases. Am J Obstet Gynecol 147(8):861–869. Scopus. https://doi.org/10.1016/0002-9378(83)90236-3
63. Kaczmarek M, Nowakowski A (2003) Analysis of transient thermal processes for improved visualization of breast cancer using IR imaging. Thermal Science and Engineering Progress 2:1113–1116. Scopus
64. Kennedy DA, Lee T, Seely D (2009) A comparative review of thermography as a breast cancer screening technique. Integr Cancer Ther 8(1):9–16. https://doi.org/10.1177/1534735408326171
65. Ng EY-K (2009) A review of thermography as promising non-invasive detection modality for breast tumor. Int J Therm Sci 48(5):849–859. Scopus. https://doi.org/10.1016/j.ijthermalsci.2008.06.015
66. Lawson R (1956) Implications of surface temperatures in the diagnosis of breast cancer. Can Med Assoc J 75(4):309–310. https://www.ncbi.nlm.nih.gov/pmc/articles/PMC1824571/
67. Ekici S, Jawzal H (2020) Breast cancer diagnosis using thermography and convolutional neural networks. Med Hypotheses 137:109542. https://doi.org/10.1016/j.mehy.2019.109542
68. Ohashi Y, Uchida I (2000) Applying dynamic thermography in the diagnosis of breast cancer. IEEE Eng Med Biol Mag Q Mag Eng Med Biol Soc 19(3):42–51. https://doi.org/10.1109/51.844379
69. Long WB, Edlich RF, Winters KL, Britt LD (2005) Cold injuries. J Long-Term Eff Med Implants 15(1):67–78. https://doi.org/10.1615/jlongtermeffmedimplants.v15.i1.80
70. Dang J, Lin M, Tan C, Pham CH, Huang S, Hulsebos IF, Yenikomshian H, Gillenwater J (2021) Use of infrared thermography for assessment of burn depth and healing potential: a systematic review. J Burn Care Res Off Publ Am Burn Assoc irab108. https://doi.org/10.1093/jbcr/irab108
71. Wearn C, Lee KC, Hardwicke J, Allouni A, Bamford A, Nightingale P, Moiemen N (2018) Prospective comparative evaluation study of laser Doppler imaging and thermal imaging in the assessment of burn depth. Burns J Int Soc Burn Inj 44(1):124–133. https://doi.org/10.1016/j.burns.2017.08.004
72. Alexiou S (2014) The effect of water temperature on the human body and the swimming effort. Biol Exerc 10(2):9–23. https://research.ebsco.com/linkprocessor/plink?id=1f5a234b-b0da-3735-9a42-a1764ba8271b
73. Jimenez-Perez I, Gil-Calvo M, Vardasca R, Fernandes RJ, Vilas-Boas JP (2021) Pre-exercise skin temperature evolution is not related with 100 m front crawl performance. J Therm Biol 98:102926. https://doi.org/10.1016/j.jtherbio.2021.102926
74. Rojas-Valverde D, Gutiérrez-Vargas R, Sánchez-Ureña B, Mjaanes JM (2019) Effects on skin temperature of Marathon running in a hot humid environment: 2094 Board #250 May 30 2:00 PM – 3:30 PM. Med Sci Sports Exerc 51(6S):569. https://doi.org/10.1249/01.mss.0000562212.76160.ff
75. Aylwin P, Havenith G, Cardinale M, Lloyd A, Ihsan M, Taylor L, Adami PE, Alhammoud M, Alonso J-M, Bouscaren N, Buitrago S, Esh C, Gomez-Ezeiza J, Garrandes F, Labidi M, Lange G, Moussay S, Mtibaa K, Townsend N et al (2023) Thermoregulatory responses during road races in hot-humid conditions at the 2019 athletics world championships. J Appl Physiol 134(5):1300–1311. https://doi.org/10.1152/japplphysiol.00348.2022
76. Racinais S, Havenith G, Aylwin P, Ihsan M, Taylor L, Adami PE, Adamuz M-C, Alhammoud M, Alonso JM, Bouscaren N, Buitrago S, Cardinale M, Dyk N v, Esh CJ, Gomez-Ezeiza J, Garrandes F, Holtzhausen L, Labidi M, Lange G et al (2022) Association between thermal responses, medical events, performance, heat acclimation and health status in male and

female elite athletes during the 2019 Doha World Athletics Championships. Br J Sports Med 56(8):439–445. https://doi.org/10.1136/bjsports-2021-104569
77. Giersch GEW, Morrissey MC, Katch RK, Colburn AT, Sims ST, Stachenfeld NS, Casa DJ (2020) Menstrual cycle and thermoregulation during exercise in the heat: a systematic review and meta-analysis. J Sci Med Sport 23(12):1134–1140. https://doi.org/10.1016/j.jsams.2020.05.014
78. Fournet D, Ross L, Voelcker T, Redortier B, Havenith G (2013) Body mapping of thermoregulatory and perceptual responses of males and females running in the cold. J Therm Biol 38(6):339–344. https://doi.org/10.1016/j.jtherbio.2013.04.005
79. Eglin CM, Golden FS, Tipton MJ (2013) Cold sensitivity test for individuals with non-freezing cold injury: the effect of prior exercise. Extreme Physiol Med 2(1):16. https://doi.org/10.1186/2046-7648-2-16
80. Gerrett N, Ouzzahra Y, Coleby S, Hobbs S, Redortier B, Voelcker T, Havenith G (2014) Thermal sensitivity to warmth during rest and exercise: a sex comparison. Eur J Appl Physiol 114(7):1451–1462. https://doi.org/10.1007/s00421-014-2875-0
81. Gerrett N, Ouzzahra Y, Redortier B, Voelcker T, Havenith G (2015) Female thermal sensitivity to hot and cold during rest and exercise. Physiol Behav 152:11–19. https://doi.org/10.1016/j.physbeh.2015.08.032
82. Ouzzahra Y, Havenith G, Redortier B (2012) Regional distribution of thermal sensitivity to cold at rest and during mild exercise in males. J Therm Biol 37(7):517–523. https://doi.org/10.1016/j.jtherbio.2012.06.003
83. Jimenez-Perez I, Gil-Calvo M, Priego-Quesada JI, Aparicio I, Pérez-Soriano P, Ortiz de Anda RMC (2020) Effect of prefabricated thermoformable foot orthoses on plantar surface temperature after running: a gender comparison. J Therm Biol 91:102612. https://doi.org/10.1016/j.jtherbio.2020.102612
84. Marins JCB, Formenti D, Costa CMA, de Andrade Fernandes A, Sillero-Quintana M (2015) Circadian and gender differences in skin temperature in militaries by thermography. Infrared Phys Technol 71:322–328. https://doi.org/10.1016/j.infrared.2015.05.008
85. Bartelink ML, Wollersheim H, Theeuwes A, Van Duren D, Thien T (1990) Changes in skin blood flow during the menstrual cycle: the influence of the menstrual cycle on the peripheral circulation in healthy female volunteers. Clin Sci 78(Pt 5):527–532
86. Lei T-H, Stannard SR, Perry BG, Schlader ZJ, Cotter JD, Mündel T (2017) Influence of menstrual phase and arid vs. humid heat stress on autonomic and behavioural thermoregulation during exercise in trained but unacclimated women. J Physiol 595(9):2823–2837. https://doi.org/10.1113/JP273176
87. Cowley ES, Olenick AA, McNulty KL, Ross EZ (2021) "Invisible sportswomen": the sex data gap in sport and exercise science research. Women Sport Phys Act J 29(2):146–151. https://doi.org/10.1123/wspaj.2021-0028
88. Ferreira JJA, Mendonça LCS, Nunes LAO, Andrade Filho ACC, Rebelatto JR, Salvini TF (2008) Exercise-associated thermographic changes in young and elderly subjects. Ann Biomed Eng 36(8):1420–1427. https://doi.org/10.1007/s10439-008-9512-1
89. Petrofsky JS, Lohman E, Suh HJ, Garcia J, Anders A, Sutterfield C, Khandge C (2006) The effect of aging on conductive heat exchange in the skin at two environmental temperatures. Med Sci Monit Int Med J Exp Clin Res 12(10):CR400-408
90. Merchant RA, Morley JE, Izquierdo M (2021) Exercise, aging and frailty: guidelines for increasing function. J Nutr Health Aging 25(4):405–409. https://doi.org/10.1007/s12603-021-1590-x
91. Requena-Bueno L, Priego-Quesada JI, Jimenez-Perez I, Gil-Calvo M, Pérez-Soriano P (2020) Validation of ThermoHuman automatic thermographic software for assessing foot temperature before and after running. J Therm Biol 92:102639. https://doi.org/10.1016/j.jtherbio.2020.102639

Chapter 11
Assessment of Sports Garments Using Infrared Thermography

George Havenith and Damien Fournet

Abstract Sports garments and their properties directly influence the heat exchanges occurring at the interface between the human skin and the environment. Active skin cooling or warming through innovative apparel is increasingly developed for training and competition. Infrared thermography can provide original insights into the patterns of skin temperature distribution during exercise that can be used for clothing design using a body-mapping approach. Alternatively, infrared thermography can be used to locate the moisture (sweat) distribution in the clothing.

11.1 Introduction

Sports garments nowadays represent a necessary element for all types of physical activity across the world. Our ancestors used to run in the simplest attire fighting for survival, escaping from dangers or exhausting animals. Over the course of evolution, clothing has become a cultural as well as a safety requirement, following fashion trends, and protecting the skin against potential external aggressors. Specific equipment and clothing have always accompanied the development of ancient and modern sports.

Sports garments manufacturers are engaged in a design process combining aesthetics and various functions to meet the sports demand as well as the body's needs. Furthermore, there sometimes are sports federation rules to comply with for authorization in competing events.

Sports garments are at the interface between the human body and the environment and control the heat and mass transfers occurring at the skin surface [1].

G. Havenith (✉)
Environmental Ergonomics Research Centre, Loughborough University, Loughborough, UK
e-mail: G.Havenith@lboro.ac.uk

D. Fournet
DECATHLON SportsLab, Thermal Sciences Laboratory, Villeneuve d'Ascq, France
e-mail: damien.fournet@decathlon.com

J. I. Priego-Quesada (ed.), *Application of Infrared Thermography in Sports Science*,
https://doi.org/10.1007/978-3-031-93311-0_11

In this context, clothing interacts with physiological and physical processes impacting skin and core temperature [2] that can, with limited accuracy for core temperature, be assessed by infrared thermography. Over the last decade, much attention has been paid to regional thermoregulatory effector responses (e.g., body mapping of local skin temperatures and sweat rates) to adapt the choice and location of textile materials in the garment to better meet the body's needs [1, 3–5]. Thermal patterns obtained by infrared thermography with high spatial sensitivity are very convenient compared to contact measurements on the skin for mapping large body regions.

The assessment of sports garments, specifically their impact on the skin using this technique, enables great advances in the field of garment design and its consequence on sports performance and thermal comfort, both for recreational and professional athletes.

11.2 Sports Garments in Exercise Thermoregulation

11.2.1 The Six Basic Parameters and Heat Balance Equation

Clothing belongs to the six basic parameters [6] contributing to the overall thermal load placed on the human body in a given condition. Air temperature, radiant temperature, relative humidity (rh), and air velocity are the four parameters defining the environmental conditions. Additionally, metabolic rate (M) determines the amount of heat produced during sports, and this must be adjusted for the external work (W) performed in the outside world.

The combination of the parameters determines the potential avenues for heat dissipation toward the environment and heat gain toward the athlete's body.

The human heat balance equation integrates the different physical heat transfers and provides a calculation (11.1) of heat storage at any specific heat production. Clothing influences the intensity of the heat transfers by conduction, convection, evaporation, and radiation.

$$S = M - W - E - K - C - R \tag{11.1}$$

Equation 11.1 Heat balance equation. Where S = net heat storage (W m^{-2}), M = metabolic rate (W m^{-2}), W = external work performed (W m^{-2}, positive or negative), E = evaporative heat loss (W m^{-2}), K = conductive heat transfer (heat loss or gain) (W m^{-2}), C = convective heat transfer (heat loss or gain) (W m^{-2}), and R = radiative heat transfer (heat loss or gain) (W m^{-2})

Depending on the climate, duration, and intensity of the exercise, sports garments can minimize or maximize these heat transfers with fixed properties or specific active technologies.

The human autonomic thermoregulatory system can deal with decreasing or increasing heat storage by triggering a series of effector mechanisms (vasoconstriction/shivering and vasodilation/sweating) to maintain core temperature within a safe range. Skin temperature and the signal transmission from skin thermoreceptors to the control centres in the brain also play a role in the effective effector responses together with central thermoreceptors.

11.2.2 Basic Clothing Properties

The first basic property is clothing insulation (expressed in $m^2\ K\ W^{-1}$ or $m^2\ °C\ W^{-1}$). This corresponds to the heat resistance offered by the garment to dry heat transfer (conduction, convection, and radiation). Insulation is highly dependent on the amount of air trapped in the material (e.g., textile fibres), and this correlates well with material thickness, modulated by fibre diameter [7].

Apparel for sports can be composed of several items organized in various layers depending on the climate. It is important to consider the still air also trapped in the microclimate between the different layers of the ensemble. This microclimate can be thick or thin depending on the location (e.g., no air gap at the shoulders), the clothing fit, or elements to tighten clothing onto the body. Chen et al. [8] observed that thermal insulation was the largest with an air gap of 0.6 cm in windy conditions. Nielsen et al. [9] measured higher local skin temperatures with a tight-fitting compared to a loose fitting inner layer in a 5 and 20 °C environment, advocating for minimizing air disturbance due to movements to maintain insulation. Wind, posture, and body motion are factors that can indeed greatly modify insulation by compressing the garment or by pumping the air between the different layers, therefore reducing the overall heat resistance [10]. ISO 11079 provides calculations on the required clothing insulation to stay in thermal equilibrium depending on the exercise intensity [11]. For example, clothing insulation of $0.155\ m^2\ K\ W^{-1}$ (= 1 Clo unit) would be necessary in a −20 °C environment while running.

Nowadays, garment insulation for cold weather sports comes from either goose down (especially in winter jackets) or from synthetic fibres made with hollow cores or air shafts in order to trap still air [12]. Porous aluminium coating/inner lining to reflect body heat or aerogel with high thermal performance for its thickness can also be used [13].

The second basic property is clothing evaporative resistance (expressed in $m^2\ kPa\ W^{-1}$). It corresponds to the clothing's ability to transfer moisture away from the skin surface. The vaporization of liquid sweat occurs on the skin or in the clothing layers [14]. Similar to heat resistance, evaporative resistance depends on material thickness (with its enclosed air) in the case of permeable materials. Coatings, membranes, or other treatments on the garment greatly alter the transfer of vapour molecules to the environment. Some waterproof finishes close the fabric pores, thereby preventing external precipitation from wetting the body but this also traps sweat inside the microclimate. For thin garments, the fibre characteristics also affect the diffusion properties and modify vapour resistance.

Other properties related to fibre type, yarn type, and fabric construction influence heat and vapour transfers by reducing or enhancing parameters such as air permeability or moisture absorption [12]. Nielsen and Endrusick [15] measured the lowest skin temperature for underwear having an open fishnet structure in a comparison with four other polypropylene knit structures in a two-layer clothing system at 5 °C.

In many different sports, especially in cold environments, a good compromise must be found between heat and vapour resistance. Sufficient insulation must ensure adequate cold protection before the exercise or during resting periods. At moderate to high workloads, heat production can largely exceed heat loss toward the environment and sweating takes place. Vapour resistance should therefore be small to avoid excessive moisture build-up. Total clothing insulation was reduced by 20% when the moisture content of the underwear was 100% of its dry weight [16], and wet clothing can lead to after chill.

Ventilation impacts the garment microclimate and can contribute to reducing the build-up of non-evaporated and condensed moisture. This can occur through natural openings (at the neck, hands, and lower end of a shirt or jacket) or specifically designed ventilation openings such as zippers across the torso, under the arms, or at the sleeves. The combination of pit zip and side seam openings was the most effective in lowering skin temperature in outdoor jackets at 20 °C [17]. Zhang and Li [18] observed significant differences in mean skin temperature by modifying the opening designs of T-shirts worn while running in a 25 °C environment. The bellows effect provided by slits in running jackets has been found to be beneficial in terms of lowered skin temperature and improved thermal comfort [19, 20]. In contrast, in situations of lower heat production or in colder environments, the reduced insulation induced by ventilation [21] must be prevented.

Waterproof membranes can be a requirement for outdoor sports garments to cope with snow or rain. Adverse weather can potentially lead to hypothermia, especially during long-duration events with low heat production, as is the case of recreational hiking [22].

In summary, basic properties of clothing modify the heat exchanges at the skin/environment interface and therefore determine skin temperatures in a given environment. Some sports garments, however, influence skin temperature more actively and to a greater extent of cooling or warming during exercise.

11.2.3 Sports Garments and Skin Cooling

Skin temperature above 35 °C induces a perception of thermal discomfort [23] and has been found to be a critical factor in exercise performance [24]. Many cooling interventions have been explored for sports in hot weather conditions such as water baths, air exposure, cold drink ingestion, and ice application both prior to [25, 26] and after exercise [27]. The practical aspects and cost of these strategies in real use are often questioned. Skin cooling can also be induced by sports garments during exercise with a direct impact on the cardiovascular, thermal, and perceptual strain of recreational and professional sports enthusiasts.

Conductive cooling has been used through lightweight vests in different sports [28–30]. Mean skin temperature was significantly lower wearing the cooling jacket in a 32 °C 70% rh environment with an improved exercise time to exhaustion in male athletes [29]. Palm skin temperature was reduced by up to 1.5 °C wearing an ice vest (Flexi Cold Vest, Interspiro Ab, Lidingö, Sweden) made from water-filled cooling elements chilled at −20 °C before the exercise [30]. Neuromuscular fatigue and ventilatory responses were also lowered compared to a control trial with no vest in a 30 °C 40% rh environment leading to an increase of over 20% in running time. The seven male participants reported improved thermal comfort throughout the exercise protocol [30]. During a 5 km running time trial in a 25 °C, 55% rh environment, master athletes also experienced improved thermal comfort with a cooling vest, but the performance was similar to the control condition [28]. The HyperKewl vest (TechNiche, Vista, California, USA) soaked with water and placed in the refrigerator (6 °C > 8 h) was worn during the whole time trial. It induced significantly lower skin temperatures, especially at the trunk with up to a 2 °C difference during the first half of the trial [28].

The efficiency of the cooling interventions relates to the surface area covered by the cooling garments as well as the ability to ensure an effective fit. In previous examples, the body surface area covered was approximately 0.20 m^2.

Conductive cooling has also been specifically targeted at the neck region. Despite its relatively small surface, the neck is an area of high alliesthesial thermosensitivity [31, 32] and can also be cooled effectively with minimal disruption to sporting actions or attire [33]. Several studies have used a commercially available cooling collar (model CCX, Black Ice LLC, Lakeland, TN) frozen for 24–28 h (at −80 °C) in order to highlight its benefits for performance (time trials, repeated sprints) and perceived thermal strain in 30–32 °C 50–70% rh [33–36]. Neck skin temperature was reduced by up to 18 °C during the first stage of exercise.

The integration of phase change materials (PCM) into adjustable cooling packs for jackets represents another option to promote conductive cooling [37–39]. Skin temperature with a PCM vest can be reduced by up to 3–7 °C compared to without a vest in hot environments [37, 39].

Convective cooling was combined with PCM to enhance skin cooling in a hybrid system [40]. Four ventilation fans were incorporated into the clothing system with a maximum airflow of 0.012 m^3/s. Local skin temperatures were significantly reduced especially in the upper body (maximum reduction from 1 °C to 7 °C) compared to the control condition during a walking protocol in a 36 °C, 60% rh. The hybrid system was also effective in improving thermal and wetness perceptions in the upper and lower body during both exercise and recovery periods [40]. Evaporative cooling was used through a lightweight garment made of hydrophilic fibres able to sustain water evaporation when wetted and consequently provide skin cooling [41]. During a running protocol in a 30 °C, 44% rh, this garment produced significantly lower mean and local skin temperatures by up to 2 and 5 °C, respectively. Cooling the skin before the exercise elicited reduced sweat loss, whereas cooling during exercise reduced heart rate and the rate of perceived exertion. In both interventions, thermal comfort, thermal sensation, and wetness perception were improved by mild evaporative cooling on the torso [41].

In the above cooling techniques, the lower skin temperatures enable a greater core-to-skin temperature gradient for dissipating heat from deeper regions of the body. Cardiovascular strain is reduced because less of the total cardiac output needs to be directed towards the skin, allowing more blood to be directed to active skeletal muscle.

All the above studies used only contact sensors to measure skin temperature in various body regions as well as mean weighed skin temperature calculations. While being extremely useful for monitoring the changes in skin temperature over time [42], contact sensors cannot indicate the total extent of skin cooling over body regions. Moreover, point-to-point variation in local skin temperature can vary by as much as 7 °C over 5 cm [43]. Infrared thermography therefore represents a complementary tool to assess the efficiency of skin cooling especially because surface area is a key factor in the summation of autonomic and behavioural responses.

11.2.4 Sports Garments and Skin Warming

In cold environments, heat production during exercise is often sufficient to maintain total heat balance. However, extremities can be at risk of extensive cooling due to vasoconstriction. The clothing strategy is classically to choose adequate insulation with multilayer garments including footwear, gloves, and hats [44]. Nevertheless, freedom of movement is key to many sports, and increased bulk with increasing insulation should be avoided. Most of the examples below are more related to extreme industrial applications, but the principles are now used in sports garments for enhanced local and overall thermal comfort.

Conductive warming through electrical heating was applied on the hands (heating wire on the fingers and back of hands covered by Arctic mitts) and separately on the torso (10 Kapton insulated flexible heaters) [45]. Finger skin temperatures were maintained at 28–35 °C in both conditions, but torso heating (at 42 °C) contributed to an eight-time higher finger blood flow in a −25 °C environment.

In a less extreme context (2 °C, 80% rh) and using a walking protocol, Song and Wang [40] observed 2 °C higher skin temperatures at the stomach and back using an electrically heated vest (seven heating pads) compared to the same unheated vest. This was also true with chemically heated garments composed of 14 non-reusable body warmers (iron powder, activated carbon, water, vermiculite, and salt) for approximately the same surface as the heating pads. Both heated conditions induced improved whole-body and local thermal sensations.

Conductive warming using PCM garments has been explored with simulation [46] or manikin studies [44] but no human testing to date.

Conductive and convective warming can be obtained through liquid warming garments such as the one designed to be worn underneath a spacesuit, with tubing

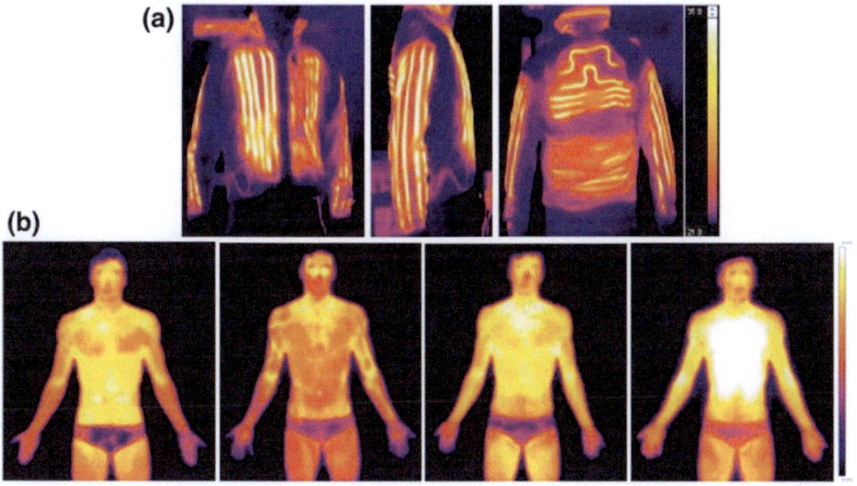

Fig. 11.1 (**a**) Thermography of the heated jacket. (**b**) A sample of upper body thermograms from left to right: before warm-up, after warm-up, and after the recovery period for the control (left two images) and for the heating condition (right two images). Temperature range 21–37 °C. (Figure reproduced with permission Wilkins and Havenith [49])

over all body regions. This has proven to be effective in elevating skin temperatures, especially in the fingers [47].

Some sports garments can also be worn during or at the end of the warmup phase or during recovery periods such as halftime of team sports events. Faulkner et al. [48] demonstrated the benefits of pre-exercise conductive warming of lower leg muscles with heated trousers. This intervention increased local skin temperature and most importantly reduced the fall in muscle temperature (by approximately 1 °C at 3 cm depth), therefore preserving optimal power contractile properties for subsequent exercise performance. Similarly, Wilkins and Havenith [49] showed that a heated jacket worn between a swimming warm-up and the actual race can increase skin temperature and improve sprint swimming performance (Fig. 11.1).

Similar to cold conditions, infrared thermography could become useful to assess the spatial extent of skin warming in order to evaluate more precisely the efficiency of heating interventions or specific active garments during exercise.

11.3 Infrared Thermography Assessing the Influence of Sports Garments

The microenvironment between the skin and the innermost layer of clothing affects skin temperature which, in turn, affects thermoregulatory responses. Skin temperatures should accurately assess the effectiveness of a clothing ensemble [12].

Moreover, core temperature is rarely modified by sports garments (except for protective sports garments) and depends to a considerable extent upon metabolic heat production. The concept of "thermal envelope" has then been suggested to evaluate the influence of sports garments in ecologically valid sports-related exercise protocols [50]. It refers to upper and lower limits in mean body temperature (from mean skin and core temperatures) elicited by clothing assemblies that could be worn under a given environmental condition. Measuring skin temperature in various locations, especially in cool environments [51], is therefore highly important for this approach. Though infrared thermography cannot track skin temperatures over time in clothed subjects, it provides the benefit of measuring large body surfaces with high spatial resolution. Previous research has pointed out that skin temperatures and distribution immediately after undressing are the same as those which existed under the clothing [52, 53] though it has also been shown that stopping the activity for taking the IR measurements can affect outcomes due to the strong change in convective heat loss when changing from running to standstill for the IR imaging [54, 55]. Nevertheless, with careful data collection and standardized undressing procedures, this opens a wide area of testing the influence of many kinds of sports garments. The current knowledge in this field is presented with key examples in specific sports.

11.3.1 Assessment of Running Wear

Many studies have investigated the skin temperature distribution of semi-nude runners using infrared thermography [5, 56–59]. Apart from the pioneering work from Clark et al. [57], no studies reported whole-body thermograms as they focused on specific body parts to understand the underlying mechanisms modifying skin temperatures between physiological (skin blood flow, muscle activation, and body composition) and external factors (air and radiant temperature, air velocity, and relative humidity). Fournet et al. [4] provided the first average whole-body thermograms based on a specific image processing technique [60] able to combine several thermograms of different participants with a variety of body dimensions into a single, representative group image.

A quantitative and qualitative data analysis was provided on male and female runners with a special focus on the link between skin temperatures and skinfold thickness variations [60].

Using the same approach, Fournet [52] explored the thermal patterns of clothed runners at distinct stages of the exercise protocol. Twelve physically active males performed a 40-min running bout at 70% VO_{2max} in a 5 °C, 50% rh environment with frontal wind at 5 m s^{-1}. This was preceded by 5 min standing prior to exercise and followed by 10 min post-exercise recovery also standing on the treadmill with 2.8 m s^{-1} frontal wind. Participants were wearing a one-layer tight-fitting winter long-sleeved top and tights (Kalenji Isolate 4000, DECATHLON, France). Infrared thermography assessment was obtained at four different stages: prior to exercise

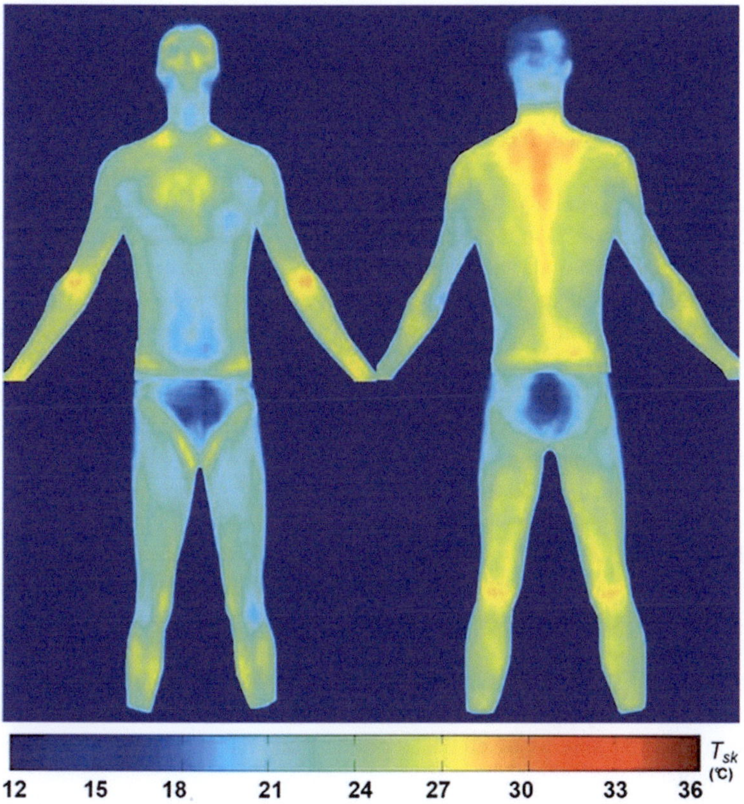

Fig. 11.2 Anterior (*left*) and posterior (*right*) group averaged whole-body maps of skin temperature after 40 min of running at 70% VO_{2max} in a 5 °C, 50% rh environment with uniform clothing insulation. (Figure adapted with permission from Fournet [52])

(PRE), after 10 min of running (RUN10), after 40 min (RUN40) of running, and after 10 min of recovery (POST); following a strictly controlled and short undressing procedure away from the wind. Mean whole-body thermograms ($n = 12$) for the group are presented at stage RUN40 for the anterior and posterior body (Fig. 11.2).

Skin temperature underneath clothing was colder in regions such as the abdomen, pectorals, thighs, triceps, patella, neck, and cheeks. They corresponded to regions deeply or poorly perfused, with large fat or muscle insulation, highly exposed to convective air cooling. The warm regions were located in the upper chest, spine, inner thighs, calves, elbow crease, popliteal fossa, and orbits. They corresponded to regions with either high perfusion, low fat or muscle insulation, and the natural creases of the human body where cross radiation exists between close surfaces or where surfaces are directly in contact.

A Y-shape of colder skin temperatures was noticeable over the anterior torso (pectoral and abdomen) and a Y-shape of warmer temperatures over the posterior torso (upper back and spine). Fournet [52] provided the general hypothesis of the compromise between the superficial blood flow and body characteristics (morphology and composition) to explain the skin temperature variations across the body. Hot spots in the upper chest, neck, and upper back were strikingly in line with brown fat deposits in infants. As brown fat has been observed in adults, especially in lean persons exposed to cold [61], it could be postulated that these regions contributed to specific warm spots due to their high vascularization and not necessarily due to their associated local thermogenesis.

Interestingly, mean skin temperature was similar in the clothed condition at 5 °C compared to the semi-nude condition at 10 °C [4]. The thermal patterns were almost identical between the conditions except for the warmer skin temperature at the upper chest and forearms (clothed) and at the calves and posterior legs (semi-nude). This could imply that heat dissipation is favoured close to heat production in the nude moving lower limbs. On the other hand, heat could be conveyed back to the heart and then dissipated when quitting the aortic arch (upper chest) more intensively in a clothed condition.

Roberts et al. [5] investigated the thermoregulatory responses of seven males wearing different base-layer garments during an intermittent running protocol. The base-layer hot was claimed to keep the wearer comfortably cool during exercise in a warm environment, whereas the base-layer cold garment was claimed to keep the wearer warm in cold conditions. These two garments with good wicking properties were compared to a 100% cotton base layer and bare-chested condition. Two running bouts lasting 23 min were repeated in a 20 °C, 50% rh environment on a treadmill with speeds varying from 6, 12, and 15 km h^{-1}. Mean, maximum, and minimum skin temperatures were assessed based on the front and back torso area covered by the garments.

At the end of the exercise, core temperature was similar between clothing conditions. On the other hand, the mean skin temperature over the torso was the lowest in the bare-chested condition (28.5 ± 0.8 °C) followed by the base-layer hot (29.1 ± 0.4 °C). These were significantly lower than in the base-layer cold (29.8 ± 0.8 °C) and the 100% cotton garment (29.8 ± 0.8 °C). The thermoregulatory responses were associated with improved thermal comfort in bare-chest and base-layer hot conditions compared to base-layer cold and cotton. Roberts et al. [5] postulated that this could be due to the low moisture retention in the lightweight base-layer hot (small yarn diameter, more interstices), therefore reducing skin wettedness, an indicator of thermal discomfort [7].

Some similarities in the thermal patterns were found such as the larger areas of higher temperature over the sternal region and the external oblique muscles of the abdomen. Thermal peaks could be observed in each condition, and this is in line with previous work on semi-nude participants [58, 62] in connection with cutaneous perforator veins transferring blood from deep active regions [63]. The differences in evaporation and skin contact, especially when garments became saturated, could explain the significant differences between the conditions [5].

To date, this has been the only study exploring different clothing compositions using infrared thermography to evaluate its impact on skin temperatures and the overall thermal pattern after running.

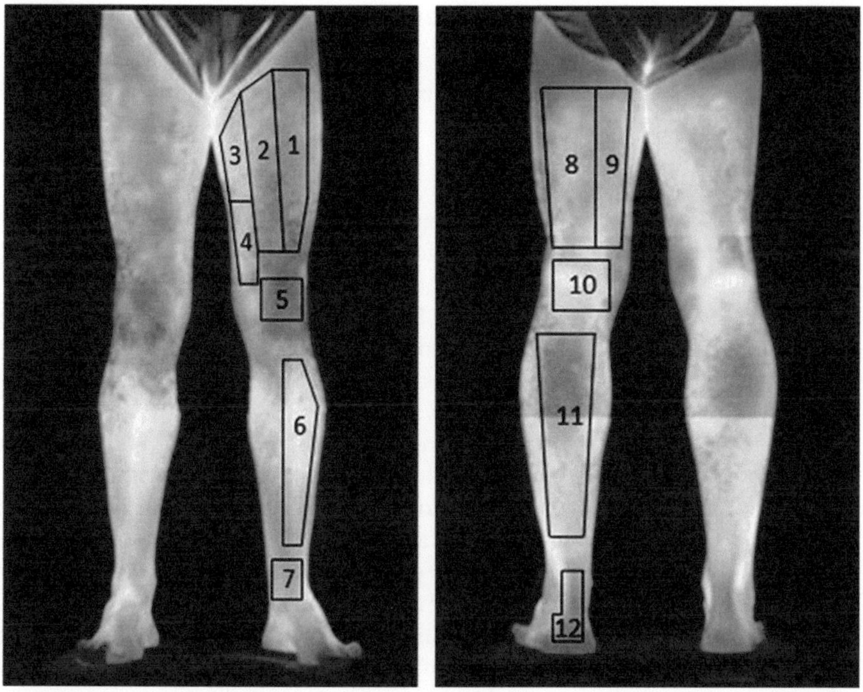

Fig. 11.3 Regions of interest for the evaluation of graduated compression stockings. *1* Vastus Lateralis. *2* Recture femoris. *3* Abductor. *4* Vastus medialis. *5* Knee. *6* Tibialis anterior. *7* Ankle anterior. *8* Biceps femoris. *9* Semitendinosus. *10* Popliteal. *11* Gastrocnemius. *12* Achilles. (Figure reproduced with permission from Priego Quesada et al. [64])

Priego Quesada et al. [64] conducted a specific study to point out the effects of graduated compression stockings on skin temperature after running. Forty-two participants (29 males and 13 females) performed two running tests in a 24 °C, 50% rh lasting 30 min (10 min of warm up and 20 min at 75% of maximal aerobic speed) with and without compression in the lower leg. They hypothesized a direct influence of the covered body regions with extra insulation as well as a potential effect on areas with no garment contact. A quantitative analysis was performed based on a definition of several regions of interest (Fig. 11.3).

It was shown that running with compression stockings increased local skin temperatures by 0.04–0.91 °C which could be due to the garment insulation. Moreover, skin temperatures in regions not covered by the stockings (vastus lateralis, abductor, and semitendinosus) increased by 0.01–0.52 °C, and it could be explained by a greater superficial perfusion induced by the lower limb coverage.

The knowledge of the thermal patterns under running apparel [5, 52, 64] could be extended to a myriad of clothing conditions and exercise types (continuous, intermittent, sprints, and short/long distances) in order to map the similarities and differences in skin temperature distribution.

11.3.2 Assessment of Sports Bras

The influence of sports bras has mainly been studied in the field of biomechanics with the aim of reducing breast movement as well as exercise-related breast pain [65, 66]. This led to recommendations for the use of well-fitted and supportive sports bras while exercising. Ayres et al. [67] were the first to specifically investigate the influence of this clothing layer and composition on skin temperatures.

Eight females with C-cup breasts performed a short-duration exercise protocol (20 min) in a 27 °C, 46% rh environment [67]. The protocol included a warm-up walk at 4 km h^{-1} (1% gradient) followed by five countermovement jumps and five maximum effort agility "T" tests followed by 8 min of running at 10 km h^{-1}. They were either wearing a composite sports bra (65% nylon, 22% polyester, 13% elastane, Under Armour, Stability, USA) or a polyester sports bra (100% polyester, Under Armour, Strength, USA). The bare-breast and abdomen skin temperatures were assessed using infrared thermography before and immediately after exercise. Reflective markers were taped onto the skin in order to standardize the quantitative analysis with different regions of interest as shown in Fig. 11.4. The temperature of the left and right breast was averaged for each image. Results indicated that breast skin temperature was reduced to a lesser extent than abdominal skin temperatures (−0.6 vs. −2.1 °C) in both bra conditions demonstrating the reduced evaporative heat loss caused by the bra. This effect of insulation provided by the bra on breast skin temperature was also observed by Fournet et al. [4] during a running protocol in females.

In Ayres et al.'s study [67], bra composition had a significant influence on skin temperatures with a larger overall decrease over the exercise period in the polyester (−1.5 °C) compared to the composite sports bra (−0.9 °C). While bra comfort and bra fit were similar between conditions, participants reported greater thermal comfort in the polyester than in the composite bra. Yet, there was no significant correlation between thermal comfort and pre-post skin temperatures, with large variability in comfort responses. Ayres et al. [67] also pointed out that both bras displayed a

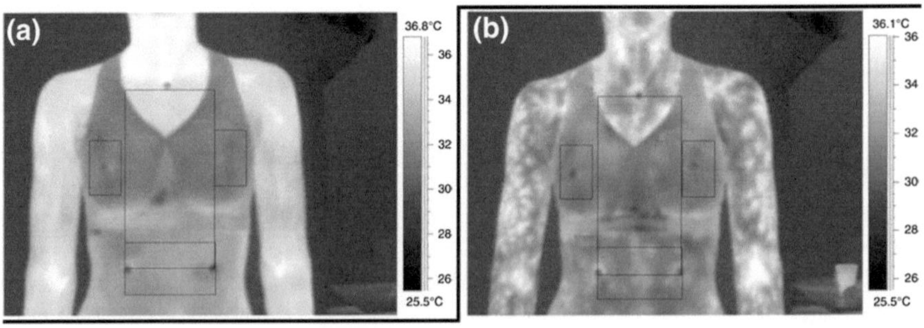

Fig. 11.4 Thermograms of bra and skin temperatures of a participant wearing the composite bra before (**a**) and after (**b**) exercise in the thermoneutral condition. Regions of interest are highlighted. (Figure reproduced with permission from Ayres et al. [67])

similar shape and coverage. Nevertheless, there was a mesh section in the composite bra at the back, and this corresponded approximately to an open uncovered section in the polyester bra.

Several future directions in bra ergonomics were suggested with the influence of additional clothing layers and the contribution of other physiological factors (hydration, heart rate, and core temperature) on breast thermal comfort. The results also highlighted the importance of optimizing the design of sports bras in order to facilitate the cooling ability of the skin. These changes can improve thermal comfort which may also lead to differences in athletic performance [67].

11.3.3 Assessment of Hiking Ensembles

Hiking is a popular recreational activity with large participant numbers varying widely in terms of age, gender, and hiking experience [68]. In severe weather, hypothermia can occur during hill walking, and this has been associated with reduced clothing insulation (cold, wet, and windy weather) and progressive fatigue due to the inability to compensate for heat loss with heat production [69–71]. Adequate clothing is therefore required in changing outdoor settings (water repellent and minimal condensation) along with a good pacing strategy to allow for sufficient energy expenditure. Apart from Pugh [72], no studies have assessed body temperatures and associated thermal comfort of hikers facing non-adverse conditions. Fournet et al. [73] replicated the thermal strain of a normal day hike up a hill in cool conditions within the laboratory and integrated infrared thermography to assess skin temperatures under clothing over the whole body at different stages.

Eight males and eight females performed a simulated morning walk 2*30 min of uphill walking (15% gradient) at 55% VO_{2max} followed by 15 min seated rest at the top (with 10 km h^{-1} wind) followed by 30 min of downhill walking (15% gradient) at 20% VO_{2max} [73]. In the first half of the ascent, participants were wearing trousers (Forclaz 900, Quechua, Decathlon, France), T-shirts (Forclaz Quechua, Decathlon, France), fleece (Florclaz 50, Quechua, Decathlon, France), and sports bra for females (Sportance, Kalenji, Decathlon, France), as well as backpack with 10% body mass (Forclaz 50 L, Quechua, Decathlon, France). In the second phase of the ascent, they removed the fleece layer. Environmental conditions were changed from 5 °C, 50% rh to 15 °C, 60% rh during the first 30 min to mimic real-life conditions in a condensed manner.

Mean skin temperature was significantly lower for females compared to males (−0.7 °C) during the uphill and resting stages [73]. Locally, skin temperature changes were more pronounced in females compared to males, except for the legs. Notably, anterior arm skin temperatures fell by 3 °C during the uphill phase in females and by 2 °C in males. Females had significantly colder thermal sensations at the start, a less humid wetness perception after 30 min of ascent, and they had larger thermal discomfort at rest when exposed to wind.

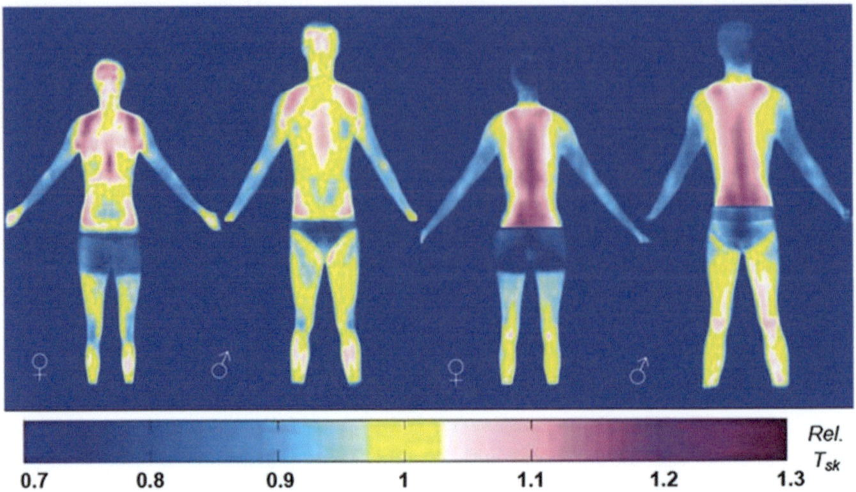

Fig. 11.5 Anterior (*left*) and posterior (*right*) groups averaged relative whole-body maps of skin temperature for eight females (♀) and eight males (♂) after 60 min of uphill walking at 55% VO_{2max} in a 15 °C, 60% rh environment with uniform clothing insulation. Thermograms are normalized to mean skin temperature, with a value of 1 equal to mean skin temperature in both groups. For example, 0.7 corresponds to 0.7 times the mean skin temperature. (Figure adapted with permission from Fournet [52])

Group-averaged thermograms were obtained at each stage using the image processing method described by Fournet et al. [60]. The relative maps of skin temperature distribution enabled a direct comparison between males and females despite differences in mean skin temperature (Fig. 11.5).

Patterns of skin temperature distribution were highly similar between males and females. This is in line with a study on unclothed male and female runners [4]. Interestingly, no correlation between the temperature distribution with local skinfold thickness was observed, suggesting that other factors than fat distribution are responsible for the skin temperature distribution.

Main features, observed consistently in the thermal patterns, were related to the warmer upper body versus lower body, a V or Y shape of warmer skin temperature over the anterior torso, the warmer lower leg overactive muscles, the colder arms due to the absence of clothing insulation (short sleeve T-shirt during the second ascent phase). The influence of the backpack was clearly noticeable with higher skin temperatures in localized body regions in contact with the shoulder straps, the belt (with abdominal pads), and the upper/lower back which seemed more pronounced in females (Fig. 11.5).

Interestingly, males performed an additional trial without a backpack at a similar heat production. The quantitative analysis (based on specific control points) highlighted the magnitude of these local effects reaching a maximum of 3 °C higher skin temperatures in the lower back for the backpack condition [52].

The use of infrared thermography for the assessment of sports garments described in the literature remains limited to a small number of studies. In certain thermo-physiological studies where rapid temperature changes are expected, it seems inadequate unless the area of interest is not located under clothing. On the other hand, infrared thermography and specific processing methods can help in the evaluation of clothing interventions such as warming or cooling jackets [49, 74], different garment compositions [5, 67], the addition or subtraction of clothing layers [52, 64], or other sports equipment [52, 75]. The main advantage of mapping large body surfaces is rarely exploited as quantitative analysis are often limited to defined regions of interest (e.g., Ayres et al. [67]) not necessarily considering the natural pattern of skin temperature distribution. The understanding of these patterns in connection with clothing properties, the type of sports, and external determinants (e.g., wind) can be extremely useful to better adapt the design, the choice of material of sports garments using a body-mapping approach for improved thermal comfort, and exercise performance.

11.4 Infrared Thermography and Body-Mapping Sports Garments

11.4.1 Design and Evaluation of Body-Mapping Garments

A body-mapping approach refers to the knowledge of biological and/or sensorial responses associated with several body regions exposed to the same stimulus or condition. Havenith et al. [1] described the principles of this body-mapping approach in the field of thermal comfort and clothing ergonomics. Regional clothing adaptation can be made according to the regional knowledge of skin temperature, sweat production/skin wettedness, and the regional skin sensitivities to temperature and wettedness [1]. Many body-mapping studies on these parameters have been performed using different techniques, with local thermal stimulations for sensitivity mapping, [76–79], local sweat collections with absorbent pads for sweat mapping [1, 80, 81], simultaneous contact sensors [31, 82], and infrared thermography for skin temperature mapping [4, 52, 60, 73]. The combination of the different body maps and their relevance according to the exercise type and environmental conditions is a determinant for clothing manufacturers in the development of sports clothing. This may lead to improved thermal-wet perceptions, more efficient heat loss or preservation, and reduced sweat loss which can all contribute to exercise performance. Wu et al. [83] attempted to use this kind of approach by combining skin temperatures and several thermal-wet perceptions in connection with the evaluation of ten kinds of hygroscopic fibres. They developed a T-shirt assembled with four fabrics matching regional requirements.

The choice and position of fabrics can be based on various body maps. Yet, in the overall garment construction, garment fit plays a crucial role by modifying the heat

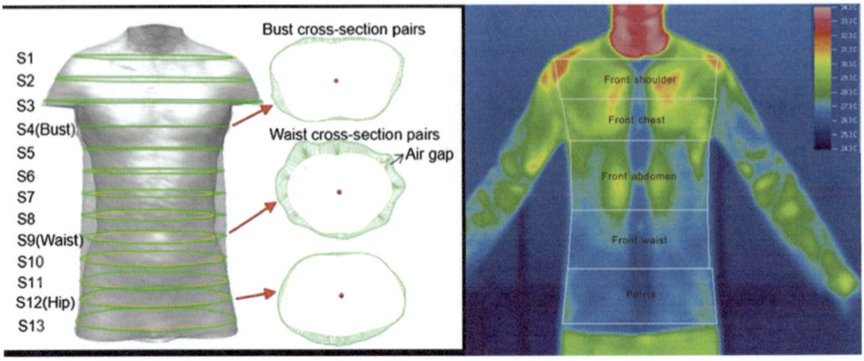

Fig. 11.6 3D scanner analysis with the selection of cross-section pairs (*left*) and clothing surface temperatures (*right*) with regions of interest for the same garment. (Figure reproduced with permission from Zhang et al. [86])

and mass transfers between the skin and the environment. Some studies have therefore performed quantification of the volume and size of air gaps determined by garment fit [84–86]. They have highlighted the uneven distribution of air gap layers using 3D body scanners. Zhang et al. [86] combined this assessment with infrared thermography of clothing surface temperatures [86] (Fig. 11.6).

They found that air gaps usually have larger thicknesses in concave areas (lumbar region) and smaller thicknesses in convex areas (hips and shoulders) formed by the human skeleton and muscles. Clothing surface temperature decreases with increasing air gap thickness up to a critical thickness of 15 mm, above which natural convection occurs [86]. Based on this local knowledge, a revised pattern was designed with additional cover cloth on the shoulder and waist darts in the back waist region. This led to an improved thermal insulation of the new garment as measured on a thermal manikin.

In the five-level system for the analysis of the physiological properties of textiles and garments [87], manikin testing represents a key level in the assessment of sportswear. It provides overall heat and vapour resistances as well as localized resistances depending on the number of body segments. Kicklighter et al. [88] reported that the scientific knowledge of body-mapping sportswear engineering has lagged far behind commercial practices of these garments.

A few studies have analysed the thermal properties of body-mapping sportswear on sweating manikins [18, 89, 90] as well as their implications for physiological and perceptual responses through human testing [4, 18, 83, 86].

The presence of mesh or openings at the two vertical side seams was found to be ideal for ventilative cooling based on manikin testing in a constant temperature mode [18, 89]. Wang et al. [90] used a manikin in thermoregulatory mode and demonstrated that body-mapping sportswear did not provide global advantages (mean skin and core temperature), but local improvements in heat and vapour transfer were beneficial for local thermal perceptions, especially with a modular system.

Wu et al. [83] measured lower thermal and stickiness sensations using a body-mapping T-shirt in 25 females at very low exercise levels in a 24 °C 55% rh environment. Zhang and Li [18] observed lower skin temperatures with mesh fabrics at the front torso (vertical) for six males running at 40% VO_{2max} in a 25 °C, 50% rh environment. There were no differences in perceptual responses between all body-mapping garments varying in mesh location, but thermal comfort was significantly improved compared to a no-mesh control condition [18]. Lastly, Wang et al. [86] reported significantly improved physiological responses (lower mean skin and core temperature, mean torso temperature, and heart rate) and local thermal sensations with a modular tight-fitting body-mapping kit worn by eight males running at 10 km h^{-1} in a 30 °C 40% rh environment.

In warm to hot environments, these body-mapping studies focused on adapting the wicking, air permeability and vapour resistance of fabrics assembled in sports garments based on knowledge of sweat production, distribution, and efficient garment construction. Skin temperatures were measured by local contact sensors potentially over-estimating or under-estimating the impact of specific fabrics locally [43]. Infrared thermography could bring more insights into the real efficiency of body-mapping sports garments and provide directions into improving their benefits for thermal comfort and exercise performance.

11.4.2 From Thermograms to Body-Mapping Garments

Infrared thermography can be used for whole-body temperature mapping [3, 4, 52, 60, 73], and this knowledge can be translated into body-mapping sportswear. Havenith et al. [1] suggested that placing extra insulation on the specific cold body regions could be beneficial for overall thermal comfort while targeting the warm regions could prevent unwanted heat loss. Knowledge of the natural whole-body distribution of temperature is required for this kind of strategy towards improved garments, especially for cold-weather sports or activities. Furthermore, this distribution can also serve in the heat to maximize heat loss or for targeted cooling interventions though regional skin temperatures become almost uniform above 30 °C air temperature during exercise [91].

Domina et al. [92] described a method based on 3D scanning and infrared thermography to provide individual thermal profiles. Image registration techniques are used to transform 2D thermograms onto 3D scans of the same person (Fig. 11.7). They suggested that this could lead to body-mapping garments customized by function (heat dissipation or preservation) and by fit necessary for designers and product developers. Domina et al. [92] mentioned potential improvements in performance as well as cost savings due to more specific placing of fabrics or finishes. Preliminary results using this method (body-mapping garment designed for the upper body) were promising for improved thermal comfort and reduced skin temperature changes during a short duration exercise (Harvard Step test) on two males in a 2 °C but not in a 29 °C environment [93].

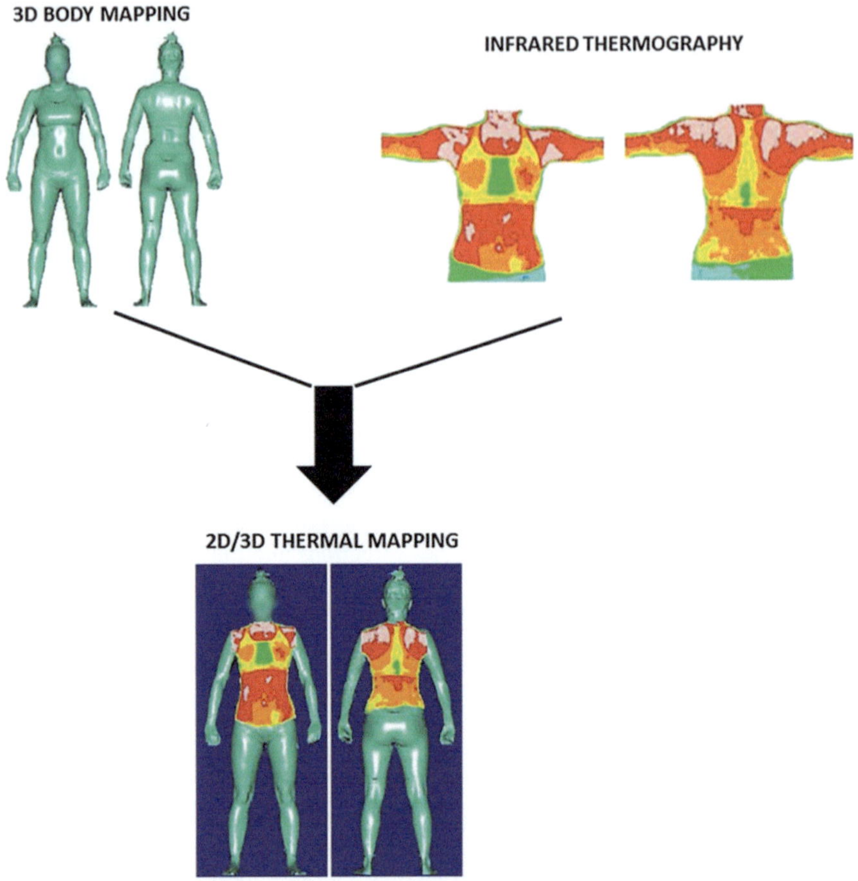

Fig. 11.7 2D/3D thermal mapping method described in the study of Domina et al. [92]. (Figure adapted with permission from Domina et al. [92])

Fournet et al. [3] investigated for the first time the suggestion by Havenith et al. [1]. Body-mapping sports garments were designed based on the natural distribution of skin temperature distribution under clothing obtained by infrared thermography while running in the cold (see Fig. 11.2). Thermal patterns were divided into two isotherms of skin temperatures to determine the body regions where clothing insulation was modified. Two levels of insulation (high and low) were applied within the running ensemble and compared to an ensemble where moderate insulation was uniformly distributed (Clothing U). Clothing C was designed to add insulation in the naturally cold regions (and reduced insulation elsewhere) and Clothing W added insulation in the naturally warm regions (Fig. 11.8). All clothing ensembles had similar overall clothing insulation (0.7 Clo). Twelve males performed a 40-min running bout at 70% VO_{2max} in a 5 °C, 50% rh environment. Overall, thermoregulatory

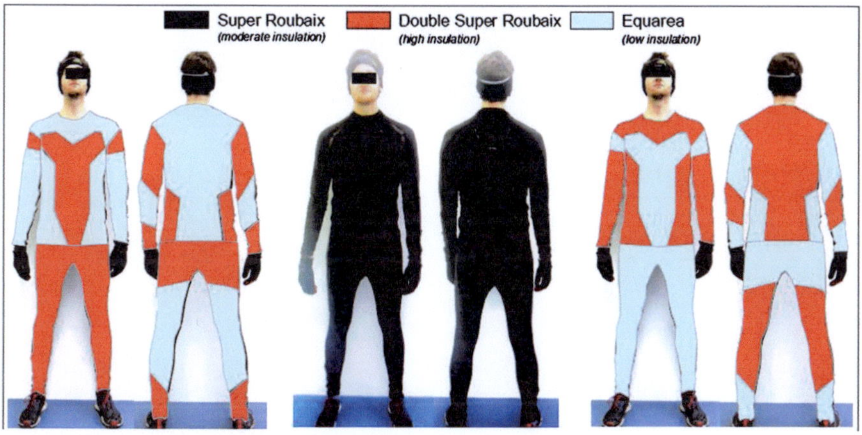

Fig. 11.8 Schematic presentation of the three clothing ensembles (C, U, and W). Normal colour was black for all ensembles. (Figure adapted with permission from Fournet [52])

responses were not affected by the type of clothing (similar mean skin and core temperature). Regional skin temperatures assessed by infrared thermography were strongly affected due to the different distributions of insulation. Clothing U led to a natural skin temperature distribution, whereas Clothing W amplified this distribution pattern, while Clothing C significantly attenuated skin temperature variations across the whole body (creating a more homogenous pattern). High insulation areas induced 2–2.5 °C higher regional skin temperatures compared to thinly insulated conditions irrespective of placing the insulation on naturally "warm" or "cold" regions (Fig. 11.9). High insulation in W had a significant negative impact on thermal comfort in the upper back and arms as they were perceived warmer and wetter compared to clothing C and U. Overall, there were more positive comfort ratings for C suggesting the benefits of reducing skin temperature contrasts between regions [3].

Further studies are required to understand the influence of body-mapping sportswear in the cold. Manipulation of local heat resistances could be mixed with the manipulation of vapour resistances to maximize comfort and performance in moderate to high-intensity exercise where sweat production and skin wettedness can be detrimental [7]. Fournet [52] also presented modular systems allowing easy and quick behavioural adjustments of the local heat loss by the wearer to consider the wide variety of weather conditions (especially in outdoor sports) and the potential variations in exercise intensity due to pacing or the sports demand.

Infrared thermography was the perfect tool to assess the efficiency of the clothing intervention and can be promising in both evaluating normal skin temperature patterns and deviations from these patterns with body-mapping sportswear [3].

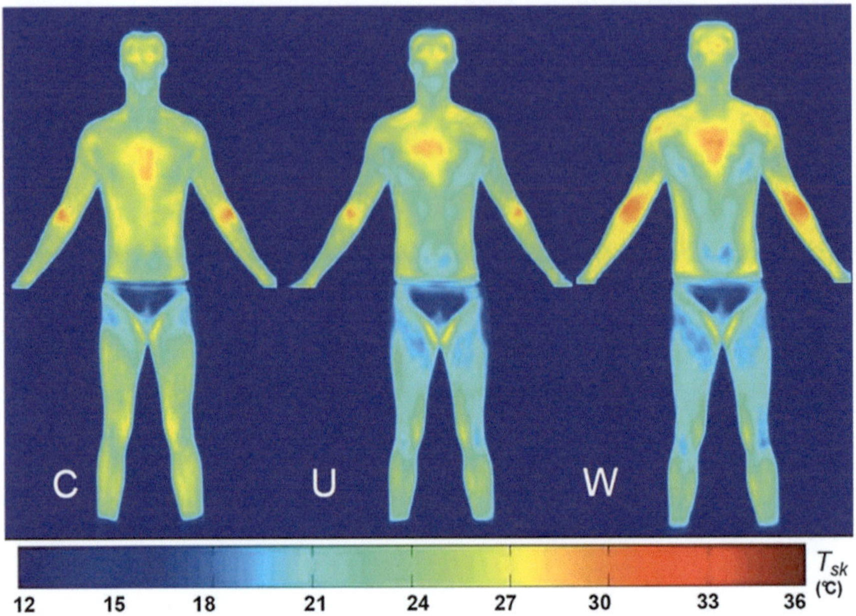

Fig. 11.9 Anterior group-averaged whole body maps of skin temperature for clothing C (higher insulation in cold regions), clothing U (uniform moderate insulation), and clothing W (higher insulation in the warm regions) for 12 males after 40 min of running at 70% VO_{2max} in a 5 °C environment. (Figure adapted with permission from Fournet [52])

11.4.3 Body Mapping Sweat Accumulation by IR Imaging

While most infrared imaging research related to body mapping and the development of specialized body-mapping garments has focussed on skin temperature distribution, attempts have been made to use infrared imaging to consider the distribution of sweat accumulation in the clothing during exercise. When clothing gets wet, it tends to cool down due to the evaporative cooling, which can be captured by IR imaging. The problem is that while the clothing is on the body, the clothing temperature is affected by the underlying skin temperature, local convection, evaporation, and ventilation, which makes the interpretation of IR images of the clothing itself problematic. Raccuglia et al. [94] tried to evaluate the usability of IR imaging for the determination of sweat distribution in clothing (running T-shirts). They solved the issue of the influence of spatial skin temperature variations by taking the clothing off the body after different exercise durations and putting it on a wooden template. There, the temperature distribution in the shirt will only be affected by moisture evaporation. They validated the method by cutting up the T-shirt after the IR imaging and weighing different sections (see Fig. 11.10). Due to the destructive nature of the validation, they had to run many experiments of different duration to see what happened over time. The results showed that the method worked well in

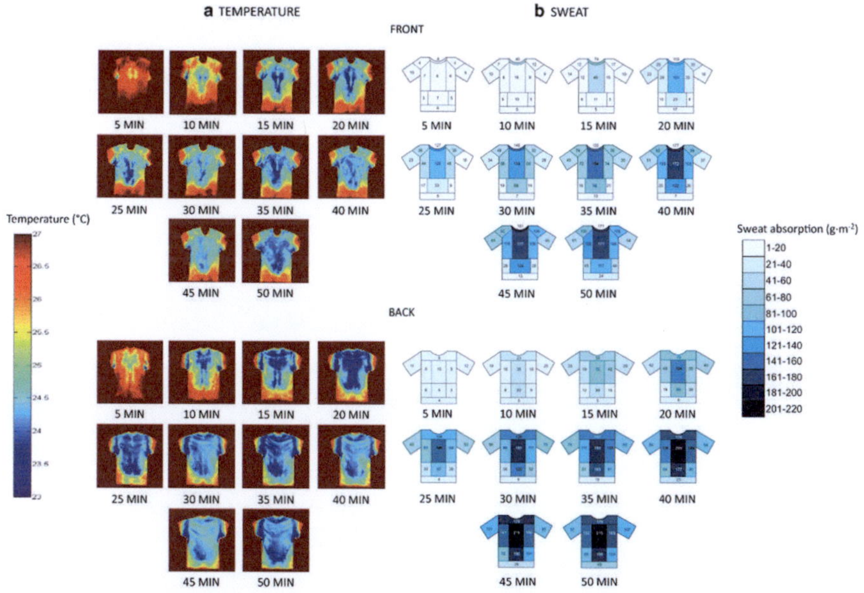

Fig. 11.10 Average (eight participants) maps of temperature (panel a, left) and sweat accumulation (panel b, right) distribution across the front side of the garments, over 50 min of running exercise. Data for each time point were obtained from ten different duration running trials. (Figure from [94]; published under Creative Commons CC BY license)

terms of qualitative determination of the sweat accumulation distribution, but that it had limited use to determine sweat accumulation quantitatively. Molly Subhash et al. [95] used the same approach in determining armpit sweat accumulation in garments in relation to the use of antiperspirants. In addition, using high-definition temporal recordings of a small area of skin, they successfully looked at sweat gland activation with IR imaging.

11.4.4 Adjusting Body-Mapping Garments to Different Populations

With body-mapping sports clothing design trying to adapt clothing properties across the body to the spatial distribution of skin temperatures and sweating patterns over the body, a higher level of physiological functionality can be reached compared to normal sports clothing. To go one step further, one can design differentially for specific groups in the population, like males versus females, for different age groups, or people with disabilities. Data on group differences in sweating with regards to age have been recently published for older (68 ± 5 years) and young (24 ± 2 years) adults [96] and for children [97].

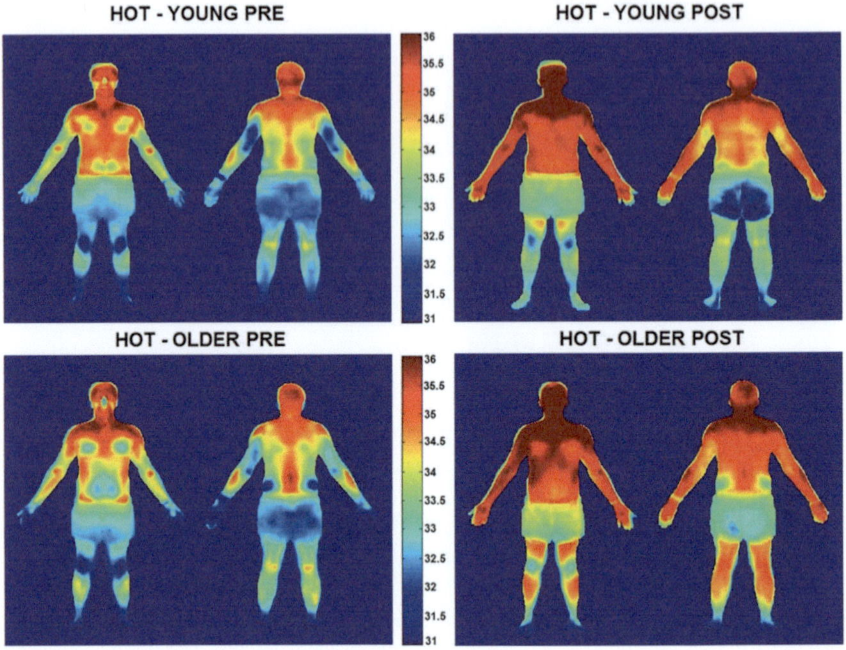

Fig. 11.11 Pre and post-rest period (32 °C and 50% RH), group averaged (n = 10 in the young (18–30 years), and n = 10 in the older (60–80 years) group) body maps of absolute skin temperature (°C). Temperature scale is adapted based on the overall mean skin temperature of the condition (~ 34 °C) ± 4 °C. (Figure from Coull et al. [96] (published under Creative Commons CC BY license))

Figure 11.11 shows the temperature distribution in young and older adults before and after a rest period in the heat, showing the temperature changes induced by the heat exposure, and the differences between the age groups. The older males showed higher leg skin temperatures than the young, which reflects the reduced sweating in the leg area in ageing. With the sweating withdrawing from the extremities with ageing [98], there will be a shift from vapour transfer to convective cooling with age, which can be considered in the garment fabric choices. Similarly, Fig. 11.12 shows the skin temperature distribution of pre-versus mid-pubertal children during exercise [99], showing the much higher skin temperatures pre-puberty. This difference is linked to lower sweat production in pre-pubertal children [97] and illustrates the higher dependence of the younger children on convective heat loss, rather than evaporative. This has led to the development of an age-specific running T-shirt for younger children by Adidas® that focusses more on convective losses.

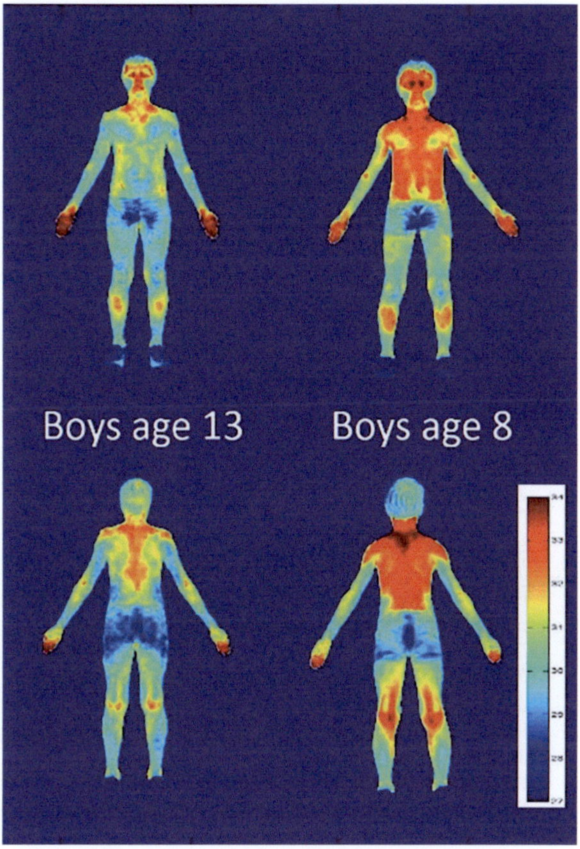

Fig. 11.12 Skin temperature patterns during exercise at the same relative heart rate in pre- and mid-pubertal children, showing the higher skin temperatures pre-puberty (Havenith et al. [99]; published under Creative Commons CC BY license)

11.5 Conclusion

Infrared thermography has been used only to a small extent in the field of sportswear ergonomics. There are yet numerous examples where this method could be used in addition to contact sensors for the measurement of skin temperatures. Some challenges need to be overcome regarding processing time, method, and speed of clothing removal, as well as standardized processing techniques for the analysis of thermograms. Nevertheless, the individual qualitative and quantitative information obtained by infrared thermography can be extremely valuable to assess the efficiency of sports garments at the skin interface (warming, cooling, and reducing temperature contrasts), consequently leading to improvements in thermal comfort and/or exercise performance.

Athletes or coaches may benefit from the development of external devices (such as drones with integrated infrared cameras) able to monitor the uncovered body parts (hands and face) in sporting situations in order to track/predict emotional and physiological changes [54, 55, 100]. With increasing wearable sensors integrated into clothing, one can imagine sports garments that could automatically adapt to these changes taking also into account weather conditions, heat production, and skin temperature/wettedness.

Infrared thermography can also help in the development of more customized sports garments where individual spatial temperature patterns can be useful for targeted heat loss or preservation. High-resolution thermograms can display the exact positions and variations of muscle/skin perforators where substantial heat is being released. Favouring the extraction of heat or cooling intensely, these spots may prove beneficial for temperature regulation.

References

1. Havenith G, Smith CJ, Fukazawa T (2008) The skin interface—meeting point of physiology and clothing science. J Fiber Bioeng Inf 1:93–98. https://doi.org/10.3993/jfbi09200802
2. Foster J, Lloyd AB, Havenith G (2021) Non-contact infrared assessment of human body temperature: the journal temperature toolbox. Temperature 8(4):306–319. https://doi.org/10.1080/23328940.2021.1899546
3. Fournet D, Redortier B, Havenith G (2015) Can body-mapped garments improve thermal comfort for sport in the cold? Extreme Physiol Med 4:A74
4. Fournet D, Ross L, Voelcker T et al (2013a) Body mapping of thermoregulatory and perceptual responses of males and females running in the cold. J Therm Biol 38:339–344. https://doi.org/10.1016/j.jtherbio.2013.04.005
5. Roberts BC, Waller TM, Caine MP (2007) Thermoregulatory response to base layer garments during treadmill exercise. Int J Sports Sci Eng 1:29–38
6. Parsons K (2002) Human thermal environments: the effects of hot, moderate, and cold environments on human health, comfort and performance, 2nd edn. CRC Press
7. Havenith G (2003) Interaction of clothing and thermoregulation. Exog Dermatol 1:221–230, https://doi.org/10.1159/000068802
8. Chen YS, Fan J, Qian X, Zhang W (2004) Effect of garment fit on thermal insulation and evaporative resistance. Text Res J 74:742–748
9. Nielsen R, Gavhed DC, Nilsson H (1989) Thermal function of a clothing ensemble during work: dependency on inner clothing layer fit. Ergonomics 32:1581–1594. https://doi.org/10.1080/00140138908966927
10. Havenith G, Heus R, Lotens WA (1990) Resultant clothing insulation: a function of body movement, posture, wind, clothing fit and ensemble thickness. Ergonomics 33:67–84. https://doi.org/10.1080/00140139008927094
11. ISO (2007) Ergonomics of the thermal environment—determination and interpretation of cold stress when using required clothing insulation (IREQ) and local cooling effects. Int. Organ, Stand, p 11079
12. Pascoe DD, Bellingar TA, McCluskey BS (1994) Clothing and exercise. II. Influence of clothing during exercise/work in environmental extremes. Sports Med 18:94–108
13. Jin L, Hong K, Yoon K (2013) Effect of aerogel on thermal protective performance of firefighter clothing. J Fiber Bioeng Inf 6:315–324
14. Havenith G, Bröde P, den Hartog E et al (2013) Evaporative cooling: effective latent heat of evaporation in relation to evaporation distance from the skin. J Appl Physiol (1985) 114:778–785. https://doi.org/10.1152/japplphysiol.01271.2012
15. Nielsen R, Endrusick TL (1990) Thermoregulatory responses to intermittent exercise are influenced by knit structure of underwear. Eur J Appl Physiol 60:15–25

16. Wang F, Shi W, Lu Y et al (2016) Effects of moisture content and clothing fit on clothing apparent "wet" thermal insulation: a thermal manikin study. Text Res J 86:57–63
17. Ruckman JE, Murray R, Choi HS (1999) Engineering of clothing systems for improved thermophysiological comfort: the effect of openings. Int J Cloth Sci Technol 11:37–52
18. Zhang X, Li J (2010) Effects of clothing ventilative designs on thermoregulatory responses during exercise. In: 2010 international conference on biomedical engineering and computer science. IEEE, pp 1–4
19. Lim J-H, Roh E-K, Yoo H-S, Kim E (2009) Ventilation and comfort sensation by slit positions of running wear jackets. J Korean Soc Cloth Text 33:1794–1805
20. Yeon S-M, Kim H-E (2005) Effect of slit ventilation system in sportswear on physiological responses. Fash Text Res J 7:75–80
21. Bouskill LM, Havenith G, Kuklane K et al (2002) Relationship between clothing ventilation and thermal insulation. AIHA J 63:262–268, https://doi.org/10.1080/15428110208984712
22. Thompson RL, Hayward JS (1996) Wet-cold exposure and hypothermia: thermal and metabolic responses to prolonged exercise in rain. J Appl Physiol (1985) 81:1128–1137
23. Gagge AP, Stolwijk JA, Saltin B (1969) Comfort and thermal sensations and associated physiological responses during exercise at various ambient temperatures. Environ Res 2:209–229
24. Sawka MN, Cheuvront SN, Kenefick RW (2012) High skin temperature and hypohydration impair aerobic performance. Exp Physiol 97:327–332. https://doi.org/10.1113/expphysiol.2011.061002
25. Marino FE (2002) Methods, advantages, and limitations of body cooling for exercise performance. Br J Sports Med 36:89–94
26. Ross M, Abbiss C, Laursen P et al (2013) Precooling methods and their effects on athletic performance: a systematic review and practical applications. Sports Med 43:207–225. https://doi.org/10.1007/s40279-012-0014-9
27. Poppendieck W, Faude O, Wegmann M, Meyer T (2013) Cooling and performance recovery of trained athletes: a meta-analytical review. Int J Sports Physiol Perform 8:227–242
28. Eijsvogels TMH, Bongers CCWG, Veltmeijer MTW et al (2014) Cooling during exercise in temperate conditions: impact on performance and thermoregulation. Int J Sports Med 35:840–846. https://doi.org/10.1055/s-0034-1368723
29. Hasegawa H, Takatori T, Komura T, Yamasaki M (2005) Wearing a cooling jacket during exercise reduces thermal strain and improves endurance exercise performance in a warm environment. J Strength Cond Res 19:122–128. https://doi.org/10.1519/14503.1
30. Luomala MJ, Oksa J, Salmi JA et al (2012) Adding a cooling vest during cycling improves performance in warm and humid conditions. J Therm Biol 37:47–55
31. Cotter JD, Taylor NAS (2005) The distribution of cutaneous sudomotor and alliesthesial thermosensitivity in mildly heat-stressed humans: an open-loop approach. J Physiol 565:335–345. https://doi.org/10.1113/jphysiol.2004.081562
32. Nakamura M, Yoda T, Crawshaw LI et al (2013) Relative importance of different surface regions for thermal comfort in humans. Eur J Appl Physiol 113:63–76. https://doi.org/10.1007/s00421-012-2406-9
33. Tyler CJ, Sunderland C (2011a) Neck cooling and running performance in the heat: single versus repeated application. Med Sci Sports Exerc 43:2388–2395. https://doi.org/10.1249/MSS.0b013e318222ef72
34. Minniti A, Tyler CJ, Sunderland C (2011) Effects of a cooling collar on affect, ratings of perceived exertion, and running performance in the heat. Eur J Sport Sci 11:419–429
35. Sunderland C, Stevens R, Everson B, Tyler CJ (2015) Neck-cooling improves repeated sprint performance in the heat. Front Physiol 6:314. https://doi.org/10.3389/fphys.2015.00314
36. Tyler CJ, Sunderland C (2011b) Cooling the neck region during exercise in the heat. J Athl Train 46:61–68. https://doi.org/10.4085/1062-6050-46.1.61
37. Gao C, Kuklane K, Holmér I (2011) Cooling vests with phase change materials: the effects of melting temperature on heat strain alleviation in an extremely hot environment. Eur J Appl Physiol 111:1207–1216. https://doi.org/10.1007/s00421-010-1748-4
38. House JR, Lunt HC, Taylor R et al (2013) The impact of a phase-change cooling vest on heat strain and the effect of different cooling pack melting temperatures. Eur J Appl Physiol 113:1223–1231. https://doi.org/10.1007/s00421-012-2534-2

39. Yazdanirad S, Dehghan H (2016) Designing of the cooling vest from paraffin compounds and evaluation of its impact under laboratory hot conditions. Int J Prev Med 7:47
40. Song W, Wang F (2015) The hybrid personal cooling system (PCS) could effectively reduce the heat strain while exercising in a hot and moderate humid environment. Ergonomics:1–10. https://doi.org/10.1080/00140139.2015.1105305
41. Filingeri D, Fournet D, Hodder S, Havenith G (2015) Mild evaporative cooling applied to the torso provides thermoregulatory benefits during running in the heat. Scand J Med Sci Sports 25(Suppl 1):200–210. https://doi.org/10.1111/sms.12322
42. Bach AJE, Stewart IB, Disher AE, Costello JT (2015) A comparison between conductive and infrared devices for measuring mean skin temperature at rest, during exercise in the heat, and recovery. PLoS One 10:e0117907. https://doi.org/10.1371/journal.pone.0117907
43. Frim J, Livingstone SD, Reed LD et al (1990) Body composition and skin temperature variation. J Appl Physiol (1985) 68:540–543
44. Wang F, Gao C, Kuklane K, Holmér I (2010) A review of technology of personal heating garments. Int J Occup Saf Ergon 16:387–404. https://doi.org/10.1080/10803548.2010.11076854
45. Brajkovic D, Ducharme MB (2003) Finger dexterity, skin temperature, and blood flow during auxiliary heating in the cold. J Appl Physiol (1985) 95:758–770. https://doi.org/10.1152/japplphysiol.00051.2003
46. Wang S, Li Y, Tokura H et al (2006) Computer simulation of multi-phase coupled heat and moisture transfer in clothing assembly with a phase change material in a cold environment. In: International conference technologies e-learning digital entertainment. Springer, pp 1103–1106
47. Koscheyev VS, Leon GR, Coca A, Treviño RC (2006) Physiological design of a space suit cooling/warming garment and thermal control as keys to improve astronaut comfort, performance, and safety. Habitation 11:15–25
48. Faulkner SH, Ferguson RA, Gerrett N et al (2013) Reducing muscle temperature drop after warm-up improves sprint cycling performance. Med Sci Sports Exerc 45:359–365. https://doi.org/10.1249/MSS.0b013e31826fba7f
49. Wilkins EL, Havenith G (2016) External heating garments used post-warm-up improve upper body power and elite sprint swimming performance. Proc Inst Mech Eng Part P J Sports Eng Technol. https://doi.org/10.1177/1754337116650322
50. Corbett J, Barwood MJ, Tipton MJ (2015) Physiological cost and thermal envelope: a novel approach to cycle garment evaluation during a representative protocol. Scand J Med Sci Sports 25:152–158. https://doi.org/10.1111/sms.12176
51. Nielsen R, Nielsen B (1984) Measurement of mean skin temperature of clothed persons in cool environments. Eur J Appl Physiol 53:231–236
52. Fournet D (2013) Skin temperature variations in the cold. Loughborough University, https://hdl.handle.net/2134/13551
53. Livingstone SD, Reed LD, Nolan RW, Cattroll SW (1988) Measurement of torso skin temperature under clothing. Eur J Appl Physiol 57:225–229
54. Aylwin PE, Racinais S, Bermon S, Lloyd A, Hodder S, Havenith G (2021) The use of infrared thermography for the dynamic measurement of skin temperature of moving athletes during competition; methodological issues. Physiol Meas 42(8):084004. https://doi.org/10.1088/1361-6579/ac1872
55. Aylwin P, Havenith G, Cardinale M, Lloyd A, Ihsan M, Taylor L, Adami PE, Alhammoud M, Alonso J-M, Bouscaren N, Buitrago S, Esh C, Gomez-Ezeiza J, Garrandes F, Labidi M, Lange G, Moussay S, Mtibaa K, Townsend N et al (2023) Thermoregulatory responses during road races in hot-humid conditions at the 2019 Athletics World Championships. J Appl Physiol 134(5):1300–1311. https://doi.org/10.1152/japplphysiol.00348.2022
56. Cena K, Clark JA (1976) Proceedings: thermographic observations of skin temperatures of trained and untrained runners. J Physiol 257:8P–9P
57. Clark RP, Mullan BJ, Pugh LG (1977) Skin temperature during running—a study using infrared colour thermography. J Physiol 267:53–62

58. Merla A, Mattei PA, Di Donato L, Romani GL (2010) Thermal imaging of cutaneous temperature modifications in runners during graded exercise. Ann Biomed Eng 38:158–163. https://doi.org/10.1007/s10439-009-9809-8
59. Veghte JH, Adams WC, Bernauer EM (1979) Temperature changes during exercise measured by thermography. Aviat Space Environ Med 50:708–713
60. Fournet D, Redortier B, Havenith G (2012) A method for whole-body skin temperature mapping in humans. Thermol Int 22:157–159. https://repository.lboro.ac.uk/ndownloader/files/16958690/1
61. van Marken Lichtenbelt WD, Vanhommerig JW, Smulders NM et al (2009) Cold-activated brown adipose tissue in healthy men. N Engl J Med 360:1500–1508. https://doi.org/10.1056/NEJMoa0808718
62. Hunold S, Mietzsch E, Werner J (1992) Thermographic studies on patterns of skin temperature after exercise. Eur J Appl Physiol 65:550–554
63. Binzoni T, Leung T, Delpy DT et al (2004) Mapping human skeletal muscle perforator vessels using a quantum well infrared photodetector (QWIP) might explain the variability of NIRS and LDF measurements. Phys Med Biol 49:N165–N173
64. Priego Quesada JI, Lucas-Cuevas AG, Gil-Calvo M et al (2015) Effects of graduated compression stockings on skin temperature after running. J Therm Biol 52:130–136. https://doi.org/10.1016/j.jtherbio.2015.06.005
65. Mason BR, Page KA, Fallon K (1999) An analysis of movement and discomfort of the female breast during exercise and the effects of breast support in three cases. J Sci Med Sport Sports Med Aust 2:134–144
66. White JL, Scurr JC, Smith NA (2009) The effect of breast support on kinetics during overground running performance. Ergonomics 52:492–498. https://doi.org/10.1080/00140130802707907
67. Ayres B, White J, Hedger W, Scurr J (2013) Female upper body and breast skin temperature and thermal comfort following exercise. Ergonomics 56:1194–1202. https://doi.org/10.1080/00140139.2013.789554
68. Simpson KM, Munro BJ, Steele JR (2011) Effect of load mass on posture, heart rate and subjective responses of recreational female hikers to prolonged load carriage. Appl Ergon 42:403–410. https://doi.org/10.1016/j.apergo.2010.08.018
69. Ainslie PN, Reilly T (2003) Physiology of accidental hypothermia in the mountains: a forgotten story. Br J Sports Med 37:548–550
70. Pugh LG (1964) Deaths from exposure on four Inns walking competition, March 14–15, 1964. Lancet 1:1210–1212
71. Young AJ, Castellani JW (2007) Exertional fatigue and cold exposure: mechanisms of hiker's hypothermia. Appl Physiol Nutr Metab Physiol Appliquée Nutr Métabolisme 32:793–798. https://doi.org/10.1139/H07-041
72. Pugh LG (1969) Thermal, metabolic, blood, and circulatory adjustments in prolonged outdoor exercise. Br Med J 2:657–662
73. Fournet D, Griggs KE, Redortier B, Havenith G (2013b) Sex differences in thermal strain induced by a typical hiking scenario in a cool environment. In: Cotter JD, Lucas SJE, Mundel T (eds) Proceedings of 15th international conference environment ergonomics. International Society for Environmental Ergonomics, Queenstown, pp 200–201. https://figshare.com/ndownloader/files/47008477
74. Tate M, Forster D, Mainwaring DE (2008) Influence of garment design on elite athlete cooling. Sports Technol 1:117–124
75. Dotti F, Ferri A, Moncalero M, Colonna M (2016) Thermo-physiological comfort of soft-shell back protectors under controlled environmental conditions. Appl Ergon 56:144–152. https://doi.org/10.1016/j.apergo.2016.04.002
76. Filingeri D, Fournet D, Hodder S, Havenith G (2014) Body mapping of cutaneous wetness perception across the human torso during thermo-neutral and warm environmental exposures. J Appl Physiol. https://doi.org/10.1152/japplphysiol.00535.2014

77. Gerrett N, Ouzzahra Y, Coleby S et al (2014) Thermal sensitivity to warmth during rest and exercise: a sex comparison. Eur J Appl Physiol 114:1451–1462. https://doi.org/10.1007/s00421-014-2875-0
78. Mancini F, Bauleo A, Cole J et al (2014) Whole-body mapping of spatial acuity for pain and touch. Ann Neurol 75:917–924. https://doi.org/10.1002/ana.24179
79. Ouzzahra Y, Havenith G, Redortier B (2012) Regional distribution of thermal sensitivity to cold at rest and during mild exercise in males. J Therm Biol 37:517–523. https://doi.org/10.1016/j.jtherbio.2012.06.003
80. Smith CJ, Havenith G (2012) Body mapping of sweating patterns in athletes: a sex comparison. Med Sci Sports Exerc 44:2350–2361. https://doi.org/10.1249/MSS.0b013e318267b0c4
81. Smith CJ, Havenith G (2011) Body mapping of sweating patterns in male athletes in mild exercise-induced hyperthermia. Eur J Appl Physiol 111:1391–1404. https://doi.org/10.1007/s00421-010-1744-8
82. Fukazawa T, Havenith G (2009) Differences in comfort perception in relation to local and whole-body skin wettedness. Eur J Appl Physiol 106:15–24. https://doi.org/10.1007/s00421-009-0983-z
83. Wu HY, Zhang WY, Li J (2009) Study on improving the thermal-wet comfort of clothing during exercise with an assembly of fabrics. Fibres Text East Eur 17:75
84. Jussila K, Kekäläinen M, Simonen L, Mäkinen S (2015) Determining the optimum size combination of three-layered cold protective clothing in varying wind conditions and walking speeds: thermal manikin and 3D body scanner study. J Fash Technol Text Eng 3:1–9. https://doi.org/10.4172/2329-9568.1000120
85. Psikuta A, Frackiewicz-Kaczmarek J, Frydrych IK, Rossi RM (2012) Quantitative evaluation of air gap thickness and contact area between body and garment. Text Res J. https://doi.org/10.1177/0040517512436823
86. Zhang Z, Li J, Wang Y et al (2016) Improving garment thermal insulation property by combining two non-contact measuring tools. Indian J Fibre Text Res 40:392–398
87. Umbach KH (1988) Physiological tests and evaluation models for the optimization of the performance of protective clothing. In: Mekjavic IB, Banister EW, Morrison JB (eds) Environment ergonomics. Taylor & Francis, London, pp 139–161
88. Kicklighter TH, Edsall JR, Martin M (2011) Effect of moisture-wicking garments on temperature regulation during exercise. Int J Athl Ther Train 16:9–13
89. Ho C, Fan J, Newton E, Au R (2008) Effects of athletic T-shirt designs on thermal comfort. Fibers Polym 9:503–508. https://doi.org/10.1007/s12221-008-0080-7
90. Wang F, Del Ferraro S, Molinaro V et al (2014) Assessment of body mapping sportswear using a manikin operated in constant temperature mode and thermoregulatory model control mode. Int J Biometeorol 58:1673–1682. https://doi.org/10.1007/s00484-013-0774-4
91. Werner J, Heising M, Rautenberg W, Leimann K (1985) Dynamics and topography of human temperature regulation in response to thermal and workload. Eur J Appl Physiol 53:353–358
92. Domina T, Kinnicutt P, MacGillivray M (2011) Thermal pattern variations analysed using 2D/3D mapping techniques among females. J Text Appar Technol Manag 7(1)
93. Kinnicutt P, Domina T, MacGillivray M, Lerch T (2010) Knit-in 3D mapping's effect on thermoregulation: preliminary results. J Text Inst 101:120–127
94. Raccuglia M, Heyde C, Lloyd A, Hodder S, Havenith G (2019) The use of infrared thermal imaging to measure spatial and temporal sweat retention in clothing. Int J Biometeorol 63(7):885–894. https://doi.org/10.1007/s00484-019-01701-5
95. Molly Subhash H, Ofoegbuna T, Oliveira AH, Pierce MC, Pillai S (2024) Infrared thermal imaging for assessing human perspiration and evaluating antiperspirant product efficacy. Sci Rep 14(1):24994. https://doi.org/10.1038/s41598-024-73878-8
96. Coull NA, West AM, Hodder SG, Wheeler P, Havenith G (2021) Body mapping of regional sweat distribution in young and older males. Eur J Appl Physiol. Jan;121:109–25. https://doi.org/10.1007/s00421-020-04503-5

97. Arlegui L, Smallcombe JW, Fournet D, Tolfrey K, Havenith G (2021) Body mapping of sweating patterns of pre-pubertal children during intermittent exercise in a warm environment. Eur J Appl Physiol. Dec;121:3561–76. https://doi.org/10.1007/s00421-021-04811-4
98. Inoue Y, Shibasaki M (1996) Regional differences in age-related decrements of the cutaneous vascular and sweating responses to passive heating. Eur J Appl Physiol Occup Physiol 74(1):78–84. https://doi.org/10.1007/BF00376498
99. Havenith G, Nevill ME, Morris JG, Nute ML, Hodder S (2025) Skin temperature distribution in pre- and mid pubertal boys. Manuscript in preparation
100. De Oliveira F, Moreau S, Gehin C, Dittmar A (2007) Infrared imaging analysis for thermal comfort assessment. Annual international conference of the IEEE engineering in medicine and biology society annual conference 2007:3373–3376. https://doi.org/10.1109/IEMBS.2007.4353054

Chapter 12
Thermomechanical Assessment of Sports Facilities and Equipment Using Thermography and Artificial Intelligence

Consuelo Latorre-Sánchez, Rafael Mengual Ortolá, Ricard Barberà Guillem, Sonia Gimeno Peña, and Juan Carlos González García

Abstract Sports equipment and facilities are subject to temperature variations that are reflected in other important parameters of their operation. Thermography is not only a temperature measurement technique; many mechanical and physical properties of materials have an associated thermal behavior that allows us to interpret and evaluate them. Friction, abrasion, fatigue, heat dissipation or aging can be tested using infrared cameras. In the world of sports, these and other properties give us information about the state and evolution of equipment, pavements or materials.

In addition, applying advanced artificial intelligence and machine vision techniques to thermal images allows us to extract patterns and information to characterize sporting products objectively, without contact and in real time.

12.1 Introduction

Thermography is a non-contact technique that has a wide variety of applications in the field of sports. Sports facilities are subject to temperature variations that affect their operation and interaction with users, athletes and spectators ([12, 18, 54]; [10.1016/j.buildenv.2014.03.017] Revel and Arnesano 2014). Thermal monitoring of sports surfaces and stands can ensure the safety of people and the correct use and maintenance of infrastructure. The extreme heat and cold conditions to which many facilities are subjected have important economic and safety repercussions; designing and thermally modeling cooling or heat dissipation solutions is possible using the information provided by thermography.

C. Latorre-Sánchez (✉) · R. Mengual Ortolá · R. Barberà Guillem · S. Gimeno Peña · J. C. González García
Instituto de Biomecánica de Valencia (IBV), Valencia, Spain
e-mail: Consuelo.Latorre@ibv.org

J. I. Priego-Quesada (ed.), *Application of Infrared Thermography in Sports Science*,
https://doi.org/10.1007/978-3-031-93311-0_12

In addition, sports equipment, materials and protection must be evaluated and characterized before, during and after use. Thermal cameras are very useful tools for this, both in the laboratory and in real use [6, 7, 16, 53, 60].

New computer vision techniques and artificial intelligence algorithms are allowing thermography to be incorporated into material characterization processes not only at a thermal level but also in the evaluation of mechanical properties such as wear, friction, abrasion, fatigue or aging [1, 2, 9, 22, 67, 69, 81]. Thermography is not just a temperature measurement technique; many mechanical and physical properties of materials have an associated thermal behavior that allows us to interpret and evaluate them. In addition, the application of advanced artificial intelligence (AI) and computer vision techniques [48, 62] to thermal images allows us to extract patterns and information to characterize sports products objectively, without contact and in real time [34, 53, 64, 77].

Infrared radiation technology is capable of evaluating products, measuring temperature and analyzing the thermal behavior of materials. In the quality control of products, it is a very useful tool, as it allows the detection of imperfections or deformities that can lead to breakages under operating conditions.

12.2 Methodology and Assessment Based on Thermography

All bodies above 0 degrees Kelvin (−273 °C) emit infrared radiation. Cameras that capture this radiation at frequencies where radiation and temperature behave linearly (LWIR—long wave infraRed) are called thermal cameras, and the images they provide are called thermographs or thermography. The methodology described in this chapter is applicable to the entire spectrum of infrared images which, while not enabling direct temperature measurement, makes it possible to detect thermal differences, providing heat maps and gradients for the characterization of materials and products. Their analysis and applications depend on the thermodynamic laws and principles that govern the emission, absorption and propagation of infrared radiation.

Simple concepts such as temperature, heat and thermal gradients are key to the interpretation of thermographic images (see the definition of these terms in Chap. 2 of the book), but other more complex properties are also necessary for the correct use of thermography and to understand the *directive and spectral nature of infrared radiation*. Thermography is not only a technique that provides the temperature of bodies without contact but also allows us to estimate and characterize other mechanical properties, which imply a *change in the transmission or diffusion of heat*. Processes such as expansion and compression are intuitively related to temperature, and it is logical to think that a thermal image can help us measure these characteristics. Other thermomechanical characteristics, however, are not so intuitive and require understanding the mechanisms involved in infrared radiation as well as other mechanical properties of bodies that can lead to errors in the interpretation of measurements.

Temperature is a physical quantity that measures the average kinetic energy level of the particles of a system, indicating how hot or cold is, while heat is a form of energy that is transferred between two systems due to a temperature difference. Heat can cause changes in the temperature of bodies or even in their physical state. Based on these concepts, two different methods of using thermography in the assessment of sports materials are proposed, depending on whether temperature or surface changes/gradients (due to heat transmission), are measured. These methods provide direct or indirect information on the properties of materials and equipment, respectively.

- The first method, measuring temperature, is based on thermodynamic properties and thermal radiation.
- The second method encompasses the thermomechanical coupling properties of materials and heat transfer.

In practice, it is difficult to separate the phenomena involved in temperature changes, but it is possible to estimate performance by controlling some variables in the laboratory and comparing some products with others in terms of heat, conduction, dissipation or thermoelasticity [69].

This section describes the fundamentals of thermodynamics and physical properties related to heat to introduce concepts used in the methodology of characterization of materials and sports products through thermography (Sect. 12.3). This section also addresses image processing techniques that allow information to be extracted from sports equipment and facilities and to identify changes and patterns that were not possible with traditional techniques of computer vision (Sect. 12.4).

12.3 Fundamentals and Basic Principles

This section includes the fundamentals of thermal radiation and heat as well as the parameters and properties of bodies (materials and products) that imply a change in temperature and make it possible to detect and quantify it.

12.3.1 Thermodynamics Principles

Thermographic images are based on thermodynamic principles that describe the transfer of energy in the form of heat and radiation. Thermal balance or thermodynamics equilibrium occurs when two bodies are at the same temperature, and there is no heat flow between them. When there is no thermal balance, heat is transferred through three primary mechanisms: *conduction*, *convection* and *radiation*. Conduction occurs through direct contact, convection involves the movement of a liquid or gas and radiation is the transfer of energy through emission. In thermography, conduction is the dominant factor in transient thermal problems, while radiation is the phenomenon detected by the camera. Convection plays a supporting role by modifying the thermal environment, so it influences both the surface temperature of the object and the thermal patterns captured by the infrared camera. *Infrared radiation* is situated between microwaves and visible light on the *electromagnetic spectrum* radiation. This spectrum is further divided into near-infrared (NIR), mid-infrared (MIR) and far-infrared (FIR) ranges. Depending on the atmospheric transmission windows, there are different sensors that classify short-wave IR (SWIR), mid-wave IR (MWIR) and LWIR infrared cameras.

In the part of the spectrum of the long-wave infrared (LWIR) band (7.5–14 μm), the captured radiation is linear with the temperature of the bodies, and therefore we speak of thermography and thermal cameras. The term thermography has been extended to other infrared cameras although in these ranges heat transfer rather than temperature is detected. Physical laws are fundamental for correctly interpreting thermal images and understanding the relationship between emitted radiation and the temperature of objects analyzed through thermography. More information about these laws (Stefan-Boltzmann Law, Planck's Law or Wien's Displacement) is detailed in Chap. 2 of this book. These laws enable thermography users to interpret thermal images accurately, select appropriate cameras and analyze the thermal properties of materials effectively.

12.3.2 Infrared Radiation Properties

This section presents three factors that must be considered when interpreting the results of infrared cameras, especially for the assessment of sports equipment: emissivity, the directive nature of the radiation and environmental factors.

12.3.2.1 Emissivity

Objects and surfaces are not perfect black bodies, and part of the energy they emit is due to other external radiant sources that reflect or pass through their structure. The energy balance equation allows these variables to be related by means of the principle of conservation of energy, where total energy is equal to the sum of absorbed, transmitted and reflected energy. Dividing this equation by the total energy and applying Kirchner's law, the radiometric parameters are related according to Eqs. 12.1 and 12.2.

$$\varepsilon_{\lambda} = \alpha_{\lambda} \tag{12.1}$$

$$1 = \varepsilon_{\lambda} + \tau_{\lambda} + \rho_{\lambda} \tag{12.2}$$

which links the total *emittance* (ε_λ) to the total *absorbance* (α_λ), total *transmittance* (τ_λ) and total *reflectance* (ρ_λ). The values of these parameters range from 0 to 1.

Emissivity is the capacity of a body to emit infrared radiation due to its temperature and depends on the material characteristics, being essential for accurate temperature calculations [33]. Only materials with high emissivity (close to 1) could be properly measured by thermal cameras. Knowing the emissivity of an object is crucial as it can lead to misuse of thermography on materials with low emissivity where the temperature detected by the camera is not real and corresponds to reflections of their ambient temperature or apparent temperature. It is noteworthy that this fact has not been considered due to a lack of knowledge in the scientific literature, and studies continue to be published where the results obtained in thermography are not valid because they do not consider the different emissivity of the monitored surfaces. Metals are a clear example where thermography is affected by reflections and the values it provides do not correspond to reality.

12.3.2.2 Directive Consideration of Radiation

Infrared radiation, being an electromagnetic wave, has a nature that depends on the direction of propagation, where the maximum radiation is produced perpendicular to the surface. Radiance ($W \cdot sr^{-1} \cdot m^{-2}$) refers to the amount of electromagnetic energy emitted, reflected, transmitted or received by a surface per unit area, per unit solid angle and per unit wavelength or frequency. It is a directional quantity, meaning that it depends on the direction from which it is observed.

Sometimes, holes or protrusions that generate artifacts in the detected image are not considered. Depending on the depth of the cavities or concave curvatures of a surface, multiple reflections are produced that add up to each other, increasing the temperature perceived by the camera at that point. This can be interpreted as a hotter point when, in reality, there is a thermal equilibrium throughout the object in a steady state (Fig. 12.1). Equipment with curves, holes and reflections is very difficult to characterize properly with thermography. This phenomenon is not considered in numerous scientific studies, which has caused thermography to lose credibility compared to other techniques that are more reliable, being a question of unawareness of the physical nature of radiation.

12.3.2.3 Environmental Parameters

Hidalgo García and Arco Díaz [25] resume how, in recent decades, global warming and extreme weather events have become one of the most important challenges facing humanity [35, 63] and how the modification of landscapes due to urbanized areas contributes very strongly to climate change [40, 65].

It is well-known that environmental factors such as humidity, ambient temperature and solar radiation influence thermographic data and must be considered.

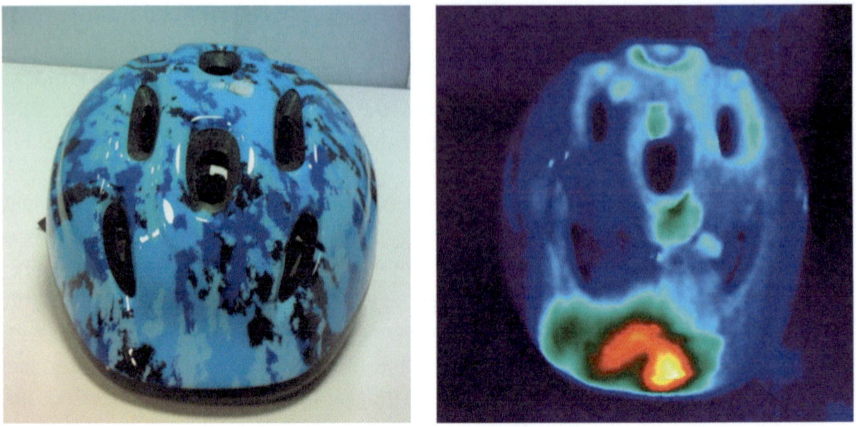

Fig. 12.1 Example of thermography that does not correspond to thermal reality. Artifacts due to reflections of the thermographer show hot spots on the helmet surface

In outdoor sports facilities and sports in open areas, there are also two concepts related to solar radiation that are especially important: urban heat island and albedo:

- *Urban heat island* (UHI) is an area, volume or region where the temperature is higher than that of the surroundings, regardless of scale [25, 65]. Numerous studies have highlighted that large temperature increases over land surface temperature can be attributed to this urban climate disturbance phenomenon [59].
- The temperature obtained by infrared remote sensing during the day depends on the water content, the evaporation and transpiration from the soil surface and other variables, such as albedo [36]. *Albedo* is generally defined as the hemisphere- and wavelength-integrated reflectivity. While reflectivity is the ratio of the amount of reflected radiation to the total incident radiation at a particular wavelength, albedo spans a range of wavelengths of interest, considering geometric effects in addition to surface effects [65]. Surface albedo (SA) is defined as the ratio between the radiant flux reflected from the Earth's surface and the incident flux and comprises the shortwave (SW domain which includes visible and NIR), where downward solar radiation is most relevant [23]. Sánchez-Zapero et al. [58], state that SA is one of the terrestrial essential climates as it controls the surface energy balance and is a sensitive indicator of environmental changes, including land degradation [20]. It also determines the fraction of solar energy absorbed by the surface and transformed into heat or latent energy. Small changes in SA could affect the global circulation [58].

Some sports facilities' surfaces and materials, such as artificial turf (Sect. 12.5.2), contribute to the UHI effect due to their limited ability to dissipate heat. Surfaces having albedo around 0.10 can produce an average surface heating greater than 60 °C in summer [11]. Thermographic studies have allowed the temperature differences between artificial and natural surfaces of sports fields and their surroundings to be quantified, emphasizing the need for sustainable design options to minimize environmental impact [33].

12.3.3 Mechanical Properties of Solids Related to Heat

This section describes the interactions between heat and deformation processes. Thermal cameras allow the detection of changes in the temperature of materials caused by the mechanical properties of the materials that generate heat. The heat flow is described by a vector field q, which is the amount of thermal energy passing through the unit surface per unit time. The heat energy Q, in calories [cal], depends on the density ρ (kg/m^3), the volume V (m^3), the specific heat (c[J/kg/°C]) and the temperature increase in °C. This thermal energy is convertible (theoretically) into mechanical energy [J].

In a cooling process, there is a heat loss by dissipation at the contact surface, radiation and convection at the free surfaces and a gain by plastic deformation in the core of the material.

The thermal changes related to many mechanical properties serve to compare and characterize physical effects and mechanical parameters, traditionally evaluated by complex measurement systems [81].

Some examples and their definitions are as follows:

- Deformation is the change in shape or size of a material under applied forces.
- Fatigue describes the weakening of a material caused by repeated loading and unloading cycles. It leads to the initiation and growth of cracks.
- Overload occurs when a material is subjected to forces beyond its capacity, leading to failure.
- Stress is the internal force per unit area within a material that arises from externally applied forces.
- Compression is the application of forces that reduce the volume of a material.
- Friction is the resistance to motion when two surfaces slide against each other.
- Abrasion is the process of scraping or wearing away a material.
- Durability refers to a material's ability to withstand wear, pressure or damage.
- Aging in materials refers to the gradual change in properties over time due to environmental factors or repeated stress. This can affect thermal and mechanical behavior.

Thermography can detect thermal signatures of overload conditions, identifying heat dissipation patterns, for example, in repetitive motions, involving cyclic loading, fatigue and eventual failure [10]. Infrared monitoring is used to study the thermal effects of such motions on material integrity and visualize stress distribution by detecting temperature variations or identifying stress and strain concentration points as well as fatigue-related processes that can lead to material destruction [9, 51].

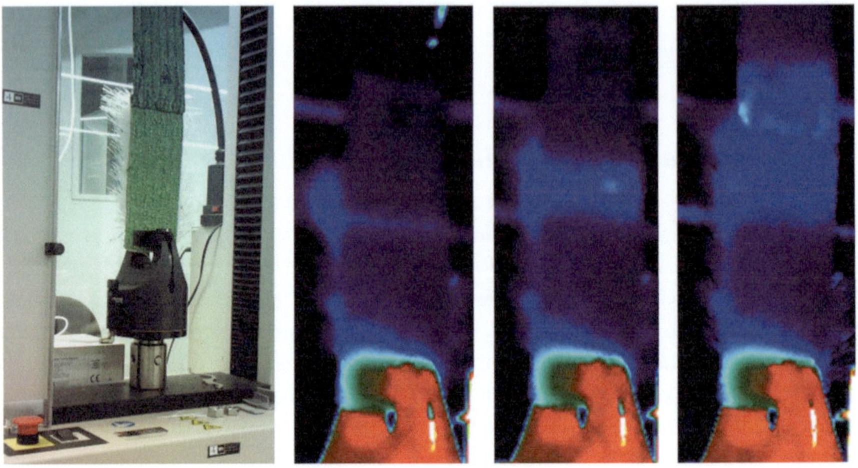

Fig. 12.2 Destructive test of artificial turf. Stress forces

Fig. 12.3 Non-destructive test on artificial turf, friction test. IBV facilities, footwear from DIADORA

Figure 12.2 shows a destructive test of artificial turf. The delaminated area increases its temperature (light blue tones) allowing different materials to be compared with thermographs in a very intuitive and qualitative way.

Figure 12.3 shows non-destructive tests on turf (rotational resistance and linear friction). The left picture corresponds to the rotational resistance test and the center picture to the linear friction in a laboratory. The right images are thermograms that highlight the temperature increase in temperature in the cleats of the football boot and the thermal signature on the turf sample. These tests provide information about the fields and also about sports footwear.

Luong [44] reported that thermography could be used to assess mechanical properties such as fatigue, stress, aging and compression. Compression is studied by observing heat emission and stress-strain behavior under compressive loads. Thermal imaging allows us to assess the behavior of objects under overload situations. Different authors [16, 51] reported that friction generates heat, which can be measured using infrared techniques to study its effects on material wear. It can help visualize the heat generated during abrasion, providing insight into material wear rates and assessing durability by monitoring thermal responses to stress and identifying potential failure and aging points.

The physical concepts that justify these properties being related to heat are mainly thermal properties: thermal conductivity, specific heat, thermal effusivity and diffusivity and thermomechanical properties: thermoelasticity, viscoplasticity and elastoplasticity. Table 12.1 summarizes them.

These concepts are related to the *thermomechanical coupling effect*, described by coupled thermal and mechanical analysis which is necessary to understand mechanical phenomena and the thermal changes they generate [72]. Equations 12.3 and 12.4 describe this interrelation.

Table 12.1 Key concepts of thermomechanical properties related to heat

Thermophysical properties	
Thermal conductivity	It quantifies the rate at which thermal energy flows through a material due to a temperature gradient, and how quickly heat spreads through a material, affecting thermal patterns observed in infrared studies [16]
Specific heat	It is the amount of heat required to raise the temperature of a unit mass of a substance by one degree. It indicates how much energy a material can store. Deformation can vary this parameter.
Thermal effusivity	It combines thermal conductivity, density and specific heat. It measures a material's ability to exchange heat with its surroundings
Thermal diffusivity	It is the thermal conductivity divided by the product of density and specific heat capacity at constant pressure. It measures the rate at which heat spreads through a material, indicating how quickly a material responds to changes in temperature. ([69, 75]. It is crucial for understanding heat transfer dynamics with thermography [16]
Thermomechanical properties	
Thermoelasticity	It studies thermomechanical effects in elastic materials, how temperature induces deformations in materials and how these deformations, in turn, affect temperature distributions [72]
Viscoplasticity	It describes the rate-dependent inelastic behavior of solids. It accounts for materials that exhibit both viscous and plastic characteristics when undergoing deformation. Viscoplastic materials show time-dependent strain under stress, meaning their deformation rate depends on the applied load and the rate at which it is applied [72]
Elastoplasticity	It describes materials that exhibit both elastic and plastic responses to applied stresses. Initially, the material deforms elastically (reversible deformation), but once the stress exceeds a certain yield point, it deforms plastically (permanent deformation). Elastoplastic models are essential for predicting material behavior under various loading conditions [72]

The *thermal energy transfer equation* for an isotropic material undergoing plastic deformation can be written as follows:

$$\rho C T' = \nabla \cdot (k \cdot \nabla T) + \beta (\sigma : \epsilon'^{p}) \tag{12.3}$$

where

ρ is the density of the material.
C is the specific heat of the material.
k is the thermal conductivity of the material.
T' is the material derivative of temperature.
β is a coefficient in the range of 0.85 and 0.95 (called Taylor-Quinney) that indicates the fraction of the plastic deformation power (σ: ϵ'p) that is transformed into heat. The remaining fraction (1-β) is stored in the material and represents the energy associated with the evolution of the microstructure of the material assuming that there are no phase changes.

Regarding the mechanical problem, in the equilibrium equations (Eq. 12.4), the *thermomechanical coupling effect* is implemented through the dependence of the relationship on temperature. Deformation can be elastic (reversible) or plastic (permanent), and thermography permits the observation of thermal effects during the process. The plastic deformation power is expressed in terms of the rigid/viscoplastic constitutive relationship [14]:

$$\rho \cdot C \cdot \dot{u} \cdot \nabla T = \nabla \cdot k \cdot \nabla T + \frac{c}{|c|} \cdot \beta \cdot \sigma_y \cdot \bar{\epsilon} \cdot \qquad a.(\text{steady state})$$

$$\rho \cdot C \cdot \left[\frac{\partial T}{\partial t} + (\dot{u} - \dot{u}_M) \cdot \nabla T \right] = \nabla \cdot k \cdot \nabla T + \frac{c}{|c|} \cdot \beta \cdot \sigma_y \cdot \bar{\epsilon} \cdot \qquad b.(\text{transient}) \tag{12.4}$$

Understanding these properties and the thermomechanical coupling is crucial for applying thermography to the evaluation of sports equipment (see Sect. 12.3.4), as they influence how materials respond to thermal and mechanical loads. Thermography allows us to detect temperature changes and relate them to properties, detecting defects, inhomogeneities and other fatigue characteristics that allow us to characterize sports equipment in a non-destructive way, such as sailboats [21].

12.3.4 Electrical Properties Related to Heat

Thermography can visualize thermal signals associated with high electrical resistance. The Joule effect is a physical phenomenon whereby electrical energy is transformed into heat when an electric current passes through a conductor. This heating is due to collisions between moving electrons and the atoms of the conducting material, which generates a dissipation of energy in the form of heat. For the thermal

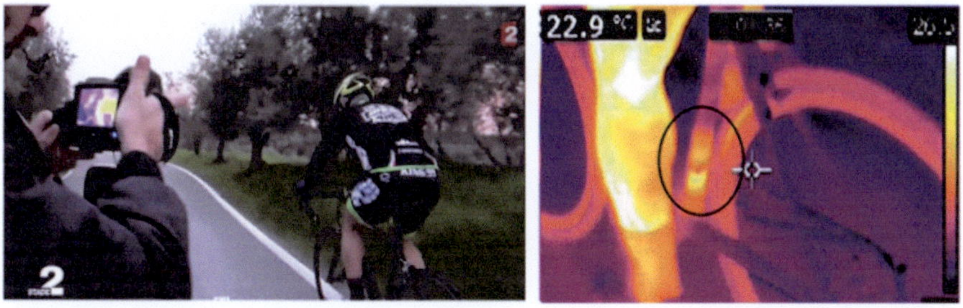

Fig. 12.4 Thermography during a cycling race. Heat from a motor inside the bicycle frame is detected due to the thermal gradient. [https://doi.org/10.1016/B978-0-08-102582-6.00002-2] (Priego Quesada and Carpes 2019)

evaluation of electrical circuits, a laser was used to induce a local thermal gradient in the test devices and with infrared emission microscopy (IREM), infrared emissions were captured. In the early 90 s, camera-based detectors replaced these techniques [31]. Thermography can be used to check the optimal performance and safety of electronic parts, ensuring that the electrical circuits and electronic components are working properly and not overheating.

Many sports equipment incorporate electronic components [12]. In the design and testing of smart sportswear or using electronic training devices, such as muscle electro-stimulators that integrated electronic components, thermal images can help monitor the thermal response to these devices, ensuring that local overheating or tissue damage does not affect the thermoregulation or create uncomfortable hot spots. Furthermore, thermographic detection of temperature increases in electronic circuits can be useful to prevent "electronic doping" or "technological doping"in those electronic devices or aid in obtaining an unfair advantage in sports competitions [https://doi.org/10.1016/B978-0-08-102582-6.00002-2] (Priego Quesada and Carpes 2019) (Fig. 12.4). In motorsports, the thermal behavior due to electronics and engines provides information of interest for the performance of vehicles complementary to mechanics (see Sect. 12.3.3).

12.4 Techniques to Evaluate Thermal Images

Thermography assessment techniques are mainly based on punctual or continuous monitoring with thermal cameras, both in the laboratory and in real indoor or outdoor conditions. However, the use of additional equipment, such as other types of visible cameras, mannequins (Fig. 12.5), climatic chambers, thermocouples, controlled radiation sources and thermal images from satellite or airborne systems such as drones, allows more information to be extracted from sports materials, products and facilities.

This section aims to serve as a guide to understanding how to apply thermography to the assessment of sports materials and facilities. It has been divided into three

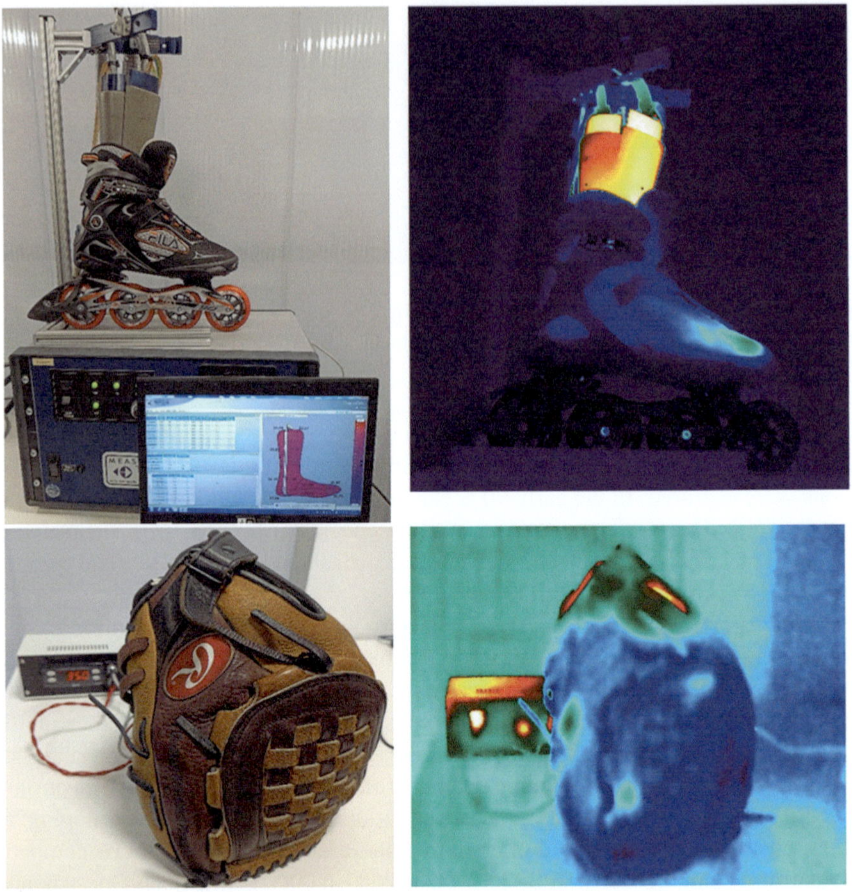

Fig. 12.5 Examples of thermography combined with other thermal equipment. (**a**) Roller skate over thermal foot mannequin foot, (**b**) softball globe with a plate hand mannequin. Active thermography to estimate thermal stress on body parts

sections: characterization and laboratory testing, thermal monitoring of sports equipment and facilities and, finally, the application of advanced artificial intelligence techniques to this sector.

12.4.1 Characterization and Lab Testing on Products and Materials

Thermal cameras are used together with other devices, allowing us to obtain information on the characteristics of the materials and equipment under test. Thermography, combined with other techniques, has become a very popular tool for non-destructive testing and evaluation (NDT&E) [62]. The increasing popularity is mainly due to the decrease in prices, the increase in infrared sensor resolution and the new image processing techniques specifically developed for this type of application [3].

This section covers the characterization and laboratory experimentation techniques based on the two thermography approaches: passive and active. Focus on active thermography; thermal non-destructive tests (TNDT) are also described, and their integration with other technologies such as ultrasonic, acoustic emission or optical sources, with mention of some examples applied to sports equipment.

12.4.1.1 Passive Thermography

Traditionally, thermography has been a passive technique, capturing radiation from bodies to measure temperature differences by measuring the emitted radiation. Nevertheless, the technique is sometimes restricted to applications with sufficient natural thermal contrast. This approach is generally a qualitative method with the primary aim of identifying discontinuities, applied as a standard quality control technique, condition monitoring or predictive maintenance [62]. Passive thermography enhances visual inspection and offers surface-level observations; however, subsurface or internal changes may require complementary methods for a comprehensive analysis.

12.4.1.2 Active Thermography

Recently, external energy sources have been incorporated into measurement methodologies to stimulate the object and generate thermal contrasts that reveal defects or intrinsic properties of the material. The energy can be applied in pulses or modulated, by contact or by radiant sources such as black bodies [29, 33, 62].

The experimental setup of active thermography is similar to passive thermography except that an external energy source is needed to produce a thermic inconsistency in the object or system. The heating or cooling energy can be generally supplied in pulsed or harmonically modulated form [61]. Laser heating, optical flash lamps, halogen lamps, hot or cold air guns, electrical heating, ultrasonic excitation, eddy currents, microwaves and other techniques are examples of external sources [22]. The assessment can take place in either transmission or reflection mode. Conventional IR thermograms can be converted into images of effusivity (in a one-sided test) and diffusivity (in a two-sided test) [69].

Figure 12.6 shows a configuration in controlled conditions in a climate chamber where the sample (tatami foam) is tested in both modes with active thermography: a black body (number 2) and a heating infrared source (number 3).

The active thermography methodology allows characterizing, thermally as well as mechanically, properties of sports products and materials indirectly through the changes that occur in their temperature due to the presence of deformations, irregularities or induced actions. It can be used for image subsurface flaws of different materials [3].

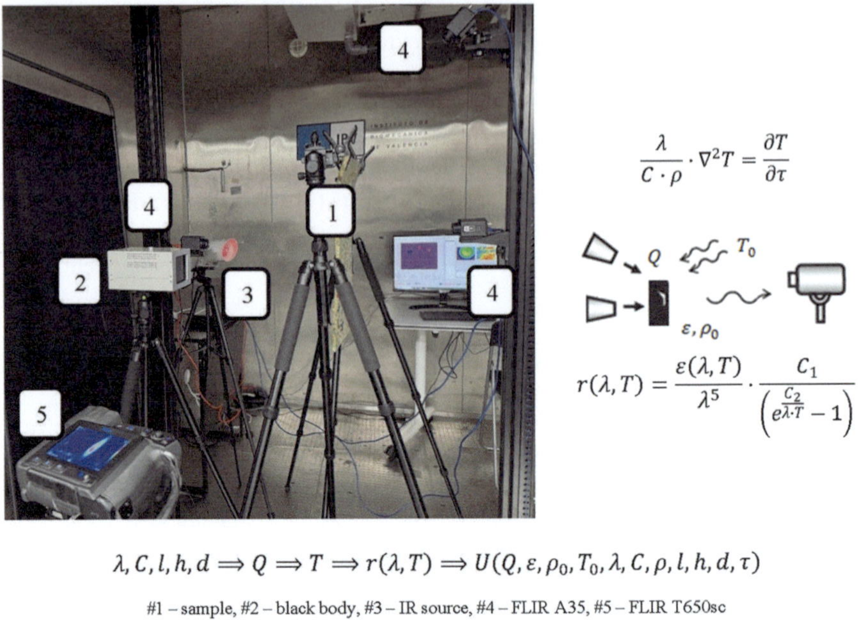

Fig. 12.6 TNDT with active thermography in a climate chamber with a black body (#2) and heat source (#3) to characterize a sports surface. Thermal chamber at IBV Facilities, equation from [69]

12.4.1.3 TNDT: Thermal Non-destructive Testing

Nowadays, thermographic diagnostics and NDT represent mature high-tech fields that combine achievements in the understanding of heat conduction, materials science, IR technology and computer data processing [69]. Thermal analysis is based on a group of techniques in which a sample property is monitored as a function of time or temperature, while the sample temperature is programmed in a specific atmosphere [26].

Active thermography offers a variety of measurement techniques and inspection methods that differ mainly in the type of excitation source used, the way the thermal energy is applied, and the data/signal processing methods.

The main non-destructive testing techniques are pulsed thermography, step heating thermography and lock-in thermography [62].

- Pulsed thermography (PT): High-power, short-duration thermal energy is applied to the surface of the object being inspected, and the thermal response of the stimulated surface is observed in a transient mode.
- Step heating thermography (SHT): SHT is similar to PT, but in contrast, a longer pulse, from a few seconds to several minutes, is used for thermal stimulation, and the surface temperature of the stimulated object is monitored during the heating and cooling phases [61].

- Lock-in thermography (LIT): LIT is an active thermography technique used to detect defects, damage, or characterize potential manufacturing issues [41]. The camera acquisition rate is synchronized with the periodic excitation, applied to the material under test. It improves signal-to-noise ratio, sensitivity and feature resolution compared to pulsed thermography. It offers significant advantages in terms of sensitivity and spatial resolution. LIT eliminates all disturbances such as surrounding reflections, local variations of the surface optical absorption and infrared emission coefficient and inhomogeneous illumination from heat sources. Another advantage compared to PT is the lower intensity which prevents excessive heating, the entire intensity being concentrated at the modulation frequency which is optimized for the specific material and intended depth range [76]. The major cost determinant of a LIT system is the IR camera used for non-contact detection of surface temperature evolution. Razani et al. [55] have demonstrated the feasibility of using cellphone attachment infrared cameras that cost significantly less for lock-in thermography investigation of biological and non-biological samples.

Vavilov [69] provides a detailed review of other combined techniques with thermography for thermal non-destructive testing and commercial systems.

Composite materials, such as carbon fiber-reinforced polymer composite (CFRP), due to the significant weight reduction at the same strength [42], are used as structural elements in aircraft, vehicles, yachts, vessels or sports equipment. These materials require high-performance mechanical properties, so their quality control is essential for safety and assessment. However, the flaws that occur during manufacturing seriously threaten security. Therefore, it is important to develop inspection techniques to assess the material. Impact damage, as it occurs at the interlaminar scale, is difficult to reveal with standard NDT methods, while active thermography PT proves to be an appropriate method for assessing such materials [6].

Gama Caetano et al., [21] report a study applied on samples made of CFRP mast of a racing sailboat. Pulsed thermography, assists in detecting subsurface defects and weak interfaces between composite layers, which may contribute to structural failures. By supporting failure analysis and condition monitoring, thermography aids in maintenance planning and enhances the reliability of high-performance sailing structures. The test equipment used a FLIR SC5600 infrared camera with a cooled detector (640 × 512 pixels) and 25 Hz acquisition frequency and an IRBox system for thermal excitation of the samples (a set of 3.4 kW halogen lamps to generate a 10 s heat pulse on the external surface of the material). The setup consisted of the heat source and the IR camera placed in front of the sample (reflection mode).

Figure 12.7 presents the PT results at the 90° position of the mast. Several heat peaks are observed distributed along the mast in the thermogram. A "high temperature" area near the sail path (left side of the thermogram) may be associated with a possible interlaminar crack (delamination).

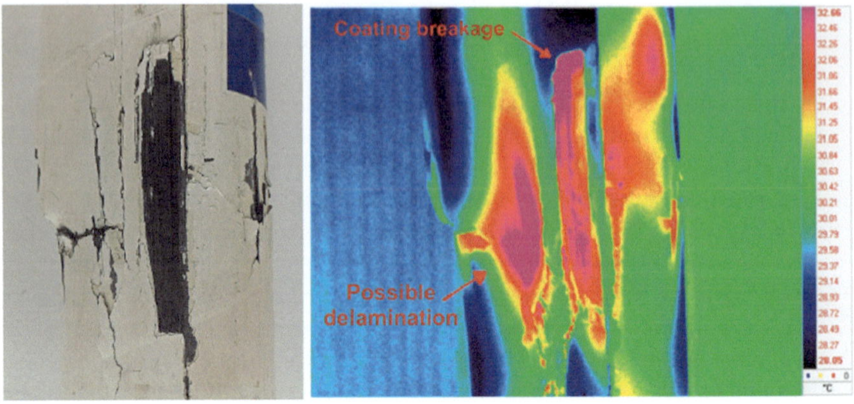

Fig. 12.7 Result of pulsed thermography for a sailboat racing mast. Photograph (left) and thermogram (right) of the analyzed area [21]

12.4.2 Thermal Monitoring in Sports

Thermal monitoring provides a dataset of thermograms that can be considered in different forms in order to be processed: a 1-D model which is independent of spatial information and where input is a vector of temperature history in terms of time for a pixel; a 2-D model which is independent of temporal information and where input is thermal images; and a 3-D model with the input of thermal images (video) series combining the 1-D and 2-D models [62]. N-D model is an advanced computer vision applied to thermography [https://doi.org/10.1007/978-3-030-20216-3_48] (Latorre-Sánchez et al. 2020) that could provide information in 3D spatial dimension with additional variables: temporal information (continuous monitoring), frequency, movement or information from other devices, such as the TNDT multimodal techniques (optical, holographic or ultrasonic data) [2, 3].

Even though most of the research in thermography is still based on 2D imaging techniques, researchers are investigating the true power and calibration of three-dimensional thermal imaging [3, 53].

Ribbens et al. [56] reported the 3D thermal monitoring applied to four billiard balls on a wooden board, with a multi-camera system around time-of-flight sensors, which are used to capture the geometric shape and location of the balls. The proposed measurement setup and methodology included an industrial robot that allows scanning of the object from multiple viewpoints and adds time as an additional fourth dimension during passive thermography experiments (Fig. 12.8). This automated system helps to compensate for the influence of directional emissivity [49].

There are a multitude of sports that already use camera systems to spatially and temporally monitor both athletes and facilities and equipment, using image triangulation from cameras spread across the court, the captures are processed by software, which is responsible for generating the 3D map of the court and the trajectory of the ball (see Sect. 12.3.4). 3D thermography applications are providing further data on

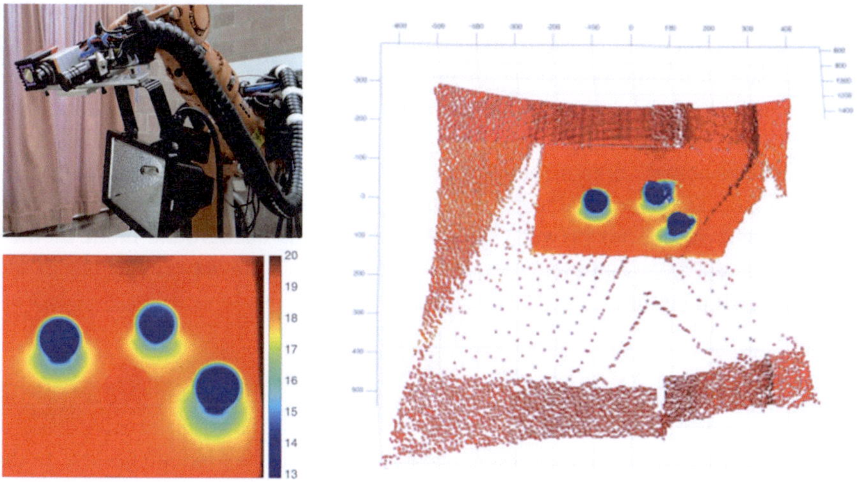

Fig. 12.8 Left top: robot, left bottom: IR2D, right: IR3D. Billard balls on a woolen panel. [56]

materials and playing techniques. New monitoring techniques in the industry are moving in a multimodal data fusion approach, thanks to the growing data processing capacity and increasingly agile AI algorithms.

12.4.3 Artificial Intelligence Applied to Infrared Images

This section closes the thermographic characterization block by summarizing the conclusions of our research on AI in sports, current applications of artificial intelligence in this field and possible future applications of artificial intelligence in professional sports.

12.4.3.1 Computer Vision Techniques

In the last decade, thermographic monitoring techniques have advanced considerably, mainly for two reasons: the fusion with other measurement techniques (see Sect. 12.3.2.1) and the evolution in image processing. Before exploring current techniques, we will start with the recent background of advanced data processing analysis applied to thermography.

Garrido et al. [22] reported different image processing methods: phase image Fourier transform (FT), cold image subtraction (CIS), principal component thermography (PCT) and thermographic signal reconstruction (TSR), to detect material defects by applying an ultrasonic excitation thermography technique (vibrothermography).

Bagavac et al. [6] presented a novel method for separating subsurface defects in cured pre-impregnated carbon fiber-reinforced polymer specimens on thermal images based on edge detection, for non-destructive evaluation. Every specimen was recorded by means of infrared thermography in reflective configuration, and raw data was processed by several techniques: TSR, fast Fourier transformation (FFT) and principal component analysis (PCA). Images were processed to determine damaged and non-damaged areas. Edge detection enables both simulated damages and boundaries between the specimen and the background in the thermal image to be detected. Edges are defined as a transition zone between damaged and non-damaged material.

12.4.3.2 AI in Thermography

The development of artificial intelligence has focused on visible images due to the wide availability of databases, cost and intuitive interpretation of results. RGB images have been used for many years to detect calibration errors and regulatory heights with computer vision techniques and AI. Incorporating infrared images (from NIR to LWIR spectrum) allows for improved detection against lighting changes and other limitations that RGB cameras have, compared to thermal ones [38]. The difficulty for thermal IA algorithms is that the amount of labeled thermal images is significantly smaller compared to visible databases.

Shrestha et al. [62] presented a review focused on thermal images and summarized artificial intelligence's rapid advancement, particularly machine learning (ML) and deep learning (DL), making thermography increasingly automated and intelligent, such as recurrent neural networks (RNNs) and convolutional neural networks (CNNs) from one to three dimensions. Shrestha et al. [62] categorize the DL algorithm into passive and active IRT:

- Deep learning in passive thermography. The use of deep learning models such as RNN, CNN, generative adversarial network (GAN), auto-encoder (AE) and restricted Boltzmann machine is often more frequent than active ones. The resolution of infrared cameras is lower than visual cameras, resulting in poor image quality and restricted use of passive infrared thermography. Several researchers have evaluated CNN [37] and GAN [78] to increase the resolution of infrared projections. For example, a DL method based on GAN was established, which improves the resolution of thermal images by integrating visual and infrared data [4]. An approach combining a GAN with a residual network has been suggested to improve the visualization of thermal and visible image fusion [78].
- Deep learning in active thermography. DL use in active is currently lower although its advancement and increasing application are expected. DL in active TNDT can be classified into subdivisions based on the data format: 1D, 2D or 3D (Table 12.2).

Infrared imaging-based machine vision (IRMV) is a common term to define the technology used to automatically inspect, detect and analyze infrared images (or

Table 12.2 DL review applied to active thermography [62]

Data format of IRT	Model	TNDT techniques and applications
1-D model	A long short-term memory recurrent neural network (LSTM-RNN)	Model based on a temperature series array has been used to automatically classify predominant defects in honeycomb materials such as adhesive build-up, peel-off and liquid ingress [28]
	CNN model	In LIT, for a thermal sequence of carbon fiber-reinforced polymer (CFRP) samples to accurately classify defective and healthy areas. Compared with PCA and Fourier transform, the model improves the resolution [13]
2-D model	2D CNN	For autonomous concrete surface crack detection based on DL integrating vision and IRT data. By spatially inspecting through unmanned vehicles equipped with a hybrid imaging technology including infrared and vision cameras as well as a continuous wave line laser [79]
	Pre-trained GoogLeNet using a transfer learning approach	To achieve automatic crack detection and visualization, the reconstructed images have been categorized into multiple patches, and each patch has been categorized
3-D model	End-to-end deep pattern learning	To provide accurate crack identification and localization, an end-to-end pattern deep region learning architecture has been suggested [27]. The temporal information was extracted from the thermal video using PCA and compressed into a single frame (matrix) with a more substantial defect feature. Due to DL is problematic for temporal data reduction, to analyze temporal and spatial data at the same time, some researchers suggest compressing the temporal data first and then processing the spatial data with one DL [24]. Temporal information was extracted from thermal video using PCA and compressed into a single frame (matrix) with larger defect features [27]
	Faster-RCNN	To detect faults from the matrix
	A technique integrating YOLO (Yolonet—you only look once) and a fully connected network (FCN)	FCN forecasts the depth of defects, whereas YOLO evaluates if there are any. The approach may be considered as a fusion of a 1D feed-forward network and a 2D YOLO model, dealing with temporal and spatial data independently [57]
	Spatial-temporal network	To analyze the fused temporal and spatial inputs, a hybrid spatial-temporal DL framework for automated thermography has been developed [43]
	An LSTM loop (three layers)	To collect the material features that reflect the character of thermal differences between the defective and healthy spots
	Visual geometry group-Unet (VGG-Unet)	To identify the out-of-standard defective area

videos) obtained by recording the intensity of infrared light emitted or reflected by observed objects [24]. This concept related thermography to artificial intelligence techniques, particularly machine learning (ML) and neural networks.

He et al. [24] reviewed the principles, cameras and thermal data of IRMV and discussed the application of deep learning on these datasets. For the segmentation of heterogeneous composite applications, He et al. [24] cited Wu (2020), who proposed a data-driven approach with a CNN and the semantic segment model U-Net, used to classify the image into defect and non-defect pixel-wisely. The rapid development of deep learning makes IRMV more and more intelligent and highly automated, thus considerably increasing its range of applications.

Vijaya Lakshmi et al. [71] have experimentally produced an artificial neural network-based analysis to improve depth resolution in non-stationary thermal wave imaging. This study describes an artificial neural network-based analysis used on quadratic frequency modulated thermal wave imaging to detect and characterize subsurface and CFRP defects, which provides improved defect detection by artificial neural network-based analysis compared to pulse compression, Hilbert phase and FFT phase analysis.

Artificial intelligence (AI) technology has evolved over the past few years, and it is evident that it is improving many sectors. Nadikattu [48] described how AI has been of great importance in the sports industry. It has played a major role in ensuring that the sector evolves from tradition to more modern forms. The article investigated the benefits of AI in the sports industry and identified some of the AI applications that are used in sports today. Also, businesses associated with sports are growing rapidly due to the huge awareness created through this technology, so it is advisable for sports businesses to stay updated.

Real-time monitoring of training workloads using wearable technology has made it possible to predict injuries in elite footballers a week before matches by combining such data with nearly 90% accuracy [12].

In F1, for example, everything from the driver to the car itself can be measured and aggregated to deliver millisecond advantages that can decide the race [73]. AI can provide comparative analysis of a variety of health parameters, which in turn can be provided by AI-enabled wearable technology [54]. Ratiu et al. [54] analyzed the advances in the use of artificial intelligence (AI) in sports biomechanics. Examples of the use of ANN in sports biomechanics are presented for javelin and discus throwing, shot put and kicking in soccer and the optimization of the serve motion in soccer.

12.4.3.3 AI in Sports

AI is applied to thermography to improve the analysis and interpretation of thermal images. This integration allows for more accurate and automated detection of thermal patterns and anomalies. For example, in the inspection of stadiums, pavements or stands, unlabeled methods (deep learning) are applied that use AI to predict thermal distributions to facilitate the identification of anomalies during thermographic inspections.

Data processing improvements have allowed thermography to be in the most ambitious lines of application of AI, in the industry in general and in the sports sector in particular. Thermography together with AI is used for early diagnosis and monitoring of facilities, allowing small temperature changes to be detected and correlated with information on the state of the pavement, grass, goals, baskets or stands. AI is used in the detection and classification of cracks in buildings through the combination of thermography and machine learning, which helps to identify pathologies and defects more efficiently and accurately than with traditional and subjective maintenance by thermographers [24].

Much of the current interest in AI is due to the role it plays in what happens on the field/court/track during matches. A key development for any sport would be the ability to use the data available in real time, thereby helping coaches, athletes and referees make decisions during the game (see Sect. 12.3.4) while also serving as a check on the health of facilities and equipment.

12.5 Applications of Thermography in Sports

This section of the chapter covers the main applications of thermography in the sports environment in three blocks. First, applications in sports facilities are addressed, with special relevance to sports surfaces made of turf, both artificial and natural.

The second block covers applications on sports materials and equipment and their potential transfer from other sectors of the industry to the sports environment. Finally, the third block describes vision technologies applied to sports.

12.5.1 Sports Facilities

12.5.1.1 Introduction

Thermography has emerged as a critical technology for assessing structural integrity, energy efficiency and thermal comfort in large sports facilities. Large venues such as stadiums, gyms and swimming pools present unique challenges due to their vast dimensions, varying occupancy patterns and complex mechanical systems. Thermal images provide facility managers with actionable insights to enhance maintenance strategies, optimize heating, ventilation and air conditioning (HVAC) systems and improve overall facility performance.

12.5.1.2 Thermal Comfort and Energy Efficiency

Thermal analysis of sports facilities plays a critical role in optimizing thermal comfort. Thermography enables the assessment of mean radiant temperature (MRT), a key parameter for evaluating comfort in large spaces such as stadiums and arenas, where uneven temperature distribution can impact both athletes and spectators. By identifying localized hotspots caused by direct sunlight, poor airflow or heat-emitting equipment, thermographic imaging informs targeted interventions to improve comfort levels [39].

Another contribution to energy conservation is by identifying inefficiencies in HVAC systems, lighting and insulation. In gyms, thermography has been successfully used to assess airflow patterns, ensuring adequate ventilation and identifying heat sources from fitness equipment. Optimizing the placement of treadmills and weightlifting machines helps distribute temperature more evenly, enhancing energy efficiency and user comfort. Similarly, in swimming pools, thermal imaging assists in balancing water and air temperatures to optimize energy consumption while maintaining comfort for swimmers and spectators.

In stadiums, thermography has been extensively used to analyze heat distribution and identify problem areas, such as metal seating exposed to direct sunlight. It has also guided insulation improvements in roofing and building envelopes, leading to enhanced energy efficiency. During live events, real-time thermographic monitoring enables dynamic HVAC adjustments, ensuring spectator comfort regardless of external weather conditions.

Thermal comfort assessment is very important in *seating areas*, due to the time that spectators spend in contact with these surfaces. It is important to have methodologies that introduce the user factor in the evaluation of them, and not only focus on regulatory criteria (for example, in the case of stadium seats, DIN EN 12727:2017-02 can be applied to check the safety of the seat, DIN-EN 13200–1:2019-05 for the general characteristics of spectator viewing spaces, DIN-EN 13200-4:2024-02 for seat characteristics). Figure 12.9 shows a test of stands under controlled and extreme conditions to compare the thermal behavior of different materials. In addition, users tested the seats at 30 °C and for 45 min (the equivalent of half-time in a football or rugby match, and approximately halfway through a basketball or handball match), and they were able to watch the sports of their choice. In this way, the aim was to simulate the real conditions of attending a sporting event in hot conditions and to evaluate thermal comfort and the thermal signature.

An additional consideration for outdoor sports facilities is the UHI effect (see Sect. 12.3.2.2), a phenomenon where densely built areas retain heat, raising local temperatures [80]. This is particularly relevant where elevated temperatures can affect both athlete performance and spectator experience. Thermography is particularly effective in identifying these heat-retaining zones, providing critical data to mitigate their impact through strategies such as the use of reflective materials, green roofing and optimized landscaping. Addressing the UHI effect in stadium environments, thermographic analyses have informed the implementation of reflective

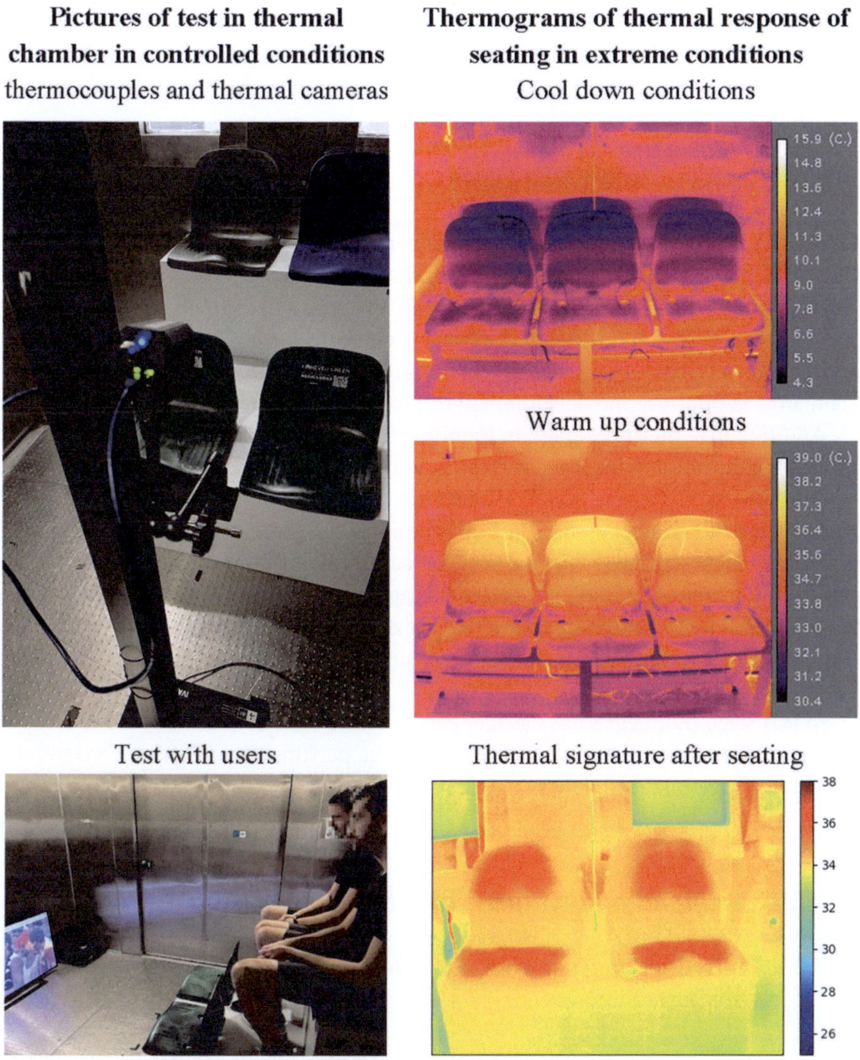

Fig. 12.9 Characterization of recycled material of seats (provided by FANBASE) and evaluated in controlled conditions in a climate chamber. Assessment in extreme climate conditions thermal comfort with users (IBV Facilities)

roofing materials, shaded spectator zones, vegetation to reduce localized heat build-up [39], and the overall energy demand for cooling systems [33]. Revel and Arnesano (2014) [10.1016/j.buildenv.2014.03.017] applied thermography to the design of energy-efficient ventilation systems to mitigate the effects of the UHI phenomenon in outdoor pools.

Moreover, integrating thermal images with modern facility management systems allows for continuous monitoring and optimization of environmental conditions

[39] and contributes to sustainability goals by reducing the carbon footprint of sports facilities [33].

12.5.1.3 Applications to Structural Diagnosis and Preventive Maintenance

Lee et al. [39] describe how infrared thermography is widely applied in sports facilities, offering valuable benefits in maintenance and structural diagnostics. Thermography facilitates the monitoring of material conditions, enabling facility managers to assess the wear and aging of components such as synthetic tracks, sports courts or roofing elements. This ensures consistent performance and safety standards, supporting the development of preventive maintenance plans [33].

Additionally, thermography is highly effective in detecting delamination and moisture infiltration beneath surfaces such as tiles, concrete facades and synthetic flooring. These hidden defects, if left unaddressed, can compromise structural integrity and lead to costly repairs. In high-humidity environments such as swimming pools and locker rooms, infrared images can pinpoint condensation-prone areas, preventing material degradation and reducing maintenance costs [33]. For instance, in swimming pools and gyms, detecting areas prone to condensation or water infiltration can prevent accidents and ensure the longevity of the facility's infrastructure.

In this way, by integrating thermal cameras into routine maintenance protocols, sports facility managers can ensure consistent performance, enhance safety and optimize operational efficiency.

12.5.2 Artificial and Natural Turf Sports Surfaces

12.5.2.1 Introduction: Importance of Thermal Management in Sports Surfaces

Sports surfaces, including artificial and natural turf, are essential for sports such as football, hockey and tennis. However, their thermal behavior presents challenges for performance, safety and the environment. According to Petrass et al. (2015) [10.1177/1754337114553692], artificial turf can reach temperatures over 70 °C in warm climates due to its synthetic materials, increasing the risk of burns and reducing performance. Natural turf regulates temperature better through evapotranspiration but also faces issues in extreme conditions such as drought or frost [32].

Effective thermal management relies on strategies such as controlled irrigation, reflective materials and innovative designs. Thermography plays a key role by providing precise, non-invasive thermal analysis (Petrass et al. 2015) [10.1177/1754337114553692]. Infrared cameras permit to creation of detailed thermal maps, identifying *heat accumulation areas* and evaluating *cooling strategies* for artificial turf. It also compares different infill materials, such as rubber and organic options, to optimize surface design.

12.5.2.2 Evaluation of Friction and Abrasion on Artificial Turf

The risk of skin abrasion on sports fields, particularly artificial turf, increases as surface temperature rises, intensifying friction upon contact. Petrass et al. (2015) [10.1177/1754337114553692] highlight that this effect is especially concerning in high-intensity sports, where athletes frequently interact with the ground, making them more vulnerable to thermal burns and skin irritations. Recent studies, including those by Braun et al. (2024) [10.1002/csc2.21383], have shown that surface temperatures above 50 °C significantly increase the risk of abrasion, affecting both the safety and comfort of players.

By using thermal imaging, it is possible to identify areas with greater heat accumulation and friction, providing a detailed analysis of thermal conditions during contact. Additionally, thermography enables the quantification of temperature distribution in real-time, which is useful for assessing the performance of different materials used in artificial turf, such as rubber or organic infills, in reducing the risk of abrasion.

The use of thermography has driven the development of strategies to improve player safety on artificial turf surfaces. These improvements include:

- *Design of advanced materials*: Petrass et al. (2015) [10.1177/1754337114553692] emphasize the importance of incorporating infills with optimized thermal properties, such as high-reflectivity or low-thermal-conductivity materials, to effectively reduce surface temperatures on artificial turf.
- *Constant thermal monitoring*: Braun et al. (2024) [10.1002/csc2.21383] highlight the importance of real-time thermal monitoring systems, which enable the early detection of hazardous conditions, helping to mitigate risks before they become a serious threat to athletes.
- *Thermal maintenance protocols*: Petrass et al. (2015) [10.1177/1754337114553692] describe thermal maintenance protocols that rely on thermographic data, including strategies such as controlled irrigation and thermal coatings. These approaches help reduce heat accumulation and decrease the risk of injuries.

The manual of the *Fédération Internationale de Football Association* (FIFA) includes a specific test for evaluating abrasion on artificial turf, which is a key industry standard. This test simulates the contact between the surface and the skin under various temperature and friction conditions, allowing the determination of the surface's abrasiveness. Thermography complements such evaluations by offering a detailed thermal analysis, facilitating the identification of critical areas and validating materials that meet international safety requirements [https://digitalhub.fifa.com/m/7cfc3106515eb3e2/original/FIFAQuality-Programme-for-Football-Turf-Test-Manual-I-Test-Methods-April-2024-edition.pdf] (FIFA 2024). Figure 12.10 shows an example of this type of test, carried out at homologated FIFA lab.

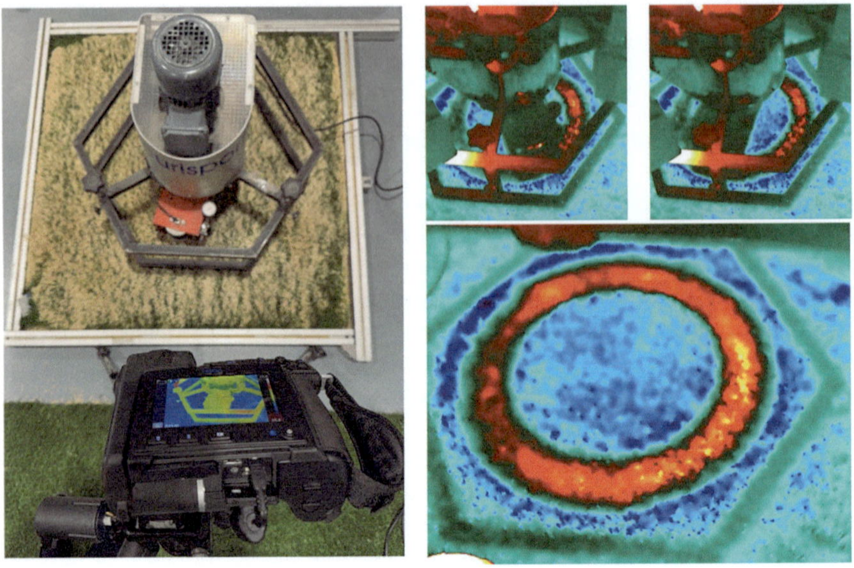

Fig. 12.10 Evaluation of friction and abrasion on artificial turf in the laboratory. (IBV Sport LAb Facilities)

12.5.2.3 Applications of Thermography for Surface Maintenance

Thermography has emerged as a key tool for analyzing and improving the thermal conditions of sports surfaces, enabling precise, data-driven management of cooling measures applied to both artificial and natural turf. One of the most relevant applications is real-time thermal monitoring, which allows for the identification of high-heat accumulation areas on sports surfaces. This capability facilitates the implementation of targeted cooling strategies, such as localized irrigation, and the evaluation of the efficiency of reflective materials or thermal coatings designed to reduce surface temperatures [39].

Moreover, infrared thermography provides key data to optimize water usage for surface cooling, particularly for artificial turf. Through detailed analysis of thermal images, it is possible to determine the specific moments and zones that require irrigation, reducing water waste and maximizing the effectiveness of cooling strategies. Petrass et al. (2015) [10.1177/1754337114553692] indicated that intermittent and controlled irrigation can reduce surface temperatures by up to 20%, depending on climatic conditions and the structural design of the field. This underscores the role of thermography in optimizing water management and promoting a more efficient use of this vital resource [32]. Moreover, thermography plays a fundamental role in assessing the environmental impact of cooling strategies. By monitoring the thermal efficiency of materials and water consumption, it is possible to develop sustainable practices that balance sports performance needs with the conservation of natural resources. For example, combining controlled irrigation with high-reflectivity materials can significantly reduce water dependency, which is particularly beneficial in drought-prone regions (Petrass et al. 2015) [10.1177/1754337114553692]. Additionally, data obtained through thermography allows for the modeling of long-term scenarios to optimize thermal conditions and minimize the environmental impact of these surfaces.

12.5.2.4 Evaluation and Design of Infill Materials for Artificial Turf Sports Surfaces

Thermography enables precise analysis of how temperature and humidity are distributed across the surface, providing essential information to optimize the selection of infill materials and reduce the risk of overheating [80].

Infill materials such as sand, rubber and organic compounds play a crucial role in the thermal regulation of these surfaces. For instance, black rubber, commonly used for its durability, exhibits a high heat absorption capacity, which can lead to extremely high surface temperatures under sunny conditions. In contrast, more innovative infill materials, such as organic options (cork or coconut fibers) or those coated with high-reflectivity pigments, have proven effective in minimizing heat accumulation, thereby improving both comfort and safety in sports use. However, each material presents specific challenges, ranging from costs to durability, making a detailed analysis of their thermal behavior indispensable (Braun et al. 2024) [10.1002/csc2.21383].

Scientific literature supports the use of thermography as a key tool for material selection. McNitt and Landschoot (2003) [10.2135/cropsci2003.9570] have demonstrated that thermography effectively compares different infill compositions and their heat dissipation capacities. Petrass et al. (2015) [10.1177/1754337114553692] also applied thermography to demonstrate that incorporating nanomaterials, such as titanium oxide particles, into traditional infills had significant potential for mitigating accumulated heat.

Figure 12.11 shows a pair of thermograms of an artificial turf sample with organic infills, monitored with active thermography (external heat source simulating solar radiation). Moisture characterization adding controlled water quantity and monitoring thermal behavior provide a qualitative analysis of their performance.

In this context, FIFA has played a significant role in evaluating and regulating the quality of sports surfaces through its *Quality Programme for Artificial Turf*. This program establishes strict standards, including specific tests to measure surface heat accumulation. One such test is the "FIFA Heat Test," designed to evaluate how infill materials influence surface temperature under high solar radiation conditions. Thermography is employed in these tests to generate thermal maps that identify critical areas and allow for comparisons of the effectiveness of different materials [https://digitalhub.fifa.com/m/7cfc3106515eb3e2/original/FIFA-Quality-Programme-for-Football-Turf-Test-Manual-I-Test-Methods-April-2024-edition.pdf] (FIFA 2024) (Fig. 12.12. left). Figure 12.12, right and Fig. 12.13 correspond to the "Splash Test" which measures the dispersion of the filling after an angled bounce

Fig. 12.11 Characterization of moisture properties of artificial turf with organic components by active thermography. Sample of artificial turf (with organic components of wood, coconut and grape seed fibers) in the lab. (IBV Facilities)

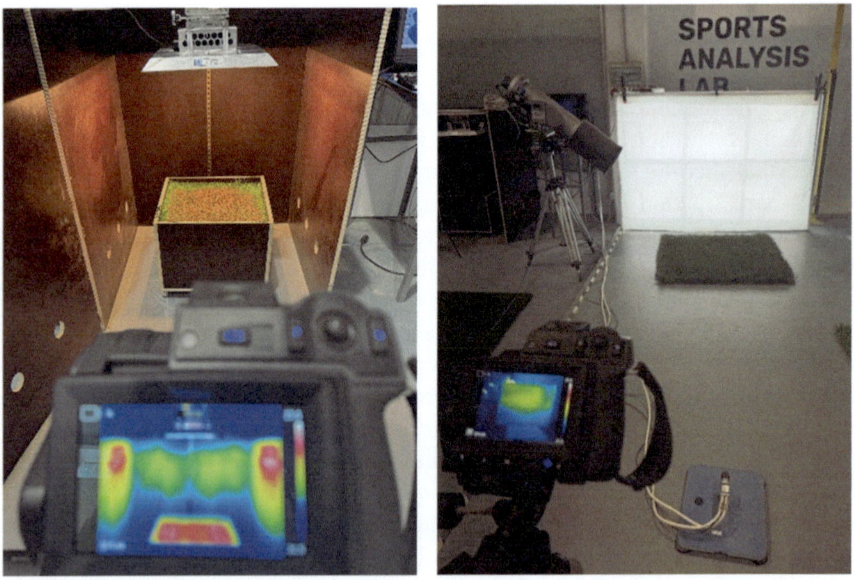

Fig. 12.12 "FIFA Heat Test" and "Splash Test". [https://digitalhub.fifa.com/m/7cfc3106515eb3e2/original/FIFA-Quality-Programme-for-Football-Turf-Test-Manual-I-Test-Methods-April-2024-edition.pdf] (FIFA 2024)

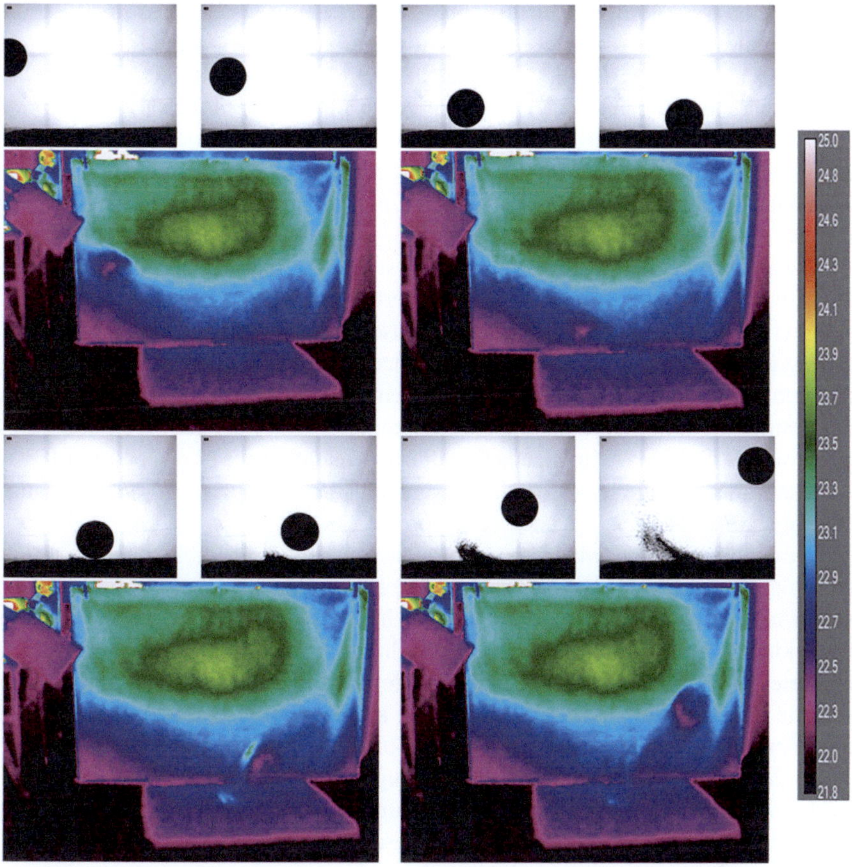

Fig. 12.13 Monitoring with thermal cameras and high-speed cameras, the impact of a soccer ball on a sample of artificial turf. Splash test. [https://digitalhub.fifa.com/m/7cfc3106515eb3e2/original/FIFA-Quality-Programme-for-Football-Turf-Test-Manual-I-Test-Methods-April-2024-edition.pdf] (FIFA 2024)

of a balloon is launched from a cannon at a given speed and angle. The measurement is carried out by processing images obtained with a high-speed camera. The thermal camera also provides the thermal signature of the ball after the impact.

12.5.2.5 Thermography Applied to Natural Turf Sports Surfaces

Thermal analysis has been used with success in the evaluation of turf surface temperature, providing valuable insights into its health, water-use efficiency and response to environmental conditions. Jones (2004) [https://doi.org/10.1016/S0065-2296(04)41003-9] stated that one of the main benefits of thermography in this context is its ability to identify areas experiencing water stress or irrigation-related issues. On sports surfaces, where turf uniformity is essential for athlete performance and safety, thermal cameras detect temperature variations that reveal zones with insufficient water or soil compaction. Jones (2004) [https://doi.org/10.1016/S0065-2296(04)41003-9] also explained that warmer areas often indicate reduced turf transpiration due to damaged roots or uneven water distribution. This analysis allows for early intervention, optimizing irrigation and maintaining the quality of the playing field.

Thermography also allows for seasonal studies of natural turf sports surfaces, analyzing how climatic conditions impact performance throughout the year. During summer months, high temperatures can cause thermal stress on the turf, compromising its transpiration capacity and resistance to wear caused by intensive training and competitions. Conversely, low winter temperatures slow biological processes, increasing susceptibility to diseases and reducing recovery capacity after use. These observations provide essential data for adjusting maintenance practices according to each season's specific needs, such as irrigation frequency, aeration or fertilization (Braun et al. 2024).

In extreme climates, where sports surfaces are more vulnerable to deterioration, thermography plays a key role in facilitating the selection of turf varieties better suited to such conditions. For instance, in hot and dry regions, thermal analysis helps identify species that tolerate heat and drought stress, while in cold climates, it evaluates turf resistance to frost. Furthermore, thermography can be used to design specific management strategies, such as the application of temporary shading, the use of protective covers or the incorporation of soil amendments to improve thermal properties Another critical aspect of managing natural turf sports surfaces is detecting freezing conditions, as subzero temperatures can compromise field functionality and increase the risk of athlete injuries. Thermography allows real-time identification of freezing patterns, highlighting vulnerable areas where the turf may become rigid or lose traction. These patterns are often influenced by local microclimates, soil properties or differences in turf density [https://doi.org/10.1016/S0065-2296(04)41003-9] (Jones 2004).

12.5.3 Sports Equipment

Infrared imaging is widely utilized across various industries due to its non-invasive nature and real-time thermal imaging capabilities. Its primary applications include defect detection ([10.21611/qirt.2020.012] Fang et al. 2021; [67]; [10.3390/s23218780] Zhao et al. 2023), predictive maintenance ([10.1109/TII.2018.2884738] (Alfredo Osornio-Rios et al. 2019; [47, 70]), performance optimization ([50]) and condition monitoring [10.1109/TII.2018.2884738] (Alfredo Osornio-Rios et al. 2019; [60]). These applications enhance operational efficiency, improve safety standards and contribute to cost-effectiveness across different sectors. Specifically, thermography plays a key role in detecting defects in materials and electronic systems,

Table 12.3 Summary of potential applications of infrared thermography in sports equipment

Category	Industrial application	Potential sports equipment application
Defect detection	Electronic components and systems	Identifying overheating issues in electronic sports devices (e.g. fitness trackers and smart wearables)
	Safety	Detecting potential overheating hazards in wearable technology for athletes
	Quality control	Ensuring consistent material properties in sports gear manufacturing
Predictive maintenance	Preventive maintenance	Early detection of material fatigue in high-impact sports equipment (e.g. hockey sticks, golf clubs and football helmets)
	Automotive and transportation equipment	Monitoring tire temperatures in motorsports or brake systems in cycling for enhanced safety
Performance optimization	Performance enhancement	Analyzing thermal distribution in equipment to optimize material performance (e.g. tennis rackets and football boots)
	Sportswear and apparel	Assessing thermoregulation properties in athletic clothing to improve comfort and performance
	Energy conservation	Detecting inefficient thermal properties in protective sports gear (e.g. helmets and shin guards)
Condition monitoring	Equipment monitoring	Monitoring the wear and performance of sports machinery (e.g. treadmills and rowing machines)
	Training environments	Evaluating airflow and insulation in high-performance sports facilities to enhance conditioning strategies
	Structural integrity assessment	Assessing the durability and safety of sports arenas and training structures

Source: Own elaboration based on industrial and sports applications of thermography

enabling preventive maintenance for industrial equipment, optimizing performance through energy conservation and quality control and continuously monitoring critical systems such as electrical infrastructure and mechanical components.

Within the sports sector, thermography has been predominantly applied to assess surfaces, sportswear and physical performance ([10.1088/1748-9326/ac32fb] Guyer et al. 2021; [45];]; [10.1088/1361-6579/ac0fbd] Perpetuini et al. 2021; [10.1371/journal.pone.0211197] Priego et al. 2019) (Sect. 12.5.3.5). However, its application in sports equipment remains relatively limited, primarily focusing on thermomechanical coupling characterization. Despite this, given its proven effectiveness in related domains, its broader integration into sports equipment assessment and optimization may expand in the near future.

Building on these industrial and sports applications, Table 12.3 categorizes potential uses of infrared thermography in sports equipment, aligning them with key functional areas.

Expanding on this idea, the use of infrared thermography in sports equipment offers valuable opportunities to enhance safety and maintenance. This technology can help identify early signs of wear, monitor temperature-related changes and

Fig. 12.14 TNDT—passive thermography test under compression to evaluate the valve of an official FFCV T-4 Ball (Football 8), provided by LUANVI

support effective maintenance practices. The following sections examine specific applications where thermography has been applied in sports equipment.

12.5.3.1 Football, Soccer and Ball Sports

Thermal imaging is applied in football and other ball sports to detect structural changes in balls, such as panel detachment, uneven material distribution and wear-related thermal variations [44]. This application supports consistency in ball construction and performance [74]. Additionally, thermal images contribute to energy restitution analysis, examining how balls absorb and release energy upon impact and informing research on materials and construction techniques that influence elasticity and trajectory control ([44]; Priego Quesada and Carpes 2019). Figure 12.14 shows an example of a non-destructive test that uses compression and thermography to make quality control of different products (IBV facilities, football sample belonging to the *Federació de Futbol de la Comunitat Valenciana* (FFCV)).

12.5.3.2 Tennis

In tennis, thermal cameras are used to assess racket string wear, providing insights into stress distribution and the development of microcracks over time [https://doi.org/10.1016/B978-0-08-102582-6.00002-2] (Priego Quesada and Carpes 2019).

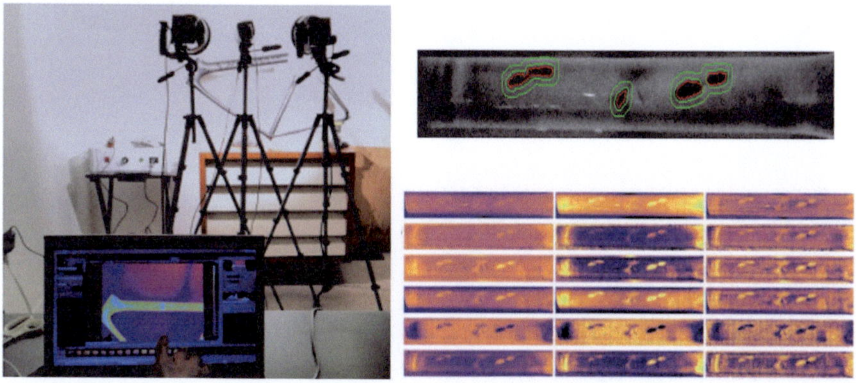

Fig. 12.15 Inspection method to detect impact damage on a bike [68]

It also examines the friction between strings and its impact on material durability and wear patterns [44], assisting in the development of maintenance strategies that extend racket lifespan and enhance gameplay performance. Similar tests as Fig. 12.14 are carried out for tennis and padel balls, to measure compression forces and detect defects from manufacture.

12.5.3.3 Racing Sailboat

Thermographic techniques contribute to the structural assessment of racing sailboats by detecting defects in composite materials [21]. Infrared imaging, along with other non-destructive testing methods such as radiography and computed tomography, helps identify damage locations and material weaknesses in carbon fiber-reinforced polymer masts. Figure 12.7 shows an example of a non-destructive test on a sailboat.

12.5.3.4 Cycling

Thermography in cycling is utilized for multiple applications, including detecting hidden motors by identifying heat anomalies that may indicate embedded electric propulsion systems and technological doping (Fig. 12.4) [https://doi.org/10.1016/B978-0-08-102582-6.00002-2] (Priego Quesada and Carpes 2019). Additionally, it aids in evaluating saddle pressure distribution, offering insights into pressure variations that support studies on bike fit and rider comfort. These insights help inform adjustments aimed at enhancing comfort and minimizing discomfort over extended periods. Furthermore, thermography serves as an inspection method based on active thermography, a well-established technique successfully applied to other materials [68]. Figure 12.15 shows the obtained regions that correspond to defects due to

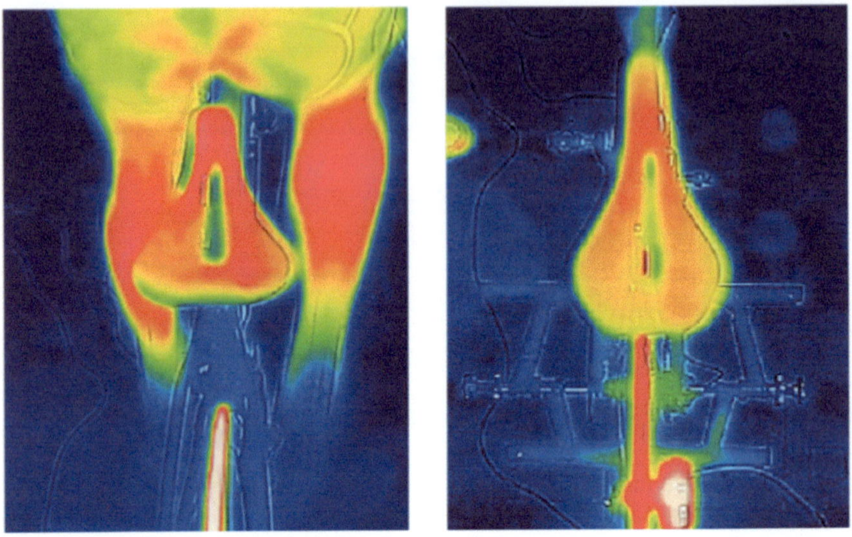

Fig. 12.16 Assessment of the temperature distribution over the saddle during cycling. [https://doi.org/10.1016/B978-0-08-102582-6.00002-2] (Priego Quesada and Carpes 2019)

impact damage. The test of IR inspection was performed with active thermography. Figure 12.16 shows the thermal signature over the saddle.

12.5.3.5 Motorsports

Thermal imaging is used in motorsports to examine tire performance. Infrared cameras generate thermal maps of tire surfaces, assessing temperature distribution to aid in tire pressure monitoring, traction control and race strategy adjustments ([45]; Priego Quesada and Carpes 2019). Furthermore, thermography supports real-time condition monitoring by identifying overheating and pressure variations, contributing to tire management and reducing the risk of performance issues.

Just like many of its Formula One competitors, Red Bull Racing uses wind tunnels to study and test the aerodynamics of its vehicles. Although wind tunnel testing is very useful, the process is complex, and the testing itself can lead to accelerated component wear. To better protect its investment, Red Bull Racing relies on FLIR thermal imaging cameras to monitor temperatures inside the wind tunnel (Fig. 12.17).

12.5.3.6 Equestrian Sports

In equestrian sports, thermography assists in saddle evaluation by analyzing pressure points and heat distribution that may indicate improper fit or stress areas affecting both horse and rider [https://doi.org/10.1016/B978-0-08-102582-6.00002-2]

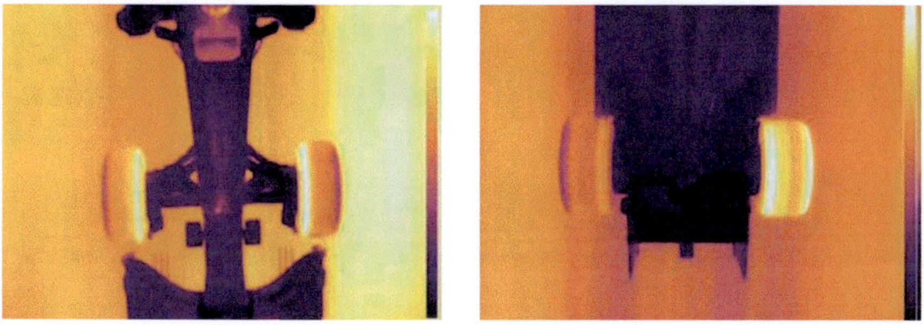

Fig. 12.17 George Trigg, thermal imaging to recognize the model and allow engineers to identify possible problems as a result of overheating in an early stage. (Image from FLIR and Appliter Thermography)

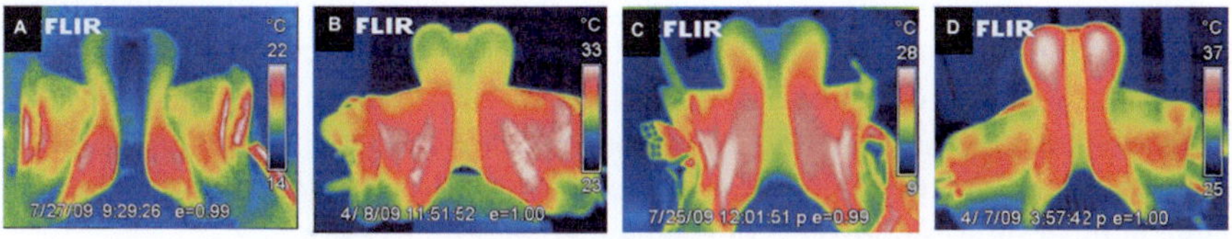

Fig. 12.18 Contact area (%) between saddle panels and thoracolumbar region of horses identified by thermography. (**a**) 25%, (**b**) 50%, (**c**) 75%, (**d**) 100%. ([10.1016/j.jevs.2011.05.011] Arruda et al. 2011)

(Priego Quesada and Carpes 2019). Over time, thermal imaging is also used to monitor saddle wear patterns, helping ensure appropriate weight distribution and preventing discomfort [10.1016/j.jevs.2011.05.011] (Arruda et al. 2011) (Fig. 12.18).

12.6 Vision Technologies and Sports

The world of sports has integrated technology into the design, manufacture and evaluation of materials but also into the development of sporting activities, taking advantage of the benefit of monitoring matches and using images for a multitude of applications in real time.

This is why cameras are currently essential in top-class sporting events, to improve spectator perception and for decision-making, from the "photo finish" to "slow motion replay" [19].

Fig. 12.19 Hot spot of ball and bat contact as "bright spot." (Images obtained from https://www.bbgsports.com/hotspot.html (02 June 2025 checked))

12.6.1 Decision-Making Technologies in Sports

Visualization and measurement technologies are changing the relationship between spectators and referees at sporting events. The umpire decision review system (DRS) is a technology-based system used to assist match officials with their decision-making. This system is known as the "third referee" [17, 18] and serves to improve the accuracy of umpiring decisions while also allowing the collection of a large number of useful images for training IA algorithms [48]:

- "*Hawk-Eye*" (www.hawkeyeinnovations.co.uk) is a reconstructed track device (RTD) system that uses visible light cameras to track the ball's trajectory. The spatial and temporal coordinates of these pixels are represented numerically, and a statistical algorithm reconstructs the flight and point of impact of the ball and crucial features of the playing area [15, 18].
- "*Hot Spot*" is an infrared imaging system used in cricket to determine whether the ball has struck the batsman, bat or pad. When the cricket ball comes in contact with the bat or the protective gear of the batsman, the friction due to this impact locally generates heat which is a result of energy transfer from ball to bat. Figure 12.19 shows several thermograms where the impacts show the thermal contrast due to the heat transfer (gray palette). This heat signature is captured over a number of sequential frames by an infrared video camera and computationally processed [46], with the use of several cameras located at the loggerhead of the ground. The system uses technology developed in the military for tank and jet fighter tracking. These cameras sense and measure heat from friction generated by a collision, such as a ball on the pad, a ball on the bat, a ball on the ground or a ball on the glove. BBG Sports was the inventor of "Hot Spot," integrating hardware, software and network components to create innovative, turnkey, real-time solutions for sports broadcasting since its debut in 2006 cricket was revolutionized [15].

The HotSpot technology is accurate but expensive. The first versions presented some errors when the ball hits the gloves, due to the silicone tape reducing the friction and a mark will not appear on the bat. Like video replays, low-intensity impacts that generate little friction are also likely that the clarity of their output is affected

Fig. 12.20 Foxtenn proposal for WPT. Simulation of combined RGB, IR (right corner) and high-speed camera system

by ambient conditions. Other factors that may also explain the relatively low take-up of HotSpot, especially outside cricket, are the high costs of the cameras, the need for specialist operators and the limited scope for re-using the data generated. In 2012, BBG Sport introduced a new generation of Hot Spot using high-performance thermal imaging cameras that improved sensitivity and much less motion blur. Following the success of this updated Hot Spot system, BBG Sport and Selex signed an exclusivity agreement for the supply of SLX-Hawk cameras for Hot Spot in cricket and other sports [5, 8, 15, 17, 66].

Spectators, referees and regulatory bodies have a legitimate interest in reducing uncertainty and eliminating serious errors. Data analysis technologies based on AI and pattern recognition are helping to correct the limitations of these systems. AI, in particular with infrared imaging, requires a series of recommendations to implement sports measurement technologies [18]. Hawk-Eye has continued to improve in line with developments in computer technology, providing additional information [66]. The same technology (based on Hawk-Eye imaging) is used in other sports in different ways, such as Auto-Ref [18, 30], video assistant referee (VAR), football video support (FVS) or Foxtenn for tennis and padel.

In addition to video refereeing, future implementations will be the basis for a broad statistical record of data such as the speed of the hit, the heat map of each athlete, the hitting zones, the distribution of time in the offensive or defensive phase and their positioning on the court, the main rebound zones or the net coverage. Figure 12.20 shows a simulated proposal of the Foxtenn system for the World Padel Tour (WPT) that includes a thermal image fused with high-speed cameras.

Some of these machine vision and camera-based systems are used in international competitions, and the limitations they present can be overcome in the short term with new AI techniques and cameras with higher spectral and spatial resolution and lower cost. The reduction in costs of thermal cameras, more accessible and with better performance, will facilitate their inclusion and integration with other vision technologies [48]. Thermography will reduce the uncertainty inherent in other vision technologies working in dual mode.

12.7 Conclusions and Future Research

In recent years, thermography has evolved into an invaluable tool in the sports sector, thanks to advancements in artificial intelligence and data processing and the development of more accessible, high-resolution thermal cameras. The application of computer vision techniques in thermography expands the analysis capabilities, improves the accuracy in the detection of artefacts and anomalies and optimizes processes in sports equipment and facilities maintenance. The automation of these thermographic analyses reduces the economic and time costs. This evolution has broadened its applications in several key areas:

- *Thermomechanical Characterization of Sports Equipment and Materials*: Utilizing thermography, in passive and active configuration, detailed assessments of the thermomechanical coupling properties of sports equipment and materials can be conducted. Thermography is a valuable tool in the field of sports equipment, with applications including performance monitoring, defect detection and safety assessments. However, challenges, such as the lack of standardized analysis methods, and variability in protocols limit its broader application. Proper calibration and consistent methodologies are important to ensure accurate and reliable results. Additionally, this technique is most effective when used alongside traditional diagnostic methods to provide a more comprehensive assessment. Recent advancements, such as the integration of machine learning and deep learning algorithms, are improving infrared image efficiency and accuracy in defect detection and quality evaluation. These developments, combined with continued research and technological improvements, could further expand its applications in sports science.
- *Real-Time Thermal Monitoring of Sports Facilities*: The integration of thermographic systems allows for continuous, real-time surveillance of sports facilities. This proactive approach aids in the early detection of potential issues, facilitating timely maintenance and ensuring the safety of both athletes and spectators. With growing technological progress and interdisciplinary collaboration, thermal cameras have the potential to make a significant contribution to the evaluation and development of sports facilities. Infrared analysis is a powerful tool for the evaluation, maintenance and management of sports facilities to identify thermal anomalies, structural weaknesses and energy inefficiencies translated into tangible benefits, including enhanced safety, comfort and sustainability. Kim et al. [33] suggested that to fully leverage its potential, it is essential to adopt an integrated approach that combines thermal comfort analysis, energy efficiency optimization and preventive maintenance planning. Advanced algorithms and artificial intelligence could refine anomaly detection while combining thermography with complimentary inspection techniques, such as ultrasonic or acoustic testing, could enhance the detection of subsurface defects. Additionally, dynamic monitoring systems that could provide continuous thermographic data could enable real-time environmental adjustments, further optimizing facility performance. Another promising avenue for research involves exploring the thermal

behavior of innovative construction materials used in modern sports facilities. By understanding how these materials interact with environmental factors, thermographic applications can be expanded to include advanced design and construction strategies. C. Yang et al. [80] suggested that addressing urban heat islands through thermographic studies and implementing mitigation measures, such as increased vegetation or reflective materials, could significantly enhance *thermal comfort* while reducing energy consumption. By addressing these research priorities, thermography will continue to play a central role in the efficient and sustainable management of sports facilities.

- *Fusion of Real-Time Vision Technologies, 3D and Tracking Systems*: Combining thermography with other imaging modalities enhances the depth and accuracy of data collected. This fusion provides comprehensive insights into various parameters, leading to improved decision-making processes in training and facility management. Additionally, tracking systems for balls, pucks or other equipment components enhance the analysis of in-game dynamics, contributing to performance optimization and decision-making. The advancement of AI, 3D techniques and the Internet of Things (IoT) opens up an opportunity to implement such intelligent surveillance systems [1]. The AI in the case of sports is primarily centered on player tracking, match prediction and opponent behavior analysis that help in making the training methodology and the overall gameplay strategies better [5].

These developments underline the critical role of thermography, bolstered by artificial intelligence and cutting-edge technologies, in raising the standards of safety, performance and efficiency in the sports industry.

References

1. Ahmed AA, Echi M (2021) Hawk-eye: an AI-powered threat detector for intelligent surveillance cameras. IEEE Access 9:63283–63293. https://doi.org/10.1109/ACCESS.2021.3074319
2. Akhloufi MA, Verney B (2013) Fusion framework for 3D inspection and thermal NDT. 2013-01-2171. https://doi.org/10.4271/2013-01-2171
3. Akhloufi MA, Verney B (2015) Multimodal registration and fusion for 3D thermal imaging. Math Probl Eng 2015:1–14. https://doi.org/10.1155/2015/450101
4. Almasri F, Debeir O (2019) Multimodal sensor fusion in single thermal image super-resolution. In: Computer vision–ACCV 2018 workshops: 14th Asian conference on computer vision, Perth, Australia, December 2–6, 2018, Revised Selected Papers 14, pp 418–433
5. Antalamarad NM, Upadhye J (2024) Role of artificial intelligence (AI) in sports. ITM Web of Conf 68:01004. https://doi.org/10.1051/itmconf/20246801004
6. Bagavac P, Krstulović-Opara L, Domazet Ž (2018) Pulse phase thermography: impact damage retrieval. Mater Today Proc 5(13):26578–26583. https://doi.org/10.1016/j.matpr.2018.08.119
7. Bagavac P, Krstulović-Opara L, Domazet Ž (2021) Enhancing IR thermographic inspection of subsurface defects by using the technique of edge detection. Russ J Nondestruct Test 57(7):609–618. https://doi.org/10.1134/S1061830921070020
8. Baker I, Hipwood L, Maxey C, Weller H, Thorne P (2012) High-performance, low-cost IR detector technology. SPIE Newsroom, 15

9. Berthel B, Wattrisse B, Chrysochoos A, Galtier A (2007) Thermographic analysis of fatigue dissipation properties of steel sheets. Strain 43(3):273–279. https://doi.org/10.1111/j.1475-1305.2007.00349.x
10. Blokhin D, Sheinin V (2021) Thermomechanical effects in different geomaterials in limiting behavior of cyclic loading. IOP Conf Ser Earth Environ Sci 773(1):012055. https://doi.org/10.1088/1755-1315/773/1/012055
11. Bonafoni S, Baldinelli G, Rotili A, Verducci P (2017) Albedo and surface temperature relation in urban areas: analysis with different sensors. In: 2017 Joint Urban Remote Sensing Event (JURSE), pp 1–4. https://doi.org/10.1109/JURSE.2017.7924612
12. Brady C, Tuyls K, Omidshafiei S (2021) AI for sports, 1.ª ed. CRC Press. https://doi.org/10.1201/9781003196532
13. Cao Y, Dong Y, Cao Y, Yang J, Yang MY (2020) Two-stream convolutional neural network for non-destructive subsurface defect detection via similarity comparison of lock-in thermography signals. NDT E Int 112:102246. https://doi.org/10.1016/j.ndteint.2020.102246
14. Cavaliere MA, Goldschmit MB, Dvorkin EN (2004) Implementación Del Acoplamiento Termo-Mecánico En Procesos De Conformado Masivo De Metales Utilizando El Método De Las Pseudo-Concentraciones. Mecánica Computacional:2333–2352
15. Choudhary D, Kumar S (2021) Impact of the technology of cricket
16. Chrysochoos A (2012) Infrared thermography applied to the analysis of material behavior: a brief overview. Quant InfraRed Thermogr J 9(2):193–208. https://doi.org/10.1080/17686733.2012.746069
17. Cole S (2014) Has computer technology enhanced international cricket? Comp Everyone 42
18. Collins H, Evans R (2012) Sport-decision aids and the "CSI-effect": why cricket uses hawk-eye well and tennis uses it badly. Public Underst Sci 21(8):904–921. https://doi.org/10.1177/0963662511407991
19. Dc S (2022) Artificial intelligence in sport: an ethical issue. Unity J 3(01):27–39. https://doi.org/10.3126/unityj.v3i01.43313
20. Dickinson RE (1995) Land processes in climate models. Remote Sens Environ 51(1):27–38. https://doi.org/10.1016/0034-4257(94)00062-R
21. Gama Caetano D, Scandiuzzi Valença De Castro D, Giron Camerini C, Kotik HG, Alho A (2023) Failure analysis of a CFRP mast of a racing sailboat. Eng Fail Anal 149:107263. https://doi.org/10.1016/j.engfailanal.2023.107263
22. Garrido I, Lagüela S, Otero R, Arias P (2020) Thermographic methodologies used in infrastructure inspection: A review—post-processing procedures. Appl Energy 266:114857. https://doi.org/10.1016/j.apenergy.2020.114857
23. Gueymard CA, Lara-Fanego V, Sengupta M, Xie Y (2019) Surface albedo and reflectance: review of definitions, angular and spectral effects, and intercomparison of major data sources in support of advanced solar irradiance modeling over the Americas. Sol Energy 182:194–212. https://doi.org/10.1016/j.solener.2019.02.040
24. He Y, Deng B, Wang H, Cheng L, Zhou K, Cai S, Ciampa F (2021) Infrared machine vision and infrared thermography with deep learning: a review. Infrared Phys Technol 116:103754. https://doi.org/10.1016/j.infrared.2021.103754
25. Hidalgo García D, Arco Díaz J (2021) Modeling of the urban Heat Island on local climatic zones of a city using sentinel 3 images: urban determining factors. Urban Clim 37:100840. https://doi.org/10.1016/j.uclim.2021.100840
26. Hill JO, International Confederation for Thermal Analysis (1991) For better thermal analysis and calorimetry. International Confederation for Thermal Analysis. https://books.google.es/books?id=R7B4AAAACAAJ
27. Hu J, Xu W, Gao B, Tian GY, Wang Y, Wu Y, Yin Y, Chen J (2018) Pattern deep region learning for crack detection in thermography diagnosis system. Metals 8(8):612. https://doi.org/10.3390/met8080612

28. Hu C, Duan Y, Liu S, Yan Y, Tao N, Osman A, Ibarra-Castanedo C, Sfarra S, Chen D, Zhang C (2019) LSTM-RNN-based defect classification in honeycomb structures using infrared thermography. Infrared Phys Technol 102:103032. https://doi.org/10.1016/j.infrared.2019.103032
29. Hwang S, An Y-K, Kim J-M, Sohn H (2019) Monitoring and instantaneous evaluation of fatigue crack using integrated passive and active laser thermography. Opt Lasers Eng 119:9–17. https://doi.org/10.1016/j.optlaseng.2019.02.001
30. Jayalath LM (2021) Hawk eye technology used in cricket. South Asian Res J Eng Technol 3(2):55–67. https://doi.org/10.36346/sarjet.2021.v03i02.002
31. Kachare OR (2020) Fault isolation & failure analysis techniques for the development of robust USB2 ESD protection. EOS & ESD fault isolation & fault analysis techniques on USB2 lanes. PhD Thesis. California State University, Sacramento
32. Kanaan A, Sevostianova E, Leinauer B, Sevostianov I (2020) Water requirements for cooling artificial turf. J Irrig Drain Eng 146(10):05020004. https://doi.org/10.1061/(ASCE)IR.1943-4774.0001506
33. Kim H, Lamichhane N, Kim C, Shrestha R (2023) Innovations in building diagnostics and condition monitoring: a comprehensive review of infrared thermography applications. Buildings 13(11):2829. https://doi.org/10.3390/buildings13112829
34. Korman P, Zieliński J, Krzysztof K, Straburzyńska-Lupa A (2016) Possible uses of infrared thermography in sport. https://api.semanticscholar.org/CorpusID:204290988
35. Kovats S, Ebi KL, Menne B (2003) Methods of assessing human health vulnerability and public health adaptation to climate change. World Health Organization. Regional Office for Europe
36. Krapez J-C, Olioso A (2011) A combination of temperature, vegetation indexes and albedo, as obtained by airborne hyperspectral remote sensing, for the evaluation of soil moisture. Quant InfraRed Thermogr J 8(2):187–200. https://doi.org/10.3166/qirt.8.187-200
37. Kwasniewska A, Ruminski J, Szankin M, Kaczmarek M (2020) Super-resolved thermal imagery for high-accuracy facial areas detection and analysis. Eng Appl Artif Intell 87:103263. https://doi.org/10.1016/j.engappai.2019.103263
38. Latorre Sánchez C, Antonio Solves J, Sanchiz Navarro J, Bayona Salvador R, Laparra J, Palomares N, Solaz J (2022) Methodology based on 3D thermal scanner and AI integration to model thermal comfort and ergonomics. In: 13th international conference on applied human factors and ergonomics (AHFE 2022). https://doi.org/10.54941/ahfe1001896
39. Lee D-S, Kim E-J, Cho Y-H, Kang J-W, Jo J-H (2019) A field study on application of infrared thermography for estimating mean radiant temperatures in large stadiums. Energ Buildings 202:109360. https://doi.org/10.1016/j.enbuild.2019.109360
40. Li J, Song C, Cao L, Zhu F, Meng X, Wu J (2011) Impacts of landscape structure on surface urban heat islands: a case study of Shanghai, China. Remote Sens Environ 115(12):3249–3263. https://doi.org/10.1016/j.rse.2011.07.008
41. Liu J, Yang W, Dai J (2010) Research on thermal wave processing of lock-in thermography based on analyzing image sequences for NDT. Infrared Phys Technol 53(5):348–357. https://doi.org/10.1016/j.infrared.2010.06.002
42. Liu B, Zhang H, Fernandes H, Maldague X (2016) Experimental evaluation of pulsed thermography, lock-in thermography and vibrothermography on foreign object defect (FOD) in CFRP. Sensors 16(5):743. https://doi.org/10.3390/s16050743
43. Luo Q, Gao B, Woo WL, Yang Y (2019) Temporal and spatial deep learning network for infrared thermal defect detection. NDT E Int 108:102164. https://doi.org/10.1016/j.ndteint.2019.102164
44. Luong MP (2007) Non-destructive testing of sports equipment: the use of infrared thermography. In: Materials in sports equipment. Elsevier, pp 35–59. https://doi.org/10.1533/9781845693664.1.35
45. Luong MP (2017) Assessment of equipment using infrared thermography in sports. In: Priego Quesada JI (ed) Application of infrared thermography in sports science. Springer International Publishing, pp 185–209. https://doi.org/10.1007/978-3-319-47410-6_8

46. Maeda T, Wang Y, Raskar R, Kadambi A (2019) Thermal non-line-of-sight imaging. In: 2019 IEEE International Conference on Computational Photography (ICCP), pp 1–11. https://doi.org/10.1109/ICCPHOT.2019.8747343
47. Massaro A, Galiano A (2020) Infrared thermography for intelligent robotic systems in research industry inspections: thermography in industry processes. In: Habib MK (ed) Advances in computational intelligence and robotics. IGI Global, pp 98–125. https://doi.org/10.4018/978-1-7998-0137-5.ch005
48. Nadikattu RR (2020) Implementation of new ways of artificial intelligence in sports. SSRN Electron J. https://doi.org/10.2139/ssrn.3620017
49. Peeters J, Ribbens B, Dirckx JJJ, Steenackers G (2016) Determining directional emissivity: numerical estimation and experimental validation by using infrared thermography. Infrared Phys Technol 77:344–350. https://doi.org/10.1016/j.infrared.2016.06.016
50. Plastropoulos, A., Adiuku, N., Avdelidis, N. P., Ibarra-Castanedo, C., Maldague, X., & Servais, P. (2023). A short review on infrared thermography: history, philosophy, approaches, standards, and application examples. En N. P. Avdelidis (Ed.), Thermosense: thermal infrared applications XLV (3). SPIE. https://doi.org/10.1117/12.2664586
51. Plekhov O, Naimark O (2013) Statistical model of submicrocrack evolution under cyclic loading. In: Proceedings of the 13th international conference on fracture, pp 16–21
52. Priego Quesada JI (ed) (2017) Application of infrared thermography in sports science. Springer International Publishing. https://doi.org/10.1007/978-3-319-47410-6
53. Rangel J, Soldan S, (2014) 3D thermal imaging: fusion of thermography and depth cameras. In: Proceedings of the 2014 international conference on quantitative InfraRed thermography. 2014 Quantitative InfraRed Thermography. https://doi.org/10.21611/qirt.2014.035
54. Ratiu OG, Badau D, Carstea CG, Badau A, Paraschiv F (2010) Artificial intelligence (AI) in sports. In: Proceedings of the 9th WSEAS international conference on artificial intelligence, knowledge engineering and data bases, pp 93–97
55. Razani M, Parkhimchyk A, Tabatabaei N (2018) Lock-in thermography using a cellphone attachment infrared camera. AIP Adv 8(3):035305. https://doi.org/10.1063/1.5021601
56. Ribbens B, Peeters J, Bogaerts B, Steenackers G, Sels S, Penne R, Van Barel G (2016) 4D active and passive thermography measurement system using a KUKA KR16 robot and time-of-flight imaging. Proceedings of the 2016 international conference on quantitative InfraRed thermography. In: 2016 quantitative InfraRed thermography. https://doi.org/10.21611/qirt.2016.106
57. Saeed N, King N, Said Z, Omar MA (2019) Automatic defects detection in CFRP thermograms, using convolutional neural networks and transfer learning. Infrared Phys Technol 102:103048. https://doi.org/10.1016/j.infrared.2019.103048
58. Sánchez-Zapero J, Camacho F, Martínez-Sánchez E, Gorroño J, León-Tavares J, Benhadj I, Toté C, Swinnen E, Muñoz-Sabater J (2023) Global estimates of surface albedo from Sentinel-3 OLCI and SLSTR data for Copernicus climate change service: algorithm and preliminary validation. Remote Sens Environ 287:113460. https://doi.org/10.1016/j.rse.2023.113460
59. Santamouris M (2020) Recent progress on urban overheating and heat Island research. Integrated assessment of the energy, environmental, vulnerability and health impact. Synergies with the global climate change. Energ Buildings 207:109482. https://doi.org/10.1016/j.enbuild.2019.109482
60. Sarawade AA, Charniya NN (2018) Infrared thermography and its applications: a review. In: 2018 3rd international conference on communication and electronics systems (ICCES), pp 280–285. https://doi.org/10.1109/CESYS.2018.8723875
61. Shrestha R, Kim W (2018) Non-destructive testing and evaluation of materials using active thermography and enhancement of signal to noise ratio through data fusion. Infrared Phys Technol 94:78–84. https://doi.org/10.1016/j.infrared.2018.08.027
62. Shrestha R, Moradi M, Sfarra S, Kim W (2024) Infrared thermography: philosophy, approaches, analysis—processing, and guidelines. In: Non-destructive material characterization methods. Elsevier, pp 249–279. https://doi.org/10.1016/B978-0-323-91150-4.00017-3

63. Song J, Chen W, Zhang J, Huang K, Hou B, Prishchepov AV (2020) Effects of building density on land surface temperature in China: spatial patterns and determinants. Landsc Urban Plan 198:103794. https://doi.org/10.1016/j.landurbplan.2020.103794
64. Szurko A, Kasprzyk-Kucewicz T, Cholewka A, Kazior M, Sieroń K, Stanek A, Morawiec T (2022) Thermovision as a tool for athletes to verify the symmetry of work of individual muscle segments. Int J Environ Res Public Health 19(14):8490. https://doi.org/10.3390/ijerph19148490
65. Taha H (2004) Heat islands and energy
66. Ting S, Chilukuri MV (2009) Novel pattern recognition technique for an intelligent cricket decision making system. In: 2009 IEEE Instrumentation and Measurement Technology Conference, pp 978–982. https://doi.org/10.1109/IMTC.2009.5168594
67. Usamentiaga R, Venegas P, Guerediaga J, Vega L, Molleda J, Bulnes F (2014) Infrared thermography for temperature measurement and non-destructive testing. Sensors 14(7):12305–12348, Article 7. https://doi.org/10.3390/s140712305
68. Usamentiaga R, Ibarra-Castanedo C, Klein M, Maldague X, Peeters J, Sanchez-Beato A (2017) Nondestructive evaluation of carbon fiber bicycle frames using infrared thermography. Sensors 17(11):2679, Article 11. https://doi.org/10.3390/s17112679
69. Vavilov V (2014) Thermal NDT: historical milestones, state-of-the-art and trends. Quant InfraRed Thermogr J 11(1):66–83. https://doi.org/10.1080/17686733.2014.897016
70. Venegas P, Ivorra E, Ortega M, Sáez De Ocáriz I (2022) Towards the automation of infrared thermography inspections for industrial maintenance applications. Sensors 22(2):613. https://doi.org/10.3390/s22020613
71. Vijaya Lakshmi A, Ghali VS, Suresh B, Vesala GT, Naik RB (2021) A thermographic system for quantitative and automated subsurface detail visualization. In: Mukhopadhyay CK, Mulaveesala R (eds) Advances in non-destructive evaluation. Springer, pp 27–39. https://doi.org/10.1007/978-981-16-0186-6_4
72. Wagoner RH, Chenot JL (2001) Metal forming Analysis, 1.ª ed. Cambridge University Press. https://doi.org/10.1017/CBO9781139087070
73. Watts M, Watkins S (2014) Aerodynamic structure and development of formula 1 racing car wakes. SAE Int J Passeng Cars Mech Syst 7(3):1096–1105. https://doi.org/10.4271/2014-01-0600
74. Wiart N, Kelley J, James D, Allen T (2011) Effect of temperature on the dynamic properties of soccer balls. Proc Instit Mech Eng Part P J Sports Eng Technol 225(4):189–198. https://doi.org/10.1177/1754337111411644
75. Wilcox LM, Donnell KM (2023) Thermal diffusivity materials characterization via active microwave thermography. *2023 IEEE international instrumentation and measurement technology conference (I2MTC)*, 1-6. https://doi.org/10.1109/I2MTC53148.2023.10175955
76. Wu D, Busse G (1998) Lock-in thermography for nondestructive evaluation of materials. Revue générale de thermique 37(8):693–703
77. Xie Y, Sattari K, Zhang C, Lin J (2023) Toward autonomous laboratories: convergence of artificial intelligence and experimental automation. Prog Mater Sci 132:101043. https://doi.org/10.1016/j.pmatsci.2022.101043
78. Xu D, Wang Y, Xu S, Zhu K, Zhang N, Zhang X (2020) Infrared and visible image fusion with a generative adversarial network and a residual network. Appl Sci 10(2):554. https://doi.org/10.3390/app10020554
79. Yang J, Wang W, Lin G, Li Q, Sun Y, Sun Y (2019) Infrared thermal imaging-based crack detection using deep learning. IEEE Access 7:182060–182077. https://doi.org/10.1109/ACCESS.2019.2958264
80. Yang C, Zhu W, Sun J, Xu X, Wang R, Lu Y, Zhang S, Zhou W (2021) Assessing the effects of 2D/3D urban morphology on the 3D urban thermal environment by using multi-source remote sensing data and UAV measurements: A case study of the snow-climate city of Changchun, China. J Clean Prod 321:128956. https://doi.org/10.1016/j.jclepro.2021.128956

81. Zhou R, Su H, Wen Z (2022) Experimental study on leakage detection of grassed earth dam by passive infrared thermography. NDT E Int 126:102583. https://doi.org/10.1016/j.ndteint.2021.102583
82. Zhou L, Li J, Yang S, Alkhalifah T, Marzouki R (2025) Application of Carrera unified formulation and hybrid AI-composition with CNN and ReliefF feature selection: presenting some useful suggestions for improving the stability of football and sport equipment via advanced nanocomposites. Mech Adv Mater Struct 1-20:1. https://doi.org/10.1080/15376494.2024.2442493

Chapter 13
Infrared Thermography: A Possible Role in Psychophysiology of Sports?

Damiano Formenti, Arcangelo Merla, and David Perpetuini

Abstract Infrared thermography (IRT) is an upcoming, promising methodology in the field of psychophysiology. Mental and emotional components of behaviour play an important role in the determination of human performance in sporting competition scenarios. Driven by sympathetic nerve activity, affective and emotional states derive from muscular activity, skin blood flow, and perspiration patterns in specific body parts. The goal of this chapter is to introduce the assessment of emotional states and computational psychophysiology through thermal infrared imaging in sports and exercise.

13.1 Introduction: The Importance of Psychophysiology in Sports and Exercise

In today's highly competitive world of sports events, athletes and sportsmen of various levels are looking for ways to get bigger, stronger, and faster, with the aim of excelling in competitions. Technological advances in equipment also play an important role in rising performance. However, the physical and physiological components of the performance represent only a few of the factors necessary for rising success in sports competitions. In fact, it is well-accepted in the literature that the psychological side of sport (i.e., motivation, concentration, and mental toughness) is also critical to success [1, 2]. As regard, several important athletes from various disciplines regularly acknowledge the importance of the psychology of sport and the need to work on their mental capacities. The last decades have been

D. Formenti
Department of Biotechnology and Life Sciences (DBSV), University of Insubria, Varese, Italy
e-mail: damiano.formenti@uninsubria.it

A. Merla · D. Perpetuini (✉)
Department of Engineering and Geology (INGEO), University "G. d'Annunzio" of Chieti-Pescara, Pescara, Italy
e-mail: arcangelo.merla@unich.it; david.perpetuini@unich.it

J. I. Priego-Quesada (ed.), *Application of Infrared Thermography in Sports Science*,
https://doi.org/10.1007/978-3-031-93311-0_13

characterized by an increase in the involvement of sports psychologists by both individual athletes and professional teams and federations. This is thanks to the fact that growing attention has been devoted to the study of the interaction between the brain and the body in the field of sport and exercise [2].

The relationship between the brain and body has always been a subject of interest in research. The science aimed to study such brain-body relationships is denoted as psychophysiology. Psychophysiology has been defined by Sternbach [3] as the inference of psychological and emotional states from an examination of physiological measures. According to this definition, the electrical signals derived from physiological measures are recorded from the body surface of subjects under investigation [3].

Looking at the development of psychophysiology back in history, it seems that the main idea of psychophysiology has been that every physiological modification is accompanied by a parallel change reflected in the mental and/or emotional state [4]. Specifically, authors formulated this psychophysiological principle: "Every change in the physiological state is accompanied by an appropriate change in the mental-emotional state, conscious or unconscious, and conversely, every change in the mental-emotional state, conscious or unconscious, is accompanied by an appropriate change in the physiological state." [5]. This concept has been extensively studied and developed in the last decades, thus supporting the notion that any physical or emotional stresses are associated with various physiological and psychological responses [6]. This is evident in conditions in which the human body is subjected to both psychological and physiological stresses, such as during any type of physical exercise, with particular emphasis during competitions.

Among a variety of indicators of stress, elevated heart rate and high state anxiety are considered two of the most important parameters to monitor stress status. The literature indicates that the effects of physical and emotional stress on heart rate response have been investigated among various athletes and disciplines. For example, heart rate responses to physical and emotional stress have been reported for automobile drivers [7], archeries [8], and parachuting athletes [9]. However, most of the studies that investigated the response to the stress of parachuting focused on physiological or psychological indices of stress (such as heart rate and anxiety, respectively) [10]. Only very few studies adopted a combined approach, matching the two types of measure. For example, the study by Falk and Bar-Eli [10] examined the psychophysiological stress responses to parachuting using the two common indicators of stress such as heart rate and state anxiety. This approach allowed the authors to investigate the interaction between the two indices, thus supporting the central idea of the psychophysiological principle, formulated by Green et al. [5]. Measuring physiological variables, technologies and instruments play an important role in psychophysiology, thus gathering information on the processes occurring in the brain and the body of an exercising subject. In this sense, the aim of this chapter is to briefly describe the methods actually used in psychophysiological studies in the sport and exercise field, with particular interest in thermal biofeedback. As regards, we propose infrared thermography as a potential methodology for studying emotional responses in the field of assessment of psychophysiological states in sport and exercise by means of the thermal signature of such states.

13.2 Biofeedback Modalities in Sport and Exercise

13.2.1 Overview of the Biofeedback Modalities in Sport and Exercise

Scientists and practitioners investigating the psychophysiological response to exercise of the human body can apply three different levels of measurement: verbal measurement (e.g., expressions of subjective experience), motor measurement (e.g., particular movements of the body, facial expressions, and observed behaviour during training or competitions), and physiological (e.g., electromyography activity, galvanic skin response, heart rate, and skin temperature) [11].

The act of monitoring the body's physiological responses of an individual and displaying this information is denoted as biofeedback [12]. In this sense, biofeedback helps the subject to detect and control physiological functions by using monitoring devices. The basic concept is to provide individuals with information about the inner status of their bodies, brains, and their interactions [13].

Although the development of new technologies and instrumentations especially in the last decades, investigations in the field of psychophysiology and biofeedback are usually conducted in well-controlled conditions, such as laboratory settings. Biofeedback devices and related modalities are usually used to train people to better control their physiological processes within a controlled laboratory setting, rather than for being used during real situations, such as competitions. As regards, biofeedback research has traditionally focused mainly on closed-skill sports, rather than on open-skill sports [14].

In general, biofeedback is considered by the scientific community a useful tool for stress management and control. However, demonstrating a direct relationship between biofeedback training and performance is not simple because of the variety of factors that contribute to the outcome of the performance [11, 14–17]. As regards, studies that have been conducted on this issue with the aim to test the efficacy of a period of biofeedback training have indicated that training with biofeedback contributes to decreasing both physiological and self-determined stress levels [11, 18], which may be indirectly beneficial for performance.

A major application for biofeedback is detecting and helping in the management of psychophysiological arousal, in particular in those situations characterized by arousal. Examples are situations in which athletes are requested to perform tasks in a real competition scenario. The situation may exacerbate when the task is crucial for the result of competition, and in turn, its outcome determines the conditions of winners or losers [19]. These conditions may be found in both team and individual sports where concentration and psychological states play a fundamental role. Specifically, the condition during which an athlete is under pressure has been defined as choking [20].

The relationship between arousal states and performance has been described and explained by sports psychology using the inverted-U hypothesis [21]. This theory states that at low arousal levels, performance will be below par because the exerciser or athlete is not activated enough from a psychological point of view. As arousal increases, performance will reach an optimal point which in consequence results in the best performance. However, further excess increases in arousal cause a following decrement in performance. In this way, this theory can be represented by an inverted U. In fact, high performance will be reached with the optimal level of arousal, whereas lesser performance with either low or very high arousal.

Physiological processes associated with a status of over-arousal and choking include different modalities of responses, such as tension on skeletal muscle, electrodermal activity, and peripheral vasoconstriction. These are the most used biofeedback modalities [11, 12].

The term "biofeedback modalities" refers in particular to different types of instrumentation used for recording and monitoring different physiological signals. Several biofeedback modalities have been used in sports and exercise, depending on the situation, each one with its own characteristics. Examples are the measurement of muscle activity and tension by electromyography, the measurement of electrodermal and sweat gland activities, the measurement of the brain's electrical activity by electroencephalography, the measurement of heart rate and heart rate variability by electrocardiography or heart rate monitor, and the measurement of peripheral skin temperature as an index of skin blood flow (see Table 13.1 for a resume of the most common biofeedback modalities). As a result, biofeedback training with one or more modalities combined together have been used both by researchers and practitioners to study the possible improvement in the performance of athletes in different disciplines via psychoregulation mechanisms [2, 17, 18, 22, 23].

Before presenting the thermal biofeedback and in the following sections proposing infrared thermography (IRT) as a potential method for investigating psychophysiological states in sport, we dedicate space to two other biofeedback modalities that, for their nature of being applied on the surface of the body, are the more ones related to the thermal biofeedback.

Table 13.1 Most common biofeedback modalities, together with the measured variables and the sensors

Biofeedback modalities	Measures	Sensor
Electromyography	Muscle activity, tension, and dysfunction	Electrodes on muscle area of tension
Electrodermal activity	Galvanic skin response and sweat gland activity	Contact sensors on fingers or hand palms
Electroencephalography	Electrical activity of the brain	Electrodes embedded in a cap placed on the head
Electrocardiography or heart rate monitor	Heart rate and heart rate variability	Sensors and heart rate belt on the thorax
Thermal	Skin temperature	Contact sensors or infrared thermography
Functional near-infrared spectroscopy	Cerebral hemodynamics related to brain activity	Optodes embedded on a cap placed on the head

In the following paragraphs, we therefore provide a brief explanation and proper examples of electromyography activity biofeedback and electrodermal activity biofeedback. These biofeedback modalities are usually used in combination with the thermal biofeedback.

13.2.2 Electromyography Activity Biofeedback

Measurement of the electrical activity preceding and occurring during muscle contraction is called electromyography. Electromyography quantifies the electrical energy discharged by the motor nerve that induces a muscle contraction [24]. Generally, surface electrodes are positioned on the skin over the muscle of interest. Electromyography biofeedback consists of recording and showing the electrical activity of the target muscle to the subject [25]. Specifically, quantification of electrical activity can be shown to the subject either with visual signals or with auditory signals. Furthermore, the combination of both visual and auditory information is possible [11]. In this way, electromyography provides the subject with a reliable indication of muscle tension. The most common application of electromyography biofeedback is to evaluate the efficacy of periods of electromyography biofeedback training. The aim is therefore to verify that a control of self-regulation has been achieved for a given criterion.

An interesting example of an electromyography biofeedback application can be found in the study of Blumenstein and colleagues [26]. In that study, a combination of three psychoregulative procedures of relaxation and excitation with biofeedback were provided to subjects with the purpose of investigating their role in physiological performance variables. A sample of 39 students was assigned to three groups of psychoregulatory treatment (i.e., autogenic and imagery training; music and imagery training; and autogenic, music, and imagery training). Moreover, a placebo group and a control group were also included in the study. Imagery training consisted of a 100 m run task. The treatment and control conditions lasted 13 sessions of 20 min each. Only in the last six sessions, training was accompanied by electromyography biofeedback. It was found that the combination of biofeedback with autogenic, imagery, and music training produced a significant augmenting effect on physiological components and athletic performance [26].

13.2.3 Electrodermal Activity Biofeedback

Another widely used biofeedback modality is concerned with the electrical activity of the skin surface. It is not easy to determine whether sweat glands were active or to quantify the sweat produced [12]. However, since sweat contains salts making it electrically conductive, sweaty skin is more conductive to electricity than dry skin. Therefore, skin conductance activity reflects well sweat gland activity. This phenomenon is known as electrodermal activity, and it has historically been known as galvanic skin response.

Galvanic skin response is based on the fact that sweating is controlled by the autonomic nervous system, and skin conductance is an indication of psychological or physiological arousal. When the autonomic nervous system is highly aroused, it reflects its activity by increasing sweat gland activity, which in turn increases skin conductance. In this way, skin conductance can be considered a reliable measure of emotional and sympathetic responses to emotional stimuli of an affective nature [27]. Galvanic skin response is recognized as a method to gain objective access to psychophysiological arousal. Electrodermal activity is usually monitored and quantified by contact biofeedback instruments. A galvanic skin response device applies a very small electrical pressure (voltage) to the skin, measuring the amount of electrical current that the skin will allow to pass. A galvanic skin response device is typically applied on the fingers or the palmar surface of the hand (where there are many sweat glands).

Galvanic skin response biofeedback may be used on its own or in conjunction with other mental and relaxation techniques to institute a relaxation response that might be used to combat the high status of pre-competition anxiety that may limit performance [17].

A recent paper by Pusenjak and colleagues [17] examined whether an 8-week biofeedback training period (a combination of various biofeedback exercises, such as breathing, muscle, galvanic skin response, and temperature) could improve the psychophysiological control over competitive anxiety and in turn, enhance athletic performance. Participants were divided into two groups: experimental group ($n = 18$) and control group ($n = 21$). Psychophysiological responses (including galvanic skin response) to stress tests were assessed before and after the 8-week biofeedback training period. It was found that a greater number of participants in the experimental group were able to successfully control their psychophysiological parameters, with respect to other subjects of the control group. Specifically, better regulation of galvanic skin response in the stress tests was found in the experimental group, thus supporting the usefulness of biofeedback training [17].

13.2.4 Thermal Biofeedback

One of the most popular biofeedback modalities is thermal feedback (often referred to as skin temperature). This feedback, as for the others, is usually transformed and expressed to the subjects in the form of audio and visual signals, which reflect skin temperature modifications [11].

In general, thermal biofeedback provides information on the peripheral circulation. This is because skin temperature is a function of skin blood flow [28, 29]. Temperature feedback is usually recorded on the surface extremities of the body (such as distal parts of limbs) [30] but can be also recorded in other parts of the body, such as limb muscles, trunk, or face.

Thermal biofeedback can be used in combination with other biofeedback modalities, such as electrodermal biofeedback (i.e., galvanic skin response) and electromyography biofeedback [11, 12]. In this way, integrating the information of different natures coming from various biofeedback modalities may be beneficial for the subjects, which have the possibility to evaluate their behaviour establishing relationships between different biofeedback modalities signals.

The mechanisms on which the thermal biofeedback is based are related to the sympathetic activity of the central nervous system. It is well-known that skin temperature is influenced by both external and internal factors of the body (see Chap. 4 of this book for a detailed description of these factors). Among the internal factors, the cardiovascular mechanisms affecting skin temperature are closely related to the sympathetic activity of the autonomic nervous system [31].

The autonomic nervous system is activated when the subject experiences particular emotions, thus causing an increase in arousal. When the autonomic nervous system is activated, the smooth muscles surrounding the blood vessels near the skin surface are likely to contract, thus resulting in superficial vasoconstriction. This vasoconstriction causes a decrease in skin blood flow, which in turn accompanies a decrease in skin temperature. Conversely, an increase in skin temperature is caused by vasodilation that results from the relaxation of sympathetic activity, and in particular from the relaxation of the autonomic nervous system [31].

Biofeedback training finds a role in the link between skin temperature modulation, and activation of the autonomic nervous system. Learning to modulate peripheral skin temperatures has been widely investigated through the application of thermal biofeedback training. Thermal biofeedback aims to teach individuals to consciously and voluntarily adapt their skin temperature, without medication or external application [30]. The efficacy of thermal biofeedback training has been investigated and proven. For instance, self-regulation of peripheral skin temperatures by people affected by Raynaud's phenomenon has been shown to be particularly beneficial since their ability to control and adapt the skin temperature of limb extremities is often scarce [32]. Moreover, individuals with migraine headaches also have shown improvements after learning to relax by increasing distal skin temperatures via thermal biofeedback training [33]. With particular regard to sport and exercise, studies involving thermal biofeedback modalities are scarce. However, it is worth noticing that thermal biofeedback training was applied and studied for self-regulation of skin temperature extremities in sports activities characterized by cold weather (i.e., skiing, ice-hockey, ice-skating, curling, football, and mountaineering) [30, 34].

An interesting study was conducted by Kappes and Chapman [30]. The authors examined the effects of indoor versus outdoor thermal biofeedback training on the self-regulation of digital skin temperature, measured by contact sensors, in outdoor sports. Moreover, the accuracy of self-estimation of skin temperature after the training period has been also investigated. A sample of 25 individuals was randomly divided into three groups: indoor subjects trained exclusively indoors, outdoor subjects trained exclusively outdoors, and control subjects, which did not receive any training. Subjects of the two experimental groups trained twice a week for four consecutive weeks. Pre and post-tests were performed in an unheated tent outdoors. Results indicated that the outdoor-trained group performed better than the two other groups. Interestingly, there was no significant difference in overall temperature between groups, whereas all subjects overestimated their temperatures when requested to give an estimation. The authors concluded by proposing the notion that learning to control the skin temperature of distal parts of the body in cold environments may depend on environmental context [30].

13.2.5 Electroencephalography and Functional Near-Infrared Spectroscopy Biofeedback

The integration of electroencephalography (EEG) and functional near-infrared spectroscopy (fNIRS) in sports has emerged as a promising frontier for enhancing athletic performance and mental resilience. EEG biofeedback has been shown to significantly improve athletes' cognitive functions, such as attention and reaction time to various stimuli, which are critical in high-stakes sports environments. For instance, a study on judo athletes demonstrated that EEG biofeedback training could enhance visual reaction times, suggesting that athletes can learn to produce and sustain neural patterns associated with improved performance [35]. This training enables athletes to self-regulate their brain activity, leading to better focus and reduced anxiety during competitions [36]. Furthermore, it has been effectively utilized to enhance motivation and self-confidence in combat sports, where quick decision-making under pressure is essential. The ability to modulate brain activity through EEG biofeedback allows performance enhancement favouring tactical flexibility during competitions [37]. In addition, fNIRS offers a complementary approach for monitoring brain activity in sports activities. This non-invasive imaging technique measures changes in hemoglobin concentrations, providing insights into cerebral oxygenation and hemodynamics associated with neural activity [38]. The portability and tolerance of fNIRS to movement, thanks to recent technological advancement, make it particularly suitable for real-time monitoring of athletes during training and competition [39]. In fact, research has indicated that fNIRS can be effectively used to assess cognitive demands and emotional states in athletes, thereby informing training protocols to enhance both physical and mental performance [40]. For example, fNIRS has been employed to investigate the effects of physical exertion on cognitive function, revealing how different training regimens can influence brain activity patterns [41]. The combination of EEG and fNIRS in a multimodal approach allows for a more comprehensive understanding of the psychophysiological factors, delivering information on both electrical and hemodynamic activities and providing insights into the neurovascular coupling. Moreover, the application of these technologies extends beyond performance enhancement to injury prevention and rehabilitation. By utilizing EEG and fNIRS, coaches and sports scientists can gain insights into athletes' mental states and cognitive load during training, which can help in identifying potential risks for injuries related to mental and physical fatigue or lack of focus [42]. As research continues to evolve, the integration of EEG and fNIRS in sports applications holds the potential to revolutionize how athletes train, compete, and recover, ultimately leading to enhanced performance and well-being.

13.2.6 Electrocardiography or Heart Rate Monitor Biofeedback

Electrocardiography (ECG) and heart rate variability (HRV) biofeedback represent critical advancements in the monitoring and management of cardiovascular health. The significance of HRV is underscored by its ability to reflect the balance between sympathetic and parasympathetic nervous system activities, which are pivotal in regulating various metabolic processes. The evolution of biofeedback techniques, particularly HRV biofeedback, has been recognized as a therapeutic intervention aimed at enhancing autonomic regulation and overall well-being. HRV biofeedback has emerged as a promising modality for improving psychophysiological outcomes, particularly in high-stress environments such as competitive sports [43]. Research indicates that HRV biofeedback can facilitate self-regulation of physiological responses, thereby enhancing performance and reducing anxiety among athletes [23]. This technique not only fosters awareness of physiological states but also empowers individuals to modify their autonomic responses through guided breathing and relaxation techniques. Recent advancements in technology have further enhanced the practicality of ECG and HRV monitoring through the development of innovative devices and materials. For instance, flexible, conductive fabric-backed microneedle electrodes have been introduced, providing a comfortable and effective means of capturing electrophysiological signals for continuous monitoring [44]. These developments are essential for long-term ECG monitoring, which is crucial for the early detection of cardiovascular disorders [45]. Furthermore, the advent of in-ear and wearable devices has facilitated unobtrusive monitoring of heart rhythms, enabling real-time data collection without the constraints of traditional ECG setups [46]. The implications of ECG and HRV biofeedback extend into the realm of emotional well-being, by providing real-time feedback on physiological states. Individuals can gain insights into their stress responses and learn techniques to enhance their emotional regulation. Overall, the integration of biofeedback also with virtual reality has shown promise in enhancing performance in sports by allowing athletes to visualize and regulate their physiological states during competition [47].

13.3 Infrared Thermography as a Potential Ecological Method for Assessing Psychophysiological States in Sport

Classical instruments employed for monitoring psychophysiological and emotional states through the thermal signal of autonomic activity usually require contact sensors or devices (such as thermistors and thermocouples), thus resulting somehow invasive and potentially biasing the estimation of the state of the subject. This is because the compliant participation of the subject in the measurement is required (please see Chap. 7 for further details on contact thermal sensors and IRT comparison).

In the last decade, IRT has been proposed as a potential method for non-invasive and ecological recording of autonomic nervous system activity [48]. IRT, in fact, allows the contactless and completely non-invasive recording of the skin temperature through the measurement of the thermal radiation of the body [49–51].

Bioheat exchange is controlled by the autonomic nervous system, which plays a fundamental role in the modulation of various physiological processes, such as heart rate, breathing, tissue metabolism, perspiration, respiration, and skin blood perfusion, particularly during exercise. Therefore, the autonomic nervous system can be considered an effective tool for observations of emotional responses and affective states [50]. The potentialities of IRT as a method for non-invasive and ecological recording of autonomic nervous system activity have been widely demonstrated, as IRT is now well-recognized by the scientific community of psychophysiology [49, 50, 52, 53].

Studies have demonstrated that IRT was able to characterize competing subdivisions of the autonomic nervous system [52, 54, 55]. Thermal imaging in the field of psychophysiology is usually performed on the facial area of individuals. This is because the face is the body part usually exposed to social communication and interaction with the external world. The reliability and validity of IRT in the detection of psychophysiological states were investigated by comparing data recorded simultaneously by thermal imaging and by golden standard methods, such as electrocardiography, piezoelectric thorax stripe for breathing monitoring, nasal thermistors, and galvanic skin response. As regards the latter, studies have demonstrated that IRT and galvanic skin response have a similar detection power [56, 57]. An almost exclusive feature of IRT is its non-invasiveness. This feature makes IRT a powerful tool for research in stress field. In a recent study, Engert and colleagues investigated the ability of IRT to detect conditions of stress [58]. Thermal imprints were compared to established stress markers (i.e., heart rate, heart rate variability, finger skin temperature, α-amylase, and cortisol) in healthy subjects participating in two standardized laboratory stress tests: the cold pressor test [59] and the trier social stress test [60]. Although the thermal imprints of different regions of the face and stress marker outcomes demonstrated poor correlations, the thermal responses correlated with stress-induced mood changes. On the contrary, the established stress markers did not correlate with stress-induced mood modifications. Overall, the authors concluded supporting the notion that IRT can be a useful tool for the estimation of sympathetic activity in the field of human stress and emotional research.

Moreover, Di Giacinto et al. [56] examined fear conditioning in posttraumatic stress disorder using both IRT and galvanic skin response. The authors examined fear processing in posttraumatic stress disorder patients with mild symptoms and in individuals who did not develop symptoms (both groups were victims of a bank robbery), through the study of fear-conditioned response. Based on the positive results of the study, it was concluded that the analysis of facial thermal response assessed by IRT during the conditioning paradigm is a promising psychometric method of investigation, even in the case of low levels of posttraumatic stress disorder symptom severity [56].

Moreover, IRT was also proposed as a potential tool to create an atlas of the thermal expression of emotional states by using a proper classification algorithm [61, 62]. This would be based on the characterization of the thermal signal in facial regions of autonomic valence (nose or nose tip, perioral or maxillary areas, periorbital, supraorbital areas, and forehead), thus monitoring the modulation of the autonomic activity.

Ioannou et al. [49] summarized 23 experimental papers that employed IRT for investigations of affective and emotional nature. Moreover, the authors provided a detailed portrait of the adopted experimental protocols and the thermal changes that occurred in selected regions of interest [49].

Interestingly, the majority of the studies have observed that, when the autonomic nervous system is activated in response to a specific emotion or affective stimulus, the temperature of the nose tip (indicated by the authors as the most sensible facial area) tended to decrease, attributing this effect to vasoconstriction mechanisms and to the restriction of the skin blood flow.

Similar to what happened to the nose area, a decrease in skin temperature has been observed also on the upper lip or maxillary area, reflecting sweat gland activation. Conversely, an increment in skin temperature has been observed on the forehead and in the region between the eyes and the nose. Overall, a specific direction of the skin temperature change has been found for specific emotion and stimuli. To clarify this point, we report here a modified table of that paper, which provides an overview of the direction of the skin temperature variation in the considered regions of interest in response to different emotions [49] (Table 13.2).

13.3.1 Methodological Aspects of IRT in Psychophysiology

As indicated by the review of Ioannou et al. [49], the most used region of the body for investigating thermal expression of emotional states is the face [52, 56–58]. Facial thermal prints provide a useful channel for studying psychophysiological states. To extract information of affective nature from a thermal image series, regions of interest (ROIs) are selected in specific regions of the face. The most common region used in psychophysiological studies is the nose or nose tip.

Table 13.2 Overview of the direction of the skin temperature variation in different regions of interest in response to different emotions

	Emotions							
Regions	Stress	Fear	Startle	Sexual arousal	Anxiety	Joy	Pain	Guilt
Nose	↓	↓		↑		↓		↓
Cheeks			↓					
Periorbital			↑	↑	↑			
Supraorbital			↑		↑			
Forehead	↓↑	↓		↑	↑		↓	
Maxillary	↓	↓	↓				↓	↓
Neck–carotid			↑					
Tail	↓							
Fingers/palm		↓					↓	
Lips/mouth				↑				

Adapted and modified from Ioannou et al. [49]

In fact, it has been shown that the nose tip temperature depends on the activation of the sympathetic nervous system in response to emotional stimuli and reflects sympathetic alpha-adrenergic vasomotor effects [55, 56, 62–64]. Furthermore, sympathetic stimulation of the blood vessels in the nose can also have smaller vasodilatory effects via the action of cholinergic and beta-adrenergic receptors [65]. Figure 13.1 shows an example of a thermal image with a male subject's face, and the region of interest positioned on the nose tip. As regards the choice of the nasal area, Di Giacinto and colleagues [56] explained why they decided to focus analysis on the nose tip in the caption of Fig. 13.2 of that paper. We report here this figure with relative caption modified (Fig. 13.2). In particular, they recorded thermal images pre- and post-acoustic stimulation. A general skin temperature decrease could be observed over the whole face. However, they stressed that the skin temperature decrease was particularly appreciable on the nose tip, where blue areas can be easily spotted. Therefore, temperature data were extracted from a region of interest on the nose tip.

However, apart from the nose, the literature reported also other regions that have been studied, such as the forehead, the inner cantus of the eye (peri-orbital area), the maxillary area and the lips, the cheeks, and carotid, as well as fingers. An overview of the studies in the literature using IRT to estimate autonomic nervous system activity is provided in Table 13.3.

Figure 13.3 shows a schematic representation of the most common ROIs used in psychophysiological state assessment, together with a vascular representation of the major subcutaneous vessels affecting the skin temperature of the face.

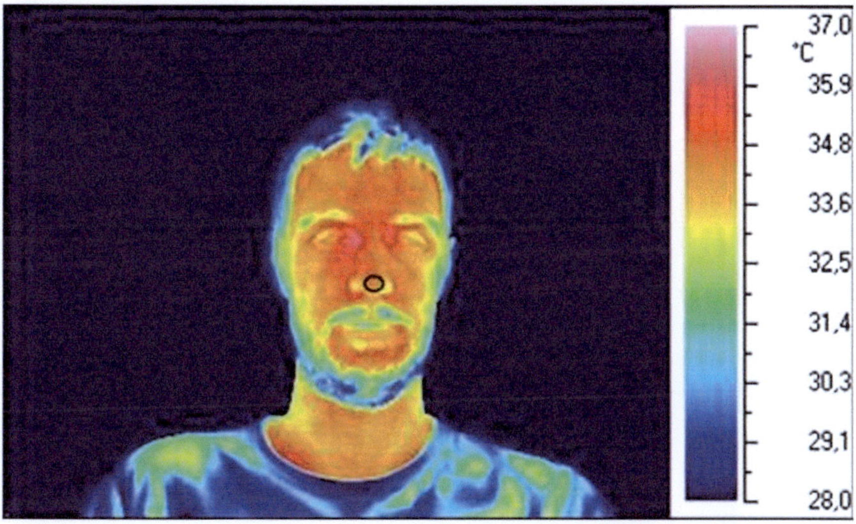

Fig. 13.1 Thermal image of a representative subject's face. The *black circle* indicates the nose tip region usually selected as a region of interest in studies assessing psychophysiological states

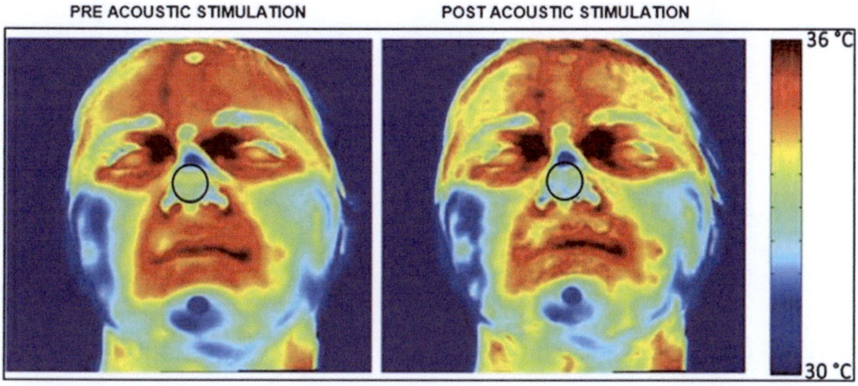

Fig. 13.2 Facial thermal images in a representative subject, before the acoustic stimulation (*left*) and soon after the acoustic stimulation (*right*). A general temperature decrease can be observed over the whole face. In particular, while the cheeks seem not to change their average temperature values and pattern, the nose tip, perioral, maxillary and forehead regions clearly present a temperature decrement due to the appearance of colder spots. The temperature decrease is particularly appreciable on the nose tip, where blue areas can be easily spotted. The *circles* over the face are paper markers put on to facilitate the tracking of the region of interest along the procedure. The black-contour circular region on the nose tip is the region of interest used to extract the temperature data. Figure adapted and modified from Giacinto et al. [56]

Table 13.3 Overview of the studies in psychophysiology using IRT to estimate autonomic nervous system activity

Author	Year	N	Emotion	Experimental paradigm	Baseline	Regions of interest
Gioia et al.	[66]	25	Stress	Stroop test	Rest	Various on the face
Calvin and Duffy	[67]	33	Stress	Driving/mental loading task	Rest	Forehead, nose
Pavlidis et al.	[68]	17	Stress	Laparoscopic drill training	Natural landscapes	Perinasal
Puri et al.	[69]	12	Stress	Stroop test	Rest	Supraorbital vessels
Kang et al.	[70]	9	Stress	Alphabet arithmetic task	Rest	Forehead, nose
Mizukami et al.	[71]	34 (pairs)	Stress	Mother–infant separation stress/stranger exposure	Held by mother	Forehead
Kistler et al.	[31]	20	Fear	Horror movie	Prestimulation	Fingers
Merla and Romani	[55]	10	Fear	Electric stimulation and trigger	Prestimulation	Face, palm
Perpetuini et al.	[72]	53	Startle	White noise	Rest	Nose
Shastri et al.	[73]	10	Startle	Glass breaking, phone ringing	Mental task-counting circles	Periorbital, supraorbital, maxillary
Naemura et al.	[74]	52	Startle	White noise (45–100 dB)	Comparison between groups	Nasal region
Pavlidis et al.	[48]	6	Startle	Loud noise (60 dB)	Rest, sit quietly in a dark room	Periorbital area, cheeks, neck area
Gane et al.	[75]	11	Startle	Loud noise (102 dB)	Image-matching task	Periorbital
Ebisch et al.	[76]	12 (dyads)	Empathy	Toy mishap	Playing with toys	Face: nose, maxillary
Manini et al.	[77]	18 (dyads)	Empathy	Toy mishap	Playing with toys	Face: nose, maxillary
Ioannou et al.	[52]	15	Guilt	Toy mishap	Playing with toys	Nose
Hahn et al.	[78]	16	Sexual arousal	Touch on high intimate regions	Neutral face presentation	Nose, lip, periorbital
Merla and Romani	[55]	10	Embarrassment	Presence of strangers while performing a mental task	Prestimulation	Maxillary, face, palm
Nakanishi and Matsumura	[64]	12	Laughter	Playing	Prestimulation/acclimatization	Nose, forehead, cheek

Pavlidis et al.	[79]	12	Anxiety	Mock interrogation	Prestimulation	Face
Tsiamyrtzis et al.	[80]	39	Anxiety	Mock interrogation	Prestimulation	Periorbital vessels
Zhu et al.	[81]	38	Anxiety	Mock interrogation	Prestimulation	Supraorbital vessels

The table indicates the baseline and the experimental approach followed to induce the emotion of interest. The last column indicates the region of interest in which thermal observations were made. Adapted and modified from Ioannou et al. [49]

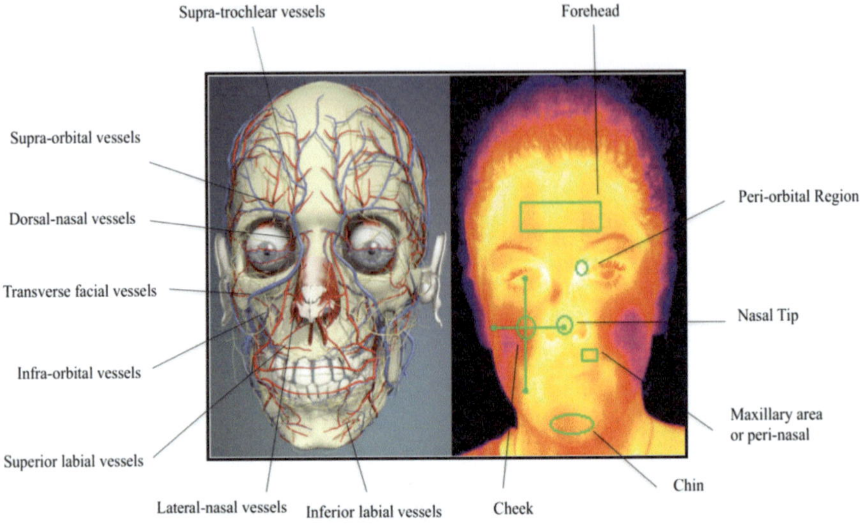

Fig. 13.3 Vascular representation of the major subcutaneous vessels affecting the skin temperature of the face (*left panel*). Thermal representation of the face with the most common ROIs (*right panel*). (Figure modified from Ioannou et al. [49])

13.3.2 Future Directions of Research in Sport Psychophysiology Using IRT

All in all, although the actual increasing use of IRT for investigating affective and emotional nature with various experimental paradigms, none of the studies summarized and presented in the review by Ioannou et al. [49] employed conditions involving sport and exercise activities (Table 13.3). In fact, to the best of our knowledge, studies that have applied IRT in psychophysiological state assessment in the sport and exercise scenario are scarce.

However, analysing and extrapolating the findings of the examined studies, in particular focusing on the study by Ioannou and colleagues [49], together with the methodologies employed, it is reason to hypothesize that IRT may be a useful tool in the field of emotional research in sport and exercise settings. Specifically, we suggest the use of IRT in those activities that cause null or limited activation of thermoregulatory processes, such as golf, curling, archery, and shooting, where the concentration and the psychological components are maximized, thus in turn maximizing the activation of the autonomic nervous system.

Furthermore, the application of IRT to study the autonomic nervous system activation in other kinds of sports activities is also possible. For example, game sports activities can be beneficial in the use of IRT for assessing psychophysiological states. This application may be restricted to particular phases of the game, such as penalty kicks in football and free throws in basketball, when thermoregulatory processes as well as movements of the subjects, especially the face, are negligible.

Nevertheless, cautions are needed in planning studies on these sports activities because of the activation of thermoregulatory processes, which may act as confounding factors of the thermal signal coming from autonomic nervous system activation, thus creating possible difficult interpretations of the thermal data. Moreover, difficulty in its application can be related to the experimental setting, given the fact that strategies should be adopted to position the thermal camera in a non-laboratory setting and a non-controlled environment.

Using IRT to monitor the psychophysiological condition of the athletes during the execution of physical exercise could facilitate the integration of intelligent systems into sports and exercise science with the goal of enhancing performance, improving training methodologies, and facilitating rehabilitation processes. The concept of human-machine interaction (HMI) encompasses a range of interactions between humans and machines, often facilitated by interfaces that allow for real-time communication and feedback. This interaction is particularly relevant in sports, where athletes rely on technology for performance analysis and feedback. Notably, recent advancements in wearable technology have transformed HMI in sports [82]. Such technologies can monitor athletes' movements in real time, providing valuable data for performance optimization, and the application of machine learning (ML) algorithms can allow for accurate assessments of athletes' movements, enabling tailored training programs that address individual needs. Moreover, HMI technologies also play a significant role in rehabilitation and injury prevention. In fact, the ability to monitor athletes' movements and thermal responses in real time allows for early detection of potential injuries and improvement of rehabilitation outcomes.

Concerning the employment of IRT for HMI solutions in sports science, one of the primary applications is the monitoring of muscular activity and fatigue. Perpetuini et al. discussed how data-driven ML approaches applied to thermal imaging data can estimate muscular activity and fatigue levels, demonstrating the technology's potential for real-time monitoring during exercise [83]. This capability is particularly beneficial for athletes, as it allows for immediate feedback on their physical state, enabling coaches and trainers to adjust training regimens. Furthermore, the ability to assess temperature during exercise can provide insights into the onset of fatigue, which is crucial for optimizing performance and preventing injuries. Perpetuini et al. further elaborated on the use of thermal imaging to analyse superficial vascular changes induced by physical activity, which can inform coaches about an athlete's recovery status and readiness for subsequent training sessions [84]. This application is particularly relevant in high-intensity sports, where understanding the body's thermal responses can help in designing effective training programs that can minimize the risk of overtraining and injury.

13.4 Benefits and Limitations of Infrared Thermography for Assessing Psychophysiological States in Sport

As for every tool, IRT has both benefits and limitations that should be considered before planning any kind of psychophysiological experiments in the sport and exercise field. In this paragraph, we provide a general overview of the actual literature about the most significant benefits and limitations characterizing the use of IRT in psychophysiology research, with particular interest in psychophysiology of sport.

Over the years, especially in the last decades, the sensitivity and resolution of the thermal camera have dramatically improved [85]. In fact, physiological events that would be invisible to the naked eye, such as the activation of perspiration pores, can now be detected and documented by use of thermal infrared imaging [86].

Infrared cameras allow wireless recordings of skin temperature, from a distance, without interfering with the subject's behaviour and the experimental setting despite the possible subject's movements [49, 77, 85, 87]. Likewise, also the development of software for analysing thermal images and extracting temperature values has increased the possibility of using IRT in sports and exercise studies. In fact, thanks to the solutions provided by tracking the regions of interest, it is possible to follow the body area under investigation avoiding any contact with the subjects, and without the use of any markers on the skin [88]. This great potential can contribute to the diffusion of IRT not only in psychophysiology research but also in sport and exercise research, where the movement of subjects makes difficult the acquisition of thermal images and their following analysis.

Another advantage lies in the fact that, with respect to galvanic skin response, IRT provides the same physiological recording efficiency reflecting accurately the autonomic nature of the psychological phenomenon during the experimental procedure without any direct contact. In fact, it has been noticed that galvanic skin response electrolytes can change conductivity over time [89], both spontaneously during baseline recordings [90] or during non-steady state phase reflecting stimuli [91].

Moreover, galvanic skin responses are so sensitive to emotional stimuli that reach maximum activity levels in conditions of different degrees of intensity, making them indistinguishable from each other, which is a problem that IRT overcomes [63]. This notion was supported also by Kreibig [92] who stated that studies investigating autonomic nervous system response to emotion have been impeded by the quasi-exclusive use of common physiological measures, such as heart rate and electrodermal activity. These variables have been used as sole indicators of the activation state of the organism [92].

On the other hand, it is necessary to highlight that, although the advantages offered by IRT, limitations are also present and should be known by researchers who are going to study emotional states using IRT.

First, it is worth noticing that the thermal signal of the skin in response to a general stimulus is quite slow, also with respect to the skin conductance. Therefore, the slowness of the thermal signal, which occurs as the result of the microvascular change, perspiration, or muscular activity, maybe a limitation of IRT. As regards, Kuraoka and Nakamura [63] in their paper stressed the fact that IRT has a longer latency with respect to skin conductance in response to a stimulus. Specifically, it has been shown that the earliest that temperature changes can be observed is within 10 s from the stimulus, whereas changes in skin conductance are evident in only 3 s from the stimulus [52, 63].

Most studies investigating emotional responses with IRT have focused on the skin temperature of the nose, which is considered a region of the face particularly sensible and reliable for identifying thermal signals of psychological states [63, 93]. However, despite the large use of the nose as a region of interest, controversies exist in the literature about the cause of the observed temperature change in the nasal area. In fact, the changes in the skin temperature of the nose have been attributed to the act of breathing, especially with heavy breathing, and not only to the activation of the autonomic nervous system [48].

Therefore, it is difficult to exclude the fact that nasal skin temperature changes are not due to airflow patterns or blood flow modifications associated with muscular activity of the facial muscles. Nevertheless, nose temperature changes have been clearly shown to respond to stimuli of an affective and emotional nature [64], thus providing evidence to support monitoring of the thermal signal of the nasal area as a reliable indication of autonomic nervous system activation.

Finally, it is worth noticing that IRT was quite expensive tool for monitoring skin temperature and microcirculation, thus making its prevalence in studies scarce. However, especially in the last decade, development in infrared instrumentation technology of both thermal cameras and related computer software, as well as the consequent lowering prices, have contributed to the diffusion of thermography in psychophysiology. In detail, one of the most notable advancements in thermal camera technology is the increase in spatial and temporal resolution. Modern thermal cameras offer high spatial resolutions of up to 1280 × 1024 pixels and temporal resolutions with full-frame frequency rates reaching up to 150 Hz [88]. This level of detail allows for more precise thermal mapping of the skin, enabling researchers and practitioners to detect subtle changes in the temperature pattern that may correlate with psychophysiological states such as stress, anxiety, and fatigue. The high thermal sensitivity of these cameras, which can be as low as 15 mK at 30 °C, further enhances their ability to capture variations in skin temperature, providing a more accurate representation of an athlete's physiological condition. Moreover, the computational requirements for processing thermal IR imaging data are comparable to those needed for visible-band imaging data, facilitating the integration of IRT into automated systems for remote monitoring of physiological activity [94]. In addition, ML approaches have been developed to define ROIs on the body together with tracking algorithms able to follow the different ROIs through the thermograms of thermal videos, further enhancing the potentiality of IRT in sports science [95, 96].

13.5 Computation of Psychophysiological Signals from Infrared Thermography

Conventional approaches for assessing psychophysiological states are based on the measurements of physiological variables reflecting the sympathetic activity, in particular the autonomic nervous system activity. As regards, most of the physiological variables measured in psychophysiological studies usually require the application of contact sensors, such as electrodes, probes, and belts. It is therefore evident that these sensors may have also limitations because of their nature to be in direct contact with the body of subjects. Moreover, in complex physiological studies, many variables may be monitored at the same time to study their relationship in response to a stimulus. In this sense, more variables are to be assessed, and more contact sensors are to be positioned on the body of subjects at the same time. This certainly has negative implications for the comfort of the subjects during the experimental protocol, from both physical and psychological points of view.

Recently, IRT has been proposed as a useful solution for the non-invasive and ecological recording of thermal signatures indicating activity of the autonomic nervous system. Since the face is naturally exposed to social communication and interaction, thermal imaging for psychophysiology is generally performed by recording images of the subject's face [49, 50, 77].

In order to exceed limitations related to the use of thermal contact sensors, computational psychophysiology based on the thermal imaging modelling approach has been proposed and introduced in the literature [88]. Therefore, IRT has found an application also in the computation of various physiological signals from its thermal signals.

Different psychophysiological signals can be identified and indirectly calculated through analysis of thermal signatures of the face in response to emotional and stress stimuli. Specifically, it has been demonstrated that the application of bioheat transfer models to thermal signals by IRT allows computing different physiological variables, such as heart rate, breathing rate, and cutaneous blood perfusion. In this section, we present briefly the possibility of computing heart rate, breathing rate, and cutaneous perfusion from the analysis of a series of thermal images. For more details on the models presented, we suggest to refer to the review by Cardone et al. [88].

One of the most used signals in psychophysiology is heart rate. Analysis of thermal signals provided by IRT allows the computation and the quantification of the heart rate at a distance in a completely non-invasive way thanks to the modelling of the pulsatile propagation of skin blood flow [54, 97, 98]. In fact, the heart contraction during the ventricular systole induces a pressure wave, which propagates through the arterial tissue. Arterial pulse reflects the heart activity thus providing a measure of cardiac interbeat intervals, heart rate, and its variability [99]. The method proposed is based on the information on the thermal signal of major superficial vessels (e.g., carotid, supraorbital, and temporal), recorded through a highly sensitive thermographic system (Fig. 13.4).

Breathing consists of continuous inspiration and expiration cycles during which heat is exchanged with the external environment by airflow through the nostrils. These exchanges create a quasiperiodic thermal signal in the proximity of the nostrils that varies between high (expiration phase) and low (inspiration phase) skin temperature values. In fact, it has been shown that, instead of the classical thermal contact sensors, this phenomenon can be captured and monitored by IRT at a distance, with an accuracy of 96.43% [100]. The estimation of breathing rate through thermal imaging was found to be very accurate, as proved by studies that have

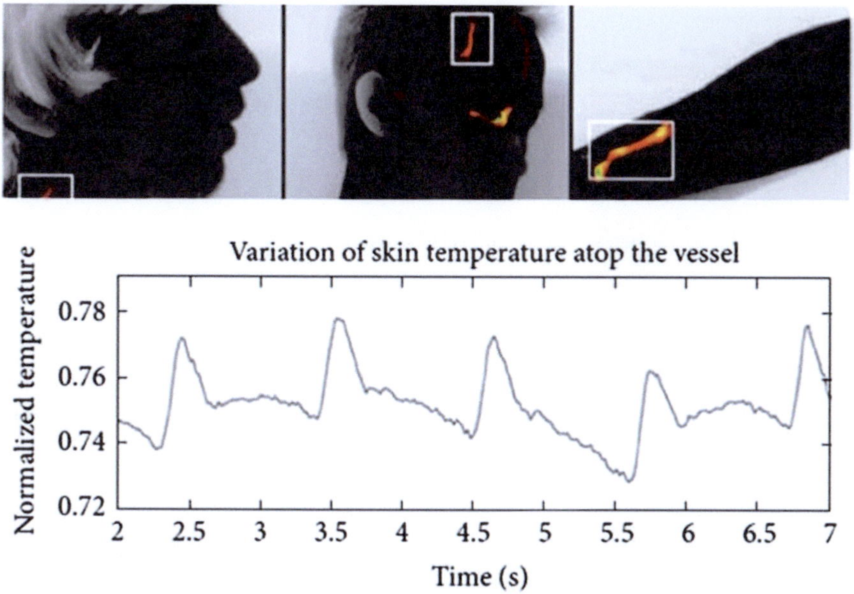

Fig. 13.4 Heart rate computation from thermal images data. Recording point on the carotid arteriovenous complex, the frontotemporal region, and the wrist (*upper panel*). Skin temperature dynamic after removing frequency signals lower than 0.67 Hz (40 beats min^{-1}) and higher than 1.67 Hz (100 beats min^{-1}). (Figure adapted and modified from Garbey et al. [54])

compared with respiratory signals recorded by conventional sensors [101, 102]. An example of the computation of breathing rate from thermal signal is shown in Fig. 13.5.

It has been demonstrated that also cutaneous perfusion can be computed from a high-frequency series of thermal images recorded on a specific region of the body [103]. This is because cutaneous temperature modification within a short time is mainly due to the heat gain/loss by convection attributable to the blood flow of cutaneous blood vessels and the heat conducted by subcutaneous tissue. The model showed that cutaneous blood flow depends mostly on the time derivative of the cutaneous temperature and on the difference between the temperatures of the cutaneous layers and the inner tissues [103]. The method proposed by Merla and colleagues has been validated by comparison with laser Doppler imagery [103]. This study produced evidence that it is possible to transform a series of raw thermal images into a series of cutaneous blood flow images (Fig. 13.6).

The above-mentioned studies support the value of IRT for estimating and monitoring physiological variables such as heart rate, breathing rate, and cutaneous blood perfusion. Reliability and validity were proven by comparing data simultaneously recorded by IRT and by golden standard methods, such as piezoelectric pulse meter for heart rate monitoring, piezoelectric thorax stripe for breathing monitoring, nasal thermistors, and laser Doppler imagery. Therefore, we feel to support the use of IRT as a powerful and ecological tool for computational physiology.

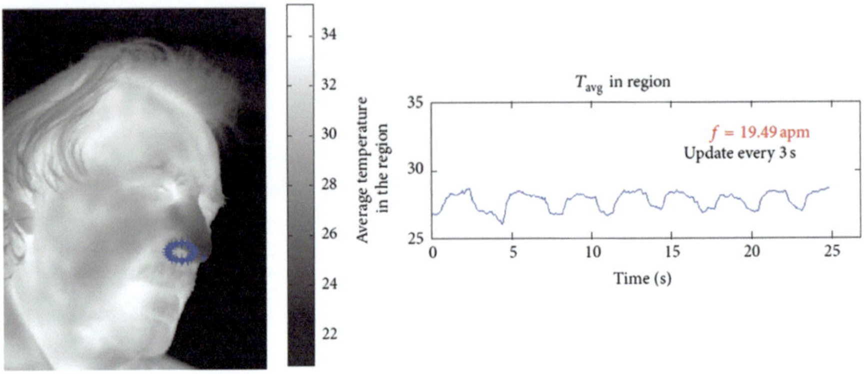

Fig. 13.5 Thermal image of a subject with the track of the airflow on a region of interest on the nostril (*left panel*). Dynamic average temperature of the region of interest close to the nostril. (Figure adapted and modified from Cardone et al. [88])

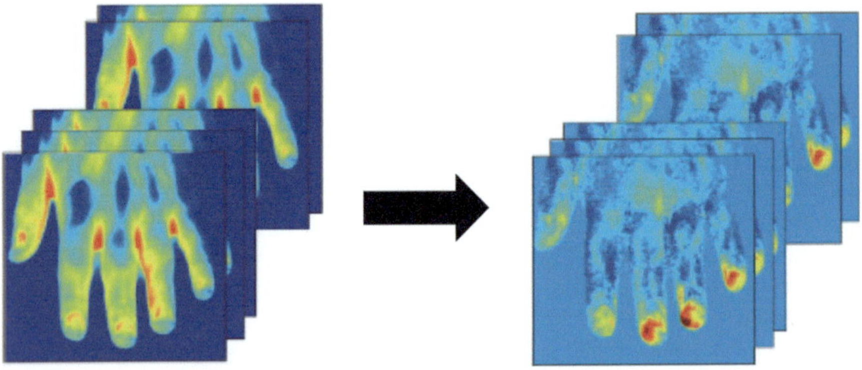

Fig. 13.6 From thermal image series to cutaneous blood flow images. Thermal images are converted into a series of thermal images—cutaneous blood flow images by applying computational models for bioheat exchange. (Figure adapted and modified from Merla et al. [103])

However, researchers interested in estimating physiological variables from a series of thermal images should be cautious in generalizing these findings to the sport and exercise field. In fact, the studies that have assessed the value of IRT for the estimation of other physiological variables did not include exercise conditions. Exercise creates a non-steady state condition during which phenomena of various nature happen in the human body.

Therefore, in addition to the above paragraph on future directions of research, we would like to stimulate researchers in planning and conducting investigations aiming to compare data simultaneously recorded by IRT and by golden standard methods for assessing physiological variables under non-steady state exercise conditions. For example, we suggest the study of the comparison of data recorded by IRT and golden standard methods during well-controlled laboratory human performance.

Incremental maximal exercise and submaximal steady-state exercise at different intensities on a cycle ergometer and treadmill can be used as exercise stimuli triggering non-steady-state conditions.

Moreover, it has been demonstrated the possibility of employing ML algorithms to estimate physiological signals from thermal imaging data. For instance, Perpetuini and colleagues demonstrated the feasibility of estimating EMG parameters, such as the average rectified value and the median frequency, from IRT [83], whereas Di Credico et al. estimated HRV parameters from facial thermal videos [104]. These findings highlighted the importance of employing contactless methods to deliver physiological metrics in a non-invasive and contactless manner in sports and clinical applications.

The findings of these future studies can contribute to fill a gap in the literature, which may extend the use of IRT in sport and exercise psychophysiology, not only for measuring skin temperature, estimating skin blood flow, and estimating the autonomic nervous system activity but also for measuring other physiological variables, such as heart rate and breathing rate, in a completely non-invasive way.

13.6 Conclusions

In this chapter, we presented a general overview of the literature about the potentialities of IRT in the assessment of psychophysiological states in the sport and exercise scenario.

Despite the lack of studies in the literature, IRT may provide alternative ways for addressing questions regarding autonomic nervous system function in response to exercise and emotions that have so far been unexplored. In particular, considering the present literature about the application of IRT in psychophysiology, it is reasonable to hypothesize that IRT may become a useful tool in the field of emotional research in sports and exercise settings.

In light of the above-mentioned evidence, we feel that IRT is a novel wireless, ecological methodology for quantifying emotional responses in the sport and exercise field. In particular, we recommend its use in those sports activities characterized by low to null thermoregulatory responses. Specifically, we feel to support the use of IRT in those activities that cause null or limited activation of thermoregulatory mechanisms, such as golf, curling, archery, and shooting.

Moreover, IRT application to other kinds of sports activities should also be considered. This is particularly promising in game sports activities, especially in specific phases of the game where the movements of the subjects and the thermoregulatory processes are negligible and limited (such as penalty kicks in football and free throws in basketball).

All in all, the maturity and the feasibility achieved in recent years by IRT in the general field of psychophysiology suggest its use even in the more specific field of sport psychophysiology. However, researchers and practitioners should be aware of the existence of limitations for using IRT, especially in a real-world sporting scenario. Cautions must be adopted by researchers and practitioners in interpreting data to avoid the attribution of any psychological valence to merely thermoregulatory or acclimatization processes.

References

1. Hanin YL (2000) Emotions in sport. Human Kinetics, Champaign
2. Zaichkowsky L (2012) Psychophysiology and neuroscience in sport: introduction to the special issue. J Clin Sport Psychol 6:1
3. Sternbach RA (1966) Principles of psychophysiology: an introductory text and readings. Academic Press, New York
4. Andreassi JL (2000) Psychophysiology: human behavior & physiological response. Psychology Press, Mahwah
5. Green E, Green A, Walters ED (1999) Voluntary control of internal states: psychological and physiological. Subtle Energ Energy Med 10:71–88
6. Neiss R (1988) Reconceptualizing arousal: psychobiological states in motor performance. Psychol Bull 103:345–366
7. Schwaberger G (1987) Heart rate, metabolic and hormonal responses to maximal psycho-emotional and physical stress in motor car racing drivers. Int Arch Occup Environ Health 59:579–604
8. Robazza C, Bortoli L, Nougier V (1999) Emotions, heart rate and performance in archery. A case study. J Sports Med Phys Fitness 39:169–176
9. Hynynen E, Konttinen N, Rusko H (2009) Heart rate variability and stress hormones in novice and experienced parachutists anticipating a jump. Aviat Space Environ Med 80:976–980
10. Falk B, Bar-Eli M (1995) The psycho-physiological response to parachuting among novice and experienced parachutists. Aviat Space Environ Med 66:114–117
11. Blumenstein B, Bar-Eli M, Tenenbaum G (2002) Brain and body in sport and exercise: biofeedback applications in performance enhancement. Wiley, Chichester
12. Schwartz MS, Andrasik F (2016) Biofeedback, a practitioner's guide, 4th edn. Guilford Publications, New York
13. Birbaumer N, Kimmel HD (1979) Biofeedback and self-regulation. Wiley, Hillsdale
14. Sandweiss JH, Wolf SL (2013) Biofeedback and sports science. Springer Science & Business Media, New York
15. Donald M, Werthner P (2015) Special issue: evidence-based applications of biofeedback and neurofeedback in sport. Biofeedback 43:51–53
16. Petruzzello SJ, Landers DM, Salazar W (1991) Biofeedback and sport/exercise performance: applications and limitations. Behav Ther 22:379–392
17. Pusenjak N, Grad A, Tusak M et al (2015) Can biofeedback training of psychophysiological responses enhance athletes' sport performance? A practitioner's perspective. Phys Sportsmed 43:287–299. https://doi.org/10.1080/00913847.2015.1069169
18. Zaichkowsky LD, Fuchs CZ (1988) Biofeedback applications in exercise and athletic performance. Exerc Sport Sci Rev 16:381–421
19. Vickers JN, Williams AM (2007) Performing under pressure: the effects of physiological arousal, cognitive anxiety, and gaze control in biathlon. J Mot Behav 39:381–394. https://doi.org/10.3200/JMBR.39.5.381-394
20. Wallace HM, Baumeister RF, Vohs KD (2005) Audience support and choking under pressure: a home disadvantage? J Sports Sci 23:429–438. https://doi.org/10.1080/02640410400021666
21. Williams J (2009) Applied sport psychology: personal growth to peak performance. McGraw-Hill Education, New York

22. Park JL, Fairweather MM, Donaldson DI (2015) Making the case for mobile cognition: EEG and sports performance. Neurosci Biobehav Rev 52:117–130. https://doi.org/10.1016/j.neubiorev.2015.02.014
23. Paul M, Garg K (2012) The effect of heart rate variability biofeedback on performance psychology of basketball players. Appl Psychophysiol Biofeedback 37:131–144. https://doi.org/10.1007/s10484-012-9185-2
24. Rainoldi A, Moritani T, Boccia G (2016) EMG in exercise physiology and sports. In: Merletti R, Farina D (eds) Surface electromyography physiology, engineering, and applications. Wiley, New York, pp 501–539
25. Giggins OM, Persson UM, Caulfield B (2013) Biofeedback in rehabilitation. J Neuroeng Rehabil 10:60. https://doi.org/10.1186/1743-0003-10-60
26. Blumenstein B, Bar-Eli M, Tenenbaum G (1995) The augmenting role of biofeedback: effects of autogenic, imagery and music training on physiological indices and athletic performance. J Sports Sci 13:343–354. https://doi.org/10.1080/02640419508732248
27. Carlson NR, Birkett MA (2016) Physiology of behavior. Pearson Education, Upper Saddle River
28. Schlager O, Gschwandtner ME, Herberg K et al (2010) Correlation of infrared thermography and skin perfusion in Raynaud patients and in healthy controls. Microvasc Res 80:54–57. https://doi.org/10.1016/j.mvr.2010.01.010
29. Swain ID, Grant LJ (1989) Methods of measuring skin blood flow. Phys Med Biol 34:151–175
30. Kappes BM, Chapman SJ (1984) The effects of indoor versus outdoor thermal biofeedback training in cold-weather sports. J Sport Psychol 6:305–311
31. Kistler A, Mariauzouls C, von Berlepsch K (1998) Fingertip temperature as an indicator for sympathetic responses. Int J Psychophysiol 29:35–41
32. Keefe FJ, Surwit RS, Pilon RN (1980) Biofeedback, autogenic training, and progressive relaxation in the treatment of Raynaud's disease: a comparative study. J Appl Behav Anal 13:3–11. https://doi.org/10.1901/jaba.1980.13-3
33. Khazan IZ (2013) The clinical handbook of biofeedback: a step-by-step guide for training and practice with mindfulness. Wiley, New York
34. Donald MW, Hovmand J (1981) Autoregulation of skin temperature with feedback-assisted relaxation of the target limb, and controlled variation in local air temperature. Percept Mot Skills 53:799–809. https://doi.org/10.2466/pms.1981.53.3.799
35. Prończuk M, Trybek G, Terbalyan A, Markowski J, Pilch J, Krzysztofik M, Kostrzewa M, Mostowik A, Maszczyk A (2023) The effects of EEG biofeedback training on visual reaction time in judo athletes. J Hum Kinet 89:247–258. https://doi.org/10.5114/jhk/174272
36. Kolayis H (2012) Using EEG biofeedback in karate: the relationship among anxiety, motivation and brain waves. School Phys Educ Sport 8(1):13–18
37. Rydzik Ł, Wąsacz W, Ambroży T, Javdaneh N, Brydak K, Kopańska M (2023) The use of neurofeedback in sports training. Syst Rev Brain Sci 13(4):Article 4. https://doi.org/10.3390/brainsci13040660
38. Yücel MA, Selb JJ, Huppert TJ, Franceschini MA, Boas DA (2017) Functional near infrared spectroscopy: enabling routine functional brain imaging. Curr Opin Biomed Eng 4:78–86. https://doi.org/10.1016/j.cobme.2017.09.011
39. Pinti P, Tachtsidis I, Hamilton A, Hirsch J, Aichelburg C, Gilbert S, Burgess PW (2020) The present and future use of functional near-infrared spectroscopy (fNIRS) for cognitive neuroscience. Ann N Y Acad Sci 1464(1):5–29. https://doi.org/10.1111/nyas.13948
40. Tan SJ, Kerr G, Sullivan JP, Peake JM (2019) A brief review of the application of Neuroergonomics in skilled cognition during expert sports performance. Front Hum Neurosci 13. https://doi.org/10.3389/fnhum.2019.00278
41. Herold F, Wiegel P, Scholkmann F, Müller NG (2018) Applications of Functional Near-Infrared Spectroscopy (fNIRS) neuroimaging in exercise–cognition science: a systematic, methodology-focused review. J Clin Med 7(12):Article 12. https://doi.org/10.3390/jcm7120466

42. Graczyk M, Pąchalska M, Ziółkowski A, Mańko G, Łukaszewska B, Kochanowicz K, Mirski A, Kropotov ID (2014) Neurofeedback training for peak performance. Ann Agric Environ Med 21(4):871–875. https://doi.org/10.5604/12321966.1129950
43. Shaffer F, Crawford J, Moss D (2013) Bcia launches a heart rate variability biofeedback certificate of completion. Biofeedback 41(1):4–6
44. Singh OP, Bocchino A, Guillerm T, Hu Y, Stam F, O'Mahony C (2024) Flexible, conductive fabric-backed, microneedle electrodes for electrophysiological monitoring. Adv Mater Technol 9(3):2301606. https://doi.org/10.1002/admt.202301606
45. Nigusse AB, Malengier B, Mengistie DA, Tseghai GB, Van Langenhove L (2020) Development of washable silver printed textile electrodes for long-term ECG monitoring. Sensors 20(21):Article 21. https://doi.org/10.3390/s20216233
46. van der Heijden P, Gilbert C, Jafari S, Lucchini MA (2024) Multi-channel soft dry electrodes for electrocardiography acquisition in the Ear Region. Sensors 24(2):Article 2. https://doi.org/10.3390/s24020420
47. Lagos L, Vaschillo E, Vaschillo B, Lehrer P, Bates M, Pandina R (2010) Virtual reality–assisted heart rate variability biofeedback as a strategy to improve golf performance: a case study. Biofeedback 39(1):15–20. https://doi.org/10.5298/1081-5937-39.1.11
48. Pavlidis I, Levine J, Baukol P (2001) Thermal image analysis for anxiety detection. In: Proceedings of 2001 international conference on image processing. IEEE, pp 315–318
49. Ioannou S, Gallese V, Merla A (2014) Thermal infrared imaging in psychophysiology: potentialities and limits. Psychophysiology 51:951–963. https://doi.org/10.1111/psyp.12243
50. Merla A (2014) Thermal expression of intersubjectivity offers new possibilities to human-machine and technologically mediated interactions. Front Psychol 5:802. https://doi.org/10.3389/fpsyg.2014.00802
51. Wilson AN, Gupta KA, Koduru BH, Kumar A, Jha A, Cenkeramaddi LR (2023) Recent advances in thermal imaging and its applications using machine learning: a review. IEEE Sens J 23(4):3395–3407. https://doi.org/10.1109/JSEN.2023.3234335
52. Ioannou S, Ebisch S, Aureli T et al (2013) The autonomic signature of guilt in children: a thermal infrared imaging study. PLoS One 8:e79440. https://doi.org/10.1371/journal.pone.0079440
53. Tavares IM, Vardasca R, Cera N, Pereira R, Nimbi FM, Lisy D, Janssen E, Nobre PJ (2018) A review of infrared thermography as applied to human sexual psychophysiology. Int J Psychophysiol 133:28–40. https://doi.org/10.1016/j.ijpsycho.2018.09.001
54. Garbey M, Sun N, Merla A, Pavlidis I (2007) Contact-free measurement of cardiac pulse based on the analysis of thermal imagery. IEEE Trans Biomed Eng 54:1418–1426
55. Merla A, Romani GL (2007) Thermal signatures of emotional arousal: a functional infrared imaging study. In: 2007 29th annual international conference of the IEEE engineering in medicine and biology society. IEEE, pp 247–249
56. Di Giacinto A, Brunetti M, Sepede G et al (2014) Thermal signature of fear conditioning in mild post traumatic stress disorder. Neuroscience 266:216–223. https://doi.org/10.1016/j.neuroscience.2014.02.009
57. Shastri D, Merla A, Tsiamyrtzis P, Pavlidis I (2009) Imaging facial signs of neurophysiological responses. IEEE Trans Biomed Eng 56:477–484
58. Engert V, Merla A, Grant JA et al (2014) Exploring the use of thermal infrared imaging in human stress research. PLoS One 9:e90782. https://doi.org/10.1371/journal.pone.0090782
59. Hines EA, Brown GE (1932) A standard stimulus for measuring vasomotor reactions: its application in the study of hypertension. In: Proceedings of Staff Meet Mayo Clinic, p 332
60. Kirschbaum C, Pirke KM, Hellhammer DH (1993) The "trier social stress test"—a tool for investigating psychobiological stress responses in a laboratory setting. Neuropsychobiology 28:76–81
61. Khan MM, Ward RD, Ingleby M (2009) Classifying pretended and evoked facial expressions of positive and negative affective states using infrared measurement of skin temperature. ACM Trans Appl Percept 6:6

62. Nhan BR, Chau T (2010) Classifying affective states using thermal infrared imaging of the human face. IEEE Trans Biomed Eng 57:979–987. https://doi.org/10.1109/TBME.2009.2035926
63. Kuraoka K, Nakamura K (2011) The use of nasal skin temperature measurements in studying emotion in macaque monkeys. Physiol Behav 102:347–355. https://doi.org/10.1016/j.physbeh.2010.11.029
64. Nakanishi R, Imai-Matsumura K (2008) Facial skin temperature decreases in infants with joyful expression. Infant Behav Dev 31:137–144. https://doi.org/10.1016/j.infbeh.2007.09.001
65. Johnson J, Minson CT, Kellogg DL (2011) Cutaneous vasodilator and vasoconstrictor mechanisms in temperature regulation. Compr Physiol. https://doi.org/10.1002/cphy.c130015
66. Gioia F, Greco A, Callara AL, Scilingo EP (2022) Towards a contactless stress classification using thermal imaging. Sensors 22(3):Article 3. https://doi.org/10.3390/s22030976
67. Or CK, Duffy VG (2007) Development of a facial skin temperature-based methodology for non-intrusive mental workload measurement. Occup Ergon 7(2):83–94
68. Pavlidis I, Tsiamyrtzis P, Shastri D, Wesley A, Zhou Y, Lindner P et al (2012) Fast by nature how stress patterns define human experience and performance in dexterous tasks. Sci Rep 2(1):305
69. Puri C, Olson L, Pavlidis I, Levine J, Starren J (2005) StressCam: non-contact measurement of users' emotional states through thermal imaging. In: CHI'05 extended abstracts on human factors in computing systems, pp 1725–1728
70. Kang J, McGinley JA, McFadyen G, Babski-Reeves K (2006) Determining learning level and effective training times using thermography. In: Proceedings of army science conference, Orlando, Florida, USA
71. Mizukami K, Kobayashi N, Ishii T, Iwata H (1990) First selective attachment begins in early infancy: a study using telethermography. Infant Behav Dev 13(3):257–271
72. Perpetuini D, Cardone D, Filippini C, Chiarelli AM, Merla A (2019) Modelling impulse response function of functional infrared imaging for general linear model analysis of autonomic activity. Sensors 19(4):Article 4. https://doi.org/10.3390/s19040849
73. Shastri D, Papadakis M, Tsiamyrtzis P, Bass B, Pavlidis I (2012) Perinasal imaging of physiological stress and its affective potential. IEEE Trans Affect Comput 3(3):366–378
74. Naemura A, Tsuda K, Suzuki N (1993) Effects of loud noise on nasal skin temperature. Shinrigaku kenkyu Jpn J Psychol 64(1):51–54
75. Gane L, Power S, Kushki A, Chau T (2011) Thermal imaging of the periorbital regions during the presentation of an auditory startle stimulus. PLoS One 6(11):e27268
76. Ebisch SJ, Aureli T, Bafunno D, Cardone D, Romani GL, Merla A (2012) Mother and child in synchrony: thermal facial imprints of autonomic contagion. Biol Psychol 89(1):123–129
77. Manini B, Cardone D, Ebisch SJH et al (2013) Mom feels what her child feels: thermal signatures of vicarious autonomic response while watching children in a stressful situation. Front Hum Neurosci 7:299. https://doi.org/10.3389/fnhum.2013.00299
78. Hahn AC, Whitehead RD, Albrecht M, Lefevre CE, Perrett DI (2012) Hot or not? Thermal reactions to social contact. Biol Lett 8(5):864–867
79. Pavlidis I, Eberhardt NL, Levine JA (2002) Seeing through the face of deception. Nature 415(6867):35–35
80. Tsiamyrtzis P, Dowdall J, Shastri D, Pavlidis IT, Frank MG, Ekman P (2007) Imaging facial physiology for the detection of deceit. Int J Comput Vision 71:197–214
81. Zhu Z, Tsiamyrtzis P, Pavlidis I (2007) Forehead thermal signature extraction in lie detection. In: 2007 29th annual international conference of the IEEE engineering in medicine and biology society. IEEE, pp 243–246
82. Chen F, Song X, Fu J, Liang J, Zhou J, Cai J, Zhang Y, Zhu M, Ding Y, Jiang J, Chen Z, Qi Y, Zhou Z, Huang Q, Zhang Y, Zheng Z (2024) An all-fibrous, permeable, adhesive, and stretchable self-powered electronic skin for sign language recognition. J Mater Chem A 12(44):30298–30308. https://doi.org/10.1039/D4TA05825A

83. Perpetuini D, Formenti D, Cardone D, Trecroci A, Rossi A, Di Credico A, Merati G, Alberti G, Di Baldassarre A, Merla A (2023) Can data-driven supervised machine learning approaches applied to infrared thermal imaging data estimate muscular activity and fatigue? Sensors 23(2):Article 2. https://doi.org/10.3390/s23020832
84. Perpetuini D, Formenti D, Cardone D, Filippini C, Merla A (2021) Regions of interest selection and thermal imaging data analysis in sports and exercise science: a narrative review. Physiol Meas 42(8):08TR01. https://doi.org/10.1088/1361-6579/ac0fbd
85. Vianna DML, Carrive P (2005) Changes in cutaneous and body temperature during and after conditioned fear to context in the rat. Eur J Neurosci 21:2505–2512. https://doi.org/10.1111/j.1460-9568.2005.04073.x
86. Vainer BG (2005) FPA-based infrared thermography as applied to the study of cutaneous perspiration and stimulated vascular response in humans. Phys Med Biol 50:R63. https://doi.org/10.1088/0031-9155/50/23/R01
87. Anbar M (2002) Assessment of physiologic and pathologic radiative heat dissipation using dynamic infrared imaging. Ann N Y Acad Sci 972:111–118
88. Cardone D, Pinti P, Merla A (2015) Thermal infrared imaging-based computational psychophysiology for psychometrics. Comput Math Methods Med 2015:984353. https://doi.org/10.1155/2015/984353
89. Wagner HL (1988) Social psychophysiology and emotion: theory and clinical applications. John Wiley & Sons Australia Limited, Milton
90. Laine CM, Spitler KM, Mosher CP, Gothard KM (2009) Behavioral triggers of skin conductance responses and their neural correlates in the primate amygdala. J Neurophysiol 101:1749–1754. https://doi.org/10.1152/jn.91110.2008
91. Amiez C, Procyk E, Honoré J et al (2003) Reward anticipation, cognition, and electrodermal activity in the conditioned monkey. Exp Brain Res 149:267–275. https://doi.org/10.1007/s00221-002-1353-9
92. Kreibig SD (2010) Autonomic nervous system activity in emotion: a review. Biol Psychol 84:394–421. https://doi.org/10.1016/j.biopsycho.2010.03.010
93. Nakayama K, Goto S, Kuraoka K, Nakamura K (2005) Decrease in nasal temperature of rhesus monkeys (Macaca mulatta) in negative emotional state. Physiol Behav 84:783–790. https://doi.org/10.1016/j.physbeh.2005.03.009
94. Cardone D, Merla A (2017) New frontiers for applications of thermal infrared imaging devices: computational psychophysiology in the neurosciences. Sensors 17(5):Article 5. https://doi.org/10.3390/s17051042
95. Cardone D, Spadolini E, Perpetuini D, Filippini C, Chiarelli AM, Merla A (2021) Automated warping procedure for facial thermal imaging based on features identification in the visible domain. Infrared Phys Technol 112:103595. https://doi.org/10.1016/j.infrared.2020.103595
96. Requena-Bueno L, Priego-Quesada JI, Jimenez-Perez I, Gil-Calvo M, Perez-Soriano P (2020) Validation of ThermoHuman automatic thermographic software for assessing foot temperature before and after running. J Therm Biol:102639. https://doi.org/10.1016/j.jtherbio.2020.102639
97. Chekmenev SY, Farag AA, Essock EA (2007) Thermal imaging of the superficial temporal artery: an arterial pulse recovery model. In: 2007 IEEE conference on computer vision and pattern recognition. IEEE, pp 1–6
98. Sun N, Pavlidis I, Garbey M, Fei J (2006) Harvesting the thermal cardiac pulse signal. In: International conference on medical image computing and computer-assisted intervention. Springer, Berlin, pp 569–576
99. Schmid C, Soatto S, Tomasi C (2006) 2006 IEEE computer society conference on computer vision and pattern recognition workshops: CVPR workshops. IEEE, Piscataway
100. Murthy R, Pavlidis I (2006) Noncontact measurement of breathing function. IEEE Eng Med Biol Mag Q Mag Eng Med Biol Soc 25:57–67

101. Lewis GF, Gatto RG, Porges SW (2011) A novel method for extracting respiration rate and relative tidal volume from infrared thermography. Psychophysiology 48:877–887. https://doi.org/10.1111/j.1469-8986.2010.01167.x
102. Murthy JN, van Jaarsveld J, Fei J et al (2009) Thermal infrared imaging: a novel method to monitor airflow during polysomnography. Sleep 32:1521–1527
103. Merla A, Di Donato L, Romani GL et al (2008) Comparison of thermal infrared and laser doppler imaging in the assessment of cutaneous tissue perfusion in scleroderma patients and healthy controls. Int J Immunopathol Pharmacol 21:679–686
104. Di Credico A, Perpetuini D, Izzicupo P, Gaggi G, Cardone D, Filippini C, Merla A, Ghinassi B, Di Baldassarre A (2022) Estimation of heart rate variability parameters by machine learning approaches applied to facial infrared thermal imaging. Front Cardiovasc Med 9. https://doi.org/10.3389/fcvm.2022.893374

Chapter 14
Foot Temperature Assessment

Marina Gil-Calvo, Irene Jimenez-Perez, Pedro Pérez-Soriano, and Jose Ignacio Priego-Quesada

Abstract Thermographic studies of the foot can be very useful in three different ways: preventing injury, analyzing sporting technique, and assessing the effects of footwear and clothing. This chapter aims to discuss several methodological points concerning the thermal analysis of the foot using infrared thermography, as well as to discuss how it can be applied to the science of sports, both in areas already researched and those as yet uninvestigated.

14.1 Introduction: Foot Anatomy and Physiology

To understand the possibilities that the analysis of foot temperature through infrared thermography presents to us, we first of all have to know the characteristics of that region of the body. The foot presents a very particular anatomy and physiology for several reasons. It has its own vascularization and thermoregulation, as it is one of the most peripheral regions of the body. Moreover, the feet support all the weight of

M. Gil-Calvo (✉)
Faculty of Physical Activity and Sports Sciences, Universidad de León, León, Spain

Research Group in Sport Biomechanics (GIBD), Department of Physical Education and Sports, University of Valencia, Valencia, Spain
e-mail: magic@unileon.es

I. Jimenez-Perez · J. I. Priego-Quesada
Research Group in Sport Biomechanics (GIBD), Department of Physical Education and Sports, University of Valencia, Valencia, Spain

Biophysics and Medical Physics Group, Department of Physiology, University of Valencia, Valencia, Spain
e-mail: j.ignacio.priego@uv.es

P. Pérez-Soriano
Research Group in Sport Biomechanics (GIBD), Department of Physical Education and Sports, University of Valencia, Valencia, Spain
e-mail: pedro.perez-soriano@uv.es

J. I. Priego-Quesada (ed.), *Application of Infrared Thermography in Sports Science*,
https://doi.org/10.1007/978-3-031-93311-0_14

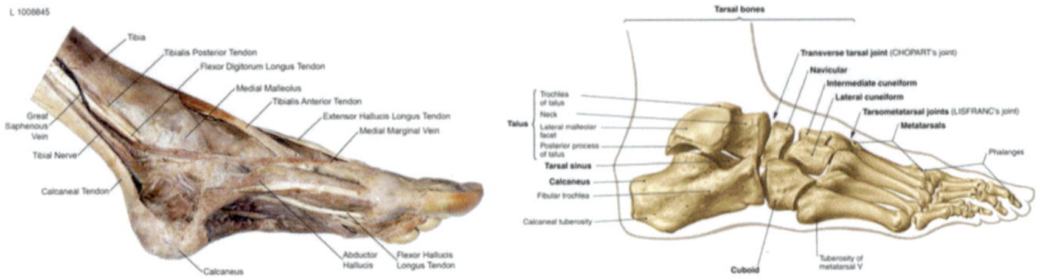

Fig. 14.1 Muscles and tendons on the medial side of the foot (left) and lateral view of the foot skeleton (right). (Figure obtained from [144])

an individual and are subject to considerable forces during physical exercise. We shall, therefore, now introduce the anatomic and thermophysiological structure of the foot (Fig. 14.1).

The foot acts as a structural platform that can bear all the different loads and movements of the human body [1]. It consists of 28 bones (including the sesamoids), the movements of which are all interrelated [1]. Bone tissue behaves as a good heat sink due to its porous structure and represents between 28% and 31% of the foot mass [2]. Furthermore, the bones of the foot are laid out in the form of longitudinal and cross-sectional arcs that provide greater support than other layouts would [3] (Fig. 14.1).

The structure of the foot is similar to that of the hand, but there are certain differences that allow it to support weight and adapt to different surfaces and speeds of locomotion. For that reason, the big toe of the foot is less mobile but has a greater structural solidity [3].

There are two groups of muscles related to foot movement. On the one hand, there are the extrinsic muscles of the foot, which are located in the leg and exercise their action by pulling on the tendons that are inserted into the ankle and foot [3]. These are responsible for making the movements of dorsiflexion, plantar flexion, inversion, and eversion of the foot [3]. These movements, especially the eversion of the ankle, have been widely studied through biomechanics applied to sport, as they are the ones most associated with the risk of injury [3, 4]. On the other hand, there are the intrinsic muscles of the foot. These muscles are responsible for the fine movements of the toes such as flexion, extension, adduction, and abduction [3].

When the muscles are active, they can release great amounts of energy. In the foot, the extrinsic muscles, located in the leg, are responsible for bigger movements, so there is no great heat source within the foot and, therefore, the thermal energy generated comes from other sources [2]. This fact, together with the higher presence of the heat dispersal structures that the foot possesses (such as the sweat glands or bone tissue), actually facilitates heat loss in that region, so increasing susceptibility to injuries caused by cold [5].

Blood flow is rarely stable in the feet as the thermal changes affect many cutaneous blood vessels [6–8]. Blood supply to the foot comes from arteries that enter behind the malleolus and along the dorsal surface of the foot [9]. Medial plantar artery enters through the medial malleolus, across the sole, and provides blood to the intrinsic muscles of the foot, to the skin of the medial plantar region and the medial three toes [9]. The external plantar artery feeds blood to the calcaneus and adjacent muscles, to the plantar tarsus and to the tarsometatarsal joints and the lateral zone of the sole of the foot [9]. In the dorsal zone are the medial and lateral tarsal arteries, the arcuate artery, and the first dorsal metatarsal artery, which takes the blood to the first and second toes [9]. For their part, the second, third, and fourth dorsal metatarsal arteries each supply blood to their own toe and the next one [9]. In the foot, moreover, the draining of the blood takes place from the toes toward the dorsal venous arch (nearer the surface) and the plantar venous arch (deeper) [9].

Venous return takes place through the tibial vein and the foot fascia, which mix with the blood from the dorsal surface before leaving the saphenous veins [2].

The normal state of the veins on the soles of the feet is that of vasoconstriction, and consequently, the accumulation of venous blood is minimum, a passive dilation of the veins being produced in hot situations [10, 11]. In contrast, there are mechanisms of active vasodilation in the dorsal region that are activated in hot situations [12], such as, for example, during exposure to heat or undertaking physical exercise, where vasodilation, blood flow, and perspiration increase, as they are essential for dispersing heat [13, 14]. If, on the contrary, the human body cools down, vasoconstriction is activated, most of the blood flow goes to the central areas of the body, and reduces the arterial flow to the limbs [2]. In mountain sports where temperatures can be very low and, therefore, great vasoconstriction takes place, cases of foot amputation as a result of freezing are well known [15].

As commented on earlier, perspiration is one of the most important heat-dissipating mechanisms in hot surroundings [14]. The sole of the foot has around 467 cm^2 of sweat glands, whereas the dorsal surface of the foot has only 119 cm^2, the sole, therefore, has an 80% concentration as opposed to 20% on the top [2]. Moreover, feet and hands lose more water than the rest of the body (2–4 times more) through the loss of transepidermal water. The dorsal surface of the foot loses *0.05 $mg\ cm^{-2}\ min^{-1}$, whereas the sole loses 0.10 $mg\ cm^{-2}\ min^{-1}$ [16].

The foot plays a fundamental role in human thermoregulation as a thermal radiator whenever there is more than sufficient heat transported by the blood to these regions [17]. Thus, knowledge of foot temperature can help us to understand the effects that footwear and exercise have, both on the risk and prevention of injuries, and on sporting performance. In this sense, infrared thermography is now considered a valid and reliable technique when it comes to discovering the surface temperature of the skin, provided that the factors that may affect its measurement are controlled and an appropriate methodology to prevent errors of measurement is followed.

14.2 Methodological Aspects Related to Foot Temperature Assessment

As foot temperature is very unstable due to factors that will be explained later on, it is essential to be able to control a series of methodological aspects linked to the assessment of their temperature. That methodological control helps to minimize errors in measuring foot temperature, thus increasing the validity and reliability of infrared thermography.

14.2.1 Advantages and Limitations of Infrared Thermography in Assessing Foot Temperature

Measuring foot surface temperature by means of infrared thermography brings with it a series of advantages and limitations. As mentioned earlier, infrared thermography is a quick, accurate, noninvasive, and objective method for measuring foot surface temperature [18, 19]. One of the main advantages of infrared thermography is that the measurements are taken at a distance, in such a way that it allows us to see the distribution of temperature in a noninvasive manner and at no risk to the patients [20, 21]. These characteristics make thermography a highly recommended tool when it comes to assessing the temperature of any region of the body, including the soles of the feet.

Furthermore, by measuring the surface temperature of the soles of the feet, a great amount of information can be obtained on the physical, circulatory, and thermoregulatory state of the individual being assessed. However, it is important to bear in mind that both skin temperature and limb temperature are very dependent on underlying circulation and the metabolic tissue rate [22]. This may involve interindividual differences or differences caused by the influence of environmental conditions [23–25].

Similarly, the use of thermography enables us to define the location of the area where an inflammation and/or an injury has taken place [26, 149]. The ability to detect injuries or inflammations through infrared thermography is because inflammatory response is characterized by an increase in the permeability of the blood vessels, which leads to an increased blood flow, thus altering the body's heat pattern [26].

Despite that, the dynamic nature of foot temperature regulation and the variability of its reaction to the very same stimulus, with regard to the initial temperature, have to be borne in mind [21, 27]. Here, there are multiple factors that make the temperature of the feet vary, such as environmental conditions, the moment of measuring and the physiological conditions of the region that is to be measured, so it is very complicated to establish a significant relationship between the foot and clinical diagnosis or sport intervention [28].

The main advantages as far as other injury-detecting instruments are concerned, such as magnetic resonances or scanners, are that infrared thermography does not emit radiation, it works from a distance, it is portable, and is low-cost [20, 29, 30]. Nevertheless, as far as foot temperature analysis goes, it has one clear limitation and that it is not possible to monitor foot temperature during the foot's daily activity. Being an imaging technique, it is necessary to undertake measurements in laboratory conditions and in most cases, undressed and outside the foot's normal habitat, the shoe. This limitation is very evident in the sports setting where, when undertaking physical exercise, athletes usually have their feet covered by socks and footwear, which makes infrared thermography measurements on the soles of the feet, in a dynamic way, impossible.

Regarding sports sciences, thermographic analysis of the foot can offer numerous applications of great interest, as will be explained in this chapter. However, its use in this field is quite recent, so further studies are required that may help to clarify its potential. In the clinical field, the main limitation of thermography is the insufficiency of databases with which to diagnose any type of injury, as, to date, we can only detect the production or recovery from an inflammation and/or injury, but it is still not possible to diagnose it accurately.

However, there has been great progress in recent years in the detection of diabetic foot through machine learning [137, 148], which can be extrapolated in the future to other injuries and pathologies. Nevertheless, currently, therefore, the best way of using thermography is in combination with other techniques, instead of as a replacement of them [29, 31]. By doing so, it can reveal physiological changes that appear before clinical signs appear and enable early diagnosis and intervention [26, 30, 32].

Moreover, on some occasions, it can provide relevant information on other physiological variables that conventional diagnosis techniques (such as magnetic resonance, scanners, etc.) do not take into account [30].

14.2.2 Reproducibility

One of the determining aspects in measuring the temperature of the soles of the feet is the low reproducibility of measurements in that region. Skin temperature is more stable in the trunk than in the limbs, where there is high variability that seems to depend on endogenous factors, such as the sympathetic nervous system or the environmental temperature [33].

Furthermore, in the most distal regions of the human body, such as the feet, peripheral circulation is weaker and there is a constant struggle between the mechanisms of vasoconstriction and vasodilation that regulate the blood flow of the skin, so increasing the variability of skin surface temperature and reducing the reproducibility of thermographic measurements in that region [33]. It is for that reason, more important than ever, that we must be able of controlling any extrinsic factor (such as preventing alcohol, tobacco, and medicinal intake or avoiding physical exercise before measuring, etc. as stated in Chap. 4 of this book) that may affect the infrared thermography measurements of the feet, so as to reduce the variability of measurements in these regions [33].

Regarding sports studies, where physical exercise is usually undertaken and temperature measured before and after, it would be interesting if future studies could compare the reproducibility of the variation in temperature (difference between pre- and postexercise) with absolute temperatures. It is possible that, although the reproducibility of the foot is low [33], the effect of exercise may be more constant.

Another factor that could influence the reproducibility of the temperature measurements of the soles of the feet is the thermographic camera used and its technical characteristics. In this sense, the authors of this chapter demonstrated in a previous study that when small ROIs (i.e., hallux or toes) are analyzed, it is more challenging to perform a detailed analysis of them [152]. When a camera with low resolution is used, the variations in the results are greater due to the difficulty in selecting the specific ROI to be analyzed [152].

In this context, when the aim of the thermal analysis is to use small regions of interest, it is recommended to use cameras with the highest possible thermographic resolution ($\geq$320 × 240 pixels), low spatial resolution (IFOV; $\leq$1.35 mrad), and good thermal sensitivity (NETD; $\leq$0.05 °C). In addition, these recommendations could be extrapolated to the rest of the ROIs delimited in the soles of the feet due to their small size compared to other regions of the body, such as the trunk or thighs. Finally, excellent inter-rater and intra-rater reliability was observed when the criteria of the ROIs' definition were well established [156].

14.2.3 Thermal Imaging of the Feet

There seems to be a consensus about the position for taking thermographic images of the soles of the feet. The images are taken following a period of thermal equilibrium (or adaptation to room temperature), where the participant, who is subject to thermal imaging, is positioned sitting down with their legs extended (Fig. 14.2).

When the image is to be taken, it is recommendable to place a matt black antireflection panel behind the feet to minimize the effects of the reflected temperature of the surroundings, and to isolate the image of the foot sole and not capture other regions of the body such as the legs (Fig. 14.3a) [19, 34]. Nevertheless, it is important to bear in mind that the feet (together with the hands) are regions of the human body that present the highest values of radiant heat loss. Hence, these regions at rest (mainly the fingers and toes) usually present very similar temperatures to the surrounding room temperature where measuring takes place, and it is often complicated to visualize the fingers or toes as they are of such similar temperature to the surrounding temperature (Fig. 14.3b).

The problems in determining ROIs are particularly relevant when using automatic software for analysis. Such tools help to reduce the time taken to analyze ROIs by up to 86%, but it has been shown that the accuracy in detecting regions is

Fig. 14.2 Recommended position for thermal imaging of the soles of the feet

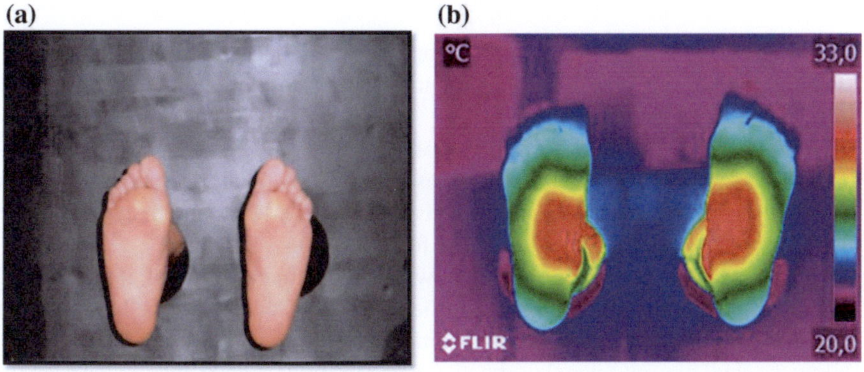

Fig. 14.3 (**a**) Antireflection panel behind the feet of a participant undergoing thermal imaging of the soles of the feet. (**b**) Example of a thermal image of toes that present temperatures similar to the surrounding room temperature. (**c**) Examples of thermal images that presented some error in the delimitations of the ROIs performed by ThermoHuman® [155]

challenging in some images [155]. These problems are especially relevant in the feet analysis because of the difficulty in identifying the coldest regions (hallux and toes), and they especially appear when the software confuses the foot with the leg (Fig. 14.3c) [155]. Therefore, it is advisable to use a panel that completely cancels the leg in the image acquisition to avoid reducing the error in the delimitation of the regions of the soles of the feet or by a higher training of the machine learning models that these automatic software use [155].

Most investigations have not reached an agreement on the time required for participants to remain in this position while adapting to the room. On the one hand, those studies where other regions of interest are measured in which the fact that the individual is seated may interfere, for example the posterior femoral zone, undertake a thermal adaptation period of between 15 and 20 min in anatomic position and once the thermal images of the body are taken, they seat the participant for a further 5 min before taking the thermal images of the soles of the feet [30, 33]. By doing so, they believe that this is sufficient time for thermal equilibrium to take place in those participants.

Besides, in studies that only focus on the soles of the feet, the participants remain in the seated position with their legs extended during the entire period of thermal adaptation [34, 35]. This period must be at least 10 min so as to adapt well to room conditions [29, 36, 142], but it could be longer if the environmental conditions outside the room are extremely cold or heat, as, in this case, the body needs a longer period to stabilize the temperature [33, 36]. Moreover, after an intense running of 30 min, it seems that 10 min are not enough to recover baseline temperatures of the soles of the feet, so that a longer period is also required to stabilize the temperature of these regions [142].

14.2.4 Regions of Interest (ROIs)

Choosing the regions of interest (ROIs) for foot assessment depends on the aims of each investigation. It seems that there is no homogenous definition of ROIs for the foot that can serve for all studies.

Ammer [37] tried to establish a protocol for thermographic analysis of the different regions of the human body, which was called the *Glamorgan Protocol*. Within the 90 regions of interest in the entire body, it established two single regions on the feet: the dorsal region and the sole of the foot (Fig. 14.4).

However, many authors do not follow the recommendations of this protocol when it comes to analyzing the sole of the foot, as, by doing so, one loses a large part of the information on what is happening in the different regions of the foot sole. Below we shall present several of the divisions that authors have undertaken both on the diabetic foot and in the medicine and physiology of sport, in an attempt to seek common ground between them.

When studying diabetic foot ulcerations, there are two main ways of dividing the foot sole. On the one hand, some authors [20, 38, 39] opt for dividing the sole of the foot in an anatomic way, which varies depending on the aims of each study. The hallux (big toe), the lesser toes, the metatarsus (which can be separate or together), and the heel typically appear separated (Fig. 14.5). In some cases, moreover, divisions in the arch and the lateral zone of the midfoot appear [39, 40].

On the other hand, some studies [41–45, 156] propose analyzing regions of the feet of diabetic subjects through the concept of angiosome, a region of the foot

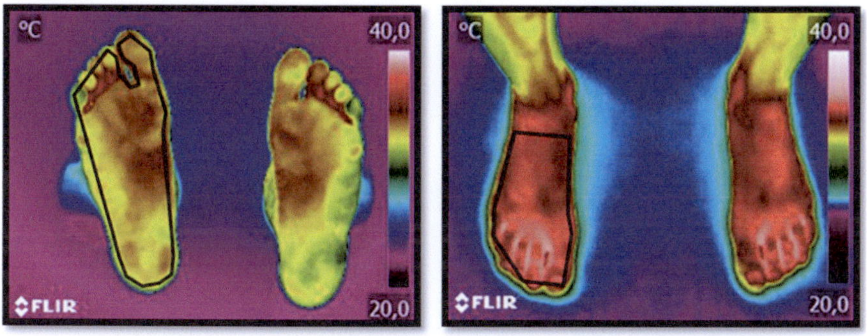

Fig. 14.4 Regions of interest on the feet according to the Glamorgan protocol [37]. The foot sole (*left*) and the dorsal region of the foot (*right*)

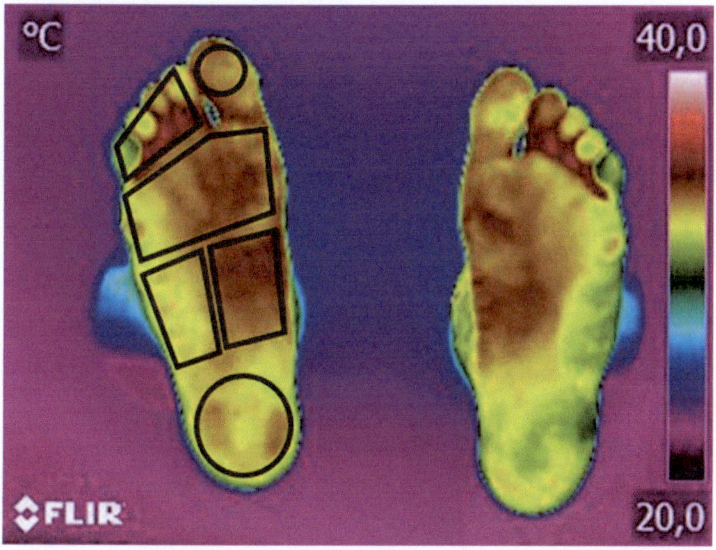

Fig. 14.5 Example of anatomical division of the diabetic foot according to Sun et al. [39]

composed of the tissues irrigated by the same artery of origin [43, 46]. These authors divide the foot sole into four zones, in accordance with the medial plantar angiosomes proposed by Attinger et al. [47]: the angiosome of the medial plantar artery (MPA), the angiosome of the lateral plantar artery (LPA), the angiosome of the medial calcaneal artery (MCA), and the angiosome of the lateral calcaneal artery (LCA) (Fig. 14.6).

In the case of medicine and physiology, the division of the regions of interest on the foot sole is more structural. Zaproudina et al. [33] undertook a division of the soles of the feet into nine regions of interest: the complete sole, three different regions of the arch and heel zone and each of the toes separately, so as to discover the reproducibility of this zone of the human body (Fig. 14.7).

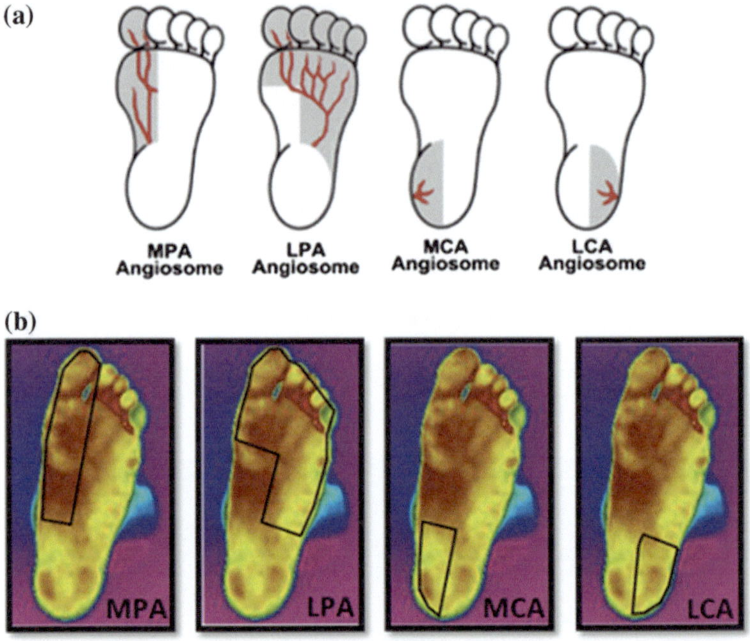

Fig. 14.6 Division of the thermographic image of the sole of the foot (**a**) according to the four angiosomes (**b**). Image A obtained from Nagase et al. [43]

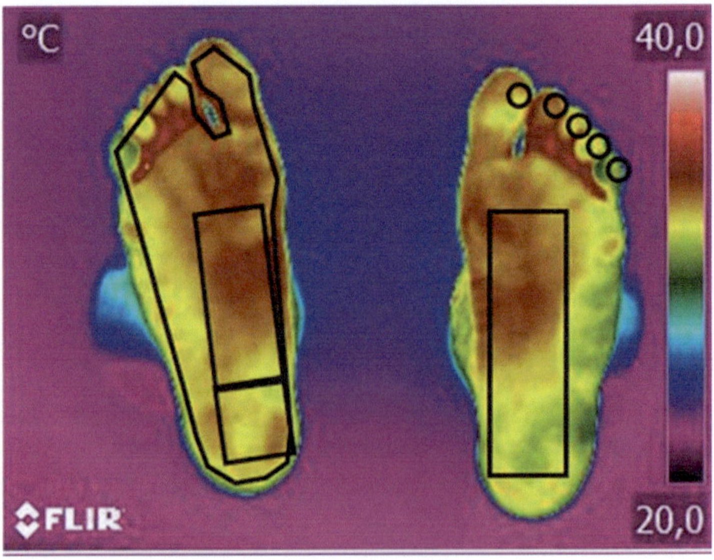

Fig. 14.7 Structural division of the foot sole according to Zaproudina et al. [33]

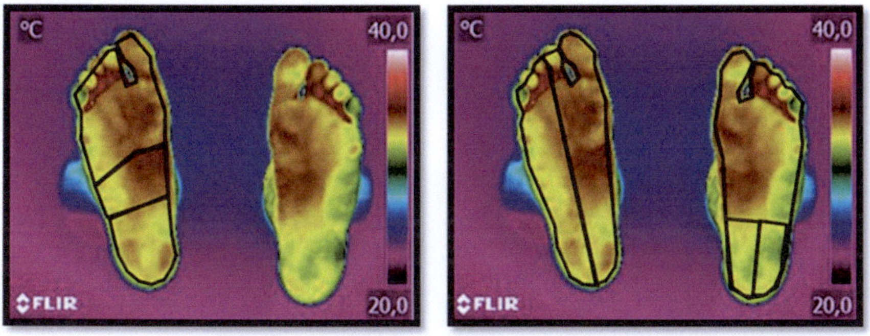

Fig. 14.8 Example of the division of the soles of the feet depending on the support provided while running according to Priego Quesada et al. [34] (*left*) and the five added regions proposed by the authors of this chapter (*right*)

In this study on the sport field [34], it was proposed that the division of the regions should follow a commonly used division for analyzing plantar pressure when running: the forefoot (50% of the surface of the foot sole), midfoot (19% of the surface of the foot sole), and rearfoot (31% remainder of surface) (Fig. 14.8). The authors of this chapter would like to add to this division a further five regions of interest for future studies: complete foot sole, medial foot sole (50% inner surface of the foot sole), lateral foot sole (50% outer surface of the foot sole), medial rearfoot (50% inner rearfoot), and lateral rearfoot (50% outer rearfoot), as shown in Fig. 14.8. This extra division will provide a deeper analysis of foot temperature with regard to exercise, as well as greater applicability in sport sciences, as will be explained in the following section.

Furthermore, if the idea of the work is to compare the results obtained on skin surface temperature of the soles of the feet with the plantar pressure data obtained using instrumented insoles or to compare these data with the automatic Thermohuman® software already implemented in the market, in that case, our recommendation is to subdivide the soles of the feet into the following ten regions of interest [146]: complete foot sole (CFS), hallux (H), toes (T), medial metatarsal (MM), central metatarsal (CM), lateral metatarsal (LM), medial midfoot (MF), lateral midfoot (LF), medial heel (MH), and lateral heel (LH) (Fig. 14.9).

14.3 Applications of Thermographic Assessment of the Foot in Sports Sciences

The use of infrared thermal imaging in sport allows us to undertake rapid assessments, both quantitatively and qualitatively. Following up the thermal response of the sports (man or woman) can provide relevant information on their health and on their performance, by contributing to the analysis of the sporting technique [48,

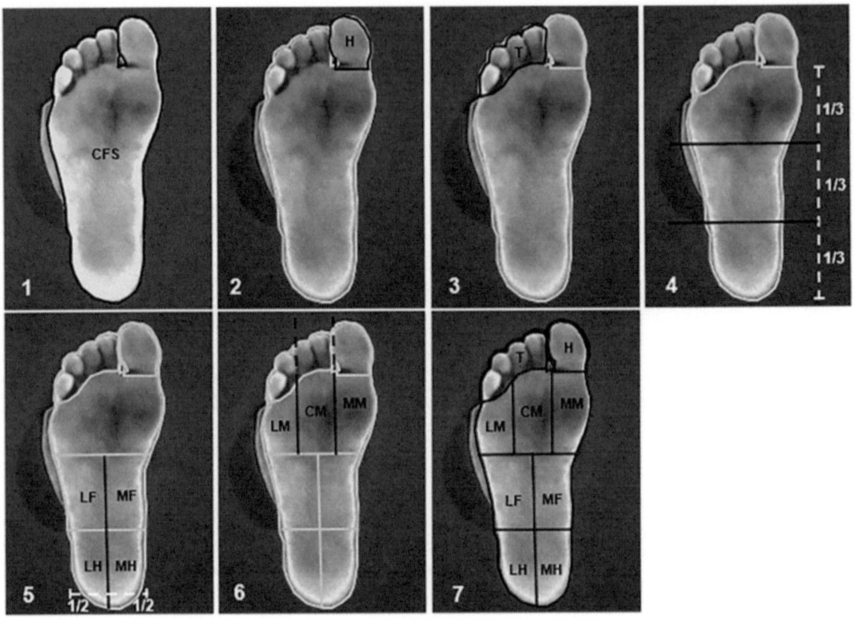

Fig. 14.9 Example of the division of the soles of the feet in ten regions according to Jimenez-Perez et al. [146]: (1) Foot sole contour; (2) Hallux delimitation; (3) delimitation of the rest of the toes; (4) determination of horizontal lines; (5) determination of the lower midline; (6) definition of vertical metatarsal lines; and (7) delimitation of nine ROIs

140]. Furthermore, knowing foot temperature may help to understand the effects of footwear and sports equipment, which are also related to the performance and health of athletes [49, 50, 161]. Hence, the following section focuses on the application of thermal imaging of the feet in terms of these three aspects: injury prevention, the analysis of sporting technique, and the assessment of footwear and sports equipment.

14.3.1 Injury Prevention

Physical and sporting activity exposes its practitioners to great physical stress, so often producing injuries related to that activity [19, 51, 52]. Injuries are one of the main problems of athletes [48], and it is not only necessary to make a correct and appropriate diagnosis and treatment of injuries, but also to work on prevention, which will allow the sportsman or woman to undertake their sport in greater comfort, well-being, and safety [53].

These injuries are accompanied by changes to skin temperature. In some cases, an increase in temperature (hyperthermia) occurs when there is inflammation or some other process that leads to an increase in blood circulation, such as infections, tendonitis, bursitis, fractured bones, etc. Whereas a fall in temperature (hypothermia) occurs as a consequence of degenerative processes, a lower vascularization, arterial or venous occlusion, nerve damage, alterations in the autonomous nervous system, etc. [54, 55].

As the human body is bilateral and almost totally symmetrical in terms of its limbs, it may be said that it possesses bilateral symmetry in the surface temperature between two contralateral body zones [56, 57]. The differences of temperature found, therefore, between two opposite zones, called asymmetries, can serve as an indicator of pathologies or dysfunctions [57, 58, 149].

Using this idea, the study of temperature asymmetries through infrared imaging has become a very valid method for use in the prevention, diagnosis, and follow-up of an injury [29, 55, 139, 143]. Moreover, this technique allows us to study the temperature of the body surface during and after movement and, therefore, detect the changes produced by exercise [19, 59]. Nevertheless, it must not be used as a diagnosis tool, but as a technique that provides additional information, complementing other tests [31, 48].

According to some studies [60, 61], asymmetries greater than 0.38 ± 0.31 °C in the dorsal region of the foot, and 0.35 ± 0.27 °C on the sole; and 0.40 ± 0.30 °C both in the dorsal and sole, respectively, can be considered abnormal and be associated with an anatomical or physiological dysfunction. However, a more recent study [57], with improvements both in applying the protocols and in the use of more modern and advanced thermographic cameras, found more accurate asymmetrical values: 0.32 ± 0.18 °C for the dorsal region of the foot and 0.28 ± 0.13 °C for the sole; smaller values than those previously accepted for any area of the body in general (0.5 ± 0.3 °C). Moreover, this research recommended avoiding the view of the entire body when undertaking thermal imaging of the feet for specific studies on the asymmetries of the foot because this can reduce the accuracy of the measurement.

One of the factors with which several different injuries are associated, especially in running, is the high plantar loads to which the foot is subject [62, 163]. It has been suggested that the increase in plantar pressures and friction in different regions of the foot may cause increments in foot skin temperatures [17, 63, 64, 153]. This supposition is based on the principle of energy conservation, where the mechanical energy produced by the repeated loads that the foot suffers in its interaction with the ground is transformed into thermal or calorific energy [17]. Hence, Shimazaki and Murata [17] found that the regions of the foot with the highest temperatures, the hallux and heel, coincided with the regions of the foot that have the highest contact loads when walking.

On the other hand, the studies of Yavuz and collaborators [64, 65] only reported a moderate linear relationship between the increase of triaxial plantar loads and the increase of foot temperature as a consequence of walking. They claimed that the changes in foot temperature are produced not only because of the plantar load but also blood circulation, the internal load, and the presence of kinetic friction in the interaction between the foot and the ground.

Another study [34] analyzed the relationship between static plantar pressure and the temperature of the soles of the feet, using thermography, following a 15 min run, with and without footwear. Although no correlations were found between the temperature and plantar pressure in any of the two conditions, they did find higher variations in temperature at the rearfoot in running without footwear, possibly as a consequence of a higher contact load, as the participants ran with a rearfoot striking pattern.

For their part, Taiar et al. [66] found higher temperature increases in the heel than in the rest of the foot when walking, whereas during running, the temperature of the sole was higher than the complete foot. Thus, they claimed that these results were due to the kinematic pattern differences between the two movements, walking versus running, and, therefore, the temperature rises are produced due to the specific movement undertaken.

The authors of this chapter conducted further research comparing plantar pressures with the surface temperature of the soles of the feet and the footwear outsole after running [147]. The results showed that plantar pressures were more strongly related to outsole footwear temperature (strong correlations) than the sole of the feet temperature (moderate correlations) in both the forefoot and rearfoot [147]. However, there was no correlation in the midfoot [147]. Thus, the analysis of footwear outsole temperature could be considered a viable alternative for understanding the behavior of plantar pressures during running [147].

To sum up, according to the above mentioned literature, although the evidence is not totally clear and still requires further investigation, it seems that the plantar temperature may be an alternative method for determining the loads or triaxial plantar shearing forces [17, 63, 64]; and, therefore, infrared thermography can provide a useful tool when it comes to assessing the mechanical stress to which the foot is subjected.

Associated with this idea, the authors of this chapter explored, using thermography, the ability of several insoles to reduce contact with ground loads. We analyzed the effect of two types of insoles (prefabricated and custom-made) on the surface temperature of the soles of the feet, after running, understanding that the use of insoles can reduce plantar pressures in several zones of the foot [67, 68]. Curiously enough, we did not find changes in plantar temperatures by using both custom-made or prefabricated foot orthoses in comparison to not wearing insoles (Fig. 14.10), possibly explained by the period (1 week) of adaptation to the insoles, where these changes may have taken place together with kinetic and kinematic adaptations (Gil-Calvo, [146]). Moreover, the multifactorial dependence of surface skin temperature and the influence of several factors such as running mechanics, mechanisms of heat dissipation, or an increase in blood flow could also neutralize the minor effects of foot orthoses on skin temperature of the soles of the feet [141].

These results were confirmed by a subsequent study in which no difference in sole temperature was found between a standard prefabricated insole and a prefabricated thermoformable insole (which adapts to the shape of the foot using a home-made microwave [146]. In this study, soles of the feet were divided into nine smaller ROIs, being the effects being similar between men and women, after 2 weeks of adaptation [146]. Furthermore, the increase in temperature of the soles of the feet after running was similar (~10 °C) in both studies [141, 146].

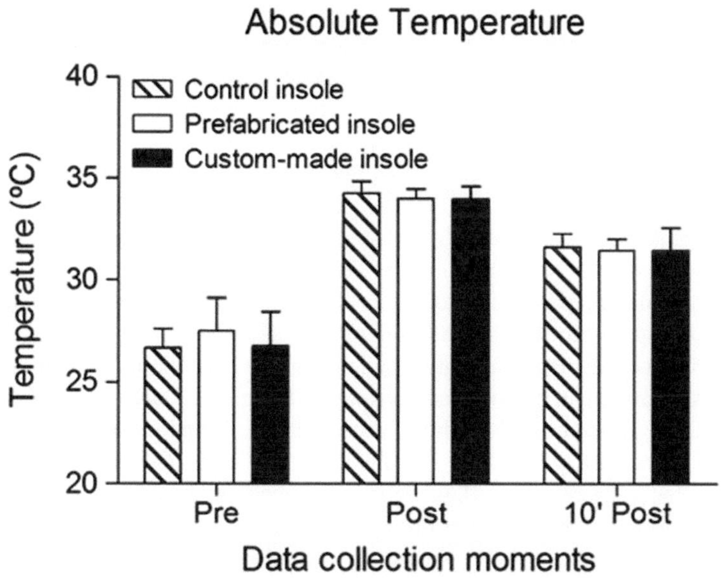

Fig. 14.10 Mean with 95% confidence interval of the surface temperature of the soles of the feet at the three moments of measuring (the exercise consisted of a 20 min run)

On the other hand, the study of Wrobel et al. [69] found lower increases in temperature in diabetic subjects, both on the forefoot and midfoot, after walking with an insole created to reduce shearing forces in comparison with a standard insole. In this study, therefore, thermography showed reductions of plantar temperature associated with lower plantar loads.

As the case of Wrobel and collaborators [69] shows, infrared imaging is a very popular tool in the diagnosis and prevention of the diabetic foot [20, 31, 70, 71]. As mentioned before, its usefulness in assessment is evident and, especially, in the prevention of injuries related to physical activity [29, 48, 55], although its application to other injuries of the foot and foot sole, other than that of the diabetic foot, has yet to be fully studied.

To date, injuries related to the feet in which thermography has mostly been used for exploration, for changes in the temperature of the skin, are as follows:

- Cutaneous Injuries

When undertaking physical activity, the body is subjected to certain repeated loads that predispose the sportsman or woman to developing an injury. Due to the fact that the skin is the body's first line of defense, it is frequently exposed to the development of dermatological lesions [72]. In the feet, in particular, the friction and repeated pressure of the skin with the footwear or sock during exercise, together with the humidity and ventilation that the foot experiences, often cause traumas to the skin, such as abrasions, chafing, calluses, blisters, black heel, or nail disorders [73–75].

Among these lesions, chafing and blisters on the feet and soles are frequent in almost all sports [73, 76], but they mostly appear in running and especially in long-distance runners [75, 77]. According to a Mailler and Adams' review of dermatological lesions in marathons [77], the most common lesions discovered on the feet were blisters (0.2–39%), followed by chafing (0.4–16%).

Although these lesions seldom hinder the athletes from continuing their activity, they can be very painful, so restricting the normal functioning of the foot and altering the biomechanics of the lower limbs, and, hence, reducing the sporting performance and increasing the risk of injury [73, 78].

During chafing, repeated friction leads to reddening and peeling of the skin [74], whereas during blistering, friction leads to the separation of the layers of the epidermis, causing local acute inflammation and the infiltration of liquid into the area [76]. This inflammation can promote an increase in the temperature of the skin, which along with the generation of a point of access, are the main indicators of the formation of blisters on the foot [72, 79]. Thus the studies of Hashmi et al. [72] and Kirkham et al. [80] investigated, using infrared imaging, the thermal changes in the skin of the foot due to friction in order to describe and quantify the physiological responses that take place in the skin of the foot before, during, and after the formation of a blister. They arrived at the conclusion that thermal imaging can be of great use in evaluating the scope of the traumatically damaged skin of the foot, both for blisters [72, 81] and other pathologies and injuries whose main characteristic is inflammation and consequently an increase in temperature in the damaged area [82].

Chafing, blisters, calluses, cuts, and all types of foot injuries are especially dangerous in people who suffer from diabetes mellitus, given that their complications can lead to the development of an ulcer, which, if not treated in time, can increase the risk of amputation [71, 82, 83].

Diabetes mellitus is a metabolic disease that is characterized by the presence of hyperglycemia caused by a lack of insulin [54]. It is one of the pandemics of the twenty first century and its prevalence is in continuous growth: according to the *International Diabetes Federation,* it is estimated that there are around 415 million people suffering from diabetes throughout the world and that this number will increase to more than 642 million in the year 2040 [84].

Among the various complications that this disease brings with it is what is called the diabetic foot, the main cause of which is a reduction in blood supply (vascular disorder) and loss of sensitivity (neuropathy) [70]. These factors, together with the repeated stress that the foot suffers in contact with the ground, can lead to the appearance of a diabetic ulcer [85, 86], very common for this disease (it arises in 15–25% of diabetic cases) [87]. Areas with probability of ulceration are associated with local increases in skin temperature due to inflammation and the enzymatic autolysis of the tissue [88, 89, 153]. Thus, checking the diabetic patient's foot temperature is very useful for the early identification of this lesion and would contribute to reducing its incidence [31, 43, 158].

Again, infrared imaging has become one of the methods with the greatest potential for assessing and diagnosing the diabetic foot [20, 70, 71, 90, 158]. Indeed, it is a very popular technique among researchers focusing on the foot and diabetic ulcers. According to several recent studies, it has been proposed as an accurate indicator of diabetes, even better than blood sugar measurements [70], and it seems to be a good substitute for measuring inflammation [82, 85]. Its already mentioned advantages as a noninvasive and noncontact technique, in comparison with other techniques, such as dermal thermometers or liquid crystal thermography [20, 31, 71], benefit the assessment of the diagnosis of the diabetic foot, as it allows us to obtain a full representation of temperature distribution in the plantar and dorsal regions [83, 157], including the medial arch, or of feet with deformities [31]; it creates no undesired pressures or the transmission of pathological organisms [43]; and its high resolution allows the detection of temperature differences, often without the need to apply image processing algorithms [31]. However, it must be stated that its use must always be as a complementary method to other techniques in medical assessment [29, 31].

The study of diabetic foot temperature by thermography has allowed us to relate certain temperatures with the risk of developing complications of the diabetic foot, such as neuropathic ulcer, Charcot arthropathy, and other complaints [83, 91]. Here,

Sun and collaborators [92] found that diabetic patients in a situation of risk have a significantly higher mean temperature of the foot (30.2 ± 1.3 °C) in comparison with healthy subjects (26.8 ± 1.8 °C). On the other hand, Bagavathiappan et al. [40] found that patients with diabetic neuropathy show a higher mean temperature of the foot (32–35 °C) than patients without neuropathy (27–30 °C).

Similarly, most investigations use asymmetry analyses with thermography by comparing the temperature of the damaged limb with the contralateral limb as control in order to determine the risk of diabetic complications [20, 71, 91]. In particular, van Netten et al. [93] found a difference in temperature of 2.2 °C between two contralateral points as the maximum cut-off value for the detection of these complications. Curiously, this value is the same as that found in previous clinical studies that used cutaneous thermometry [85, 88, 94, 95]. However, as we suggested, caution always must be taken as infrared thermography is not diagnostic and not all these asymmetries can be associated with ulcers [150].

Along these same lines, different authors [41–45] proposed asymmetry analysis on the feet of diabetics by analyzing the differences of temperature between the four contralateral angiosomes proposed by Attinger et al. [47], described earlier in Sect. 14.2.4 of this chapter.

Nevertheless, some authors [39, 92] are against diagnosis by asymmetries in systemic diseases that affect both limbs and propose standardizing the absolute temperature of the foot sole with respect to the temperature of the forehead. They take the temperature of the forehead as a standard index, given that they claim that this is positively correlated with common central temperatures, is devoid of derivations from thermoregulatory flow and of countercurrent mechanisms that could significantly alter skin temperature [28], and finally because this temperature varies little.

Finally, other studies use more complex methods based on automated detection through the analysis of infrared thermal images, using algorithms, to identify the risk of ulceration [71, 82, 83, 96, 148].

- Bone Injuries

The validity of infrared thermal imaging as a diagnostic tool for fractures due to stress has been conclusively shown as a result of the inflammatory process that accompanies them [54]. According to Goodman et al. [97], thermography can provide an accurate and noninvasive way of distinguishing a fracture from stress from other injuries with similar pain symptoms among runners. Besides its application in the follow-up on recovery from the injury, it allows the sportsman/woman to take up his or her activity again in the shortest and safest period of time.

However, its use has been little studied despite the fact that these injuries are produced quite frequently as a consequence of doing sport, especially in the metatarsals [98]. Thus, DiBenedetto [99], in 2002, used thermography to evaluate the risk of suffering a fracture by stress in the foot during basic military training.

Recently, the use of infrared thermography in the emergency room showed that fractures are associated with asymmetries higher than 0.9 °C [138]. Niehof and collaborators [100] reported that this technique can be useful for complementing the diagnosis of complex regional pain syndrome (chronic pain that often compromises a limb, possibly as a consequence of a dysfunction of the nervous system) in patients with fractures in the feet, taking thermographic recordings in the dorsal and plantar regions of the feet.

- Muscular Injuries

Muscular injuries, among which are muscle cramps, contractures, strains, and contusions, are the most frequent injuries among sportsmen/women after osteoarticular injuries [101]. Generally, they have an important inflammatory component that is easily detectable by a local thermal variation [102], so they can be assessed by means of measuring temperature through infrared thermography technique, as several studies have shown [102–105].

The detection of muscular pain from delayed onset muscle soreness (DOMS) has even been investigated with this technique as the Chap. 6 of this book shows. However, there is lack of studies of muscular injuries produced in the feet using thermography, when it may be of great interest in such common sporting injuries as plantar fasciitis [53].

One interesting study [30] assessed low back pain using thermal imaging on the soles of the feet. Curiously, this investigation found a correlation between temperature changes in the plantar surface of the feet and the intensity of low back pain, because the salient innervation of the low back region transports information to the soles of the feet by means of the sympathetic system.

- Joint Injuries

Foot sprains usually occur in the midfoot and the first metatarsophalangeal joint, but are not very common amongst sportsmen/women, whereas ankle sprains, closely related to the foot, are among the most common injuries in sport [106, 107]. Even so, there have been few scientific investigations into the application of infrared imaging in the study of any type of ankle joint injury [108, 151] and even less of the foot. According to current literature, Oliveira et al. [108] were the first to use this technique in the assessment of an ankle sprain to study its potential as a complementary tool for diagnosing degrees of this injury. Moreover, asymmetries of 1.2 ± 0.9 °C are recently observed in tendinopathies [151].

In Fig. 14.11, the image on the right shows an injured foot from judo, with a sprain of the anterior talofibular ligament, and the image on the right the contralateral healthy foot, showing the regions involved in this injury according to Oliveira et al. [108]. There is a difference of 1.4 °C in temperature in the affected region, the greatest temperature value being found at the uninjured extremity. This is due to the fact that the image of the injury was taken more than 48 h after the injury took place, so, at that moment, the injury was already in the recovery process. Hence, the inflammatory component, due to the immediate incidence of the injury, had already disappeared, but one can observe a lack of blood flow to the distal zone of the injured foot, associated with a fall in the overall temperature of that extremity, which suggests a hypothermia of the right foot, due to an already known and diagnosed pathology.

Another investigation [109] applied thermography in the assessment of a sprain, this time to the knee, to explore the possible effects of a type of manual therapy in the treatment of a complete rupture of the anterior crossed ligament. Curiously the surface temperature of the skin was measured, both of the knee and the feet and, although no significant changes were found in the temperature of the feet following treatment, it was interesting that the temperature of the feet before treatment was always lower on the injured side in comparison with the healthy side, whereas the temperature in the knee region was always higher on the injured side.

- Vascular Injuries

Lastly, hardly related to physical activity and sport, Staffa and collaborators [35] assessed the effect of percutaneous transluminal angioplasty using thermography of the feet and indicated the possibility of using thermal imaging as a means of early

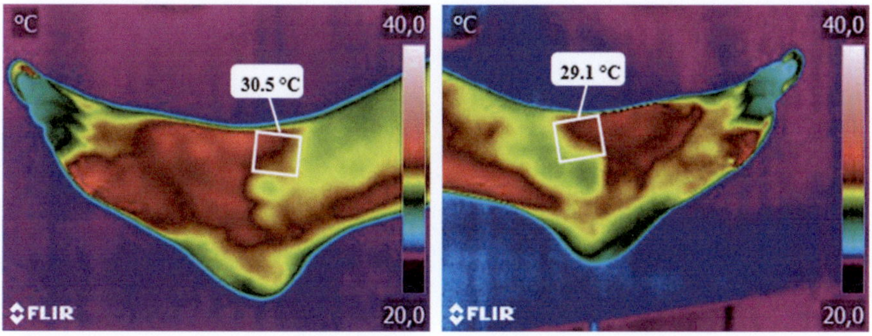

Fig. 14.11 Comparison of asymmetries between the right foot with sprain of the anterior talofibular ligament (*right*) and the healthy left foot (*left*), showing the mean temperature of the affected region

detection of the appearance of ischemic disorders of the foot, such as the diabetic foot, previously treated, or the Raynaud phenomenon.

The Raynaud phenomenon has for a long time been evaluated using infrared thermography [110–114], being especially focused on the hands [115, 116].

It is characterized by the manifestation of episodic vasospasms in the fingers and toes as a consequence of exposure to cold temperatures or strong emotions [117]. During these vascular spasms, a narrowing of the blood vessels takes place which reduces the blood supply to the fingers and toes and produces a drastic reduction in temperature [112, 117].

Specifically, the study of Lim et al. [116] identified, through measuring the gradient of foot temperature using infrared thermography, a difference of temperature equal to or greater than 3.11 °C between the dorsal of the foot and the big toe, as the cut-off value that may indicate the presence of this disease (Fig. 14.12).

Both Lim et al. [116] and Tse et al. [114] reported that this technique can be very useful in diagnosing this phenomenon in the foot. Indeed, Tse et al. [114] claim that infrared thermography provides a more accurate and objective method than traditional physical exploration undertaken by health professionals by identifying the asymmetries of skin temperature related to the Raynaud phenomenon.

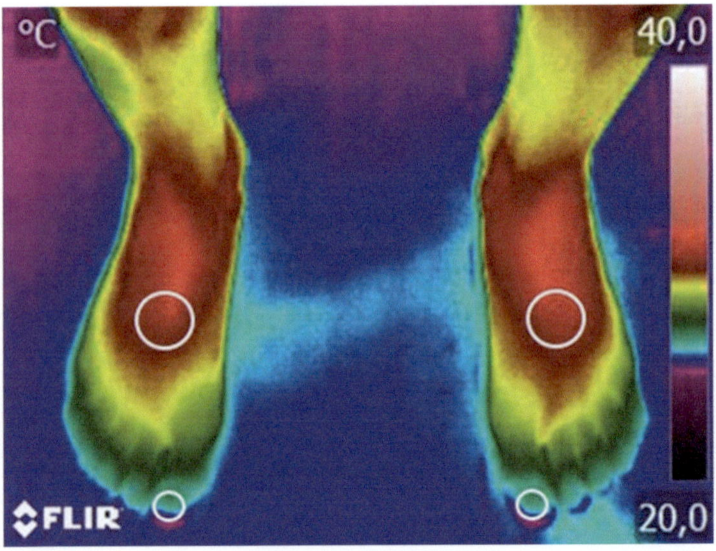

Fig. 14.12 Adaptation of the method used by Lim et al. [116] for measuring temperature gradients, comparing the dorsal surface of the foot with the big toe

14.3.2 Determining the Pattern of Running

Sporting specialization can affect the normal thermal patterns in healthy athletes [118], causing thermal asymmetries that are specific to a particular sport, such as the forearm in tennis players, the tibialis anterior in football players, the arm in volleyball and basketball players or the grasping forearm of a judo practitioner [54]. It is, therefore, important to know the thermal pattern commonly presented by the participants of a sport and the individual thermal patterns of the sportsman or woman before assessing the thermal asymmetries and venturing a possible injury [54].

Likewise, the execution of a sports technique, depending on the sport and the characteristics of the individual, can present certain thermal patterns, so it may be suggested that infrared thermography can be of great use to coaches, providing them with valuable information on the muscular activity undertaken and the technique employed as well as contributing to their improvement in training [119]. With this idea, and in relation with the foot region, the authors of this chapter undertook a study, the aim of which was to analyze the relationship between the temperature of the skin surface on the soles of the feet and the ankle eversion in running in order to determine whether infrared thermography could be a quick and simple tool for predicting the degree of ankle eversion in runners [154].

As Fig. 14.13 shows, there is a weak positive correlation between the symmetries of temperature variation (temperature difference between the medial side and the lateral side of the foot as a result of running) and the maximum ankle eversion during stance phase (MES) and a weak negative relationship with eversion at contact time. Therefore, given the weak correlations found, the previously stated hypothesis is not fulfilled by the results obtained, and the measurement of sole temperature does not seem to be an appropriate tool to determine foot eversion during running.

On the contrary, the following image (Fig. 14.14) suggests that both the type of running contact (pronation-supination) and the running foot strike pattern of an athlete (forefoot, midfoot, and rearfoot) could be studied by evaluating the temperatures of the footwear's outsoles following exercise, due also to its already proven relationship with plantar pressures as mentioned above [147], simply by taking thermographic images.

Similarly research using thermography in other populations that could also be interesting for the evaluation of this technique, such as the study of lateral dominance, the use of one extremity more than another, or the greater application of load on one foot rather than the other, and not only in running, but also in any sport where the foot meets the ground.

Building on this topic, a study performed by the authors of this chapter investigated whether inducing running asymmetry by attaching a weight to one ankle caused changes in the surface temperature of the soles of healthy runners' feet [140]. Despite the spatiotemporal asymmetry, no thermal asymmetry was observed [140]. This may be because the mechanical demands of running outweighed the effects of the induced asymmetry. Further research is needed to explore these findings in runners with pathological or natural asymmetries.

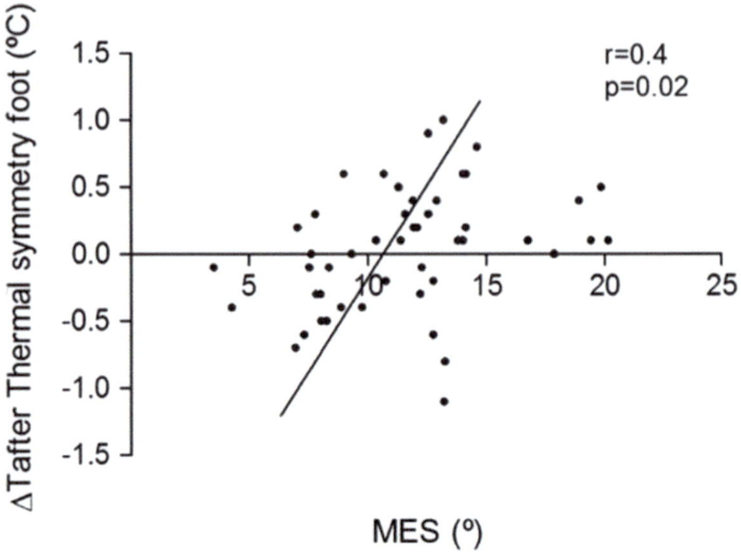

Fig. 14.13 Correlation between the maximum ankle eversion during the stance phase (MES) and the variation of the thermal symmetry of the plantar temperature (difference in temperature variation due to running between the medial and lateral sides of the foot) [154]

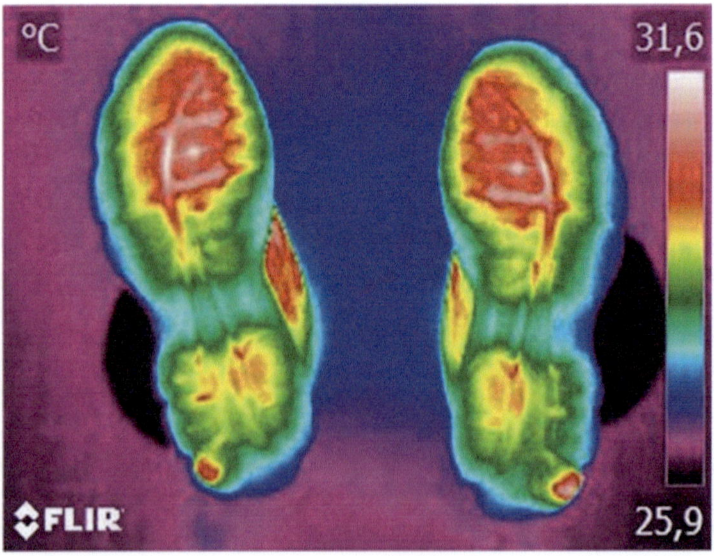

Fig. 14.14 Thermographic image of the soles of an athlete's footwear 10 min after running that could predict the type of foot contact (pronation-supination) due to the differences in temperature between regions

14.3.3 Determining the Effect of Footwear and Textiles on the Foot

Sports equipment significantly influences the sporting performance of the athlete in terms of the possible development or prevention of an injury [50, 120]. Specifically, sports footwear plays an important role in these two aspects in a great number of sports where the interaction of the foot and the ground predominates.

Currently, the sports market offers a huge selection when it comes to choosing specific footwear as there are many types of wide-ranging characteristics that differ both in structure and functionality, and in style [121]. Hence, the sportsman or woman must choose footwear bearing in mind the architecture of their feet, their sport mechanics, and the specific demands of their sports modality, as those characteristics will reduce the risk of injury and increase performance [122].

Moreover, thermal comfort is another characteristic that has to be taken into account in the design and choice of footwear, as it has great repercussions on general comfort while exercise is being undertaken and is also associated with improvement in sporting performance [123–125]. The thermal properties related to thermoregulation and thermal comfort are breathability and thermal insulation [123, 124, 161]. Consequently, footwear that allows good breathability during exercise or that in very cold situations and adverse climatology are insulating and impermeable will favor the thermal comfort of the athlete [49, 123, 161]. In addition, the type of footwear lacing could also influence foot skin temperature. Although no differences were observed between males and females, the results of an unpublished study conducted by the authors of this chapter suggest increased overall and thermal comfort with heel lock lacing.

Another garment that has an influence on the microclimate created inside the shoe and directly affects the thermoregulation of the foot during sport is the sock [159]. However, there are very few studies on the effect of socks, despite the fact that new textiles and fabrics have been designed to improve thermal equilibrium and to provide greater thermal comfort to the user [126].

In this sense, thermographic information on the foot can be very useful for the functional design of both footwear and socks; for the evaluation of the thermal properties that these garments offer to sportsmen and women, and also for observing the thermoregulation of the foot depending on the type of shoe or sock in the sports context [127].

Interestingly, in a previous work performed by the authors of this chapter, it was found that women have a generally lower basal temperature of the soles of the feet than men [146], probably due to the lower blood circulation in the peripheral areas and a higher percentage of fat in women. However, after performing 30 min of intense running, the temperatures of the soles of women's feet are equal to those of men, probably because men have greater peripheral vasodilatation and greater perspiration response in these regions [145]. This fact implies that the increase in temperature of the soles of the feet in women is higher than that of men during running. This different thermal behavior between men and women [159] should be considered by manufacturers in the design of running shoes, socks, and foot orthoses by differentiating the thermal properties between those designed for men and women.

In fact, according to the review published by Banerjee et al. [128], various applications of infrared thermography have already been reported in the textile industry, such as measuring the thermal properties of textile materials, measuring dryness and the transport of heat and humidity, and analyzing the mechanical properties or the comfort of the garment, where this technique can play a crucial role in the future.

- Breathability and Sweat

As mentioned above, the foot is a thermal radiator essential to human thermoregulation. So, when the feet are inside footwear for long periods, the temperature and humidity within the footwear tend to increase [17] and can begin to cause discomfort and irritation and even provoke the development of an injury or bacterial infection. This fact is especially relevant in summer and in warm and humid regions. Kinoshita and Bates found in their study [129] temperatures on the sole of the footwear greater than 50 °C recorded during a 40-min run on a summer's day.

These high foot and footwear temperatures, together with restrictions to ventilation as a consequence of the closed shoe, cause a high production of sweat by the foot aimed at trying to maintain body temperature around 37 °C [126]. In addition, the regions that are furthest away from the opening of the shoe and sock, i.e., the toes, receive less air movement (less ventilation), so their temperature will be highest [126]. However, a study showed how perceptions of wetness and foot discomfort did not correspond with regions of high sweat production or low skin temperature but rather seemed dominated by tactile interactions caused by foot movement within the shoe [162].

It has been shown that the foot produces great amounts of sweat in comparison to the rest of the body [130], as it has a dense region of eccrine sweat glands. These glands are distributed all over the body, but are mostly concentrated on the palms of the hand, soles of the feet and forehead. They possess an innervation of the sympathetic nervous system and their secretion is commonly called sweat, thus contributing to heat loss through evaporation [17]. Sometimes the feet and footwear produce bad smells due to the proliferation of bacteria resulting from excessive humidity derived from sweat in that area.

Nevertheless, different studies aimed to describe the production of foot sweat during exercise, in this case running, a greater amount of sweat secretion was observed in the dorsal part (~70% of the sweat production of the foot) of the feet than in the plantar region (~30% of the sweat production of the foot) [131, 162], even when there is a higher number of sweat glands in the soles of the feet [2]. Moreover, it was also found that the use of footwear limits the increase in sweat production as running intensity increases [131, 162]. Nevertheless, at some point, sweat production rates are suppressed when footwear is worn, probably because of hidromeiosis (skin swelling due to excessive hydration, closing the sweat ducts) [136].

A recent study [162] analyzing the effect of wearing shoes or barefoot during running at different intensities found that foot temperature was higher with footwear than barefoot and at a higher intensity than at lower intensity. However, foot skin temperature was related to runners' perceptual responses (i.e., greater perception of wetness and stickiness in shoes than barefoot) but could not explain the differences in sweat rates in any of the conditions tested, nor could it be associated with the regions of the feet with higher sweat rates [162]. The authors of this paper propose that the areas of development of footwear and socks should focus on the tactile interaction between these elements and the feet, with particular emphasis on the regions with the highest perceived sweating (toes and heel) concerning trying to reduce the afferents/cues generated between the different elements [162].

Thus a particularly important factor both for the comfort of sportsmen and women's feet and for the appropriate maintenance of heat dynamics during exercise is that the footwear or sock presents suitable properties of heat dissipation through convection and conduction currents between the feet and the fabric and improves breathability, i.e., the elimination of sweat through evaporation [14, 121]. For its part, the sock is fundamental for maintaining the climate of the foot, given that it transports the sweat from the foot to the upper part of the shoe so that it evaporates better [121].

In this regard, a recent study has shown that wearing socks at rest and during running does not produce major differences at the thermophysiological level, regardless of the composition of the sock fabric, but that exercising without socks leads to an increase in perceived moisture, stickiness, and thermal discomfort [160]. Therefore, socks can play a fundamental role in reducing the tactile and mechanical inputs generated between the foot and the footwear [160].

Thermal analysis of the sports shoe using infrared thermal imaging can, therefore, be very useful in identifying the areas that provide the greatest breathability to the athlete and help him or her to choose the footwear that facilitates thermoregulation, as shown in Fig. 14.15.

As can be seen in the next figure (Fig. 14.15), the specific running shoe presents higher temperatures than the mountain shoe. This is due to the fact that the running shoe is more breathable and less insulated, so facilitating the flow of temperature from the foot to the outside of the shoe. This method of analyzing footwear could become a very interesting field of research. As can be observed, the terms insulation and breathability are closely linked, so the characteristic of insulation will now be explained.

- Insulation

During exercise, the feet remain hot. However, during inactivity, the temperature of the feet falls rapidly [49]. One of the properties of footwear that helps to reduce heat loss is thermal insulation, the importance of which is accentuated when physical activity is undertaken at low temperatures or in cold and humid regions [132].

Insulation can be increased through garments or materials added to the footwear that provide local insulation, such as on the toes and heels, whose effects are clearly notable on the mean temperatures of the skin [49]. However, wearing a thick pair of

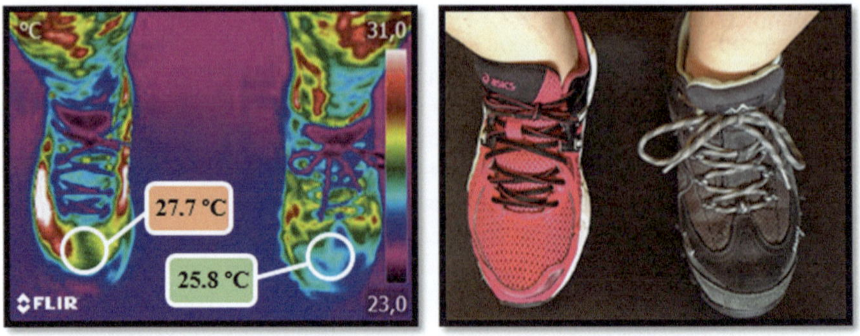

Fig. 14.15 Thermographic images of the dorsal views of the feet with two different sports shoes after undertaking a 10 min run: a specific running sports shoe (*left*): and a trail sports shoe (*right*). Mean temperatures of the regions within the circumferences are shown

socks in a tight shoe can produce the opposite effect, increasing the accumulation of humidity and so reducing the insulation of the footwear by up to 45% [132, 133]. The thermal insulation of footwear and socks can be assessed through two methodologies: directly on the feet of human beings [134], or using a thermal foot model [49, 135]. The main drawback of which is that it cannot contemplate the conditions and changes that are produced during a specific sports practice, and the alterations caused in the physiology of the foot of the user as a consequence of that practice. To date, there have been two investigations carried out on human beings that have reported increases in temperature in certain regions of the foot, during walking and running, probably as a result of the effect of footwear insulation [17, 34], respectively.

The latter, of Priego Quesada et al. [34], used, as the skin measuring technique, infrared thermography, the main advantage of which is that it does not interfere with the behavior of the athlete when in use, nor does it affect the thermoregulation of the foot as it is noninvasive. Other methods can be invasive, such as incorporating sensors on the surface of the skin of the foot, which was used by Shimazaki and Murata [17]. However, the main drawback of the infrared thermography method is that the measurements have to be taken in static positions and without footwear and socks, as it is not possible to undertake them while running in a real situation. In this sense, surface sensors have a clear advantage.

On the other hand, in various studies, the aim of which was to compare foot temperature with the use of different models of footwear and socks during running, the conclusion was reached that the perception of foot temperature by the user may differ from the real physiological temperature of the foot [121, 126]. Thus, it may be said that, when determining how a shoe or sock affects the physiological temperature of the foot, the study has to be undertaken through techniques that provide more precise information than the subjective perception of the user. Infrared thermography does this, providing speed and simplicity to the assessment. We cannot, however, ignore the perceptions of the user, given that they will be the ones to determine thermal comfort and, consequently, the performance of the athlete [126].

14.4 Conclusions

As explained in this chapter, the thermographic analysis of the foot in sports sciences presents numerous interesting applications. These applications range from the prevention of injuries and the analysis of the effects of sporting technique on foot temperature to the evaluation of sports equipment (footwear, insoles, and socks). Infrared thermography, therefore, can provide relevant information for the athlete on how to improve his or her performance and on reducing the injurys possibilities. Nevertheless, evidence is scarce in these fields and future studies are required to analyze the hypothetical potential of infrared thermography dealt with in this chapter. Methodologically, there are also issues that need to be addressed. An analysis needs to be made of which determination of the regions of interest on the foot can be of assistance to each of the different applications. Moreover, studying the reproducibility of the foot temperature variation as a result of exercise is an aspect that must be investigated.

In conclusion, thermography in the analysis of the foot in sport sciences presents great potential. Although the number of investigations in this area has increased in the recent years, it is still necessary to further investigate certain specific aspects of the thermal behavior of the foot as well as its interaction with the materials with which it is in direct or indirect contact.

References

1. Nordin M, Frankel VH (2001) Basic biomechanics of the musculoskeletal system. Lippincott Williams & Wilkins
2. Taylor NA, Machado-Moreira CA, van den Heuvel AM, Caldwell JN (2014) Hands and feet: physiological insulators, radiators and evaporators. Eur J Appl Physiol 114:2037–2060
3. Thibodeau GA, Patton KT (2007) Anatomía y fisiología. Elsevier España
4. Hreljac A (2004) Impact and overuse injuries in runners. Med Sci Sports Exerc:845–849. https://doi.org/10.1249/01.MSS.0000126803.66636.DD
5. Golden FSC, Francis TJR, Gallimore D, Pethybridge R (2013) Lessons from history: morbidity of cold injury in the Royal Marines during the Falklands Conflict of 1982. Extreme Physiol Med 2:23. https://doi.org/10.1186/2046-7648-2-23
6. Johnson JM, Kellogg DL (2010) Local thermal control of the human cutaneous circulation. J Appl Physiol (1985) 109:1229–1238. https://doi.org/10.1152/japplphysiol.00407.2010
7. Pérgola PE, Kellogg DL, Johnson JM et al (1993) Role of sympathetic nerves in the vascular effects of local temperature in human forearm skin. Am J Phys 265:H785–H792
8. Taylor WF, Johnson JM, O'Leary D, Park MK (1984) Effect of high local temperature on reflex cutaneous vasodilation. J Appl Physiol 57:191–196
9. Standring S (2015) Gray's anatomy: the anatomical basis of clinical practice. Elsevier Health Sciences
10. Blair DA, Glover WE, Roddie IC (1960) Vasomotor fibres to skin in the upper arm, calf and thigh. J Physiol 153:232–238
11. Roddie IC (2011) Circulation to skin and adipose tissue. In: Comprehensive physiology. American Physiological Society. https://doi.org/10.1002/cphy.cp020310

12. Johnson JM, Pérgola PE, Liao FK et al (1995) Skin of the dorsal aspect of human hands and fingers possesses an active vasodilator system. J Appl Physiol (1985) 78:948–954
13. Charkoudian N (2003) Skin blood flow in adult human thermoregulation: how it works, when it does not, and why. Mayo Clin Proc 78:603–612. https://doi.org/10.4065/78.5.603
14. Chicharro JL, Vaquero AF (2006) Fisiología del ejercicio/Physiology of exercise. Ed. Médica Panamericana
15. Morrison SA, Gorjanc J, Mekjavic IB (2014) Mount Everest and Makalu cold injury amputation: 40 years on. High Altitude Med Biol 15:78–83. https://doi.org/10.1089/ham.2013.1069
16. Taylor NA, Machado-Moreira CA (2013) Regional variations in transepidermal water loss, eccrine sweat gland density, sweat secretion rates and electrolyte composition in resting and exercising humans. Extreme Physiol Med 2:4. https://doi.org/10.1186/2046-7648-2-4
17. Shimazaki Y, Murata M (2015) Effect of gait on formation of thermal environment inside footwear. Appl Ergon 49:55–62. https://doi.org/10.1016/j.apergo.2015.01.007
18. de Andrade FA, dos Santos Amorim PR, Brito CJ et al (2014) Measuring skin temperature before, during and after exercise: a comparison of thermocouples and infrared thermography. Physiol Meas 35:189
19. Hildebrandt C, Zeilberger K, Ring EFJ, Raschner C (2012) The application of medical infrared thermography in sports medicine. Ultrasound 10:2
20. Mendes R, Sousa N, Almeida A et al (2015) Thermography: a technique for assessing the risk of developing diabetic foot disorders. Postgrad Med J 91:538. https://doi.org/10.1136/postgradmedj-2015-133441
21. Wasner G, Schattschneider J, Baron R (2002) Skin temperature side differences–a diagnostic tool for CRPS? Pain 98:19–26
22. Berry RJ, Kennedy AD, Scott SL et al (2003) Daily variation in the udder surface temperature of dairy cows measured by infrared thermography: potential for mastitis detection. Can J Anim Sci 83:687–693. https://doi.org/10.4141/A03-012
23. Alsaaod M, Büscher W (2012) Detection of hoof lesions using digital infrared thermography in dairy cows. J Dairy Sci 95:735–742. https://doi.org/10.3168/jds.2011-4762
24. Bowers S, Gandy S, Anderson B et al (2009) Assessment of pregnancy in the late-gestation mare using digital infrared thermography. Theriogenology 72:372–377
25. Nikkhah A, Plaizier JC, Einarson MS et al (2005) Short communication: infrared thermography and visual examination of hooves of dairy cows in two stages of lactation. J Dairy Sci 88:2749–2753. https://doi.org/10.3168/jds.S0022-0302(05)72954-4
26. Alsaaod M, Schaefer AL, Büscher W, Steiner A (2015) The role of infrared thermography as a non-invasive tool for the detection of lameness in cattle. Sensors 15:14513–14525
27. Oberle J, Elam M, Karlsson T, Wallin BG (1988) Temperature-dependent interaction between vasoconstrictor and vasodilator mechanisms in human skin. Acta Physiol Scand 132:459–469. https://doi.org/10.1111/j.1748-1716.1988.tb08353.x
28. Kakuta N, Yokoyama S, Mabuchi K (2002) Human thermal models for evaluating infrared images. Eng Med Biol Mag IEEE 21:65–72
29. Sillero-Quintana M, Fernández-Jaén T, Fernández-Cuevas I et al (2015) Infrared thermography as a support tool for screening and early diagnosis in emergencies. J Med Imaging Health Inf 5:1223–1228
30. Zaproudina N, Ming Z, Hänninen OO (2006) Plantar infrared thermography measurements and low back pain intensity. J Manip Physiol Ther 29:219–223
31. Sousa P, Felizardo V, Oliveira D et al (2015) A review of thermal methods and technologies for diabetic foot assessment. Expert Rev Med Devices 12:439–448. https://doi.org/10.1586/17434440.2015.1032251
32. Schaefer AL, Cook NJ, Luzi F et al (2013) Heat generation and the role of infrared thermography in pathological conditions. In: Thermography: current status and advances in livestock animals and in veterinary medicine, pp 69–78
33. Zaproudina N, Varmavuo V, Airaksinen O, Närhi M (2008) Reproducibility of infrared thermography measurements in healthy individuals. Physiol Meas 29:515. https://doi.org/10.1088/0967-3334/29/4/007

34. Priego Quesada JIP, Kunzler MR, da Rocha ES et al (2015) Plantar pressure and foot temperature responses to acute barefoot and shod running. Hum Mov 16:142–148. https://doi.org/10.1515/humo-2015-0040
35. Staffa E, Bernard V, Kubicek L et al (2016) Infrared thermography as option for evaluating the treatment effect of percutaneous transluminal angioplasty by patients with peripheral arterial disease. Vascular 25:42. https://doi.org/10.1177/1708538116640444
36. Marins JCB, Moreira DG, Cano SP et al (2014) Time required to stabilize thermographic images at rest. Infrared Phys Technol 65:30–35. https://doi.org/10.1016/j.infrared.2014.02.008
37. Ammer K (2008) The Glamorgan Protocol for recording and evaluation of thermal images of the human body. Thermol Int 18:125–144
38. Hall M, Shurr DG, Zimmerman MB, Saltzman CL (2004) Plantar foot surface temperatures with use of insoles. Iowa Orthop J 24:72–75
39. Sun P-C, Jao S-HE, Cheng C-K (2005) Assessing foot temperature using infrared thermography. Foot Ankle Int 26:847–853. https://doi.org/10.1177/107110070502601010
40. Bagavathiappan S, Philip J, Jayakumar T et al (2010) Correlation between plantar foot temperature and diabetic neuropathy: a case study by using an infrared thermal imaging technique. J Diabetes Sci Technol 4:1386–1392. https://doi.org/10.1177/193229681000400613
41. Bharara M, Boulger E, Grewal GS et al (2014) Applications of angiosome classification model for monitoring disease progression in the diabetic feet. In: Proceedings of the 2014 summer simulation multiconference society for computer simulation international, p 34
42. Mori T, Nagase T, Takehara K et al (2013) Morphological pattern classification system for plantar thermography of patients with diabetes. J Diabetes Sci Technol 7:1102–1112. https://doi.org/10.1177/193229681300700502
43. Nagase T, Sanada H, Takehara K et al (2011) Variations of plantar thermographic patterns in normal controls and non-ulcer diabetic patients: novel classification using angiosome concept. J Plast Reconstr Aesthet Surg 64:860–866
44. Peregrina-Barreto H, Morales-Hernández LA, Rangel-Magdaleno JJ, Vázquez-Rodríguez PD (2013) Thermal image processing for quantitative determination of temperature variations in plantar angiosomes. In: 2013 IEEE international instrumentation measurement technology conference I2MTC, pp 816–820
45. Peregrina-Barreto H, Morales-Hernandez LA, Rangel-Magdaleno JJ et al (2014) Quantitative estimation of temperature variations in plantar angiosomes: a study case for diabetic foot. Comput Math Methods Med 2014:1. https://doi.org/10.1155/2014/585306
46. Taylor GI, Palmer JH (1987) The vascular territories (angiosomes) of the body: experimental study and clinical applications. Br J Plast Surg 40:113–141
47. Attinger CE, Evans KK, Bulan E et al (2006) Angiosomes of the foot and ankle and clinical implications for limb salvage: reconstruction, incisions, and revascularization. Plast Reconstr Surg 117:261S–293S
48. Marins JCB, Fernández-Cuevas I, Arnaiz-Lastras J, et al (2015) Aplicaciones de la termografía infrarroja en el deporte: Una revisión. Revista Internacional de Medicina y Ciencias de la Actividad Fisica y del Deporte
49. Kuklane K (2004) The use of footwear insulation values measured on a thermal foot model. Int J Occup Saf Ergon 10:79–86
50. Morey G, Mademli L (2015) El calzado deportivo. In: Perez-Soriano P, Llana-Belloch S (eds) Biomecánica básica aplicada a la actividad física y el deporte. Ed. Paidotribo, pp 351–362
51. Dekker R, Kingma J, Groothoff JW et al (2000) Measurement of severity of sports injuries: an epidemiological study. Clin Rehabil 14:651–656
52. Zwerver J (Hans) (2015) Sports injuries. In: Glaudemans AWJM, Dierckx RAJO, Gielen JLMA, Zwerver J (Hans) (eds) Nuclear medicine and radiologic imaging in sports injuries. Springer Berlin Heidelberg, pp 49–67
53. Sobhani S, Dekker R, Postema K, Dijkstra PU (2013) Epidemiology of ankle and foot overuse injuries in sports: a systematic review. Scand J Med Sci Sports 23:669–686

54. Fernández-Cuevas I, Bouzas Marins JC, Arnáiz Lastras J et al (2015) Classification of factors influencing the use of infrared thermography in humans: A review. Infrared Phys Technol 71:28–55. https://doi.org/10.1016/j.infrared.2015.02.007
55. Hildebrandt C, Raschner C, Ammer K (2010) An overview of recent application of medical infrared thermography in sports medicine in Austria. Sensors 10:4700–4715
56. Vardasca R (2008) Symmetry of temperature distribution in the upper and lower extremities. Thermol Int 18:154–155
57. Vardasca R, Ring F, Plassmann P, Jones C (2012) Thermal symmetry of the upper and lower extremities in healthy subjects. Thermol Int 22:53–60
58. Vardasca R, Ring EFJ, Plassmann P, Jones CD (2007) Thermal symmetry on extremities of normal subjects. In: Conference in precedings of the first research student workshop on the faculty of advanced technology. University of Glamorgan, pp 19–24
59. Hardaker NJ, Moss AD, Richards J et al (2007) The relationship between skin surface temperature measured via non-contact thermal imaging and intra-muscular temperature of the rectus femoris muscle. Studies 13:17
60. Niu HH, Lui PW, Hu JS et al (2001) Thermal symmetry of skin temperature: normative data of normal subjects in Taiwan. Zhonghua Yi Xue Za Zhi Chin Med J Free China Ed 64:459–468
61. Uematsu S, Edwin DH, Jankel WR et al (1988) Quantification of thermal asymmetry: Part 1: normal values and reproducibility. J Neurosurg 69:552–555
62. Chuckpaiwong B, Nunley JA, Mall NA, Queen RM (2008) The effect of foot type on in-shoe plantar pressure during walking and running. Gait Posture 28:405–411
63. Brand PW (2003) Tenderizing the foot. Foot Ankle Int 24:457–461. https://doi.org/10.1177/107110070302400602
64. Yavuz M, Brem RW, Davis BL et al (2014) Temperature as a predictive tool for plantar triaxial loading. J Biomech 47:3767–3770
65. Yavuz M, Delvadia N, Atves J et al (2013) Biomechanical value of temperature in assessing plantar loading. In: Communication in the 37th annual meeting of the American Society of Biomechanics, Omaha, NE. American Society of Biomechanics
66. Taiar R, Rebay M, Vannozzi G et al (2008) Evolution of the in-shoe temperature during walking and running. Bio Eng Biomed Med:601–689
67. Lee Y-C, Lin G, Wang M-JJ (2012) Evaluating insole design with joint motion, plantar pressure and rating of perceived exertion measures. Work 41:1114–1117
68. Lucas-Cuevas AG, Pérez-Soriano P, Llana-Belloch S et al (2014) Effect of custom-made and prefabricated insoles on plantar loading parameters during running with and without fatigue. J Sports Sci 32:1712–1721. https://doi.org/10.1080/02640414.2014.915422
69. Wrobel JS, Ammanath P, Le T et al (2014) A novel shear reduction insole effect on the thermal response to walking stress, balance, and gait. J Diabetes Sci Technol 8:1151–1156
70. Lahiri BB, Bagavathiappan S, Jayakumar T, Philip J (2012) Medical applications of infrared thermography: a review. Infrared Phys Technol 55:221–235
71. Liu C, van Netten JJ, van Baal JG et al (2015) Automatic detection of diabetic foot complications with infrared thermography by asymmetric analysis. J Biomed Opt 20:026003. https://doi.org/10.1117/1.JBO.20.2.026003
72. Hashmi F, Richards BS, Forghany S et al (2013) The formation of friction blisters on the foot: the development of a laboratory-based blister creation model. Skin Res Technol 19:e479–e489. https://doi.org/10.1111/j.1600-0846.2012.00669.x
73. Basler RSW, Hunzeker CM, Garcia MA, Dexter W (2004) Athletic skin injuries. Phys Sportsmed 32:33–40. https://doi.org/10.3810/psm.2004.05.304
74. Helm MF, Helm TN, Bergfeld FW (2012) Skin problems in the long-distance runner 2500 years after the Battle of Marathon. Int J Dermatol 51:263–270
75. Purim KSM, Leite N, Purim KSM, Leite N (2014) Sports-related dermatoses among road runners in Southern Brazil. An Bras Dermatol 89:587–592. https://doi.org/10.1590/abd1806-4841.20142792

76. Honsik KA, Romeo MW, Hawley CJ et al (2007) Sideline skin and wound care for acute injuries. Curr Sports Med Rep 6:147–154
77. Mailler EA, Adams BB (2004) The wear and tear of 26.2: dermatological injuries reported on marathon day. Br J Sports Med 38:498–501. https://doi.org/10.1136/bjsm.2004.011874
78. Van Tiggelen D, Wickes S, Coorevits P et al (2009) Sock systems to prevent foot blisters and the impact on overuse injuries of the knee joint. Mil Med 174:183–189. https://doi.org/10.7205/MILMED-D-01-8508
79. Reynolds K, Darrigrand A, Roberts D et al (1995) Effects of an antiperspirant with emollients on foot-sweat accumulation and blister formation while walking in the heat. J Am Acad Dermatol 33:626–630. https://doi.org/10.1016/0190-9622(95)91283-5
80. Kirkham S, Lam S, Nester C, Hashmi F (2014) The effect of hydration on the risk of friction blister formation on the heel of the foot. Skin Res Technol 20:246–253. https://doi.org/10.1111/srt.12136
81. Hashmi F, Kirkham S, Nester C, Lam S (2016) The effect of topical anti blister products on the risk of friction blister formation on the foot. J Tissue Viability 25:167. https://doi.org/10.1016/j.jtv.2016.04.002
82. Liu C, van der Heijden F, Klein ME et al (2013) Infrared dermal thermography on diabetic feet soles to predict ulcerations: a case study. In: SPIE BiOS. International society for optics and photonics, p 85720N
83. Hernandez-Contreras D, Peregrina-Barreto H, Rangel-Magdaleno J et al (2015) Automatic classification of thermal patterns in diabetic foot based on morphological pattern spectrum. Infrared Phys Technol 73:149–157. https://doi.org/10.1016/j.infrared.2015.09.022
84. International Diabetes Federation (2015) IDF diabetes atlas, 7th edn. International Diabetes Federation, Brussels
85. Armstrong DG, Holtz-Neiderer K, Wendel C et al (2007) Skin temperature monitoring reduces the risk for diabetic foot ulceration in high-risk patients. Am J Med 120:1042–1046. https://doi.org/10.1016/j.amjmed.2007.06.028
86. Lavery LA, Armstrong DG, Vela SA et al (1998) Practical criteria for screening patients at high risk for diabetic foot ulceration. Arch Int Med 158:157–162
87. Boulton AJ, Vileikyte L, Ragnarson-Tennvall G, Apelqvist J (2005) The global burden of diabetic foot disease. Lancet 366:1719–1724. https://doi.org/10.1016/S0140-6736(05)67698-2
88. Armstrong DG, Lavery LA, Liswood PJ et al (1997) Infrared dermal thermometry for the high-risk diabetic foot. Phys Ther 77:169–175
89. Lavery LA, Higgins KR, Lanctot DR et al (2004) Home monitoring of foot skin temperatures to prevent ulceration. Diabetes Care 27:2642–2647. https://doi.org/10.2337/diacare.27.11.2642
90. Balbinot LF, Canani LH, Robinson CC et al (2012) Plantar thermography is useful in the early diagnosis of diabetic neuropathy. Clinics 67:1419–1425
91. van Netten JJ, Prijs M, van Baal JG et al (2014) Diagnostic values for skin temperature assessment to detect diabetes-related foot complications. Diabetes Technol Ther 16:714–721. https://doi.org/10.1089/dia.2014.0052
92. Sun P-C, Lin H-D, Jao S-HE et al (2006) Relationship of skin temperature to sympathetic dysfunction in diabetic at-risk feet. Diabetes Res Clin Pract 73:41–46. https://doi.org/10.1016/j.diabres.2005.12.012
93. van Netten JJ, Jannink MJA, Hijmans JM et al (2010) Long-term use of custom-made orthopedic shoes: 1.5-year follow-up study. J Rehabil Res Dev 47:643. https://doi.org/10.1682/JRRD.2010.03.0040
94. Armstrong DG, Lavery LA, Wunderlich RP, Boulton AJM (2003) Skin temperatures as a one-time screening tool do not predict future diabetic foot complications. J Am Podiatr Med Assoc 93:443–447. https://doi.org/10.7547/87507315-93-6-443
95. Armstrong DG, Lipsky BA, Polis AB, Abramson MA (2006) Does dermal thermometry predict clinical outcome in diabetic foot infection? Analysis of data from the SIDESTEP* trial. Int Wound J 3:302–307. https://doi.org/10.1111/j.1742-481X.2006.00269.x

96. van Netten JJ, van Baal JG, Liu C et al (2013) Infrared thermal imaging for automated detection of diabetic foot complications. J Diabetes Sci Technol 7:1122–1129
97. Goodman PH, Heaslet MW, Pagliano JW, Rubin BD (1985) Stress fracture diagnosis by computer-assisted thermography. Phys Sportsmed 13:114–132
98. Laird RC (2015) Acute forefoot and midfoot injuries. Clin Podiatr Med Surg 32:231–238
99. DiBenedetto M (2002) Foot evaluation by infrared imaging. Mil Med 167:384
100. Niehof SP, Beerthuizen A, Huygen FJ, Zijlstra FJ (2008) Using skin surface temperature to differentiate between complex regional pain syndrome type 1 patients after a fracture and control patients with various complaints after a fracture. Anesth Analg 106:270–277
101. Schmitt M, Guillot Y (1984) Thermography and muscular injuries in sports medicine. In: Recent advances medical thermology. Springer, pp 439–445
102. Bandeira F, Neves EB, de Moura MAM, Nohama P (2014) A termografia no apoio ao diagnóstico de lesão muscular no esporte. Rev Bras Med Esporte 20:59–64
103. Bandeira F, de Moura MAM, de Souza MA et al (2012) Pode a termografia auxiliar no diagnóstico de lesões musculares em atletas de futebol? Rev Bras Med Esporte 18:246–251
104. Barcelos EZ, Caminhas WM, Ribeiro E et al (2014) A combined method for segmentation and registration for an advanced and progressive evaluation of thermal images. Sensors 14:21950–21967. https://doi.org/10.3390/s141121950
105. Morales Ríos M, Medina Chacón E, Carnevali Fernández Á, Orozco Guillén E (2011) Termografía infrarroja y el estudio de riesgos de lesiones músculo esqueléticas. Rev Ing Ind:55–67
106. Doherty C, Delahunt E, Caulfield B et al (2014) The incidence and prevalence of ankle sprain injury: a systematic review and meta-analysis of prospective epidemiological studies. Sports Med 44:123–140
107. Tenforde AS, Yin A, Hunt KJ (2016) Foot and ankle injuries in runners. Phys Med Rehabil Clin N Am 27:121–137
108. Oliveira J, Vardasca R, Pimenta M et al (2016) Use of infrared thermography for the diagnosis and grading of sprained ankle injuries. Infrared Phys Technol 76:530–541
109. Litscher G, Ofner M, Litscher D, Others (2013) Manual Khalifa therapy in patients with completely ruptured anterior cruciate ligament in the knee: first preliminary results from thermal imaging. N Am J Med Sci 5:473
110. Ammer K (1996) Diagnosis of raynaud's phenomenon by thermography. Skin Res Technol 2:182–185
111. Cherkas LF, Carter L, Spector TD et al (2003) Use of thermographic criteria to identify Raynaud's phenomenon in a population setting. J Rheumatol 30:720–722
112. Ring F (2016) The Herschel heritage to medical thermography. J Imaging 2:13
113. Saito S, Ishii T, Kamogawa Y et al (2016) Extracorporeal shock wave therapy for digital ulcers of systemic sclerosis: a phase 2 pilot study. Tohoku J Exp Med 238:39–47
114. Tse J, Rand C, Carroll M et al (2015) Determining peripheral skin temperature: subjective versus objective measurements. Acta Paediatr 105:126–131. https://doi.org/10.1111/apa.13283
115. Chlebicka I, Matusiak U, Maj J et al (2013) Freezing fingers syndrome, primary and secondary Raynaud's phenomenon: characteristic features with hand thermography. Acta Derm Venereol 93:428–432
116. Lim MJ, Kwon SR, Jung K-H et al (2014) Digital thermography of the fingers and toes in Raynaud's phenomenon. J Korean Med Sci 29:502–506
117. Block JA, Sequeira W (2001) Raynaud's phenomenon. Lancet 357:2042–2048
118. Tauchmannova H, Gabrhel J, Cibak M (1993) Thermographic findings in different sports, their value in the prevention of soft tissue injuries. Themol Osterr 3:91–95
119. Chudecka M, Lubkowska A, Leźnicka K, Krupecki K (2015) The use of thermal imaging in the evaluation of the symmetry of muscle activity in various types of exercises (symmetrical and asymmetrical). J Hum Kinet 49:141–147

120. Stefanyshyn DJ, Wannop JW (2015) Biomechanics research and sport equipment development. Sports Eng 18:191–202. https://doi.org/10.1007/s12283-015-0183-5
121. Barkley RM, Bumgarner MR, Poss EM, Senchina DS (2011) Physiological versus perceived foot temperature, and perceived comfort, during treadmill running in shoes and socks of various constructions. Am J Undergrad Res 10:7–14
122. Yamashita MH (2005) Evaluation and selection of shoe wear and orthoses for the runner. Phys Med Rehabil Clin N Am 16:801–829
123. Covill D, Guan ZW, Bailey M, Raval H (2011) Development of thermal models of footwear using finite element analysis. Proc Inst Mech Eng H 225:268–281. https://doi.org/10.1243/09544119JEIM860
124. Fan J, Tsang HW (2008) Effect of clothing thermal properties on the thermal comfort sensation during active sports. Text Res J 78:111–118
125. Luo G, Stergiou P, Worobets J et al (2009) Improved footwear comfort reduces oxygen consumption during running. Footwear Sci 1:25–29
126. Purvis A, Tunstall H (2004) Effects of sock type on foot skin temperature and thermal demand during exercise. Ergonomics 47:1657–1668. https://doi.org/10.1080/00140130412331290880
127. Shen WW, Gu YD, Bull AM (2013) Characteristic of foot surface temperature variety during continual low-intensity exercise. Biotechnol Indian J 7:352–356
128. Banerjee D, Chattopadhyay SK, Tuli S (2013) Infrared thermography in material research–a review of textile applications. Indian J Fiber Text Res 38:427–437
129. Kinoshita H, Bates BT (1996) The effect of environmental temperature on the properties of running shoes. J Appl Biomech 12:258–268
130. Park SJ, Tamura T (1992) Distribution of evaporation rate on human body surface. Ann Physiol Anthropol 11:593–609
131. Smith CJ, Machado-Moreira CA, Plant G et al (2013) Design data for footwear: sweating distribution on the human foot. Int J Cloth Sci Technol 25:43–58
132. Kuklane K, Geng Q, Holmér I (1998) Effect of footwear insulation on thermal responses in the cold. Int J Occup Saf Ergon 4:137–152
133. Kuklane K, Gavhed D, Fredriksson K (2001) A field study in dairy farms: thermal condition of feet. Int J Ind Ergon 27:367–373
134. Kuklane K, Afanasieva R, Burmistrova O et al (1999) Determination of heat loss from the feet and insulation of the footwear. Int J Occup Saf Ergon 5:465–476. https://doi.org/10.1080/10803548.1999.11076432
135. Kuklane K, Ueno S, Sawada S-I, Holmér I (2009) Testing cold protection according to EN ISO 20344: is there any professional footwear that does not pass? Ann Occup Hyg 53:63–68. https://doi.org/10.1093/annhyg/men074
136. Candas V, Libert JP, Vogt JJ (1980) Effect of hidromeiosis on sweat drippage during acclimation to humid heat. Eur J Appl Physiol Occup Physiol 44(2):123–133. https://doi.org/10.1007/BF00421090
137. Castillo-Morquecho R, Guevara E, Ramirez-GarciaLuna JL, Martínez-Jiménez MA, Medina-Rangel MG, Kolosovas-Machuca ES (2024) Digital infrared thermography and machine learning for diabetic foot assessment: thermal patterns and classification. J Diabetes Metab Disord 23(2):1967–1976. https://doi.org/10.1007/s40200-024-01452-0
138. der Strasse W, auf Campos DP, Mendonça CJA, Soni JF, Mendes J, Nohama P (2022) Detecting bone lesions in the emergency room with medical infrared thermography. Biomed Eng Online 21(1):35. https://doi.org/10.1186/s12938-022-01005-7
139. Escamilla-Galindo V-L, Felipe JL, Alonso-Callejo A, Van-der-Horst R, de la Torre-Combarros A, Minafra P, Fernández-Muñoz D, Fernández-Cuevas I (2024) Return-to-play criteria based on infrared thermography during anterior cruciate ligament rehabilitation in football players. Biol Sport 42(2):161–167. https://doi.org/10.5114/biolsport.2025.144295

140. Gil-Calvo M, Herrero-Marco J, González-Peña RDJ, Perez-Soriano P, Priego-Quesada JI (2020a) Acute effect of induced asymmetrical running technique on foot skin temperature. J Therm Biol 91:102613–102620. https://doi.org/10.1016/j.jtherbio.2020.102613
141. Gil-Calvo M, Jimenez-Perez I, Priego-Quesada JI, Lucas-Cuevas ÁG, Pérez-Soriano P (2020b) Effect of custom-made and prefabricated foot orthoses on kinematic parameters during an intense prolonged run. PLoS One 15(3):e0230877. https://doi.org/10.1371/journal.pone.0230877
142. Gil-Calvo M, Priego-Quesada JI, Jimenez-Perez I, Lucas-Cuevas A, Pérez-Soriano P (2019) Effects of prefabricated and custom-made foot orthoses on skin temperature of the foot soles after running. Physiol Meas 40(5):054004. https://doi.org/10.1088/1361-6579/ab1c8c
143. Gómez-Carmona P, Fernández-Cuevas I, Sillero-Quintana M, Arnaiz-Lastras J, Navandar A (2020) Infrared thermography protocol on reducing the incidence of soccer injuries. J Sport Rehabil:1–6. https://doi.org/10.1123/jsr.2019-0056
144. Hirsch BE (2023) Chapter 1: anatomy of the foot. In: Ledoux WR, Telfer S (eds) Foot and ankle biomechanics. Academic, pp 3–44. https://doi.org/10.1016/B978-0-12-815449-6.00040-8
145. Ichinose-Kuwahara T, Inoue Y, Iseki Y, Hara S, Ogura Y, Kondo N (2010) Sex differences in the effects of physical training on sweat gland responses during a graded exercise. Exp Physiol 95(10):1026–1032. https://doi.org/10.1113/expphysiol.2010.053710
146. Jimenez-Perez I, Gil-Calvo M, Priego-Quesada JI, Aparicio I, Pérez-Soriano P, Ortiz de Anda RMC (2020) Effect of prefabricated thermoformable foot orthoses on plantar surface temperature after running: A gender comparison. J Therm Biol 91:102612. https://doi.org/10.1016/j.jtherbio.2020.102612
147. Jimenez-Perez I, Gil-Calvo M, Salvador-Palmer R, Ortiz C, de Anda RM, Pérez-Soriano P, Priego-Quesada JI (2021) Footwear outsole temperature may be more related to plantar pressure during a prolonged run than foot temperature. Physiol Meas 42(7):074004. https://doi.org/10.1088/1361-6579/ac0fbe
148. Khandakar A, Chowdhury MEH, Reaz MBI, Ali SHM, Kiranyaz S, Rahman T, Chowdhury MH, Ayari MA, Alfkey R, Bakar AAA, Malik RA, Hasan A (2022) A novel machine learning approach for severity classification of diabetic foot complications using Thermogram images. Sensors 22(11):Article 11. https://doi.org/10.3390/s22114249
149. Kumar P, Gaurav A, Rajnish RK, Sharma S, Kumar V, Aggarwal S, Patel S (2022) Applications of thermal imaging with infrared thermography in Orthopaedics. J Clin Orthop Trauma 24:101722. https://doi.org/10.1016/j.jcot.2021.101722
150. Liew H, Tang W, Plassmann P, Machin G, Simpson R, Edmonds ME, Petrova NL (2024) Infrared thermography shows that a temperature difference of 2.2°C (4°F) or greater between corresponding sites of neuropathic feet does not always Lead to a diabetic foot ulcer. J Diabetes Sci Technol. https://doi.org/10.1177/19322968241249970
151. Lopes-Martins RAB, Beloni LFLF, Santos MCG, Bueno FC, Lopes-Martins PSL, Naterstad IF, Bjordal JM, Ferreira CR, Filho ASS, Leonardo PS (2023) Infrared thermography in the diagnosis of Achilles tendinitis. A randomized double-blind pilot clinical trial. Manual Ther Posturol Rehabil J 21. https://doi.org/10.17784/mtprehabjournal.2023.21.1297
152. Machado ÁS, Priego-Quesada JI, Jimenez-Perez I, Gil-Calvo M, Carpes FP, Perez-Soriano P (2021) Influence of infrared camera model and evaluator reproducibility in the assessment of skin temperature responses to physical exercise. J Therm Biol 98:102913. https://doi.org/10.1016/j.jtherbio.2021.102913
153. Qin Q, Oe M, Ohashi Y, Shimojima Y, Imafuku M, Dai M, Nakagami G, Yamauchi T, Yeo S, Sanada H (2022) Factors associated with the local increase of skin temperature, 'Hotspot,' of callus in diabetic foot: a cross-sectional study. J Diabetes Sci Technol 16(5):1174–1182. https://doi.org/10.1177/19322968211011181
154. Quesada JIP, Gil-Calvo M, Jimenez-Perez I, Lucas-Cuevas ÁG, Pérez-Soriano P (2017) Relationship between foot eversion and thermographic foot skin temperature after running. Appl Opt 56(19):5559–5565. https://doi.org/10.1364/AO.56.005559

155. Requena-Bueno L, Priego-Quesada JI, Jimenez-Perez I, Gil-Calvo M, Perez-Soriano P (2020) Validation of ThermoHuman automatic thermographic software for assessing foot temperature before and after running. J Therm Biol:102639. https://doi.org/10.1016/j.jtherbio.2020.102639
156. Seixas A, Azevedo J, Pimenta I, Ammer K, Carvalho R, Vilas-Boas JP, Mendes J, Vardasca R (2018) Skin temperature of the foot: reliability of infrared image analysis based in the angiosome concept. Infrared Phys Technol 92:402–408. https://doi.org/10.1016/j.infrared.2018.07.002
157. Tang W, Plassmann P, Machin G, Simpson R, Edmonds ME, Petrova NL (2024) Thermography of the plantar region alone is not sufficient to detect all areas at risk of ulceration in diabetic foot patients. Int J Low Extrem Wounds. https://doi.org/10.1177/15347346241231041
158. Thakku Sivakumar D, Murray B, Moore Z, Patton D, O'Connor T, Avsar P (2024) Can thermography predict diabetic foot ulcer risk in patients with diabetes mellitus? A systematic review. J Tissue Viability 33(4):530–541. https://doi.org/10.1016/j.jtv.2024.06.018
159. Vargas NT, Chapman CL, Johnson BD, Gathercole R, Schlader ZJ (2019) Exercise intensity independently modulates thermal behavior during exercise recovery but not during exercise. J Appl Physiol 126(4):1150–1159. https://doi.org/10.1152/japplphysiol.00992.2018
160. West AM, Havenith G, Hodder S (2021) Are running socks beneficial for comfort? The role of the sock and sock fiber type on shoe microclimate and subjective evaluations. Text Res J 91(15–16):1698–1712. https://doi.org/10.1177/0040517520986511
161. West AM, Schönfisch D, Picard A, Tarrier J, Hodder S, Havenith G (2019a) Shoe microclimate: an objective characterisation and subjective evaluation. Appl Ergon 78:1–12. https://doi.org/10.1016/j.apergo.2019.01.010
162. West AM, Tarrier J, Hodder S, Havenith G (2019b) Sweat distribution and perceived wetness across the human foot: the effect of shoes and exercise intensity. Ergonomics 62(11):1450–1461. https://doi.org/10.1080/00140139.2019.1657185
163. Wilzman AR, Tenforde AS, Troy KL, Hunt K, Fogel N, Roche MD, Kraus E, Trikha R, Delp S, Fredericson M (2022) Medical and biomechanical risk factors for incident bone stress injury in collegiate runners: can plantar pressure predict injury? Orthop J Sports Med 10(6):23259671221104793. https://doi.org/10.1177/23259671221104793

Chapter 15
The Application of Infrared Thermography in Equestrian Sport

Maria Soroko-Dubrovina, Nina Čebulj-Kadunc, Mina C. G. Davies Morel, and Kevin Howell

Abstract Infrared thermography has found a broad range of applications in equestrian sport as a complementary diagnostic tool in veterinary practice and sport performance. Thermal imaging is useful for monitoring changes to a horse's body surface temperature resulting from exercise, allowing evaluation of the contribution of individual parts of the body to horse performance. Thermography can play an important role in rehabilitation, indicating an area of abnormality and so guiding where to concentrate further diagnostic imaging or treatment. It has also been used to assess the effectiveness of physical therapies.

15.1 Introduction: Detection of the Body Surface Temperature of the Horse and the Influence of the Environment on Temperature Changes

Equine sport is a unique discipline that involves two athletes, horse and human, working together as a team. Equine sport has two main sectors: racing and equestrianism. Horse racing is an immensely popular global sport divided into flat and jump

M. Soroko-Dubrovina (✉)
Institute of Animal Breeding, Wrocław University of Environmental and Life Sciences, Wrocław, Poland
e-mail: maria.soroko@upwr.edu.pl

N. Čebulj-Kadunc
Institute of Preclinical Sciences, Veterinary Faculty, University of Ljubljana, Ljubljana, Slovenia
e-mail: nina.cebulj.kadunc@vf.uni-lj.si

M. C. G. Davies Morel
Reader Emeritus, Department of Life Sciences, Aberystwyth University, Aberystwyth, UK
e-mail: mid@aber.ac.uk

K. Howell
Microvascular Diagnostics, Royal Free Hospital, London, UK

J. I. Priego-Quesada (ed.), *Application of Infrared Thermography in Sports Science*,
https://doi.org/10.1007/978-3-031-93311-0_15

racing (also referred to as National Hunt racing). In Great Britain, professional horse racing is conducted on 59 racecourses. Each racecourse is responsible for an average of 10 race meetings per annum [1].

Equestrianism involves jumping, dressage, eventing, driving, endurance, vaulting and reining disciplines. With more than 15,000 events each year, Europe is the location of more than 70% of the equestrian events held worldwide. The International Equestrian Federation (FEI) promotes equestrianism in all its forms and encourages the development of equestrian competitions throughout the world. Animal welfare, respect and health are the core values of the FEI.

Sport horses are particularly susceptible to injuries due to the high physical demand of the sport, but also due to genetic predispositions, age, conformation of the horse, type of training, rider skill, quality of surface or previous injuries [2, 3]. The contribution of each of these factors to possible injury development needs to be separately established to obtain better knowledge of injury risk and the ways to protect horses from lost training. Horse health and welfare are under human control, and therefore, increased knowledge in this area should bring about a decrease in early retirement of animals from performance sport [4]. For the improvement of the horse's health and welfare, it is important to have a tool which can predict future abnormal conditions or help in rehabilitation programmes.

Equine thermography has been established as a complementary diagnostic tool, which can detect an area of changed body surface temperature because of injury (Fig. 15.1). The broad application of thermography in veterinary medicine, rehabilitation and in assessment of welfare has been found to be very useful by many authors.

15.1.1 Normal Body Surface Temperature Distribution of the Horse

A constant temperature is a distinct feature of warm-blooded animals. To maintain it, heat generation and dissipation must balance within the body. An adult horse is assumed to have a basal rectal temperature from 37.2 °C to 38.1 °C, which increases significantly after an intensive workload to reach values near 39 °C [5]. A thermoregulatory centre located in the hypothalamus is responsible for maintaining a constant temperature by responding to incoming thermal information and determining how much heat is to be dissipated at any moment to maintain an optimum environment for all the cells of the body. Heat is produced as a result of chemical thermogenesis, taking place by tissue oxidation.

When a horse is at rest, its muscles produce approximately 50 kcal/min of heat, which is sufficient to maintain a proper internal body temperature [6]. During exercise, approximately 70–80% of the energy produced by working muscles is released as heat, and the amount of heat generated rapidly increases with work rate. A horse in trot generates 100 kcal/min, while in gallop this increases to 300 kcal/min. The

Fig. 15.1 Thermographic measurement with a handheld camera of the horse's limb

heat produced during exercise increases the temperature of the circulating blood. Blood flow increases in capillaries under the skin, where heat dissipation takes place from the skin surface. Compared to humans, muscle makes up a much larger proportion of the total body mass in the horse. Therefore, the heat produced by muscles during exercise by a horse must be lost to the environment very effectively.

Heat is dissipated at the skin surface mainly through convection, perspiration (evaporation) and infrared radiation. Evaporation is due to the molecular pressure difference of water vapour between the skin surface and environment, while the other components arise from the temperature difference between the skin surface and the environment [7]. The exact contribution of these processes during exercise is not fully understood: the balance of heat dissipation, transmission and storage may change in response to training time, gait type and speed, as well as a changing external environment. Research shows that the blood flow through the skin and evaporation are the key factors in the rate at which heat is dissipated to the environment [8, 9]. Other factors influencing the rate at which excess heat is eliminated are the horse's body surface area, hair coat thickness, the temperature and humidity of the external environment and wind speed [10].

An understanding of normal body surface temperature distribution is essential in detecting any abnormal pathologies [11, 12]. The distribution of body surface temperature across a horse is marked by a considerable inter-individual variability and is affected by environmental conditions [13–15]. In the study by Soroko et al. [16], the correlation between skin and ambient temperature was high ($r > 0.85$ at all skin sites). The variability of the temperature distribution in response to a changing ambient temperature has also been investigated. Mogg and Pollitt [17] found that low ambient temperatures (below 15 °C) caused vasoconstriction, and Palmer [13, 18] similarly reported that temperatures below 12 °C resulted in decreased blood circulation, leading to a reduction in body surface temperature in the distal parts of limbs. Ambient temperatures of 18–20 °C caused an increased, but unevenly distributed, blood flow arising from vasodilation in the distal limbs [17]. Recently, long-term studies have been conducted on Lipizzan horses that clearly indicate the influence of ambient temperature on body surface temperature [5, 19]. The resting body surface temperatures of adult Lipizzans were between 26.5 °C and 30.4 °C in May and between 22.7 °C and 28.7 °C in October. A positive correlation with ambient temperatures was also observed in resting horses.

Thermographic patterns of the healthy horse have been described [11, 20–22], with a high degree of symmetry between the left and right sides of the body [13]. This bilateral symmetry of all measured body regions before and after exercise was also confirmed in more recent studies conducted on adult Lipizzan horses during lunging [19] and riding [23] as well as on young Lipizzan fillies during lunging [5], regardless of the season. The distribution of skin surface temperature depends, in part, on thermal conduction from the veins lying at various depths within the body tissues [24–26]. Heat emanating from deeper lying arteries and internal organs is additionally conducted by the subcutaneous tissues. The distribution of the body surface temperature is determined by the arrangement and volume of the subcutaneous layer, the temperature of which depends on the metabolic activity of muscles. Body areas containing richly vascularised muscle tissue (e.g. neck, back, croup and chest) have higher surface temperatures in comparison to body areas where less muscle develops (e.g. proximal forelimb and hindlimb) or where they are not found at all (e.g. distal parts of the forelimbs and hindlimbs) (Fig. 15.2) [15].

The horse's hair coat absorbs infrared radiation emitted from the skin surface [27]. Air retained in the hair coat is a poor thermal conductor, affording efficient insulation against excessive heat loss from the skin [21]. Therefore, the hair thickness, density, length and arrangement also contribute significantly to the surface temperature distribution [28–30]. The effect of the hair coat on thermal emission

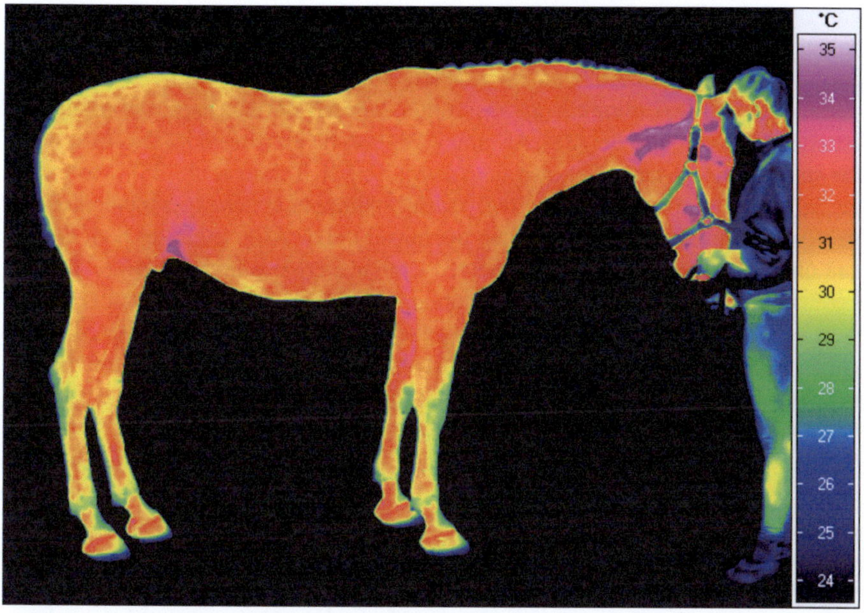

Fig. 15.2 Thermogram of the right lateral aspect of a horse. The warmest areas of the body are presented in *red*, *yellow* and *green* colours represent cooler areas and *blue* colours represent the coldest areas of the body

has many considerations. The animal's coat, as seen in a thermographic image, is simply the surface of a material with complex underlying layers. Its equilibrium temperature is determined by the balance between the loss of heat by convection to the surrounding air, heat conduction from the skin surface through the coat and also by the exchange of thermal infrared radiation to and from the environment [31]. The efficiency of infrared radiation from the coat surface is determined by the emissivity of the material. While the emissivity of the hair coat is high, it will not be precisely known for each animal and may vary across the surface, thus limiting the accuracy of temperature readings attained from thermographic imaging [32].

Clark and Cena [29] described the transmission of thermal radiation through animal coats of various colours and the related effects of environmental factors. The influence of radiant solar energy on pigmented and non-pigmented areas of the coat was also examined. There was a significant difference in body surface temperature (up to 8 °C) between hotter black pigments compared to white. It was concluded that body surface temperature differences are clearly expressed on the thermal image, but environmental factors affect the reliability of thermography for animal temperature measurements. In a later study, it was confirmed that under direct sun exposure, black areas of the skin are significantly warmer compared to unpigmented body areas. However, indoor examination did not give rise to a significant difference between pigmented and non-pigmented areas [18]. The body surface temperatures of Lipizzan horses, which are characterised by their white coat (dapple or fleabitten grey), did not differ significantly from those of other horse breeds with darker coat colours when measured indoors. These results support the relevance of measuring body surface temperature in an environment without direct sunlight [5]. It has also been suggested that solar energy absorption depends on the shape and position of areas in relation to the sun [28]. Accuracy can also be influenced by the curvature of the surface and variations in coat thickness [29, 33]. Turner et al. [33] investigated the influence of coat clipping on body surface temperature changes, demonstrating that clipped areas were warmer than unclipped areas. After exercise, both clipped and unclipped areas of the limbs showed similar increases in temperature, indicating that clipping was not necessary to obtain reliable thermographic results.

Based on the above findings, some generalisations can be made regarding the normal thermal pattern of the horse. Most studies report that the chest and the area between the hindlimbs along the ventral midline are the warmest, and the back and distal parts of the limbs are the coolest areas of the body [5, 19, 21, 23, 34]. More detailed work by Tunley and Henson [11] indicated that the dorsal midline along the back was 3 °C warmer compared to either side of the midline symmetrically, in both trained and untrained horses. Similar results were found in another study, where the back was warmer than the sides of the horse by 2 °C [35]. The temperature of the warmest areas of the distal parts of the forelimbs (the metacarpus bones, the fetlock and pastern joints) is influenced by the vascular supply. Hence, the main blood vessel areas—the palmar and digital arteries, and the palmar veins from the lateral and medial aspects—appear the warmest (Fig. 15.3). Similarly in the distal hindlimbs the warmest areas (the metatarsus bones, and the fetlock and pastern joints) are along the plantar and digital arteries, and also the plantar veins from the lateral and medial aspects (Fig. 15.4). The dorsal (cranial) and palmar/plantar aspects of the limbs are relatively cool as they are situated away from major blood vessels. The highest distal limb temperature viewed from the dorsal aspect is found in the coronary band, as it is situated close to the major arterio-venous plexus [12, 36] (Fig. 15.5). When viewed from the palmar/plantar aspect, the highest temperature is at the area between the heel bulbs (Fig. 15.6) [15].

Although the body surface temperature patterns in different horses are similar, no two horses are found to have an identical body surface temperature. The temperature distribution is affected by anatomy, individual traits in performance of the horse, injuries, training type and the level of animal adaptation to stress loads [12, 37]. It has been suggested that values of surface temperature across an individual healthy horse can range from 19 °C up to 32 °C. Flores [38] found the highest body surface temperatures (27–32 °C) on the head, neck, shoulder and flank and the lowest temperatures (24–26 °C) on the distal limbs. However, other studies have indicated slightly different temperature ranges, with the highest temperatures (25–28 °C) in the area of the head, middle part of the neck, chest and flanks, and the lowest temperatures (19–23 °C) at the distal limbs [34]. The differences in temperature ranges between these can probably be explained by the different ambient temperatures to which the horses were exposed.

Čebulj-Kadunc et al. [5, 19, 23] have recently conducted extensive long-term studies on temperature distribution in different body areas in Lipizzan horses. The

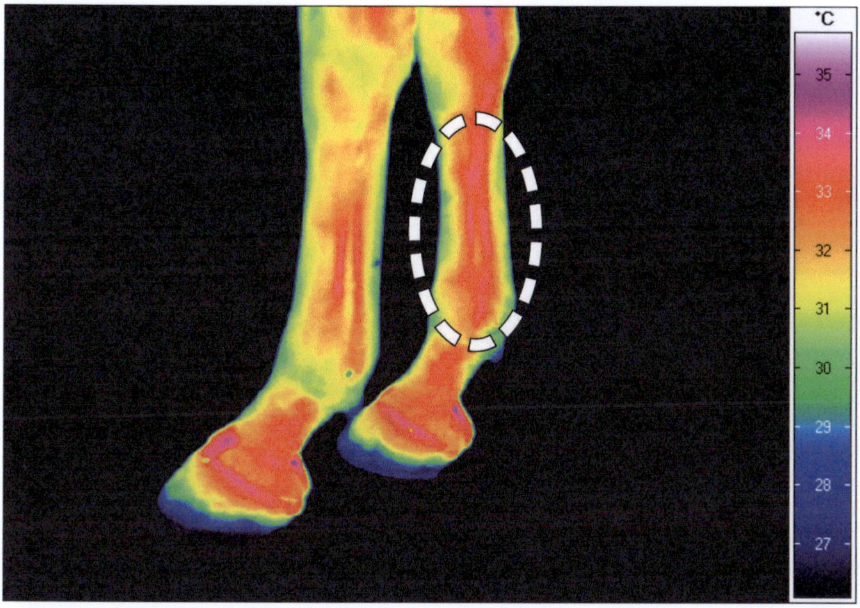

Fig. 15.3 Thermogram of the distal forelimbs (horse's *left* leg lateral and *right* leg medial aspects) from the *left* side of the horse, illustrating the higher body surface temperature associated with blood vessels (palmar artery and vein and digital artery indicated by dashed circle)

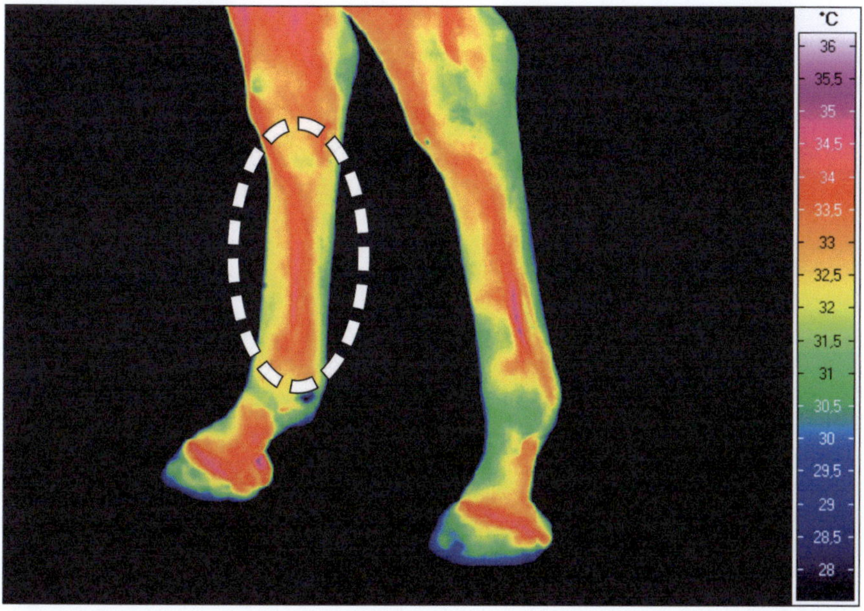

Fig. 15.4 Thermogram of the distal hindlimbs (horse's *left* leg lateral and *right* leg medial aspects) from the left side of the horse, illustrating the higher body surface temperature associated with blood vessels (plantar artery and veins and digital arteries, indicated with dashed circle)

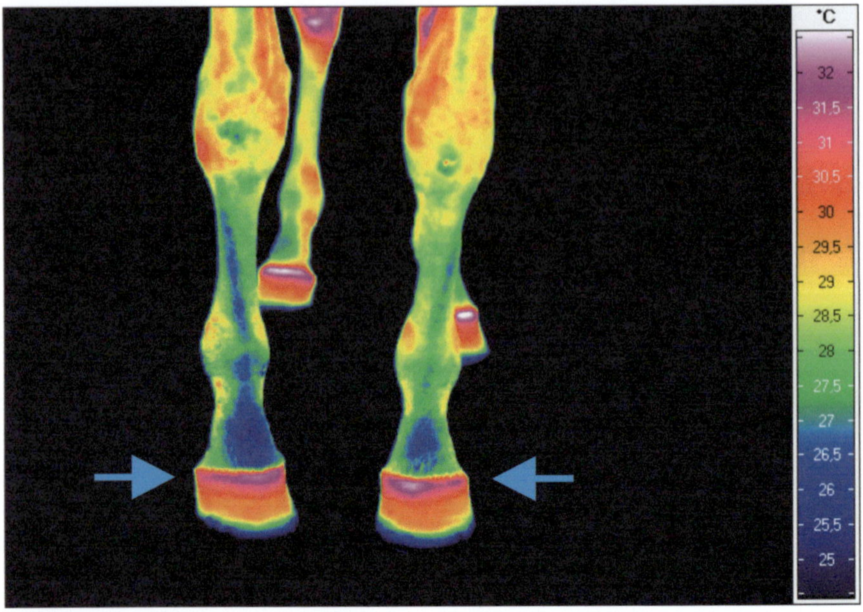

Fig. 15.5 Thermogram of the distal horse's *right* and *left* forelimbs from the dorsal aspect, illustrating the higher body surface temperature in the coronary band (indicated by arrow)

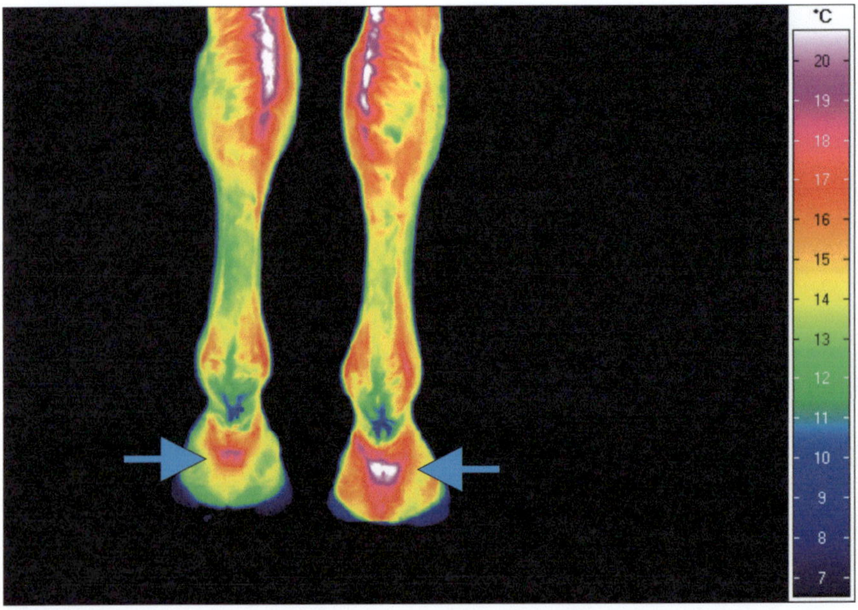

Fig. 15.6 Thermogram of the distal horse's *left* and *right* forelimbs from the palmar aspect illustrating the higher body surface temperature in the area between the heel bulbs (indicated by arrow)

resting body surface temperatures of adult Lipizzan horses used in a riding school were measured during the spring and fall months. The body surface temperatures ranged from the coldest values at the metatarsal and metacarpal regions (22–23 °C) to the warmest at the chest, neck and buttocks (27–29 °C), while the croup reached intermediate values (24–26 °C) [19]. The distribution of body surface temperatures at rest in young Lipizzan fillies [5] followed similar patterns in four consecutive exercise tests conducted at 2-week intervals during the winter months. The highest temperatures were similar in the chest, neck and buttocks (15–21 °C). Intermediate temperatures were recorded for the back and croup (12–16 °C) and the lowest for the metatarsus (11–15 °C) and metacarpus (9–13 °C). The differences between the groups of regions with the highest, medium and lowest basal temperatures within a test were significant.

In an unpublished investigation of the authors of this chapter, body surface temperatures were also compared in Lipizzan stallions tested in winter and summer at an average ambient temperature of 6.3 ± 0.5 °C in winter and 24.5 ± 0.5 °C in summer. Although the same temperature distribution was found for the regions of interest for the young Lipizzan fillies and adult riding horses, the absolute skin temperatures were significantly lower in winter than in summer. The greatest differences between the summer and winter skin temperatures were found for the distal parts of the limbs, which ranged from 11 to 14 °C in winter and from 28 to 32 °C in summer, indicating peripheral vasoconstriction in cold and vasodilation in warm environments.

15.1.2 Recommended Procedures for Equine Thermographic Examination in Accordance with Standards for Thermal Imaging in Veterinary Medicine

The proper use of thermography to evaluate body surface thermal patterns requires a controlled environment, and the physiological state of the horse must be considered in order to reduce variability and eliminate errors of interpretation [15, 39, 40]. Thermography measurement standards have been established in equine veterinary medicine [41–43]. The thermographic examination should be performed in an area sheltered from the sunlight, in the absence of air drafts and extreme variations in ambient temperature [12, 18, 44]. The temperature of the examination area should ideally be maintained between 21 °C and 26 °C. This is particularly important for the determination of body surface temperature of the distal parts of limbs, which are unstable at low ambient temperatures due to vasomotor activity [17].

At low temperatures, there is also a statistically significant influence on the bilateral symmetry of distal limb surface temperature [13]. It is recommended that the horse should be acclimatised in the examination room for 15–20 min prior to imaging. A longer period of equilibration, up to 60 min, may be required when the examined animal is transported from an extreme cold or hot environment [11, 13].

The thermographic examination should be performed when the horse is at rest, before exercise [42]. The horse must have a clean, dry hair coat and skin and should be groomed at least 1 h before the examination. Artefacts can be produced by any material on the body surface such as water (wet areas), dirt, thickened coat, irregular patterns of clipping or scars [18, 45, 46]. It is necessary that the hair coat be short, of uniform length, and lie flat against the skin to permit thermal conduction [33]. The feet should be clean, picked out and brushed to remove external contamination. Anti-inflammatory medications, vasoactive drugs, regional and local nerve blocks, sedation and tranquilisation should be avoided where possible because of their effect on superficial blood perfusion [42]. Blankets and bandages should be removed at least 30 min before examination [18].

Biological materials such as skin and hair coat have generally high emissivity, making these surfaces efficient infrared radiators. In equine thermography, a lack of accurate knowledge about the emissivity of the surfaces of the horse may limit the accuracy of temperature measurement. Emissivity is usually assumed to be 1.0, which will have a small impact on the accuracy of perhaps a few tenths of a degree Celsius [45, 47]. For thermographic examination, the horse should be presented with forelimbs and hindlimbs equally loaded and the lateral and medial aspects of the distal parts of the limbs visible from both sides [48]. Imaging of the horse should include a lateral aspect of the whole body from both sides. The distance between the horse and the camera should be about 7 metres. Images of the distal parts of the forelimbs and hindlimbs should include dorsal, palmar/plantar, lateral and medial aspects. The distance between the camera and the horse should be set at 1.5–2 metres [49]. The thermograms should also include the lateral upper part of the thoracic and pelvic areas and also the lateral aspect of the neck, head and trunk from both aspects. The distance between the camera and the horse for these parts of the body should be about 3 metres. The thoracolumbar region of the spine should be imaged from the dorsal aspect [35] from a height of approximately 2 metres and at a distance of 1.5 metres from behind the horse. Close-up thermographic images of regions of interest can be taken if necessary. Analysis of body surface temperature symmetry across contralateral body areas can be performed.

15.2 Application of Thermography as a Complementary Diagnostic Tool in Veterinary Medicine

Thermography was first introduced to veterinary medicine in the mid-1960s, when several papers reported the potential value of thermography in diagnosis. A key study presented applications of thermography for the detection of splints, bruised joints and bowed tendons [50]. Delahanty and Georgi [51] introduced the use of thermography in detecting clinical cases of squamous cell carcinoma, slab fractures of the third metacarpal bone, bone spavins and deep cervical abscesses. Since then, thermography has evolved steadily to be employed in a variety of areas in equine medicine [52, 53].

15.2.1 Thermographic Diagnosis of Soft Tissue Injury and Superficial Bone Lesions in Equine Medicine

Strömberg [46, 54, 55] was the first to use thermography in detecting and monitoring orthopaedic injuries of the distal parts of the limbs in racehorses. These studies reported that the thermographic image can predict inflammation of the superficial digital flexor tendon up to 14 days before clinical signs appear. Thermographic examination detected the increased temperature of the injured limb, even though radiography did not show any changes. It was concluded that thermography can play an essential role in preventing lameness in racehorses.

Further studies investigated the use of thermography in diagnosing pathological conditions of the limbs and back [20, 56–59]. Clinical diseases of the distal parts of the limbs that have been recognised and characterised by thermography include tendinopathy (Fig. 15.7) [57, 60], laminitis [12, 21], joint arthritis (Fig. 15.8) [22, 56, 60, 61], suspensory desmitis [21], inflammation of the stifle [62], stress fractures [21], bucked shins (Fig. 15.9) [57, 61], sole abscesses (Fig. 15.10) and navicular syndrome [33].

For back abnormalities, thermography is useful in localising spine-related diseases associated with subluxation of the third lumbar vertebrae [20], neuromuscular disease of the thoracolumbar region [35], inflammation of spinous processes (Fig. 15.11) [59] and intervertebral arthritis of the thoracolumbar spine [63, 64].

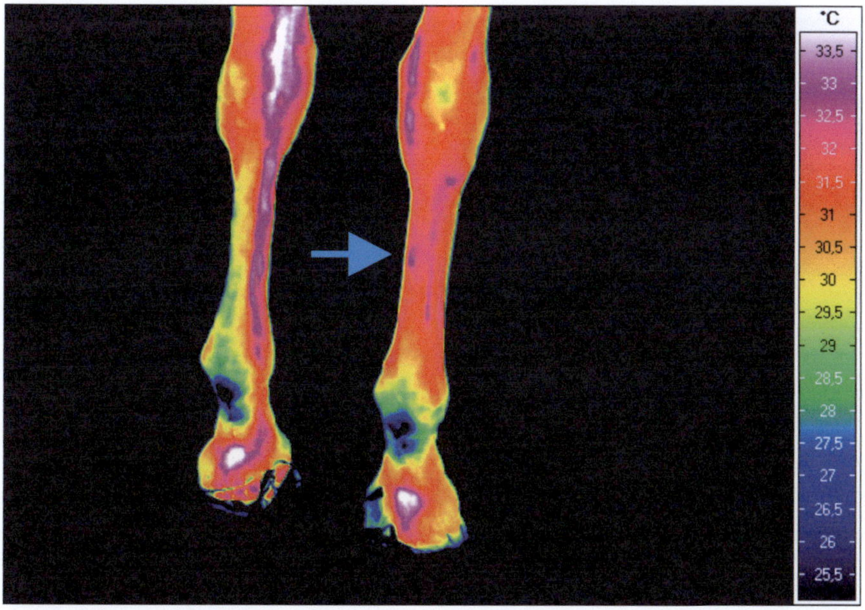

Fig. 15.7 Thermogram of the palmar aspect of the distal part of the forelimbs, illustrating tendinopathy of the superficial digital flexor tendon at the level of the third metacarpal bone of the *right* forelimb (indicated by arrow)

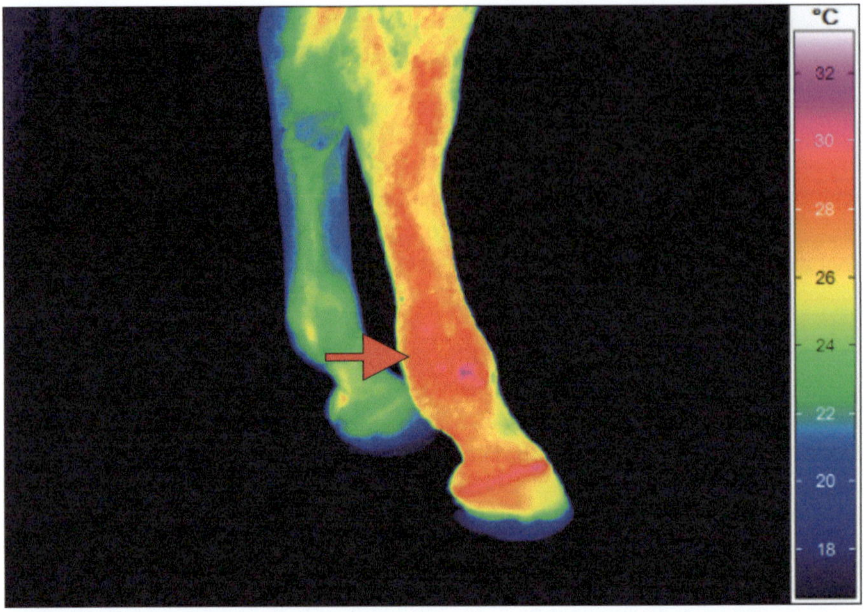

Fig. 15.8 Thermogram of the distal part of the *right* and *left* hindlimbs (horse's *right* leg lateral and *left* leg medial aspects), illustrating fetlock joint arthritis of the right hindlimb (indicated by arrow)

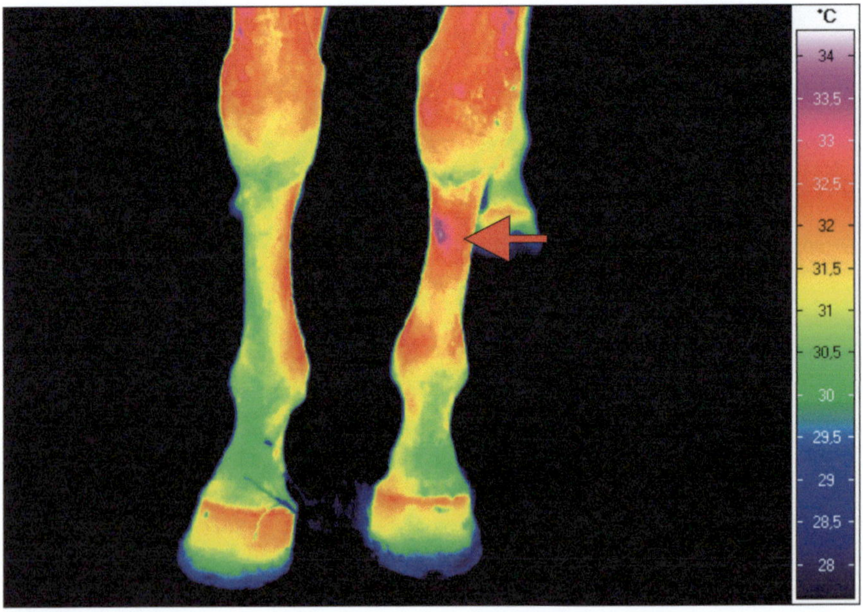

Fig. 15.9 Thermogram of the distal part of the horse's *right* and *left* forelimbs (dorsal aspect), illustrating clinical inflammation of the left third metacarpal bone (bucked shins) (indicated by arrow)

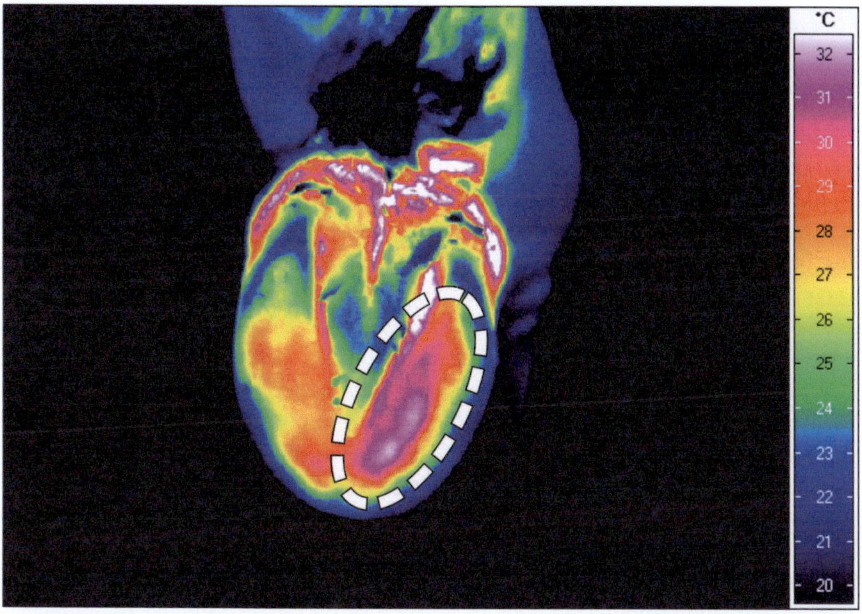

Fig. 15.10 Thermogram of the *left* hoof sole from the solar aspect, illustrating an abscess in the medial aspect of the hoof sole (right side, indicated by dashed circle). (Figure obtained from Soroko and Davies Morel [45])

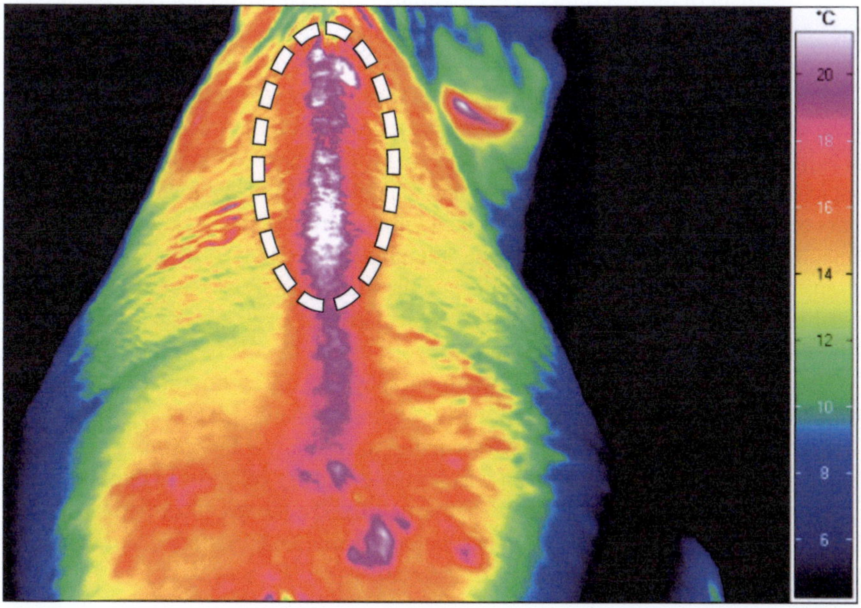

Fig. 15.11 Thermogram of the back from dorsal aspect, illustrating inflammation of the spinous processes of the thoracic vertebrae (indicated by dashed circle)

15.2.2 Thermography Complementary to Ultrasonographic and Radiographic Examinations

Thermography examination provides injury localisation and physiological inflammation information about the affected tissue. However, it is unable to specify the injury and cannot define the lesion aetiology. Therefore, the value of thermography in veterinary medicine is greatest when it is used complementary to other modalities. Much research emphasises the importance of the application of thermography as complementary to ultrasonographic and radiographic examinations in the evaluation of soft tissue injury and superficial bone lesions [12, 63, 65]. Vaden et al. [22] have shown the value of combined thermographic and radiographic examination in joint arthritis. Their study was based on 20 racehorses clinically evaluated for abnormalities in their gait due to pain in the tarsal joint. The examined horses were physically unable to perform because of early tarsal joint arthritis which was confirmed by thermography. However, no changes were detected in radiographic examinations at that stage. The study demonstrated that the early signs of arthritis can be detected by thermography prior to confirmation by radiographic examination. Later, Eddy et al. [48] correlated thermography findings with radiographic, ultrasonographic and scintigraphic methods in a group of 64 horses affected by lameness. Thermography correlated with ultrasonographic examination in 10/15 cases (66.7%), with scintigraphy in 15/20 cases (75%), but with radiography in only 15/29 cases (51.7%), suggesting that thermography may only have a limited ability to detect some of the subtle or chronic bony injuries which cause lameness. During the assessment of tendon injury, thermography examination did not correlate well to the structural reorganisation of the tendon matrix as assessed by ultrasonography [66].

In cases of back pain, thermography can localise the exact site of the lesion. In a case of subluxation of the third lumbar vertebra, where the utility of radiography was limited due to the large mass of overlying soft tissue, thermography was able to indicate the area of injury [20]. Kold and Chappell [36] investigated the effectiveness of thermography in detecting clinical inflammation of the back. Thermographic imaging was successful in identifying inflammatory activity of the superficial soft tissue overlying the thoracic vertebrae and the sacroiliac area as a result of injury. It has been suggested that thermography can also help to locate the site of pain, allowing other diagnostic imaging methods to work more effectively [64]. Fonseca et al. [63] evaluated the effectiveness of thermography and ultrasonography in combination for detecting back abnormalities in Western Quarter Horses. Thermography was used to locate the area of the lesion, and ultrasonography was used to characterise the lesion. This combination was found to be very efficient in thoracolumbar lesion diagnosis. However, not all abnormalities diagnosed ultrasonographically were detected by thermography: in particular, thermal imaging failed to detect overriding dorsal spinous processes (*kissing spine*).

15.2.3 Thermography Sensitivity for Detecting Subclinical Inflammation

Thermography may be used to diagnose subclinical inflammation before the onset of clinical signs of pathology. Strömberg [46] demonstrated the use of thermography to identify pathological changes in the superficial digital flexor tendon prior to the appearance of clinical inflammation. Vaden et al. [22] and Bowman et al. [56]) both used thermography to detect subclinical arthritic changes. Other studies also reported the use of thermography to monitor subclinical signs of inflammation in the distal parts of the limbs up to 2 weeks before the onset of clinical lameness and swelling [21, 58]. Turner et al. [58] reported the successful use of thermography in predicting distal limb tendon, fetlock, carpus and tarsus injury in Thoroughbred racehorses. The specific purpose of the study focused on predicting injuries before they became clinically evident and also on determining how well thermography harmonised with trainers' concerns. The authors concluded that thermography was a valuable tool in preventing lameness by allowing early treatment or changing the training system in response to thermographic findings. Defining more precisely the indicators of subclinical inflammation would allow quick and practical thermographic diagnosis to better protect horses from injury. Turner [21] has shown that temperature differences of more than 1 °C over 25% of the compared body area can be considered to be abnormal. In another study [57], a temperature change of 1.25 °C was considered indicative of subclinical inflammation in the distal limb. The early stages of bucked shins were detected by Turner [21] when local temperatures were higher (1–2 °C) over the dorsal third metacarpal bone compared to the surrounding areas. In human studies, a much smaller body surface temperature difference (0.4 ± 0.3 °C) between the left and right side has been reported as indicative of an abnormality [67].

15.2.4 Thermography in the Study of Neurological Disorders

In a horse with Horner's syndrome (a neurologic disease involving nerve paralysis of the sympathetic nervous system around the head and neck), Purohit et al. [68] demonstrated a local increase of body surface temperature of 2–3 °C due to vasodilation in the denervated area. Research by Waldsmith [69] and Waldsmith and Oltmann [70] indicated that regional nerve anaesthesia caused an increase in body surface temperature secondary to vasodilation from sympathetic blockade, whereas Holmes et al. [71] described local perineural anaesthesia as having no effect on body surface temperature in the forelimbs. Denoix [72], however, reported that nerve injury results in a decrease in body surface temperature.

15.3 Application of Thermography in Monitoring Rehabilitation

15.3.1 The Use of Thermography in Monitoring the Effectiveness of the Healing Process and Medicine Application

Thermography has been used in rehabilitation as a tool for monitoring the effectiveness of the healing process. Ghafir et al. [73] studied the use of intra-articular corticosteroid therapy in horses with arthritis. During the healing process, thermography quantified the response to anti-inflammatory medication. Purohit and McCoy [20] used thermography to monitor the effectiveness of anti-inflammatory therapy. Experimental horses had an inflammatory response induced in the area of the metacarpal bones, and anti-inflammatory benzydamine was injected 24 h later. Thermographic images of the distal limbs demonstrated a reduction of inflammation by indicating a decrease in temperature. When the warmth of the affected area was no longer detected by palpation and the horse appeared sound, the thermographic image continued to indicate an increased temperature. Thermography was also used to monitor corticosteroid treatment in amphotericin B-induced carpal and tarsal arthritis in ponies [56]. A group of experimental clinically healthy ponies were injected with 10 mg of amphotericin B to induce arthritis either in the carpal or tarsal joint. After arthritis induction, the joint was treated with 100 mg of corticosteroid in order to decrease the time of healing, and its effectiveness was monitored by thermographic imaging. In the case of the carpal joints, there was no significant difference in the rate of return to normal thermal patterns between the treated and untreated group. However, in the case of the tarsal joints the treated group returned to normal thermal patterns significantly faster than the control group.

The treatment of Horner's syndrome with an a2–adrenergic agonist has also been studied using thermography [68] in two horses presenting with clinical signs of the disease. Facial thermographic images were recorded before and 30 min after application of the treatment. A decrease in facial temperature of 2.8 and 3.7 °C in the two horses demonstrated the effectiveness of treatment.

Another study has indicated that thermography can be used in detecting temperature differences of the forelimb before and after isoxsuprine administration [74]. This drug is traditionally used to treat navicular disease and sesamoiditis, both of which cause an ischemic effect [75, 76]. Maximal temperature was significantly ($p < 0.05$) higher in the treated than the control group, 2.2 °C (for horses treated with 0.9 mg/kg) and 1.8 °C differences (for horses treated with 1.2 mg/kg) in the distal part of the limbs; indicating thermography to be a sensitive tool for monitoring the effectiveness of isoxsuprine administration. A similar conclusion was drawn in a study of horses with navicular disease which were orally administered with 0.6 mg/kg of isoxsuprine hydrochloride paste twice a day for 6–14 weeks. The results revealed a significant temperature increase in the distal limb which occurred from 90 to 480 min after drug administration [77]. Contrary conclusions have been presented by Harkins et al. [78], where thermography examination of the forelimbs detected a significant decrease in the superficial temperature following isoxsuprine injection 30–240 min after injection. Thermography was able to detect a temperature decrease in the forelimbs, which was apparent for no more than 4 h. Differences between the above studies could be associated with the route of administration (injection versus oral).

15.3.2 Monitoring Recovery of Muscle Strain, Inflammation and Myopathy

In physiotherapy practice, thermographic examination can help to locate the body areas potentially susceptible to injury and, in the case of an injury, to monitor the treatment progress. It helps the equine physiotherapist to determine the body areas requiring treatment and ensure quick and efficient recovery to full athletic performance.

Muscle injuries, such as muscle stress, post-exercise fatigue and azoturia, are common causes of lameness in horses and are recognised as physiological disturbances [79]. Thermography can detect temperature changes associated with muscle strain (Fig. 15.12), muscle inflammation (Fig. 15.13), caudal thigh myopathy and croup myopathy [80, 81]. Thermography may have important clinical applications in the assessment of individual muscle injuries which are difficult to diagnose [21]. Once the area of inflammation is determined, physical therapy can be applied

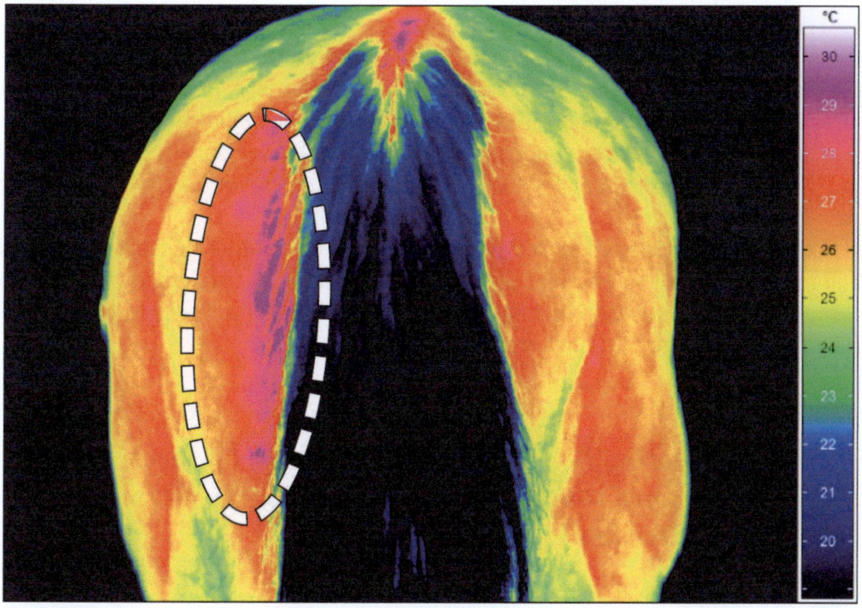

Fig. 15.12 Thermogram of the croup (caudal aspect), illustrating semitendinosus muscle strain of the *left* side of the croup (indicated by dashed circle)

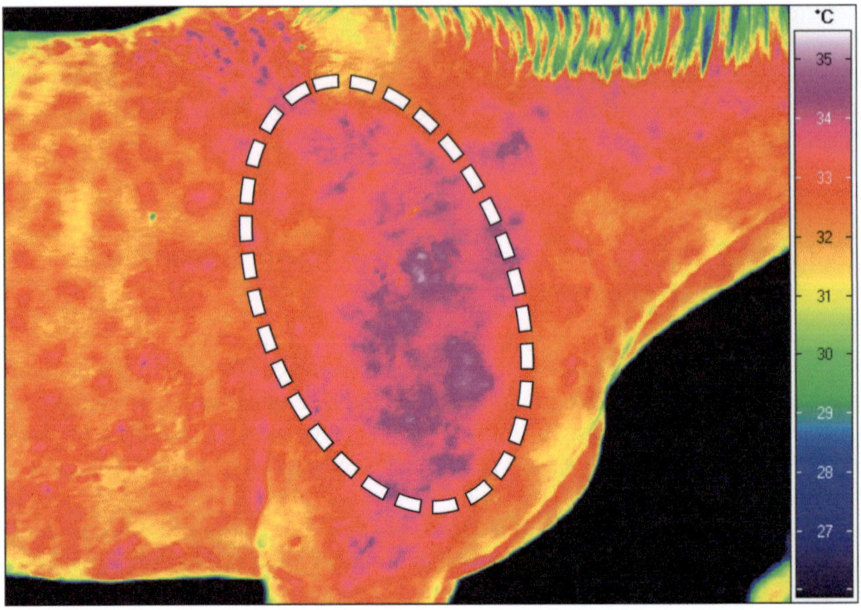

Fig. 15.13 Thermogram of the *right* side of the shoulder (lateral aspect), illustrating scapula muscle inflammation (indicated by dashed circle)

directly to that area. Then treatment and rehabilitation, such as massage or therapeutic ultrasound, can be applied more specifically [81].

The effectiveness of treatment of the thoracolumbar region has also been established by thermography. von Schweinitz [82] used thermography to monitor acupuncture therapy in a case of neuromuscular disease. Han [83] reported that an affected neurological site increased its temperature by more than 5 °C within half to three-and-a-half hours after acupuncture in a group of treated horses. Thermography has also been demonstrated to be an effective method of diagnosing chronic pain [35].

Levet et al. [84] studied the use of thermography for monitoring the development and recovery of cast sores on the distal parts of the limb. Thermography was able to detect superficial and deep dermal sores, which was useful in deciding the optimal time for cast change, avoiding serious cast complications.

15.3.3 Monitoring Body Surface Temperature Changes Associated with Rehabilitation Exercise in Horses on a Treadmill

Water treadmill exercises are helpful for horses undergoing rehabilitation processes after injuries to the back and limbs [85]. Water treadmill exercise is an aerobic activity which increases blood circulation in the working muscles [86]. Thermography has the potential to assess the changes in blood flow that reflect the working muscles undergoing rehabilitation in an aquatic environment. In the study presented by Yarnell et al. [87], thermography was used to assess surface temperature changes in the hindlimb as a measure of muscle activity and so inform rehabilitation management. The study found an increase in the body surface temperature of the hindlimb at the onset of exercise, and temperature increased until the end of exercise on the treadmill. There was a significant difference in hindlimb surface temperature when comparing exercise on a dry treadmill and with water at the level of the coffin joint and carpus joint ($p < 0.0001$).

Using dynamic thermography, another study investigated the feasibility of evaluating the heat loss from the horse's body surface during exercise on a treadmill [88]. Continuous measurement of body surface temperature change, walking (10 min), trotting (10 min), returned to walking (5 min) and during recovery (10 min), was undertaken on five horses using a thermographic camera. Four regions of interest were analysed through thermographic imaging: the neck, shoulder, chest and croup areas. It was found that, by the end of the first walk, the temperature at all regions of interest had increased, with the neck area remaining the warmest. During trot, the temperature increased for the first 5 min, with the greatest increase occurring in the chest area. During the second walk, the temperature declined in all the measured regions. By the end of the recovery period, the temperature had decreased in all regions, with the neck and chest areas demonstrating the biggest drops in temperature. It was found that heat loss from the body surface, during exercise, can be evaluated by means of dynamic body surface temperature measurement using thermography [88] (Fig. 15.14).

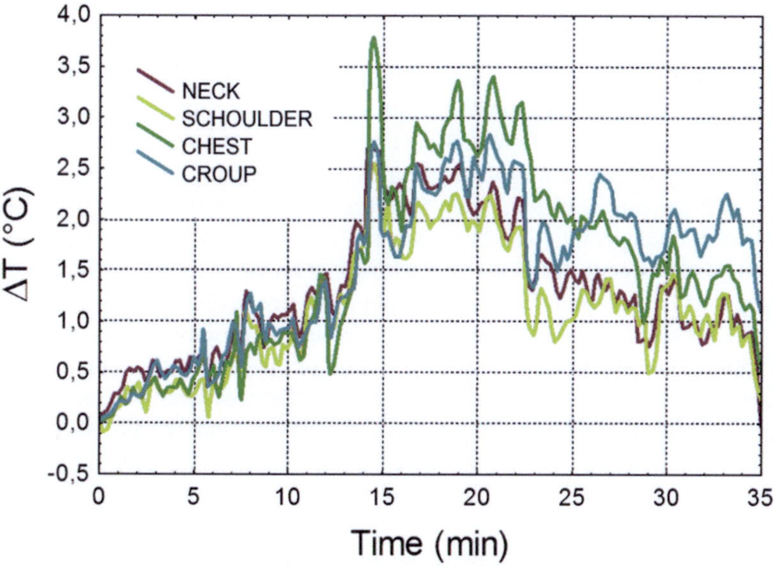

Fig. 15.14 Temperature change from baseline from all four regions: neck, shoulder, chest and croup during walk, trot, second walk and recovery period

15.3.4 Use of Thermography in Assessing Effectiveness of Physical Therapies

Thermography has been employed to evaluate the effectiveness of physical treatment methods [89]. In research by Rindler et al. [90], the impact of pulsed electromagnetic field (PEMF) therapy [91] on body surface temperature was examined using thermographic imaging on the backs of clinically healthy horses. Twenty horses received either PEMF blankets or placebo blankets on their backs for 40 min daily over 10 days. After PEMF therapy, temperatures increased by 0.7 °C compared to baseline, and this elevation persisted for 30 min. However, there were no statistically significant differences in temperature changes between the two groups. Additionally, body surface temperatures on the horses' backs showed no significant change throughout the 10-day therapy period. Thermography also assessed the PEMF BEMER system (Physical Vascular Therapy) [92] on body surface temperature in the distal parts of the forelimbs in clinically healthy horses. Horses were divided into an active BEMER group which had BEMER boots applied to the distal parts of the forelimbs, and a sham group which had BEMER boots applied but without activation of the device. Thermography did not identify changes in body surface temperature after BEMER therapy at the distal parts of the forelimbs for either group. It is possible that the distal limbs require longer BEMER therapy application in order to detect any effects on body surface temperature [93]. In a previous study, thermography assessed the influence of magnetic blankets on the body surface temperature of the back area in research and sham groups. In both groups, there was a significant increase in body surface temperature after treatment. It has also been suggested that the magnet-induced alteration in microcirculation depends on the initial vessel tone, indicating that the effect may vary depending on the condition of the treated tissue [94].

Thermography has also been utilised to observe the effects of focused extracorporeal shock wave therapy (ESWT) [95] on bone and bone-tendon junctions in clinically healthy horses. Temperature variations were measured before and after therapy in the suspensory ligament area at the metacarpus and fourth metatarsal bone, with untreated limbs serving as controls. Although treated limbs showed a slight surface temperature increase compared to control limbs, this difference was not statistically significant. It is worth noting that minor temperature fluctuations in healthy horses may stem from natural physiological metabolic activities [96]. A more recent study assessed the influence of radial extracorporeal shock wave therapy (rESWT) on the body surface temperature of the longissimus dorsi muscle in clinically healthy racing horses. Temperature of the back area was measured just before and after rESWT in a sham group and a study group. Results showed a significant increase in skin surface temperature after therapy only in the study group, confirming that rESWT increases body surface temperature overlying the longissimus dorsi muscle in clinically healthy racing Thoroughbreds [97].

Recent studies have utilised thermography to evaluate high-intensity laser therapy (HILT) [98] effects on body surface temperature in the tarsal joint area of healthy racehorses. After a single HILT treatment, the tarsal joint surface temperature rose by an average of 2.5 °C, marking the first identification of HILT's photothermal effects on a joint via thermographic imaging [99]. A follow-up study used thermography to assess HILT's impact on the superficial digital flexor tendon in the hindlimb of healthy racehorses, finding an average temperature increase of 3.5 °C post-treatment, further confirming HILT's photothermal effect on soft tissue [100].

Another study compared HILT's effects on surface temperature in pigmented versus non-pigmented skin on the forelimb fetlock joint of healthy racehorses. Results indicated that HILT increased temperature in pigmented skin, while temperature in non-pigmented skin decreased, suggesting that melanin in the epidermis significantly influences light absorption and photothermal response [101]. Additional research examined skin surface temperature changes of the carpal joint pre- and post-HILT. Findings showed that temperature increases in clipped skin were lower than in non-clipped skin (2.1 °C vs. 3.9 °C), highlighting the hair coat's role in light absorption and temperature change [102, 103].

Thermography has also been used to evaluate body surface temperature changes of horses subjected to various water cooling treatments after physical exercise. Cooling the limbs several times changed the surface body temperature from 34.2 ± 0.3 °C to 32.0 ± 0.3 °C and was more efficient than the repeated cooling of both the upper and lower body parts, leading to a temperature change from 34.6 ± 0.2 °C to 33.2 ± 0.3 °C. It was concluded that the application of cool water on the limbs three times after moderate effort seems to be sufficient to cool the horse down [104]. Recent studies have applied thermography to assess how whole-body cryostimulation affects body temperature in horses. The findings indicated that exposing horses to brief thermal stress at −140 °C for 3 min leads to a notable drop in body temperature immediately post-treatment. The extent of skin cooling varied, with the large dorsal croup showing a reduction of −1.1 °C, while the medium croup experienced a decrease of −2.8 °C. However, this research was conducted on only five horses outdoors, meaning environmental factors may have significantly impacted the results [105].

15.4 The Application of Thermography in Horse Performance

15.4.1 The Influence of Exercise on Body Surface Temperature Changes

Body surface temperature is indicative of transformations occurring in the horse due to exercise. Training entails, among other things, changes in the function of the respiratory and circulatory systems. These changes occur simultaneously in both systems and in an integrated way, as part of the body's reaction to maintain homeostasis. During training, the circulating blood volume increases, which has a beneficial effect on the stability of the circulatory system parameters and thermoregulation during exercise [106, 107]. The increased blood circulation results in increased venous return and a raised cardiac output. During exercise, muscles, tendons, ligaments and bone are subject to overloads [108]. This affects the intensity of biochemical transformations taking place in individual tissues. Muscles increase their oxygen uptake from blood flowing through blood vessels. This contributes to increased skin circulation and thus to improved dissipation of excess heat to the environment [109]. Simon et al. [110] demonstrated the influence of exercise on body surface temperature. Thermographic examination of horses was performed before and after exercise on a treadmill (Fig. 15.15). The temperature of the upper part of the body after exercise was increased by 6 °C, whereas the distal parts of the limbs were increased by 8 °C. Thermographic examinations demonstrated there was a significant temperature difference between measurements made before exercise and 15 min after exercise, and that significant differences remained evident up until 45 min after exercise. It was also found that muscular areas in the upper part of the body returned more quickly to the pre-exercise temperature, compared to the distal parts of the limbs.

In order to understand the physiological responses of Lipizzans to workloads and to establish the standards for assessing their fitness during training or performance, extensive long-term studies with graded exercise tests (10-min consecutive intervals of walk, trot and canter) have been conducted on adult Lipizzan horses used in a riding school and on young Lipizzan fillies during the initial phase of training [5, 19, 23]. The results of these studies, which are still ongoing and will be continuously supplemented by new findings, will contribute to the development of

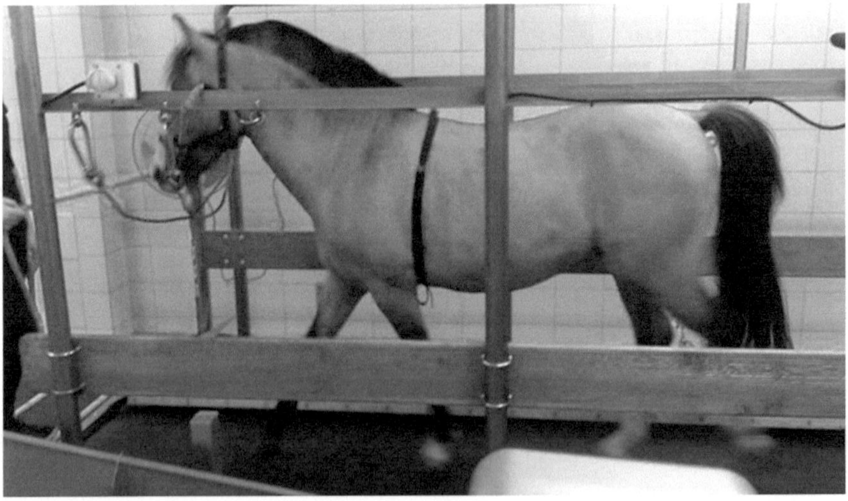

Fig. 15.15 Image of a horse working on a treadmill in walk

standards and protocols for monitoring fitness levels and training progress in Lipizzan horses and help in monitoring the health and well-being of horses in general. After exercise, which included lunging in successive 10-min intervals at a walk, trot and canter, the body surface temperatures of the young Lipizzan fillies increased significantly [5]. The areas that reached the highest temperature after exercise in all four exercise tests conducted every 2 weeks in December and January (ambient temperature range 6–11 °C) were the neck (23–29 °C), followed by the chest (22–28 °C), suggesting that the greatest accumulations of maximum temperatures are over the anatomically largest surfaces, which play a significant role in thermoregulatory function. Intermediate body surface temperatures were measured on the buttocks (20–25 °C), back (17–22 °C) and croup (15–19 °C), and the areas that warmed up the least were the metatarsus (12–15 °C) and metacarpus (11–14 °C). The different body surface temperatures of different regions within individual tests of young Lipizzan fillies indicated different muscle work of these regions, with the cranial parts of the body bearing a greater load in supporting and moving the body than the caudal parts. Similar exercise tests were also carried out on adult Lipizzan horses in spring and fall [19], at average ambient temperatures of 20 ± 0.5 °C and 11 ± 0.7 °C, respectively. The higher ambient temperatures were most likely the main reason for the higher body surface temperatures measured in these horses after exercise compared to the young Lipizzan fillies [5].

In an investigation of the authors, Lipizzan fillies were tested in four consecutive exercise tests, conducted by lunging with graded exercise load (10-min consecutive intervals of walk, trot and canter). The exercise testing started in December and was thereafter repeated every 2 weeks, at the ambient temperatures in the ranges of 6.7 ± 1.1 °C in the first exercise test to 11.1 ± 0.5 °C in the fourth exercise test. The highest average body surface temperatures in Lipizzan fillies after exercise were measured at the neck and chest (33–35 °C), followed by the buttocks and croup (30–33 °C) and metatarsus and metacarpus (25–26 °C). The highest warming (temperature difference between the values before and after exercise, Fig. 15.16) was observed for the neck (7–9 °C), followed by the chest (5–9 °C). An intermediate warming was observed for the back (3–6 °C), the croup (2.5–4.5 °C) and the buttocks (2–4 °C). The regions that warmed up the least were the metatarsus and metacarpus: the maximum warming was registered in test 1 (3.4 °C for the metatarsus and 2.4 °C for the metacarpus). The differences in warming between regions were not significant between exercise tests, with the exception of the metatarsus ($p < 0.05$). In contrast, when comparing different regions within a single test, significant differences ($p < 0.001$) in heating were confirmed (Fig. 15.16). Nevertheless, the temperature difference of the different body regions before and after the exercise remained within the ranges reported for horses under comparable workloads [111]. A similar pattern of body surface temperatures and extent of warming was also observed in adult Lipizzan horses during identical exercise tests by riding [23] and in young Lipizzan stallions (ref.).

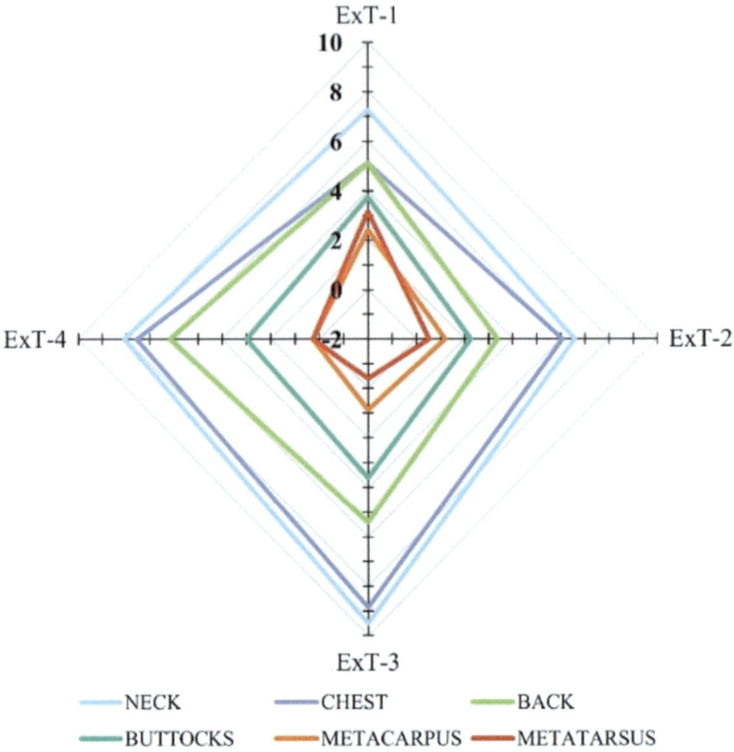

Fig. 15.16 Graphical presentation of warming (temperature differences of body surface regions before and after exercise) in young Lipizzan fillies; scaling: absolute temperatures (°C). Exercise tests 1 to 4 (ExT-1 to ExT-4) [5]

15.4.2 Changes of the Equine Musculoskeletal System During Physiological Effort Measured by Thermography

Constant overloads of the equine musculoskeletal system in response to training [112] can cause abnormalities associated with painful conditions or diseases, leading to loss of performance [4]. The physical demands on the horse are dependent on the type of training and the skills of the rider [113, 114]. During training, horses put particular strain on the distal parts of the forelimbs and back, leading to frequent injuries in these areas [4, 58, 115]. The distal bones and digital flexor tendons of the forelimbs are subjected to extreme overloads [116], which predispose the horse to injury [117]. It is therefore important to develop an understanding of the physiological response of the muscular system during exercise in order to determine the most appropriate training programme for the horse.

Soroko [118] documented abnormalities of the forelimbs associated with strains and overloads using regular thermographic examinations in racehorses. Other recent studies [34, 119–122] have investigated the use of thermography for monitoring changes in horse surface temperature resulting from exercise, and a model for temperature distribution at the surface of the forelimbs and hindlimbs has been developed. The temperature of the limb surfaces increased significantly with exercise. It was demonstrated that the precise body surface temperature depended on the training intensity and was considerably higher in forelimbs (29.3 °C in mares, 27.1 °C in stallions) in comparison to hindlimbs (28.9 °C in mares, 27.6 °C in stallions) [37]. Increased body surface temperature persisted for more than half an hour post-exercise. Measurement of the horse's body surface temperature after exercise was useful to monitor the impact of training on body surface temperature and to assess the training [123]. However, ambient temperature has to be considered in assessing the influence of exercise on body surface temperature. It was found that ambient temperature had a significant influence on surface temperature over the joints during training [124]. The fetlock joint of the forelimb and hindlimb warmed the least during training but was influenced the most by differences in ambient temperature. The surface temperature differences before and after training may be an important indicator of the thermoregulatory response to exercise in racing horses. Understanding surface temperature changes in response to regular training is necessary for future studies on diagnosing injuries of joints in young racehorses [124]. In further studies, a model of the horse's temperature before and after exercise and after recovery was determined. The rectal temperature, pulse rate and blood parameters, as well as environmental factors, were taken into consideration and correlated with body surface temperature patterns. It was concluded that body surface temperature patterns depend on exercise, and that thermography could be useful in the evaluation of the work of individual parts of the body in sport performance [119]. Assessment of body surface temperature can indicate inflamed areas that could also account for a reduction in athletic performance [58, 125]. In a recent study, the authors demonstrated that thermography remains a viable diagnostic modality to identify changes in the upper part of the body in response to physical effort [126]. It was found that exercise under the saddle contributed to asymmetric body surface temperature distribution between both sides of the body, and there was a significant increase in body surface temperature in measured areas with the highest increase in the elbow and the lowest in the base of the neck. The strongest correlation in body surface temperature increase was between the elbow and quarter, and the weakest was between the quarter and neck. Thermography technology continues to advance, thereby increasing the sensitivity of the detectors and allowing for further definition and refinement of its existing uses. Further evaluation is warranted to explore the potential of this strategy for controlling the training load of high-performance equine athletes [126].

15.4.3 The Effect of Long-Term Training on Body Surface Temperature Changes in Distal Part of the Forelimb and Back

According to Evans et al. [127], constant stress of the musculoskeletal system during intensive training increases its blood supply. In a classic thermographic study by Cena and Clark [128], the leg surface temperature of professional human athletes was significantly higher than that of college students even before exercise (31.2 °C vs. 30.3 °C respectively).

Thermography has been found to be a valuable aid in the assessment of long-term training on body surface temperature changes in racehorses [58]. It was found that the temperature of the back and the distal parts of the forelimbs in racehorses at rest was increased by long-term training [125]. In the study by Soroko et al. [121], 10 months of regular thermographic examination of racehorses allowed the identification of the most important body regions to be monitored for the effects of training. These areas included the distal parts of the limbs (carpal joints, third metacarpal bones, left fetlock joint, short pastern bone, tarsus joint), the majority of which were on the left side. Possibly, as the horses during racing and exercise were loaded on the right front limb due to the clockwise race track, when resting they were overloading the opposite side [121]. This agrees with the study by Strömberg [46] in which horses trained in a counterclockwise direction had increased left forelimb temperatures of 0.4–2.2 °C compared to the opposite side after exercise.

In further research, Soroko et al. [16] found that the thoracic vertebrae, the left third metacarpal bone and the left fetlock joint were significantly warmer in horses undergoing intensive training than in horses undergoing light training The data from this study suggest that ambient temperature, breed and training intensity level are the most important factors affecting skin temperature at the distal parts of the forelimbs and the back. Thoroughbreds were significantly warmer than Arabian and Polish Halfbreed horses at these body sites. Similar results were found in a previous study of racehorses where the distal parts of the hindlimbs of Thoroughbreds were also warmer than Arabians and Polish Halfbreed [129]. These temperature differences could be associated with varying thermal insulation of the coat in different breeds [130, 131]. Further studies demonstrated that the forelimb and hindlimb of racing mares were warmer compared to stallions [37, 129]. This may be explained by the reported higher metabolic activity in mares compared to stallions at the same level of training intensity [132].

Identifying the key thermographic diagnostic areas will help in detecting pathological conditions during training. This is particularly important in racehorses, where immediate diagnosis might help to maintain their health and condition, positively influencing their further athletic career. It is therefore suggested that prophylactic thermographic examination of key areas of the distal parts of the limbs and back should be carried out as a routine examination every 2–3 weeks for horses in regular training.

15.5 Application of Thermography in Monitoring the Use of Illegal Performance-Enhancing Procedures

15.5.1 Use of Thermography in Detecting Abnormal Temperatures Caused by Chemical and Mechanical Abrasion of Performance Horses

Thermography has found application in detecting abnormal temperature caused by chemical and mechanical abrasion of performance horses, particularly used to induce hypersensitivity in the distal parts of the limbs, for example, of gaited show horses [133]. Kaemmerer and Buntenkötter [134], as well as Stephan and Görlach [135], suggested that a thermographic image can also be used to detect the application of drugs and their effects. Turner and Scroggins [136] thermographically detected the application of an irritant to the perineal region to enhance tail lifting in a show horse. In another paper, thermography was successfully used for detecting inflammatory chemical irritants applied topically (e.g. 10% mercuric iodide) or injected subdermally (e.g. isopropyl alcohol) and induction of hypersensitisation using limb bandages containing metallic objects. Irritants were applied to the pastern area, and metallic irritants contained in limb bandages were applied to the metacarpal area to induce hypersensitivity [137]. Thermography indicated changed thermal patterns after a single application of an irritant (mercuric iodide) which persisted for 6 days. Metallic bottle caps within the leg wraps could be detected thermographically up to 24 h after application. A limitation of this application of thermography, however, is a lack of specificity: an increase in temperature may not be due to a specific drug, and equally, undetected subclinical injuries could create inflammatory patterns similar to those caused by illegal agents.

Van Hoogmoed and Synder [49] also used thermography to monitor palmar digital neurectomy, which involves cutting both palmar digital nerves to mask lameness. There was a significant temperature increase in the neurectomised limb on the first day of treatment, which persisted up to day 6 post-treatment. It is suspected that nervous interruption causes a vasodilator effect, persistent for a short period of time until the vascular system compensates, influencing a return to normal circulation. In an earlier study by Purohit and Pascoe [138], thermographic examination of neurectomised limbs indicated no temperature differences between neurectomised and non-neurectomised horses three to six weeks after the operation.

15.5.2 Use of Thermography Under International Equestrian Federation (FEI) Rules as a Complementary Diagnostic Method for the Detection of Artificially Induced Temperature Changes in Sport Horses' Limbs

The FEI approves thermography as a complementary diagnostic tool for the detection of artificially induced temperature changes in the limbs of jumping horses. Standard thermographic examination methods to diagnose hypersensitivity of the limbs appear in the FEI Veterinary Regulations [139]. These regulations contain a detailed thermographic examination protocol together with the required supporting clinical examinations. They provide the following information:

- Thermography is one of the complementary diagnostic tools used for initial limb sensitivity examination.
- The thermographic examination should include all four limbs to ascertain and assess the temperature and thermal patterns of the limbs. The horse also has a clinical examination, based on palpation (manual pressure) and observation (e.g in trot). All examinations are performed simultaneously by two examining veterinarians.
- Significant clinical findings arising from the initial limb sensitivity examination include differences between symmetrical limbs exceeding 2 °C, or both limbs showing very high or very low temperatures.
- If following the limb sensitivity examination, the two examining veterinarians agree that the Initial Limb Sensitivity Examination indicates that the horse has an abnormal limb sensitivity, the examining veterinarians will inform the person responsible (or their representative) of the findings and offer the opportunity to withdraw the horse from the event without any further consequences. If the person responsible declines to withdraw the horse, the examining veterinarians will inform the person responsible or their representative that the horse will be subject to a final limb sensitivity examination to determine whether it may continue in the competition at a time ordered by the Ground Jury according to the procedures.

Horses with hypersensitive limbs may be disqualified on the basis of horse welfare and fair play. However, limb hypersensitivity can also be produced by a range of normal occurrences, such as insect stings or accidental self-inflicted injuries. Hyposensitivity could also result from traumatic or surgical cutting of the nerves in that area of the limb (i.e. neurectomy). For this reason, thermographic examination, combined with clinical examination, is used as the first step prior to taking a horse for further investigation [139].

15.6 Application of Thermography to Assess Saddle Fit

15.6.1 Importance of the Correct Saddle Fit

Changes of the body surface temperature across the back area can be associated with pathological conditions caused by an incorrectly fitted saddle [113, 140, 141]. A saddle that does not fit the horse correctly has potentially far-reaching consequences for horse health. An ill-fitting saddle is associated with back pain, uneven sweat marks, white hairs, abnormal hair wear and soft tissue injuries or spinal stress fractures. The clinical manifestations vary from overt lameness and pain on palpation to gait alterations or behavioural changes [142–144]. The fit of the saddle for both the horse and rider must be evaluated. A correctly fitted saddle should distribute weight evenly via the panels on the horse's thoracic region, with clearance of the spinous processes by the gullet.

15.6.2 Application of Thermography in Saddle Fit Examination

The dynamic interaction between the saddle and the back of the horse can be evaluated by thermography to identify any problems with incorrect saddle fit. Thermal patterns caused by the tack while the horse is being ridden are indicative of the heat generated in areas of greater interaction with the saddle and the physiological effects of riding on the back of the horse [145]. For thermographic saddle-fit assessment, the horse should be tacked up with a saddle pad and a tightened girth. Riding should take place for at least 20 min in walk, trot and canter. Thermographic examination of the back should be performed immediately after untacking the saddle. According to Turner, "The pictures reveal the temperature fluctuations of the underside of the saddle and the back of the horse" (Fig. 15.17). The image of the back is interpreted according to the symmetry of the temperature distribution. In a correct saddle fit, the back surface temperature distribution should be the same on both sides of the spine. Locally warmer areas indicate greater interaction between the saddle and back, and cooler areas indicate less interaction. In addition, thermographic imaging of the saddle pads can assist with saddle fit.

A common example of incorrect saddle fit is a greater interaction on the back at two points: at the level of the pommel and the cantle on both sides of the back (Fig. 15.18). This indicates a "bridging saddle". Conversely, a "rocking" saddle indicates contact along the panels but not towards the pommel or cantle. Thermography has proven to be a useful tool in assessing saddle fit, evidencing asymmetry in saddle contact with the back of horses. In addition, interaction between the saddle and dorsal midline of animals was well depicted with this imaging technique [146]. Arruda et al. [140] carried out the thermographic examination of 62 saddles used on 129 jumping horses. The images obtained allowed evaluation

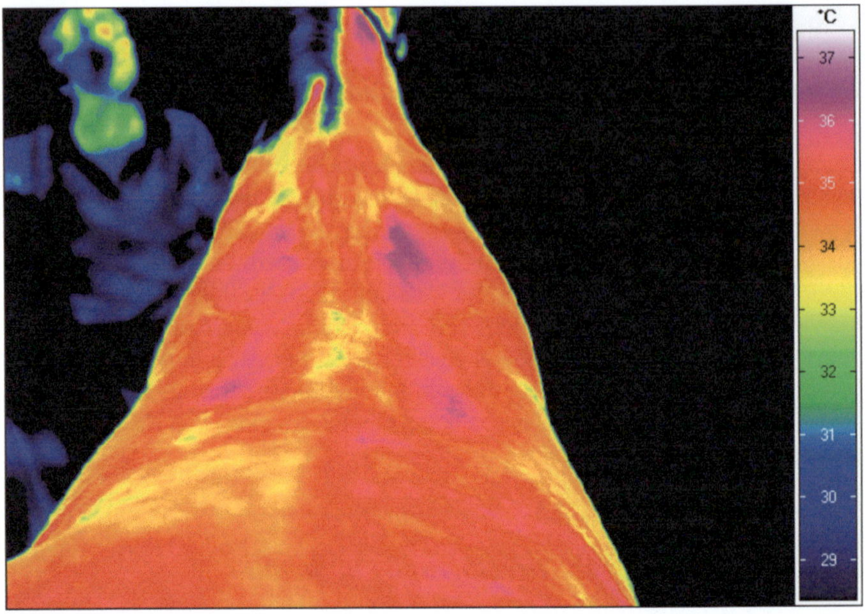

Fig. 15.17 Thermogram of the dorsal aspect of the back, taken 20 min after training in a jumping saddle. This is an example of a correct saddle fit. The back surface temperature distribution is equal on both sides of the spine

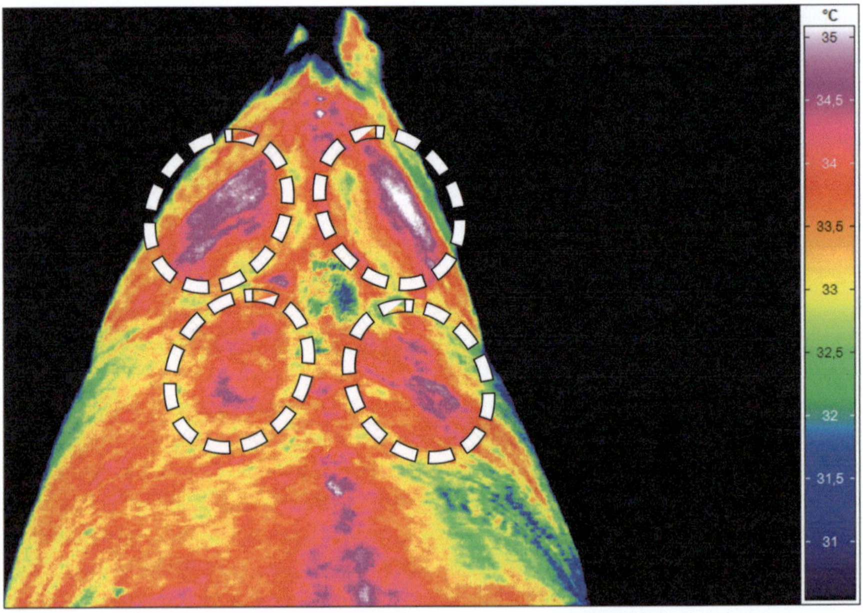

Fig. 15.18 Thermogram of the dorsal aspect of the back, taken 30 min after training in a jumping saddle. The pommel and cantle of the saddle have greater interaction on the back indicated by dashed circles. The middle part of the saddle seat has no contact with the back. This is an example of a bridging saddle. (Figure obtained from Soroko and Davies Morel [45])

of a range of features, including contact between the saddle and the spine; asymmetry in the area of interaction between the saddle panels and the back of the horse and asymmetry between the saddle panels. The main results indicated an uneven (asymmetric) pressure across the back and spine in 55.8% of horses. Saddles had direct contact with the spine in 37.2% of cases, and asymmetry between the saddle panels was evident in 62.8% of cases (Fig. 15.19).

In thermography saddle fit assessment, it must be remembered that there are many external factors that also influence the final thermographic image.

Interpreting thermograms requires meticulous analysis, as shown in a recent study that used thermography to assess how the horse, saddle and rider affect saddle fit through pressure distribution [147]. The study found that factors such as rider load, horse age and training intensity have a significant impact on pressure distribution in racing saddles, highlighting the importance of considering these factors for optimal saddle fit [147]. In a further study, the same researchers used infrared thermography to examine the effects of rider and saddle weight on saddle fit [148]. The study identified thermal patterns indicating improper saddle fit, including rocking or bridging, asymmetry between the left and right panels and front-to-back pattern inconsistencies. They highlighted specific regions of interest on the saddle panels and recommended measuring the mean temperature in six targeted areas post-training to enhance the assessment of how rider and saddle weight affect thermal distribution and saddle fit. Thermography provided a quick, easy and objective way

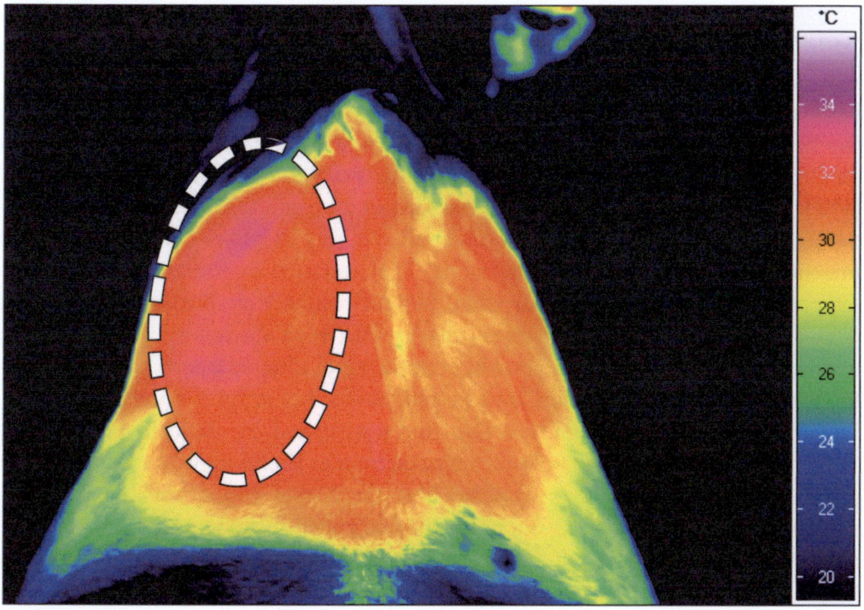

Fig. 15.19 Thermogram of the dorsal aspect of the back, taken 20 min after training in a jumping saddle. There is greater interaction on the left side between the saddle and the back (indicated by a dashed circle)

to evaluate saddle fit and analyse the load effects on thermal distribution patterns in racehorse saddles [148].

Thermography was also used to compare English and Western saddle models on Arabian horses, revealing a notable difference in mean temperatures between the right and left panel contact areas. English saddles showed higher temperatures, while the Western saddle model, despite being less favoured by endurance riders, proved more suitable for Arabian horses due to its wider panels[149].

15.7 The Application of Thermography in Measuring Stress and Assessing Horse Welfare During Training

15.7.1 Use of Thermography in Assessing Eye Temperature as a Means of Detecting Acute and Chronic Stress During Horse Performance

Horse welfare and stress-related disorders are becoming subjects of growing interest for increasing numbers of horse owners, breeders and riders. Several studies have shown that thermography represents a useful method to assess acute and chronic stress in sport horses [150, 151]. In a stressful situation the hypothalamic-pituitary-adrenal (HPA) axis is activated, along with peripheral vasodilatation due to the parasympathetic activation that follows the initial sympathetic response, causing changes in the animal's heat production and heat loss. A number of studies have shown that the temperature of the eye is a good indicator of changes in body temperature due to physiological and psychological stress. This region contains rich capillary networks innervated by the sympathetic nervous system, making it an ideal site for assessing local blood flow changes driven by autonomic mechanisms [152]. Cook et al. [153] investigated eye temperature and saliva cortisol concentration to measure adrenocortical and metabolic activity in horses before and after an injection of adrenocorticotropic hormone. They reported a correlation between eye temperature and cortisol concentration. Similar results were presented by Warren et al. [154], who found significant correlations between maximum eye temperature and both salivary and plasma cortisol concentration after adrenocorticotropic hormone injection. Further studies have investigated stress in sport horses by measuring changes in eye temperature. The study presented by Radaelli et al. [155] suggested that the evaluation of eye temperature was useful to detect physiological stress during training and to evaluate how the horses coped with the training. Thermography was an effective tool for the early identification of horses that were not fit to compete or continue in competition. In a study by Valera et al. [156], the goal was to evaluate the effectiveness of using maximum eye temperature, measured through thermography, to assess acute stress in horses during sports competitions and to compare these findings with salivary cortisol levels. The research revealed that both parameters increased immediately after the competition. However, while cortisol levels decreased 3 h post-competition, eye temperature remained elevated. Increases in eye temperature were found to correlate with increases in salivary cortisol during clipping in a study by Yarnell et al. [151]. Bartolomé et al. [157] demonstrated a correlation between increased maximum eye temperature and heart rate after jumping competitions.

In other studies, horses wearing double bridles and tight nosebands, which restrict jaw movement, exhibited increased eye temperature, indicating a physiological stress response [158]. Similarly, Fenner et al. [159] examined the effects of noseband tightness on eye temperature and found that a tight noseband combined with a double bridle significantly influenced eye temperature, further implying that overly tight nosebands cause pain or discomfort.

A later study explored the relationship between maximum eye temperature and rectal temperature in racehorses, comparing these findings with salivary cortisol concentration and heart rate at rest and after intensive exercise [160]. While all parameters increased post-exercise, the only significant correlation observed was between maximum eye temperature and rectal temperature before exercise, with limited association at other times. This raised doubts about the reliability of maximum eye temperature as an indicator of core temperature or stress in training racehorses [160].

However, Soroko et al. [161] reported a strong correlation between maximum eye temperature and plasma cortisol concentration in response to physical training, measured before, immediately after and 2 h post-training. These findings suggest that thermography can effectively monitor stress responses in horses, offering potential improvements in animal management and welfare.

15.7.2 Use of Thermography in Assessing Changes in the Body Surface Temperature as an Indicator of Stress in Response to Different Training Methods

Within international equestrian sport, there is much debate surrounding the stress associated with different training methods, in particular in relation to the head and neck position [162]. Hall et al. [163] found increased maximum eye temperature in horses lunged with a Pessoa training aid (held responsible for increasing stress during training). Becker-Birck et al. [164] studied whether hyperflexion (defined as the overflexion of the neck dorsoventrally to the point where the chin touches the pectoral region) presented an acute stress to horses. The study measured superficial body temperature at the neck, along with salivary cortisol concentration and heart rate in horses lunged with restricted (hyperflexion) and non-restricted neck positions. No acute stress responses to hyperflexion were observed.

McGreevy et al. [162] measured stress by means of eye and facial skin temperature in horses wearing nosebands with a flash strap that restricts jaw movements during training. Tightened nosebands caused an increase in eye temperature compared with the baseline values ($p = 0.016$), and the tighter the noseband was fastened, the cooler became the surface temperature of the head.

In addition to inappropriate or unfavourable training aids, horses are exposed to a variety of other sources of stress during training and competition that can affect their well-being and lead to poor performance [156].

A pilot study by Redaelli et al. [155] investigated the potential of thermography to monitor stress in endurance horses. To this aim, several physiological parameters such as heart rate, blood count, serum cortisol and temperatures of different body regions were measured using thermography in endurance horses undergoing low-, moderate- and high-intensity training. All measured parameters increased after training. In particular, eye and crown temperature showed significant positive correlations with heart rate and cortisol levels, suggesting that the eye and crown are suitable temperature measurement sites for assessing physiological stress in endurance horses to ensure that they are not being pushed beyond their limits. Another study of Negro et al. [165] performed in trotter horses demonstrated variations of eye temperatures before and after a race and their correlations to performance. A worse performance was found in horses with initial eye temperatures higher than the final measurement, suggesting the unfavourable influence of stress before the race on final competition results. Physiological stress turned into distress, lowering the race performance and potentially causing negative effects on welfare. The effectiveness of infrared thermography as a non-invasive method for stress monitoring was also explored in horses during ridden and lunging sessions [166]. Using thermography, eye temperatures were measured before and after exercise. The study also included a novel-stimulus test to assess the horses' personalities and recorded their behaviour during exercise. The results did not show any significant differences in eye temperature before and after exercise, suggesting that the horses did not experience significant physiological stress during the sessions. However, thermography could still be useful in combination with other tools for a more comprehensive assessment of horse fitness and stress levels. The study by Verdegaal et al. [167] aimed to evaluate the effectiveness of thermography in monitoring body surface temperature to detect and manage heat stress in horses during various types of field exercise. The results confirmed thermography as a useful non-invasive method for monitoring body surface temperature, even though it does not always reliably estimate core body temperature. Therefore, it was suggested that thermography can be a valuable tool for early detection of heat stress, but it should be used in conjunction with other clinical assessments and veterinary examinations for a comprehensive evaluation of a horse's thermoregulatory status. The study highlighted the need for further development of wearable monitoring technology to improve the accuracy and reliability of heat stress detection in horses.

As already mentioned, changes in serum cortisol concentration are objective and useful indicators for assessing stress in horses. Studies addressing the influence of stress caused by graded exercise load were recently performed by Čebulj-Kadunc et al. [5, 23]. The study of ten Lipizzan fillies during their initial training phase [5] aimed to determine the baseline values and fluctuations of various physiological parameters including body surface temperatures of various regions, heart rates and cortisol concentrations and their changes after a graded workload through lunging. Despite the increase in all measured values after exercise, the values including cortisol levels remained within the normal range for warm-blooded horses. A similar study [23] investigated the physiological responses of six adult Lipizzan horses to graded exercise loads during riding by a male and a female rider of different body weights and experience levels. The results also showed significant increases in all measured values after exercise, but no significant differences were found between the two riders, suggesting that rider characteristics such as body mass and experience level do not significantly affect the physiological responses of the horses including the levels of stress. The results of both studies [5, 23] most probably reflected the increased metabolic demands and heat production during physical activity which activated the hypothalamic-pituitary-adrenal axis and sympathetic nervous system and revealed positive correlations between body surface temperature, serum and saliva cortisol concentration, heart and respiratory rate, lactate concentration and some haematological parameters. However, the design of both studies did not allow for a distinction between physical stress and other forms of stress potentially endangering horses during initial training and riding. The findings highlight the importance of appropriate training practices to ensure the well-being and performance of horses as well as the need for further research in equine exercise testing, sports medicine and animal welfare, including thermography testing.

15.8 Conclusions

The power of thermography lies in its ability to provide physiological information about a "physical change" [39, 168]. Thermography specifically increases the accuracy of diagnosis by confirming inflammation and indicates in which area to concentrate further diagnosis. It can play an important complementary role in the early detection of pathology and help to prevent the financial losses associated with delayed diagnosis [12, 21, 22, 57]. The development of improved infrared technology, and better availability of this type of equipment, should contribute to more extensive use of thermography in veterinary medicine. This will be applicable not only to horses, but also to other animals.

References

1. Turner M, McCrory P, Halley W (2002) Injuries in professional horse racing in Great Britain and the Republic of Ireland during 1992–2000. Br J Sports Med 36:403–409
2. Ferraro GL (1990) Lameness diagnosis and treatment in the Thoroughbred racehorse. Vet Clin North Am Equine Pract 6:63–84
3. Mohammed HO, Hill T, Lowe J (1991) Risk factors associated with injuries in thoroughbred horses. Equine Vet J 23:445–448

4. Jeffcott LB, Rossdale PD, Freestone J et al (1982) An assessment of wastage in thoroughbred racing from conception to 4 years of age. Equine Vet J 14:185–198
5. Čebulj-Kadunc N, Frangež R, Kruljc P (2022) Fluctuations of physiological variables during conditioning of Lipizzan fillies before starting under saddle. *Animals 12*(7):Article 7. https://doi.org/10.3390/ani12070836
6. Meixner R (1979) Sauerstoff-Verbrauch und Atmungsgrossen von Pferden unter dem Reiter in verschiedenen Gangarten. W. Hohenheim
7. Di Prampero PD La temperatura corporea durante l'esercizio fiscio prolungato. Riv Cult Sport 14:6–13
8. Thiel M, Tolkmitt G, Hornicke H (1987) Body temperature changes in horses during riding: time course and effects on heart rate and respiratory frequency. In: Gillespie JR, Robinson NE (eds) Equine exercise physiology, 2nd edn. KEEP Publication, Davis, p 183
9. Kang H, Zsoldos RR, Sole-Guitart A, Narayan E, Cawdell-Smith AJ, Gaughan JB (2023) Heat stress in horses: a literature review. Int J Biometeorol 67(6):957–973. https://doi.org/10.1007/s00484-023-02467-7
10. Satchell G, McGrath M, Dixon J, Pfau T, Weller R (2015) Effects of time of day, ambient temperature and relative humidity on the repeatability of infrared thermographic imaging in horses. Equine Vet J 47(S48):13–14. https://doi.org/10.1111/evj.12486_30
11. Tunley BV, Henson FMD (2004) Reliability and repeatability of thermographic examination and the normal thermographic image of the thoracolumbar region in the horse. Equine Vet J 36:306–312
12. Turner TA (2001) Diagnostic thermography. Vet Clin North Am Equine Pract 17:95–113
13. Palmer SE (1983) Effect of ambient temperature upon the surface temperature of the equine limb. Am J Vet Res 44:1098–1101
14. Purohit RC, Bergfeld WA III, McCoy MD et al (1977) Value of clinical thermography in veterinary medicine. Auburn Vet 33:104–108
15. Soroko-Dubrovina M, Morel MCGD (2023) Equine thermography in practice, 2nd edn. CABI
16. Soroko M, Howell K, Dudek K et al (2016a) The influence of breed, age, gender, training level and ambient temperature on forelimb and back temperature in racehorses. Anim Sci J 88:347. https://doi.org/10.1111/asj.12631
17. Mogg KC, Pollitt CC (1992) Hoof and distal limb surface temperature in the normal pony under constant and changing ambient temperatures. Equine Vet J 24:134–139
18. Palmer SE (1981) Use of the portable infrared thermometer as a means of measuring limb surface temperature in the horse. Am J Vet Res 42:105–108
19. Čebulj-Kadunc N, Frangež R, Žgajnar J, Kruljc P (2019) Cardiac, respiratory and thermoregulation parameters following graded exercises in Lipizzaner horses. Veterinarski arhiv 89:11–23. https://doi.org/10.24099/vet.arhiv.0338
20. Purohit RC, McCoy MD (1980) Thermography in the diagnosis of inflammatory processes in the horse. Am J Vet Res 41:1167–1174
21. Turner TA (1991) Thermography as an aid to the clinical lameness evaluation. Vet Clin North Am Equine Pract 7:311–338
22. Vaden MF, Purohit RC, McCoy MD, Vaughan JT (1980) Thermography: a technique for subclinical diagnosis of osteoarthritis. Am J Vet Res 41:1175–1179
23. Čebulj-Kadunc N, Turk A, Kruljc P (2024) Effects of graded exercise load on variations of certain physiological parameters in Lipizzan horses during riding – a pilot study. Comp Exercise Physiol 20(2):1–13. https://doi.org/10.1163/17552559-20230049
24. Davy JR (1977) Medical applications of thermography. Phys Technol 8:54
25. Draper JW, Boag JW (1971a) Skin temperature distributions over veins and tumours. Phys Med Biol 16:645–654
26. Draper JW, Boag JW (1971b) The calculation of skin temperature distributions in thermography. Phys Med Biol 16:201–211
27. Cena K, Monteith JL (1975) Transfer processes in animal coats. I. Radiative transfer. Proc R Soc Lond B Biol Sci 188:377–393

28. Cena K, Clark JA (1973) Thermal radiation from animal coats: coat structure and measurements of radiative temperature. Phys Med Biol 18:432–443
29. Clark JA, Cena K (1977) The potential of infra-red thermography in veterinary diagnosis. Vet Rec 100:402–404
30. Meisfjord Jørgensen GH, Mejdell CM, Bøe KE (2020) Effects of hair coat characteristics on radiant surface temperature in horses. J Therm Biol 87:102474. https://doi.org/10.1016/j.jtherbio.2019.102474
31. Cena K (1974) Radiative heat loss from animal and man. In: Monteih JL, Mount LE (eds) Heat loss from animals and man. Butterworths, London, pp 34–57
32. Playà-Montmany N, Tattersall GJ (2021) Spot size, distance and emissivity errors in field applications of infrared thermography. Methods Ecol Evol 12(5):828–840. https://doi.org/10.1111/2041-210X.13563
33. Turner TA, Fessler JF, Lamp M et al (1983) Thermographic evaluation of horses with podotrochlosis. Am J Vet Res 44:535–539
34. Jodkowska E, Dudek K (2000) Badania nad symetrią temperatury powierzchni ciala koni wyścigowych [Study on symmetry of body surface temperature of race horses]. Przegl Nauk Lit Zoot 50:307–319
35. von Schweinitz DG (1999) Thermographic diagnostics in equine back pain. Vet Clin North Am Equine Pract 15:161–177
36. Kold SE, Chappell KA (1998) Use of computerised thermographic image analysis (CTIA) in equine orthopaedics: review and presentation of clinical cases. Equine Vet Educ 10:198–204
37. Jodkowska E, Dudek K, Bek-Kaczkowska I (2001) Wplyw treningu wyscigwego na temperature powierzchni ciala koni roznych ras [Effect of race training on body surface temperature of several horse breed]. Rocz Nauk Zoot 14:63–72
38. Flores CE (1978) Berührungslose Temperaturmessung an der Hautoberfläche beim Pferd. Veterinary Medicine Dissertation
39. Head MJ, Dyson S (2001) Talking the temperature of equine thermography. Vet J 162:166–167
40. Soroko M, Howell K (2018) Infrared thermography: current applications in equine medicine. J Equine Vet Sci 60:90–96.e2. https://doi.org/10.1016/j.jevs.2016.11.002
41. American Academy of Thermology. Available online: https://aathermology.org/wpcontent/uploads/2018/04/Guidelines-for-Veterinary-Thermography-2019.pdf. Accessed 8Mar 2024
42. Purohit R (2009) Standards for thermal imaging in veterinary medicine. Proceedings of the XIth European congress of thermology, thermology international Mannheim, Germany 19:99
43. Kruljc P (2023) Thermographic examination of the horse. Acta Vet Brno 73(3):289–316. https://doi.org/10.2478/acve-2023-0023
44. Westermann S, Stanek C, Schramel JP et al (2013) The effect of airflow on thermographically determined temperature of the distal forelimb of the horse. Equine Vet J 45:637–641. https://doi.org/10.1111/evj.12019
45. Soroko M, Davies Morel MCG (2016) Equine thermography in practice. CABI Publisher
46. Strömberg B (1974) The use of thermography in equine orthopedics. Vet Radiol 15:94–97
47. Howell K, Dudek K, Soroko M (2020) Thermal camera performance and image analysis repeatability in equine thermography. Infrared Phys Technol 110:103447
48. Eddy AL, Van Hoogmoed LM, Snyder JR (2001) The role of thermography in the management of equine lameness. Vet J 162:172–181. https://doi.org/10.1053/tvjl.2001.0618
49. Van Hoogmoed LM, Snyder JR (2002) Use of infrared thermography to detect injections and palmar digital neurectomy in horses. Vet J 164:129–141
50. Smith WM (1964) Applications of thermography in veterinary medicine. Ann N Y Acad Sci 121:248–254
51. Delahanty DD, Georgi JR (1965) Thermography in equine medicine. J Am Vet Med Assoc 147:235
52. da Silva TCS, de Albuquerque Mariz TM, Escodro PB (2022) Use of thermography in clinical and sports evaluations of equine animals: a review. Res Soc Dev 11(8):e13911530532

53. Čebulj-Kadunc N, Frangež R, Kruljc P (2020) Infrared thermography in equine practice. Vet Stanica 51:109-116 https://hrcak.srce.hr/file/343076Test
54. Strömberg B (1971) The normal and diseased superficial flexor tendon in race horses. A morphologic and physiologic investigation. Acta Radiol Suppl 305:1–94
55. Strömberg B (1972) Thermography of the superficial flexor tendon in race horses. Acta Radiol Diagn (Stockh) 12:295–297
56. Bowman KF, Purohit RC, Ganjam VK et al (1983) Thermographic evaluation of corticosteroid efficacy in amphotericin B-induced arthritis in ponies. Am J Vet Res 44:51–56
57. Soroko M, Henklewski R, Filipowski H, Jodkowska E (2013) The effectiveness of thermographic analysis in equine orthopedics. J Equine Vet Sci 33:760–762
58. Turner TA, Pansch J, Wilson JH (2001) Thermographic assessment of racing Thoroughbreds. In: 47th annual convention of the American association of equine practitioners, vol 47, pp 344–346
59. Turner TA, Rantanen NW, Hauser ML (1996) Alternate methods of soft tissue imaging. In: 1st proceedings of the Dubai international equine symposium, pp 165–176
60. Cetinkaya MA, Demirutku A (2012) Thermography in the assessment of equine lameness. Turk J Vet Anim Sci 36(1):43–48
61. Yanmaz LE (2018) Thermographic assessment of extremity temperature alterations of cases with bucked shin complex, splints, carpal osteoarthritis and sesamoiditis in sport horses. Erciyes Üniversitesi Veteriner Fakültesi Dergisi 15(1):41–45
62. Purohit RC, Pascoe DD, Turner TA (2006) Use of infrared imaging in veterinary medicine. In: Bronzino J (ed) Biomedical engineering handbook. CRC Press/Taylor and Francis Publication, Boca Raton, p 35
63. Fonseca BPA, Alves ALG, Nicoletti JLM et al (2006) Thermography and ultrasonography in back pain diagnosis of equine athletes. J Equine Vet Sci 26:507–516
64. Turner TA (2003) Thermography: use in equine lameness. In: Ross MW, Dyson SJ (eds) Diagnosis and management of lameness in the horse
65. Purohit RC (2008) Use of thermography in veterinary medicine. In: Cohen JM, Lee M (eds) Rehabilitation medicine and thermography. Impress Publication, Morrisville, pp 135–147
66. Hall J, Bramlage LR, Kantrowitz BM et al (1982) Correlation between contact thermography and ultrasonography in the evaluation of experimentally-induced superficial flexor tendinitis. In: Proceedings of the annual convention american association equine of practitioners, USA, pp 429–438
67. Vardasca R, Ring F, Plassmann P, Jones C (2012) Thermal symmetry of the upper and lower extremities in healthy subjects. Thermol Int 22:53–60
68. Purohit RC, McCoy MD, Bergfeld WA (1980) Thermographic diagnosis of Horner's syndrome in the horse. Am J Vet Res 41:1180–1182
69. Waldsmith JK (1992) Real-time thermography: a diagnostic tool for the equine practitioner. In: 38th annual convention of the American association of equine practitioners, vol 38, pp 455–466
70. Waldsmith JK, Oltmann JI (1994) Thermography: subclinical inflammation, diagnosis, rehabilitation, and athletic evaluation. J Equine Vet Sci 14:8–10. https://doi.org/10.1016/S0737-0806(07)80302-9
71. Holmes LC, Gaughan EM, Gorondy DA et al (2003) The effect of perineural anesthesia on infrared thermographic images of the forelimb digits of normal horses. Can Vet J 44:392–396
72. Denoix JM (1994) Diagnostic techniques for identification and documentation of tendon and ligament injuries. Vet Clin North Am Equine Pract 10:365–407
73. Ghafir Y, Art T, Lekeux P (1996) Thermographic facial pattern following an a2-adrenergic agonist injection in two horses suffering from Horner's syndrome. Equine Vet Educ 8:192–195
74. Deumer J, de Haan F, Tulp MT, van den Hoven R (1991) Effect of an isoxsuprine-resin preparation on blood flow in the equine thoracic limb. Vet Rec 129:427–429
75. Colles CM, Hickman J (1977) The arterial supply of the navicular bone and its variations in navicular disease. Equine Vet J 9:150–154

76. Turner AS, Tucker CM (1989) The evaluation of isoxsuprine hydrochloride for the treatment of navicular disease: a double blind study. Equine Vet J 21:338–341
77. Rose RJ, Allen JR, Hodgson DR, Kohnke JR (1983) Studies on isoxsuprine hydrochloride for the treatment of navicular disease. Equine Vet J 15:238–243
78. Harkins JD, Mundy GD, Stanley S et al (1996) Character and duration of pharmacological effects of intravenous isoxsuprine. Equine Vet J 28:320–326
79. Jones WE (1989) Equine sport medicine. Lea and Febiger, Philadelphia
80. Turner TA (1998) The use of thermography in lameness evaluation. In: Proceeding of the American association of equine practitioners, pp 224–226
81. Turner TA (1996) Thermography as an aid in the localization of upper hindlimb lameness. Pferdeheilkunde 12:632–634
82. von Schweinitz DG (1998) Thermographic evidence for the effectiveness of acupuncture in equine neuromuscular disease. Acupunct Med 16:14–17. https://doi.org/10.1136/aim.16.1.14
83. Han J (1987) The neuromechanical basis of pain released by acupuncture. Beijing Medical University, Beijing
84. Levet T, Martens A, Devisscher L et al (2009) Distal limb cast sores in horses: risk factors and early detection using thermography. Equine Vet J 41:18–23
85. Scott R, Nankervis K, Stringer C et al (2010) The effect of water height on stride frequency, stride length and heart rate during water treadmill exercise. Equine Vet J Suppl:662–664. https://doi.org/10.1111/j.2042-3306.2010.00194.x
86. de Graaffe Van, Fox S, Thouin L (1999) Concepts of human anatomy and physiology. McGraw-Hill
87. Yarnell K, Fleming J, Stratton TD, Brassington R (2014) Monitoring changes in skin temperature associated with exercise in horses on a water treadmill by use of infrared thermography. J Therm Biol 45:110–116. https://doi.org/10.1016/j.jtherbio.2014.08.003
88. Soroko M, Howell K, Dudek K, Wilk I, Zastrzeżyńska M, Janczarek I (2018b) A pilot study into the utility of dynamic infrared thermography for measuring body surface temperature changes during treadmill exercise in horses. J Equine Vet Sci 62:44–46. https://doi.org/10.1016/j.jevs.2017.12.010
89. Casas-Alvarado A, Ogi A, Villanueva-García D, Martínez-Burnes J, Hernández-Avalos I, Olmos-Hernández A, Mora-Medina P, Domínguez-Oliva A, Mota-Rojas D (2024) Application of infrared thermography in the rehabilitation of patients in veterinary medicine. Animals *14*(5):Article 5. https://doi.org/10.3390/ani14050696
90. Rindler N, Biermann NM, Westermann S, Buchner HHF (2014) The effect of pulsed electromagnetic field therapy on surface temperature of horses' backs. Wien Tierärztl Mon 101:137–141
91. Rostad DR (2022) Pulsed electromagnetic field therapy and its applications and usage in the equine industry. https://trace.tennessee.edu/utk_gradthes/7039/
92. Dai F, Costa ED, Giordano A, Heinzl EUL, Giongo P, Pagnozzi G, Cannas S, Minozzi G, Minero M (2022) Effects of BEMER® physical vascular therapy in horses under training. A randomized, controlled double blind study. Res Vet Sci 144:108–114. https://doi.org/10.1016/j.rvsc.2022.01.017
93. Nawrot K, Soroko-Dubrovina M, Zielińska P, Dudek K, Howell K (2024) The application of infrared thermography in the assessment of BEMER physical vascular therapy on body surface temperature in racing thoroughbreds: a preliminary study. Animals 14(11):Article 11. https://doi.org/10.3390/ani14111538
94. Edner A, Lindberg L-G, Broström H, Bergh A (2015) Does a magnetic blanket induce changes in muscular blood flow, skin temperature and muscular tension in horses? Equine Vet J 47(3):302–307. https://doi.org/10.1111/evj.12291
95. Johnson SA, Richards RB, Frisbie DD, Esselman AM, McClure SR (2023) Equine shock wave therapy – where are we now? Equine Vet J 55(4):593–606. https://doi.org/10.1111/evj.13890

96. Ringer SK, Lischer CJ, Ueltschi G (2005) Assessment of scintigraphic and thermographic changes after focused extracorporeal shock wave therapy on the origin of the suspensory ligament and the fourth metatarsal bone in horses without lameness. Am J Vet Res 66(10):1836–1842. https://doi.org/10.2460/ajvr.2005.66.1836
97. Śniegucka K, Soroko-Dubrovina M, Zielińska P, Dudek K, Žuffová K (2023) The effect of radial extracorporeal shock wave therapy (rESWT) on the skin surface temperature of the longissimus Dorsi muscle in clinically healthy racing thoroughbreds: a preliminary study. Animals 13(12):Article 12. https://doi.org/10.3390/ani13122028
98. Zielińska P, Nicpoń J, Kiełbowicz Z, Soroko M, Dudek K, Zaborski D (2020) Effects of high intensity laser therapy in the treatment of tendon and ligament injuries in performance horses. *Animals 10*(8):Article 8. https://doi.org/10.3390/ani10081327
99. Godlewska M, Soroko M, Zielińska P (2020) Assessment of vein diameter and body surface temperature after High-Intensity Laser Therapy (HILT) on the tarsal joint in healthy horses. J Equine Vet Sci 93:103198. https://doi.org/10.1016/j.jevs.2020.103198
100. Zielińska P, Soroko M, Godlewska M, Śniegucka K, Dudek K, Howell K (2022) Photothermal effects of high-intensity laser therapy on the superficial digital flexor tendon area in clinically healthy racehorses. Animals *12*(10):Article 10. https://doi.org/10.3390/ani12101253
101. Zielińska P, Soroko M, Howell K, Godlewska M, Hildebrand W, Dudek K (2021) Comparison of the effect of High-Intensity Laser Therapy (HILT) on skin surface temperature and vein diameter in pigmented and non-pigmented skin in healthy racehorses. *Animals 11*(7):Article 7. https://doi.org/10.3390/ani11071965
102. Zielińska P, Soroko M, Zwyrzykowska A, Kiełbowicz Z (2017) The use of laser biostimulation in human and animal physiotherapy–a review. Acta Vet Brno 86(1):91–96
103. Zielińska P, Soroko-Dubrovina M, Śniegucka K, Dudek K, Čebulj-Kadunc N (2023) Effects of High-Intensity Laser Therapy (HILT) on skin surface temperature and vein diameter in healthy racehorses with clipped and non-clipped coat. *Animals 13*(2):Article 2. https://doi.org/10.3390/ani13020216
104. Janczarek I, Wiśniewska A, Tkaczyk E, Wnuk-Pawlak E, Kaczmarek B, Liss-Szczepanek M, Kędzierski W (2022) Effect of different water cooling treatments on changes in rectal and surface body temperature in leisure horses after medium-intensity effort. Animals 12(4):Article 4. https://doi.org/10.3390/ani12040525
105. Bogard F, Bouchet B, Murer S, Filliard JR, Beaumont F, Polidori G (2020) Critical evaluation of whole-body cryostimulation protocol in race horses. J Equine Vet Sci 88:102944. https://doi.org/10.1016/j.jevs.2020.102944
106. Convertino VA (1991) Blood volume: its adaptation to endurance training. Med Sci Sports Exerc 23:1338–1348
107. McKeever KH (1998) Fluid balance and renal function in exercising horse. WB Sounders, Philadelphia
108. Prochno HC, Barussi FM, Bastos FZ, Weber SH, Bechara GH, Rehan IF, Michelotto PV (2020) Infrared thermography applied to monitoring musculoskeletal adaptation to training in thoroughbred race horses. J Equine Vet Sci 87:102935. https://doi.org/10.1016/j.jevs.2020.102935
109. Hinchcliff KW, Geor RJ, Kaneps AJ (2008) Equine exercise physiology. WB Sounders Elsevier
110. Simon EL, Gaughan EM, Epp T, Spire M (2006) Influence of exercise on thermographically determined surface temperatures of thoracic and pelvic limbs in horses. J Am Vet Med Assoc 229:1940–1944. https://doi.org/10.2460/javma.229.12.1940
111. Mukai K, Takahashi T, Eto D, Ohmura H, Tsubone H, Hiraga A (2007) Heart rates and blood lactate response in thoroughbred horses during a race. J Equine Sci 18(4):153–160. https://doi.org/10.1294/jes.18.153
112. Rossdale PD, Hopes R, Digby NJ, Offord K (1985) Epidemiological study of wastage among racehorses 1982 and 1983. Vet Rec 116:66–69

113. de Cocq P, van Weeren PR, Back W (2004) Effects of girth, saddle and weight on movements of the horse. Equine Vet J 36:758–763
114. Schöllhorn WI, Peham C, Licka T, Scheidl M (2006) A pattern recognition approach for the quantification of horse and rider interactions. Equine Vet J Suppl:400–405. https://doi.org/10.1111/j.2042-3306.2006.tb05576.x
115. Williams RB, Harkins LS, Hammond CJ, Wood JL (2001) Racehorse injuries, clinical problems and fatalities recorded on British racecourses from flat racing and National Hunt racing during 1996, 1997 and 1998. Equine Vet J 33:478–486
116. Ramzan PHL, Palmer L (2011) Musculoskeletal injuries in Thoroughbred racehorses: a study of three large training yards in Newmarket, UK (2005–2007). Vet J 187:325–329. https://doi.org/10.1016/j.tvjl.2009.12.019
117. Estberg L, Gardner IA, Stover SM et al (1995) Cumulative racing-speed exercise distance cluster as a risk factor for fatal musculoskeletal injury in Thoroughbred racehorses in California. Prev Vet Med 24:253–263
118. Soroko M (2011) Analyses of superficial temperature distribution of lower part of the limbs in young racing horses. Meas Autom Monit 57:1157–1160
119. Jodkowska E (2005) Temperatura powierzchni ciala jako kryterium predyspozycji wysilkowych konia [Body surface temperature as a criterion of the horse predisposition to effort]. Zesz Nauk Akad Rol:1–114
120. Jodkowska E, Rojkowski A, Sobczak Z (1990) Wplyw niektórych czynników meteorologicznych na temperaturę powierzchni skóry koni pólkriw i prymitywnych [Influence of meteorological factors on body surface temperature changes in half breed and native horses]. Zesz Nauk Akad Rol. We Wroclawiu Zoot 33
121. Soroko M, Dudek K, Howell K et al (2014) Thermographic evaluation of racehorse performance. J Equine Vet Sci 34:1076–1083
122. Soroko M, Jodkowska E, Zabłocka M (2012) The use of thermography to evaluate back musculoskeletal responses of young race horses to training. Thermol Int 22:114
123. Jodkowska E (2000) Analiza zmian temperatury na powierzchni nog koni podczas treningu wyscigowego [Analysis of temperature changes on horse legs surface during race training]. Folia Univ Agric Stetin Zoot:277–284
124. Soroko M, Górniak W, Godlewska M, Howell K (2022) The effect of training on infrared thermographic images of the forelimb and hindlimb joints of healthy racehorses. Pol J Vet Sci 25(1):83–92
125. Soroko M, Jodkowska E, Dudek K (2015) Skuteczność diagnozy termograficznej w monitorowaniu cyklu treningowego koni wyścigowych [Thermography diagnosis in monitoring annual racehorses' training cycle]. Med Wet 71:52–58
126. Soroko-Dubrovina M, Śniegucka K, Dobrowolski M, Dudek KD (2024) Application of thermography in the assessment of physical effort on body surface temperature distribution in racehorses. Pol J Vet Sci 27:221–227
127. Evans GP, Behiri JC, Vaughan LC, Bonfield W (1992) The response of equine cortical bone to loading at strain rates experienced in vivo by the galloping horse. Equine Vet J 24:125–128
128. Cena K, Clark JA (1976) Thermographic observations of skin temperatures of trained and untrained runners. J Physiol 257:8–9
129. Soroko M (2013) The effect of long-term training on racehorses' body surface temperature changes. University of Environmental and Life Sciences of Wroclaw
130. Cymbaluk NF, Christison GI (1990) Environmental effects on thermoregulation and nutrition of horses. Vet Clin North Am Equine Pract 6:355–372
131. Ousey JC, McArthur AJ, Murgatroyd PR et al (1992) Thermoregulation and total body insulation in the neonatal foal. J Therm Biol 17:1–10
132. Kędzierski W, Podolak M (2002) Wplyw treningu koni rasy arabskiej na poziom parametrów biochemicznych związanych z gospodarką węglowodanowo-lipidową [Selected hematological indices of mares from different breeds]. Med Wet 58:788–791

133. Nelson HA, Osheim DL (1975) Soring in Tennessee walking horses: detection by thermography. US Department of Agriculture, Animal and Plant Health Inspection Service, Veterinary Services Laboratory, Ames
134. Kaemmerer K, Buntenkötter S (1971) Thermovision as a measuring system in pharmacology. DTW Dtsch Tierärztl Wochenschr 78:296–298
135. Stephan E, Görlach A (1971) Measuring of surface temperatures using infrared thermography in veterinary medicine. Preliminary report. DTW Dtsch Tierärztl Wochenschr 78:330
136. Turner TA, Scoggins RD (1985) Thermographic detection of gingering in horses. J Equine Vet Sci 5:8–10
137. Van Hoogmoed L, Snyder JR, Allen AK, Waldsmith JD (2000) Use of infrared thermography to detect performance enhancing techniques in horses. Equine Vet Educ 12:102–107. https://doi.org/10.1111/j.2042-3292.2000.tb01772.x
138. Purohit RC, Pascoe DD (1997) Thermographic evaluation of peripheral neurovascular system in animal species. Thermology 7:83
139. International Equestrian Federation (2024) Limb sensitivity in equestrian competition. https://inside.fei.org/sites/default/files/2025%20Veterinary%20Regulations%20-%20marked-up.pdf. Accessed 14 Dec 2024
140. Arruda TZ, Brass KE, Flavio D (2011) Thermographic assessment of saddles used on jumping horses. J Equine Vet Sci 31:625–629
141. Harman JC (1999) Tack and saddle fit. In: Haussler KK (ed) Veterinary clinics of North America: equine practice back problems turn. Saunders Company, Philadelphia/London/Toronto/Montreal/Sydney/Tokyo, pp 247–261
142. Denoix JM (1999) Spinal biomechanics and functional anatomy. Vet Clin North Am Equine Pract 15:27–60
143. Haussler KK, Stover SM, Willits NH (1999) Pathologic changes in the lumbosacral vertebrae and pelvis in Thoroughbred racehorses. Am J Vet Res 60:143–153
144. Jeffcott LB (1999) Back problems. Historical perspective and clinical indications. Vet Clin North Am Equine Pract 15:1–12
145. Turner TA, Waldsmith JK, Wilson JH (2004) How to assess saddle fit in horses. In: Proceedings of the American association of equine practitioners, pp 196–201
146. Dantas F, Duarte MA, Marins JCB, Fonseca BPA (2019) Thermographic assessment of saddles used in Mangalarga Marchador horses. Arquivo Brasileiro de Medicina Veterinária e Zootecnia 71:1165–1170
147. Soroko M, Cwynar P, Howell K, Yarnell K, Dudek K, Zaborski D (2018a) Assessment of saddle fit in racehorses using infrared thermography. J Equine Vet Sci 63:30–34. https://doi.org/10.1016/j.jevs.2018.01.006
148. Soroko M, Zaborski D, Dudek K, Yarnell K, Górniak W, Vardasca R (2019) Evaluation of thermal pattern distributions in racehorse saddles using infrared thermography. PLoS One 14(8):e0221622. https://doi.org/10.1371/journal.pone.0221622
149. de Siqueira RF, Andrioli BM, Baumhak MJ (2019) Evaluation of two models of saddles on the back of Arabian horses through thermography. Braz J Vet Res Anim Sci 56(4):e159435
150. Hall C, Kay R, Yarnell K (2014) Assessing ridden horse behavior: professional judgment and physiological measures. J Vet Behav Clin Appl Res 9:22–29
151. Yarnell K, Hall C, Billett E (2013) An assessment of the aversive nature of an animal management procedure (clipping) using behavioral and physiological measures. Physiol Behav 118:32–39. https://doi.org/10.1016/j.physbeh.2013.05.013
152. Kim S-M, Cho G-J (2021) Validation of eye temperature assessed using infrared thermography as an indicator of welfare in horses. Appl Sci *11*(16):Article 16. https://doi.org/10.3390/app11167186
153. Cook NJ, Schaefer AL, Warren L et al (2001) Adrenocortical and metabolic responses to ACTH injection in horses: an assessment by salivary cortisol and infrared thermography of the eye. Can J Anim Sci 81:621

154. Warren LK, Cook NJ, Schaefer AL et al (2001) The use of salivary cortisol as an index of stress in horses. In: Proceedings of the17th symposium equine nutrition and physiology society
155. Redaelli V, Luzi F, Mazzola S, Bariffi GD, Zappaterra M, Nanni Costa L, Padalino B (2019) The use of infrared thermography (IRT) as stress indicator in horses trained for endurance: a pilot study. Animals *9*(3):Article 3. https://doi.org/10.3390/ani9030084
156. Valera M, Bartolomé E, Sánchez MJ et al (2012) Changes in eye temperature and stress assessment in horses during show jumping competitions. J Equine Vet Sci 32:827–830. https://doi.org/10.1016/j.jevs.2012.03.005
157. Bartolomé E, Sánchez MJ, Molina A et al (2013) Using eye temperature and heart rate for stress assessment in young horses competing in jumping competitions and its possible influence on sport performance. Anim Int J Anim Biosci 7:2044–2053. https://doi.org/10.1017/S1751731113001626
158. McGreevy P, Warren-Smith A, Guisard Y (2012b) The effect of double bridles and jaw-clamping crank nosebands on temperature of eyes and facial skin of horses. J Vet Behav 7(3):142–148. https://doi.org/10.1016/j.jveb.2011.08.001
159. Fenner K, Yoon S, White P, Starling M, McGreevy P (2016) The effect of noseband tightening on horses' behavior, eye temperature, and cardiac responses. PLoS One 11(5):e0154179. https://doi.org/10.1371/journal.pone.0154179
160. Soroko M, Howell K, Zwyrzykowska A, Dudek K, Zielińska P, Kupczyński R (2016b) Maximum eye temperature in the assessment of training in racehorses: correlations with salivary cortisol concentration, rectal temperature, and heart rate. J Equine Vet Sci 45:39–45. https://doi.org/10.1016/j.jevs.2016.06.005
161. Soroko M, Howell K, Dudek K, Waliczek A, Micek P, Flaga J (2021) Relationship between maximum eye temperature and plasma cortisol concentration in racehorses during intensive training. Pol J Vet Sci. http://psjd.icm.edu.pl/psjd/element/bwmeta1.element.oai-journals-pan-pl-120932/c/oai-journals-pan-pl-120932_full-text_9_Soroko.pdf-1
162. McGreevy P, Warren-Smith A, Guisard Y (2012a) The effect of double bridles and jaw-clamping crank nosebands on temperature of eyes and facial skin of horses. J Vet Behav Clin Appl Res 7:142–148
163. Hall C, Burton K, Maycock E, Wragg E (2011) A preliminary study into the use of infrared thermography as a means of assessing the horse's response to different training methods. J Vet Behav Clin Appl Res 6:291–292
164. Becker-Birck M, Schmidt A, Wulf M et al (2013) Cortisol release, heart rate and heart rate variability, and superficial body temperature, in horses lunged either with hyperflexion of the neck or with an extended head and neck position. J Anim Physiol Anim Nutr 97:322–330. https://doi.org/10.1111/j.1439-0396.2012.01274.x
165. Negro S, Bartolomé E, Molina A, Solé M, Gómez MD, Valera M (2018) Stress level effects on sport performance during trotting races in Spanish trotter horses. Res Vet Sci 118:86–90. https://doi.org/10.1016/j.rvsc.2018.01.017
166. Martins JN, Silva SR (2022) Use of infrared thermography to assess body temperature as a physiological stress indicator in horses during ridden and lunging sessions. *Animals 12*(23):Article 23. https://doi.org/10.3390/ani12233255
167. Verdegaal E-LJMM, Howarth GS, McWhorter TJ, Delesalle CJG (2024) Thermoregulation during Field Exercise in Horses Using Skin Temperature Monitoring. *Animals 14*(1):Article 1. https://doi.org/10.3390/ani14010136
168. Bathe PA (2007) Thermography. In: Floyd A, Mansmann R (eds) Equine podiatry. Elsevier Saunders, Saint Louis, pp 167–170

Chapter 16
Issues and Future Developments of Infrared Thermography in Sports Science

Jose Ignacio Priego-Quesada and Ricardo Vardasca

Abstract Although currently infrared thermography has a great number of applications in sports science, there are a number of different aspects that need to be investigated in order to improve the technique, the methodology, the analysis, and to increase its application areas. In this chapter, we discuss the issues with and possible developments in infrared thermography in sport science, in order to facilitate future R&D.

16.1 Introduction

As noted in previous chapters, infrared thermography has a great number of applications in sports science. Its use can be of great value in different areas such as sports medicine [1, 2], clothing design [3, 4], or physiology performance assessment [5, 6]. The use of infrared thermography in sports science has been growing continuously since the last decades. Therefore, infrared thermography has a great present in sport science.

However, it is important not to forget that the use of thermography in sports science began not long ago. For this reason, different aspects and developments need to be studied with greater detail to improve this technique in the future. Infrared

J. I. Priego-Quesada (✉)
Research Group in Sport Biomechanics (GIBD), Department of Physical Education and Sports, University of Valencia, Valencia, Spain

Biophysics and Medical Physics Group, Department of Physiology, University of Valencia, Valencia, Spain
e-mail: j.ignacio.priego@uv.es

R. Vardasca
LABIOMEP, INEGI, LAETA, University of Porto, Porto, Portugal

ISLA Santarém – Instituto Politécnico, Santarém, Portugal

LABIOMEP, CIFI2D, Faculty of Sport, University of Porto, Porto, Portugal
e-mail: ricardo.vardasca@islasantarem.pt

J. I. Priego-Quesada (ed.), *Application of Infrared Thermography in Sports Science*, https://doi.org/10.1007/978-3-031-93311-0_16

thermography presents great potential for improvement in sports science. This context implies a positive vision of the future of thermography in which the number of applications can increase, and the methodology and analysis can improve.

This chapter aims to discuss all these thermography issues that need to be investigated and improved in sports science. Through this discussion, it is expected that the present chapter will facilitate future research in academia, and the development for manufacturers of infrared thermography cameras. Moreover, this discussion may be able to discern and imagine how infrared thermography will be used in sports science in the future.

The chapter will discuss the technical aspects of thermal imagers (features, quality assurance principles, and software challenges). Additionally, the chapter will discuss aspects related to some of the main methodological problems of thermography in sports science during experimentation (e.g., sweat issue or when and how to measure) and data analysis (thermal variables, skin temperature data reference, etc.).

16.2 Infrared Camera Features

Infrared thermography cameras have improved in recent decades, resulting in a powerful and reliable technique, but always depending on the camera model [7–10]. Although cooled cameras present a better image quality and thermal sensitivity, uncooled cameras are the most used type of cameras in sports science due to their associated benefits: more accessible in price and maintenance, smaller, and easier to use [11–13]. The quality of uncooled infrared cameras is approaching the cooled instruments, which have a limited time of operation.

The most important features that researchers demand in their infrared thermography cameras are the sensors' array size, the noise equivalent temperature difference (NETD), and the repeatability of the measurement [14]. However, the minimum requirements for infrared camera features for sports and medical research are the sensor array size of 320 × 240 pixels, an NETD of <50 mK at 30 °C, and a measurement repeatability of ±2 °C (over the whole camera temperature range) or ± 2% of the overall temperature reading [15]. Some cameras present a better measurement repeatability (±1 °C or ± 1%); however, these cameras have a more inaccessible price. Although the results of some studies suggested that most of this error in accuracy is systematic, it was also observed that it is dependent on the camera used [7, 16, 17]. This means that the error between the thermographic value and the real value is always the same. Because of this, when comparisons are made of images with the same camera, the error of the difference between them does not depend on the measurement repeatability, but rather on the NETD. Although in these cases, the error of the difference between measures is lower, the measurement repeatability is maintained to determine the absolute temperature of the region of interest (ROI). Furthermore, the capacity to determine accurately the absolute temperature is of great importance for some applications. Some examples are fever screening [18] (Fig. 16.1) or the determination of high skin temperatures during high intensity exercise in hot environments in order to predict impairment of aerobic performance

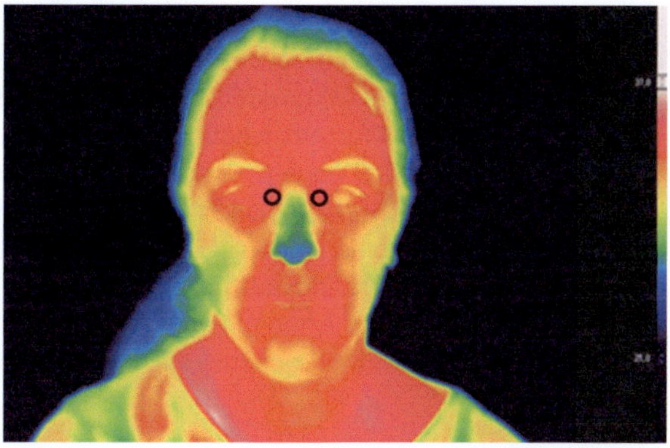

Fig. 16.1 Example of infrared thermography for determining a febrile state

[19–21]. For these reasons, the improvement of the accuracy of the infrared cameras is one of the features that researchers expect manufacturers to improve.

Otherwise, the curvature effect of the skin surface leads to an error in the measurement of the skin temperature, because it causes more thermal energy reflection [22]. Different authors have suggested the development of 3D infrared thermography by the combination of 2D thermography with other image systems (e.g., photographic systems or MRI/CT) in order to improve the accuracy and the reliability of the technique [23–26]. Since the actual existing systems of thermal imaging are 2D, the only method to create 3D images is to estimate the temperature values of pixels from another. There are three basic principles to do this: using the nearest neighbor (closest pixel), bilinear (average of 9 neighbors), and bicubic (average of a neighborhood of 16 pixels) interpolation methods [27]. Thermal imaging pixel values are radiometric and directly obtained from the energy perception of the camera sensors. If artificial new pixel values are estimated from others, then new errors are generated. From a comparison study [27], when using pixel reconstruction in thermal imaging, the best method is the most basic—the nearest neighbor—although it is not recommended for human assessments since the temperature range is very narrow and it will have a high impact on the measurement results. Despite the nice exemplificative images from the 3D thermography current developments, they are not being used currently in sports science.

Other possibility for the future is a higher use of thermographic hyperspectral cameras. These cameras integrate infrared thermography with hyperspectral imaging, enabling not only high-resolution temperature mapping but also detailed spectral analysis across multiple wavelengths [28]. Unlike traditional thermographic cameras, which provide single-band infrared images, thermographic hyperspectral cameras capture a full spectral signature of thermal emissions, allowing for a more precise identification of tissue composition, metabolic activity, and material properties [29]. This capability was applied in melanoma diagnosis, where the integration of hyperspectral and thermal imaging has shown potential to improve early detection by providing non-invasive, multi-wavelength characterization of skin lesions, aiding in the differentiation between melanoma and benign nevi [28]. In agriculture, thermographic hyperspectral cameras provide deeper insights into plant stress responses by differentiating between various physiological conditions that might not be visible with standard infrared cameras [29]. Because there is no such use in sports, we cannot know for sure whether they could provide more valuable information than current thermal imaging cameras. But perhaps these cameras can detect subtle changes in skin emissivity and muscle oxygenation, which are not discernible with conventional thermography. This additional spectral information could be crucial for assessing muscle fatigue, inflammatory responses, and thermoregulatory efficiency during exercise. Although this is pure speculation, it will be interesting to see whether these types of cameras start to be used more in sports science in the future.

16.3 Quality Assurance Principles

Every measurement method is an objective assessment; in terms of warranty, its quality, and basic quality assurance is needed. There are two types of quality assessment, within capture and periodically.

For both types of quality assurance, a calibration source is recommended, normally a black body (a device that has the ability to absorb all incident radiation at all wavelengths and only emit thermal radiation at a single temperature and known wavelength). For examination of quality assurance [11], it is recommended that one image be captured from the calibration source at the beginning of the examination and the other at the end. If the value of the ROIs of the images taken from the reference differ by more than the camera repeatability, it is recommended that the whole examination be repeated and that a verification be performed in the camera. The other option is to include the calibration source in the field of view of the thermogram taken and to verify the temperature reading within the ROI for examination quality warranty.

Two years after the acquisition of the thermal camera, it is recommended that it be sent back to the manufacturer or any certified center for recalibration, which is expensive. However, there are a few tests that can be easily performed, that allow for better knowledge of the camera and for identification of when it needs to be sent for recalibration. With these tests, you can easily verify, at low cost, the camera start-up drift, long-term drift, offset over temperature range, thermal flooding effect, and image non-uniformity [30].

The start-up drift is the time required by the camera to be stable after turning it on. The camera sensors are organized in an electronic array, when it is switched on, it needs some time before starting to capture the images due to undesirable effects such as the warming up of the internal electronics. For the image to be taken in good quality, a certain time should be refrained that warrant that equilibrium. To know this, the camera should be placed in front of a calibration source (black body) for a determinate period of time (e.g., 1 h) and images should be taken at a 1-min interval, comparing the mean temperatures of the reference ROI, the period of stabilization and possible variation can be determined. The long-term drift can be caused by ageing of the camera electronic components. In order to observe this, the start-up drift test should be repeated periodically and records of the results should be kept; this will help to identify when the camera needs to be recalibrated by the manufacturer or certified body.

Some cameras can present an error in measurements that can be introduced by the camera sensor array readout having different temperature amplifications for different temperatures (offset variation over temperature range). This should be verified every 3 months. This test can be achieved by introducing a bucket of water to the scene, and specifying different fixed temperature ranges, which should be reduced over time in reductions of 0.5 °C, by adding cold water or stirring the water to make it cool faster. By comparing the reference ROI mean temperature, it is possible to identify differences.

Thermal flooding refers to an unwanted effect in surface temperature measurements, where the presence of another object with a significantly different temperature (either cooler or warmer) in the same scene alters the true thermal reading of the target object. The warm causes flooding by stray radiance; this should be verified every month. Two buckets can be introduced in the scene at different temperatures and they can be approximated using one as reference and, verifying the reference ROI's mean temperature, the effect can be calculated. The image non-uniformity happens when the sensor cannot detect the correct radiation of the objects in the corner of the images, resulting in deficiencies of the optical path. It should be also verified periodically (typically every 2 months) through pointing the camera at a uniform wall, taking a few thermal images, drawing on them two cross-sections at the main diagonal lines and comparing their in-temperature distributions; the differences provide the non-uniformities.

The use of low-cost thermal imaging cameras is appearing in many applications [31–33]. However, all the quality aspects mentioned above are even more questionable with these cameras [8, 10]. Although it is being seen that the quality of these cameras is improving, it is necessary that in these applications, for each measurement, the validity of the data is ensured as far as possible (for example, by incorporating a black body in the image). In this sense, more development and research in this area are necessary.

16.4 Software Challenges

Thermography manufacturers provide general software for a wide range of uses. In this sense, a software package developed for human and medical assessment is lacking [11, 12], and, more specifically, for a sports science application. Additionally, manufacturers hinder the development of new software by researchers through the encryption of the thermal data in the image file. For these reasons, many thermal image processing techniques that could improve the data analysis have not yet been developed and disseminated to researchers.

The image radiometric file formats are also not uniform throughout the camera manufacturers and even within the same manufacturer, which makes thermal image integration and comparison between different systems and remote centers that have different camera models difficult [34].

Image processing techniques are essential in the application of the infrared thermography assessment [35, 36]. This section is focused on the following issues that are believed to be in the process of being solved through the development of future software:

- The determination of the regions of interest (ROIs).
- Metrics used for data analysis.
- The effect of hair on skin temperature measurement.
- The determination of the number of hot spots in a ROI.
- Machine learning approach.

The reproducibility of the determination of the ROIs in image analysis is an important factor for guarantying the accuracy of the results. The subjective and non-standardization of the determination of the ROIs could lead to a lower reproducibility and accuracy of the results [37, 38]. One of the most desired solutions is the use of software packages with an automatic ROI selection feature [3, 39–42]. These software package commonly process the thermal images by masks, which provide an outline of the ROI [36] (Fig. 16.2). A validation study with one commercial software (Thermohuman) showed an excellent reliability compared with manual determination and a 86% of time reduction in the data acquisition from ROIs determination [42]. However, these software packages are not accessible to all users because there is lack of free software packages developed by the manufacturers with this purpose [12].

Image fusion of the software packages can help in the determination of the ROIs, improving the precision and repeatability by the use of anatomical landmarks [1, 43] (Fig. 16.3). This feature also can help to better localize the affected area and the extension of the injury [1].

Alternative metrics such as entropy or percentage of pixels at different temperature ranges are appearing, although the validation of its applicability still needs to be further investigated [44]. Additionally, there are differences between people in the anatomical shapes and sizes, resulting, in most of the cases, in the ROI defined in one person having a different number of thermal pixels compared with a different person [12]. Future software packages would offer the possibility of using different analysis techniques in order to calculate the temperature of an ROI from the same number of pixels in all the participants. An example of this kind of metrics is the TMAX, a metric developed with the aim to obtain a maximum temperature value obtained by the mean of the 125 warmest pixels in the ROI [45].

On the other hand, future software packages could help to solve the issue of the human hair in skin temperature measurement. Human hair is an avascular structure with a lower emissivity, lower thermal conductivity and then lower temperature [13, 46, 47]. For this reason, the presence of hair on the skin surface could be considered an artifact in the thermal image, resulting in an underestimation of the skin

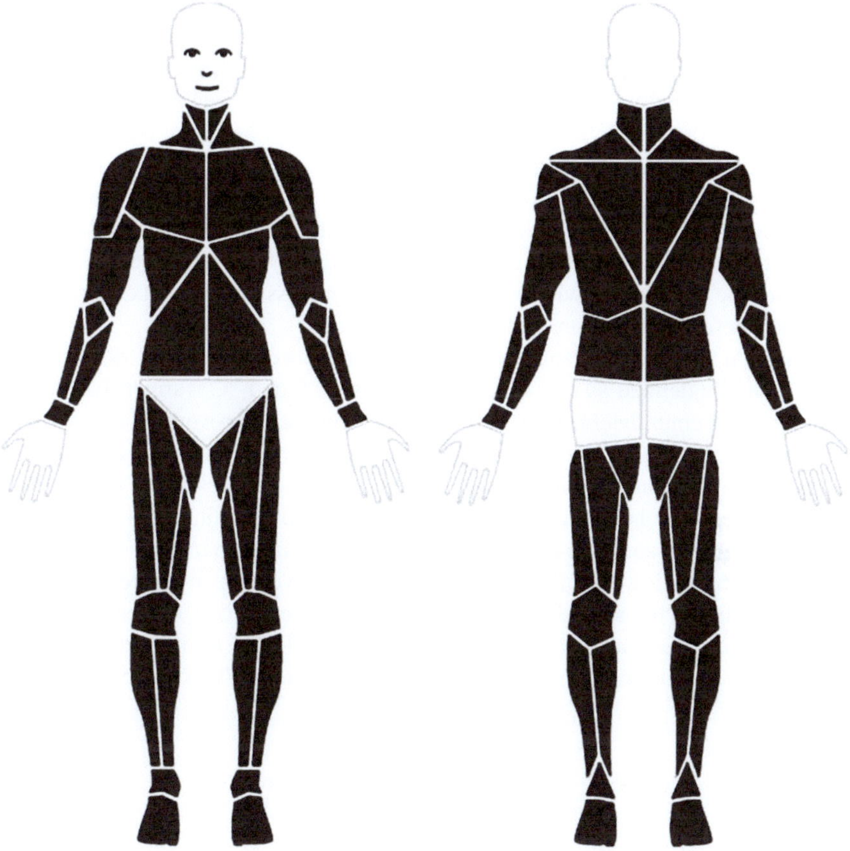

Fig. 16.2 ROI template of the ThermoHuman Cloud Software. (Figure provided by ThermoHuman Company)

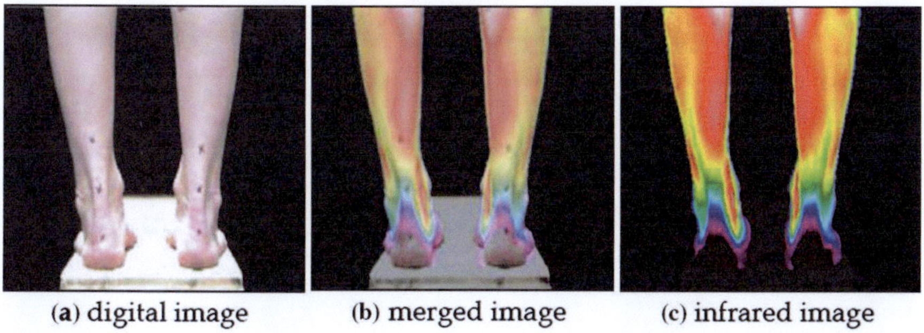

Fig. 16.3 Image fusion. (Figure obtained from Hildebrand et al. [43])

temperature [48]. Although it is common to ask for the participants to remove their body hair some days before the thermography assessment [49–52], this could be inconvenient during participant recruitment. It is probably always the best option to remove the hair, but when it is not possible, future software packages could remove the hair from the image by processing techniques, resulting in an estimated value (known error) of the skin temperature (Fig. 16.4b, c). However, for people with a high body hair density, it is possible that this processing technique would be not valid (Fig. 16.4a). Furthermore, after future development of this processing technique, it would be interesting to assess the utility of this suggested method by assessing the differences with and without this processing technique (by shaving body hair before the measurement).

Finally, a tool for the identification of the hot spots could be very valuable. Hot spots could be defined as any area that is at least 0.5 or 1 °C warmer than its surrounding and contralateral side, respectively. Another improvement could be the comparison of the ROIs' histograms [53]. Hot spots were suggested by some authors in medicine as being an interesting variable in injury assessment [43, 53–56]. Furthermore, these hot spots could be related to the perforating vessels (capillary vasodilation) [53], and then could be an interesting variable to measure during exercise (Fig. 16.5). However, there is a need for improvement in the reliability and reproducibility of hot spot identification using software [12, 53]. Moreover, entropy metric [57] can be related with the number of hot spots, but again, further research is necessary to determine that both outcomes are related. When proper software is

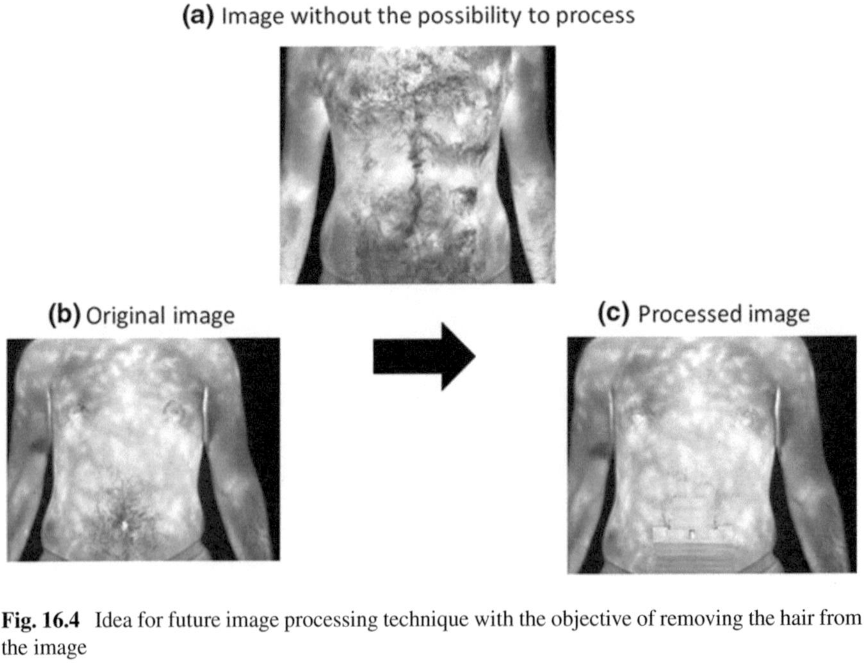

Fig. 16.4 Idea for future image processing technique with the objective of removing the hair from the image

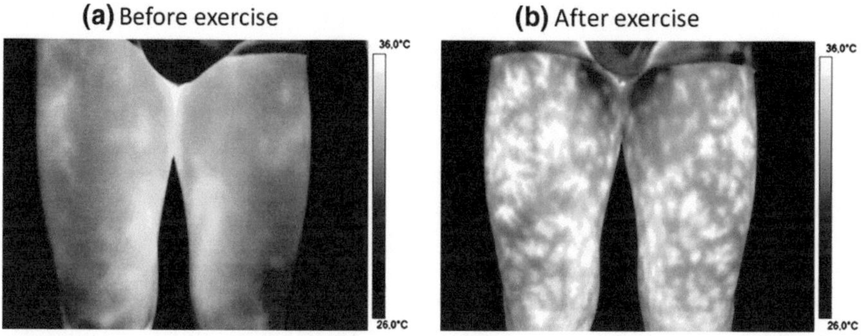

Fig. 16.5 Increase in the number of hot spots after 45 min of cycling

developed, it should be performed in studies that analyzed the hot spots during exercise.

Finally, in this section there is a heavy reliance on traditional methods of thermographic data analysis: determining ROIs and obtaining parametric data. However, the use of Artificial Intelligence methods such as Machine Learning and Deep Learning are revolutionizing the way data are analyzed. There are numerous studies showing the potential of thermography to detect pathologies based on a complete analysis of the image through Machine Learning being greater than traditional analysis [58–60]. And sport is not exempt in this sense, and it is logical to think that in some applications this type of analysis may be more interesting and valuable. Although, when using such methods, it is important to understand how they can be applied and for Machine Learning, the features must be adequately selected and for Deep Learning, the imaging inputs also have to be assertively defined, otherwise bias can be added to the analysis [61, 62].

16.5 Intraday Skin Temperature Reproducibility

Different intrinsic and extrinsic factors are necessary to control in the measurement of skin temperature [48] in order to reduce variability between the participants and between days [37]. In relation to the reproducibility of skin temperature measurement using infrared thermography, McCoy et al. [63] observed excellent reproducibility between days for infrared images from the paraspinal region. Furthermore, Zaproudina et al. [37] found moderate reproducibility in the trunk and poor reproducibility in the limbs. Zaproudina and colleagues suggested that this result was probably due to physiological variability of blood flow in the distal parts of the body [37]. Although reproducibility of skin temperature measurements using infrared thermography was assessed by these studies in baseline conditions, research into the reproducibility of thermal data after exercise is still necessary.

A study that assessed the reproducibility of skin temperature after exercise using infrared thermography was performed by Priego Quesada et al. [64]. However, the aim of this study was to determine the effect of an intervention (changing saddle height in each season) on intraday reproducibility after cycling exercise. Therefore, this study cannot be considered to be a perfect example of reproducibility research because all the tests were not performed at the same conditions; however, the results are useful to illustrate why future research is necessary. Priego Quesada et al. [64] observed good and very good reproducibility of skin temperature before exercise, and good reproducibility after exercise. However, when the authors assessed the intraday reproducibility of skin temperature variation analysis (difference between post- and pre-exercise), they observed similar good reproducibility in the trunk, but worse reproducibility in the lower limbs. The authors suggested that skin temperature variations can be more sensitive to an intervention (in this case, changes in the saddle height) than absolute temperatures. For this reason, further studies are necessary in order to investigate the reproducibility of temperature data (absolute temperature values and Δ values) after exercise.

Recent studies have further examined the reliability of infrared thermography in different conditions. Calvo-Lobo et al. [65] assessed intra- and inter-session reliability of the skin temperature triceps surae in athletes. Their findings demonstrated excellent reliability and repeatability and low standard errors of measurement. This study highlights that, despite prior concerns about variability, appropriate standardization of measurement protocols can yield highly reliable results. Additionally, Priego-Quesada et al. [66] evaluated the reproducibility of skin temperature after a cold stress test using a device that combines cryotherapy with the application of pressure in lower limbs. Their results indicated that posterior leg regions exhibited the highest reproducibility, whereas anterior thigh and posterior knee showed lower reliability. These findings reinforce the notion that regional differences in blood flow and measurement conditions can impact reproducibility and must be carefully considered in future research.

On the other hand, in the most distal regions of the human body, such as the feet, peripheral circulation is less homogenous, resulting in an increase in the variability of skin surface temperature and reducing the reproducibility of thermographic measurements in that region [37]. As was discussed in Chap. 14, the foot is one interesting region for thermography analysis with multiple possible applications in sport science. It is possible that, although the reproducibility of the foot is low [37], the effect of exercise may be more constant. Future studies are necessary to analyze the effect of exercise on intraday reproducibility of foot temperature and to compare the reproducibility of skin temperature variation with absolute temperatures.

16.6 Skin Thermoregulation Issues

The human body is homeothermic; the core temperature is maintained at a constant by the hypothalamus, and there are other body structures such as nerves, blood flow, and skin thermal sensors that play an important role in body temperature. The temperature within the body can be regulated by autonomic and behavioral means, as shown in Fig. 16.6, where the red flux represents the heat transfer path and the green represents the signal path [67].

Thermoreceptors are spread throughout the human body, internally sensing the temperature close to the organs and peripherally sensing the shell temperature under the skin dermis. Any variation in temperature values is communicated via the nervous system to the body central temperature controller, that is a group of neurons in the anterior part of the hypothalamus, known as the preoptic area. It is this area that receives nervous impulses from the thermoreceptors, mucous membranes, and other areas of the hypothalamus. Neurons from the preoptic area generate impulses at high frequency when blood temperature increases and at lower frequency when blood temperature decreases. These impulses propagate to two other parts of the hypothalamus known as the heat-losing center and the heat-promoting center, that when stimulated by the preoptic area sets into activity a series of responses that lower or rise body temperature correspondingly. When the sensory system acknowledges a decrease in core temperature, nervous impulses are sent to the preoptic area, which in turn activates the heat-promoting center through hormone production,

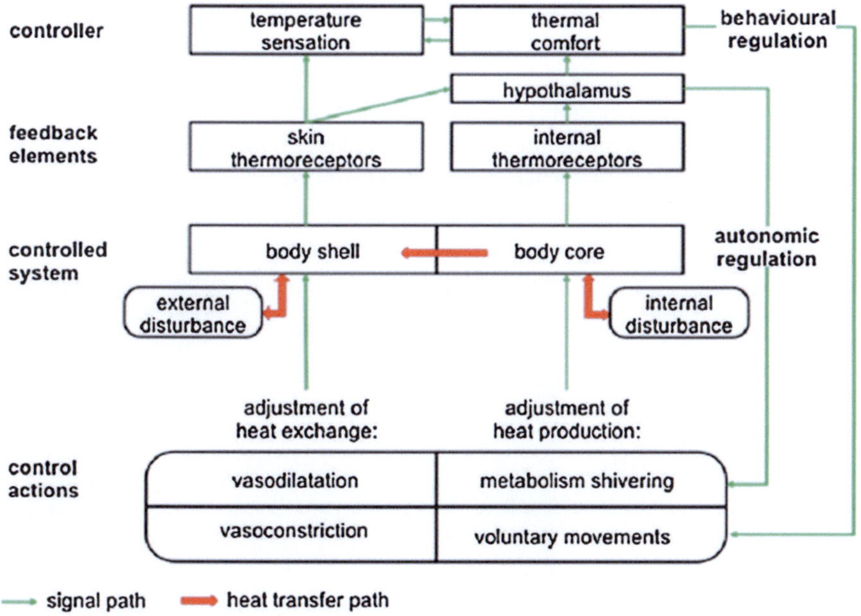

Fig. 16.6 Human body thermoregulation diagram according to Houdas and Ring [67]

activating the heat gain mechanisms such as vasoconstriction (decrease in warm blood flow), shivering (muscle contractions and stretching), and a slow increase in metabolic rate. If a core temperature increase is perceived, the thermoreceptors will send nerve signals to the preoptic area, which generates hormones that inhibit the heat-promoting center and activates the heat-losing center, which in turn activates the body temperature decrease mechanisms such as vasodilatation and sweat gland stimulation, which activate perspiration through the sympathetic nervous system [68].

Many of the thermographic studies in sports science are performed during or after exercise [5, 6, 49 50, 69–72]. Exercise may result in an increase in core temperature due to the higher metabolic activity [73, 74]. One of the main mechanisms for heat dissipation during exercise is sweating and sweat evaporation [75–77]. It is known that evaporation of sweat results in a decrease in skin temperature. This decrease in skin temperature helps in the dissipation of the core body heat, because it increases the thermal gradient between the core and the skin [20].

Although it is known that sweat evaporation is essential during exercise, the effect of sweat on thermographic measures remains unclear [78, 79] (Fig. 16.7). It was suggested that a film of water on the skin may act as a filter for infrared radiation and that this could lead to an error in the estimation of skin temperature [80]. Therefore, it was assumed that sweat produced on the skin during and after exercise could lead to errors in skin temperature calculation. For this reason, some studies have tried to remove the sweat or water from the skin without friction [81, 82] (Fig. 16.8). However, other studies have indicated that they did not remove the sweat in order to keep the natural process of sweat evaporation and because the relevance of sweat (i.e., without the film of water on the skin) on thermographic data is still unknown [64, 83, 84]. Table 16.1 summarizes the positive and negative aspects of both strategies.

Different reasons make this issue difficult to research. Not having a perfect gold standard for the skin temperature measurement makes it more difficult to know the effect of sweat [79]. In this sense, some comparison were performed between thermography and thermal contact sensors such as thermocouples [78, 79, 85, 86].

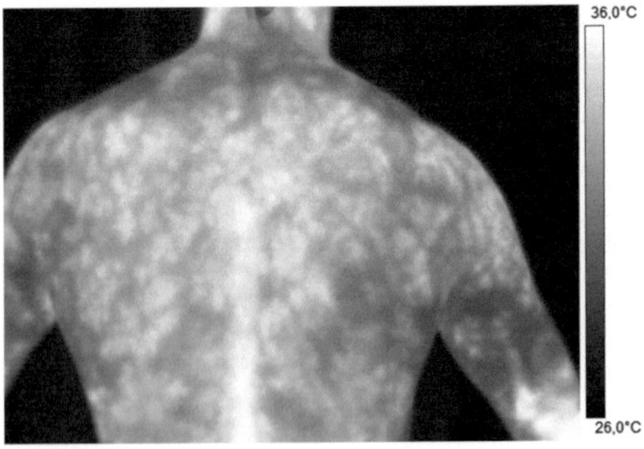

Fig. 16.7 An upper back of a cyclist immediately after cycling 45 min. Sweat was not removed and, visually, layers of sweat were not appreciated

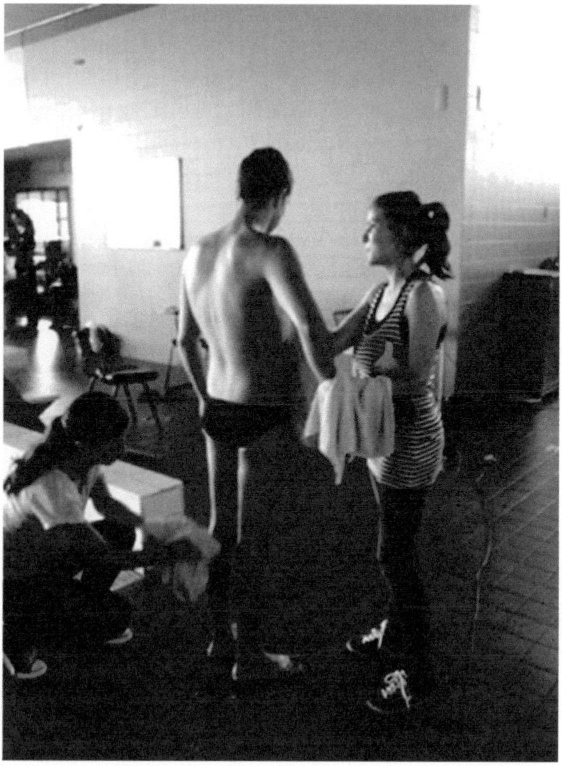

Fig. 16.8 Removal of water layers with absorbent towels without friction

Table 16.1 Advantages and disadvantages of the strategies concerning sweat in thermographic exercise studies

	Advantages	Disadvantages
Remove sweat	Remove the possible effect of sweat on thermography data	Possible friction during the removal that increase skin temperature Reduction in the capacity of sweat evaporation, thus lower capacity of skin temperature reduction
Not remove sweat	More close to reality because the natural process of sweat evaporation is not altered	Possible effect of sweat on temperature calculation that could lead to errors

However, thermal contact sensors have some other issues in skin temperature calculation. These sensors alter the skin temperature through contact with the sensor and its fixation method on convective and evaporative heat loss [78, 87, 88]. Moreover, the sensor and the attachment can also cause sweat to not evaporate and accumulate in that area, also altering the skin temperature at that specific point where it is being measured [79]. Furthermore, it is hard to know which process (error in temperature calculation by sweat, rubbing, sweat evaporation, etc.) has a greater influence on skin temperature. In this sense, one study tried to compare the differences between thermal contact sensors and infrared thermography during exercise, and during mechanical tests (hot plate system) in order to investigate the agreement between both methods during well con-trolled dry and wet heat exchange (avoiding thermoregulatory responses) [83]. The results were comparable and the authors assumed that differences between thermal contact sensors and thermography were mainly through the effect of the fixation on skin temperature. They concluded that the accumulation of sweat during cycling is not enough to form a film of water and that the sweat produced under these conditions does not affect the thermographic data [83]. However, further studies are necessary to explore this issue and provide answers to the questions below:

- Does sweat affect skin temperature? In all scenarios, none, or only where the sweat rate is very high?
- If there is an error in skin temperature calculation due to sweat, how great is this error?
- Does removal of the sweat affect skin temperature? How great is this effect?
- Is it necessary to wait for a minimum time after removing the sweat?

Different skin thermoregulation issues could be considered if the study is performed during exposure to a cold environment or to a warm environment.

Two main thermoregulatory effects resulted from cold exposure. First, the vasoconstriction of skin blood flow with the aim of reducing heat loss [89]. Second, if there is a severe cold exposure, thermogenesis via shivering, consisting of involuntary contractions of the muscles, is produced, with the aim of increasing heat production [89]. Research into the effect of cold exposure on skin temperature in sport science is especially focused on the human response to high altitude (during exercise or during specific acclimatization/training exercise protocols) (Fig. 16.9) [90–92], and in response to sport medicine treatments such as cryotherapy [93, 94]. Furthermore, in medicine, more of the studies are focused on analyzing the reasons for cold injuries [91, 95] or to assess the circulatory response in vascular diseases after the application of a cold stress [96, 97]. These tests commonly consisted of the application of cold local stress, such as the immersion of the extremities in cold water [90, 91, 95]. In these kinds of tests, the extremities are usually protected with plastic bags or gloves during the immersion [90, 91, 95–97], in order to prevent skin wetting and subsequent evaporation and cooling, as well as the effect of water on the skin on temperature estimation, as was discussed previously. However, during all the tests, environmental conditions were moderate (~20 °C) and there is nor much research simulating real cold environments. Furthermore, infrared thermography can be a useful tool to assess the thermoregulatory capacity of athletes before mountain or winter sports in cold environments, in order to improve performance and reduce the cold injury risk. Studies to assess its benefits in this kind of application are necessary.

On the other hand, in the exposure of the warm environment, the objective of human thermoregulation is heat dissipation, mainly by two mechanisms: cutaneous vasodilation and sweat evaporation [12, 89]. Research in hot environments in sport science has the main objective of simulating exercise in conditions where the

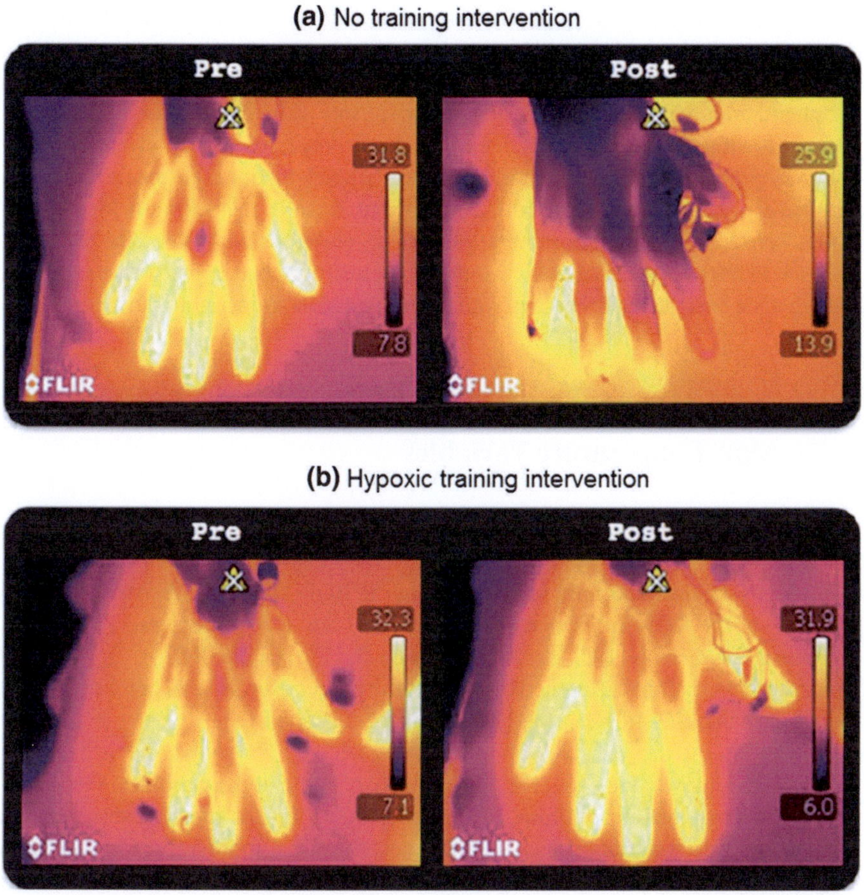

Fig. 16.9 Examples of thermographies before and after cold stress (immersion in water) in one participant without previous training (**left**) and another who had undergone a 10-day hypoxic training (right). (Figure modified from Keramidas et al. [90])

thermal stress is very high, and the risk of heat exhaustion and heat stroke is increased [19, 20, 98]. However, during room time adaptation in thermography studies, if sweat rate is very high, it could affect the thermal measurements, as was discussed previously.

In both cold and heat exposure research, the adaptations of the participants due to possible acclimatization to these environments may affect their thermoregulatory response [89, 99, 100]. In addition, if this response differs between participants, it will increase the thermal variability of the study. A methodological solution may be to make a prior period of acclimatization (between 6 and 10 days) at these temperatures for all participants. It is also important to consider that the room time adaptation before thermography measurements must be greater than that under a moderate environment [86, 101]. Moreover, it would be interesting for future studies to evaluate the ability of thermography to monitor such physiological adaptations during acclimatization periods.

Finally, it is known that in basal conditions, skin temperature depends largely on peripheral blood flow [102]. During exercise, this skin blood flow is also very important, although its effect is mixed with other processes such as sweat evaporation. Many of the studies carried out in sports science justify their results based on peripheral vasoconstriction/vasodilation without actually measuring it [6, 72, 83]. In this sense, we encourage more studies that combine skin blood flow measurements with infrared thermography would be interesting to better justify the responses and advance basic thermoregulatory knowledge of skin temperature during physical exercise.

16.7 Skin Temperature Variables

Different descriptive statistics variables can be extracted from the thermal analysis. These variables are used by different studies: the mean temperature of the ROI [3, 5, 51, 69, 103], the maximum temperature of the ROI [104–106], thermal asymmetry index of the ROI (difference between an ROI and its contralateral) [107–109], difference between one ROI that is a reference and the specific ROI [110, 111], and temperature variation (difference between during-after exercise and before) [64, 83, 84]. These variables are described in Chap. 4 of this book.

In each study, the variables of interest are selected for analysis. Furthermore, there is some knowledge about the applicability of some of the variables. For example, the thermal symmetry index is known to be useful for injury assessment [1, 108]. Temperature variation has been suggested to be a more valid measure than absolute temperatures to determine the effect of exercise [64, 83, 84]. Mean temperature was determined as the most robust and reliable metric [9, 38]. However, more studies are necessary in order to highlight which method is the most indicated in each situation. For example, the maximum temperature of the ROI is one variable used by different studies [104–106], and although it was suggested that it is more related with skin blood flow and inflammation [45], its advantages compared with the mean temperature are not clear. On the other hand, other alternative metrics were suggested that were interesting for sports science research [44, 57, 112], but again, more research is needed to investigate its application.

On the other hand, body fat tissue had an insulation capacity resulting in impairment in heat dissipation between the core and the skin [5, 113–115]. This effect of body fat tissue resulted in lower temperatures or lower variation in the skin temperature during exercise [5, 70, 114] (Fig. 16.10). For this reason, for some studies, it could be interesting to normalize the skin temperature by the fat tissue. However, more research is necessary to explore how to do this normalization. Nevertheless, the quantity of body fat tissue depends on the fitness level of the person [116] and, for this reason, it would not be recommended to normalize the skin temperature by

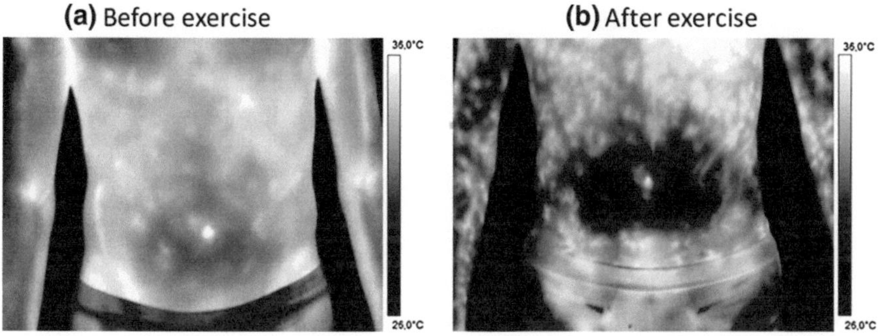

Fig. 16.10 Example of the lower skin temperature in the abdomen due to the effect of the fat tissue in this area, before (**a**) and after cycling for 45 min (**b**)

the fat tissue in studies in which the objective is to assess the effect of fitness on thermoregulation [70, 115].

16.8 Skin Temperature Data Reference Need

The idea that there is necessity for a database of thermal images or reference values of skin temperature normality has been suggested by different authors [36, 106, 117–119]. The establishment of a skin temperature reference may help to identify hyperthermic or hypothermic areas, and to easily detect some injuries or pathologies [106, 118]. With this clinical perspective, it is easy to understand why the idea to have a standard reference value is important in the medical area.

Different researchers have worked on the development of a basal thermography database. Fujimasa developed a database in 1997 with more than 20,000 binary thermal images [117]. Although the database developed was implemented in a website, this website currently seems to have ceased to be operational. Furthermore, some studies were oriented to establish normal thermal profiles for Brazilian adults [118], Polish physically active young people [120], obese women [114], and soccer players [106]. Thermal symmetry was also assessed to establish a reference value of 0.5 °C with a standard deviation of 0.3 °C [108]. There remains a necessity for reference values for medicine to be obtained [119]. Likewise, it would be interesting to establish a database for reference values that were more oriented to sports science; for example, from reference values of temperature variations produced by exercise. For an appropriate comparison with the reference values, it is important to have data for all the body areas and for each population group (data differentiated by age, sex, anthropometry, etc.) [36, 106, 118]. Furthermore, there are some environmental and methodological conditions that could alter the skin temperature that are important to take into account: room temperature, relativity humidity, body posture, time of day, etc. For medical purposes, it is also important to have data for the different pathologies. Reference values of excellent quality should be classified, taking into account all these parameters. For a sports science purpose, we need also to know reference temperature values after exercise, during exercise, and temperature variations. Reference values of excellent quality for sports science should be classified by physical fitness level, and should take into account additional information related to the exercise: type, duration, intensity, wind speed, etc. A large number of studies have been conducted in recent years in different sports and conditions, but in most of them, the samples are small and the conditions very specific.

After all this information, it is easy to understand that obtaining a skin temperature database is complicated for baseline values, and the degree of complexity increases when we think about obtaining reference values related to exercise. Instead of performing singular studies that attempt to create such databases, it may be a more accurate methodology to obtain these reference values through the use of big data analysis, by grouping all of the published studies.

16.9 Conclusions

In this chapter, we tried to present our desires in relation to future research and developments of infrared thermography in sports science. Some of these desires are possible and perhaps others are unlikely to develop. However, it is believed that these ideas could help to guide researchers and manufacturers. In order to summarize the chapter, below are the aspects that we believe should be addressed in the future R&D in sports science:

- Increase the accuracy of infrared cameras with an accessible price.
- To develop methods to increase and ensure the data quality of low-cost cameras during its application to sports science.
- Develop thermography software packages that are more orientated for solving the issues of sports science assessment.
- A higher application and investigation of 3D thermography or thermographic hyperspectral cameras in sports science.
- More research combining skin temperature and skin blood flow measurement.
- More research more the intra-day reproducibility of thermal data during and after exercise.
- Know how to deal with sweat in thermography exercise studies.
- Analyze the application of infrared thermography as a tool in the assessment of the thermoregulatory capacity of athletes before mountain or winter sports in cold environments.
- Research which skin temperature parameter is most indicated in each situation and about the application of alternative metrics that are appearing.
- To investigate more about the use of machine learning for thermal image analysis in sport applications.
- Develop a reference/database values of skin temperature for basal and exercise intervention.

References

1. Hildebrandt C, Raschner C, Ammer K (2010) An overview of recent application of medical infrared thermography in sports medicine in Austria. Sensors 10:4700–4715
2. Escamilla-Galindo V-L, Felipe JL, Alonso-Callejo A, Van-der-Horst R, de la Torre-Combarros A, Minafra P, Fernández-Muñoz D, Fernández-Cuevas I (2024) Return-to-play criteria based on infrared thermography during anterior cruciate ligament rehabilitation in football players. Biol Sport 42(2):161–167. https://doi.org/10.5114/biolsport.2025.144295
3. Fournet D, Ross L, Voelcker T et al (2013) Body mapping of thermoregulatory and perceptual responses of males and females running in the cold. J Therm Biol 38:339–344. https://doi.org/10.1016/j.jtherbio.2013.04.005
4. Raccuglia M, Heyde C, Lloyd A, Hodder S, Havenith G (2019) The use of infrared thermal imaging to measure spatial and temporal sweat retention in clothing. Int J Biometeorol 63(7):885–894. https://doi.org/10.1007/s00484-019-01701-5
5. Chudecka M, Lubkowska A (2010) Temperature changes of selected body's surfaces of handball players in the course of training estimated by thermovision, and the study of the impact of physiological and morphological factors on the skin temperature. J Therm Biol 35:379–385
6. Korman P, Kusy K, Straburzyńska-Lupa A, Kantanista A, Quintana MS, Zieliński J (2024a) Response of skin temperature, blood ammonia and lactate during incremental exercise until exhaustion in elite athletes. Sci Rep 14(1):2237. https://doi.org/10.1038/s41598-024-52374-z
7. Howell K, Dudek K, Soroko M (2020) Thermal camera performance and image analysis repeatability in equine thermography. Infrared Phys Technol 110:103447. https://doi.org/10.1016/j.infrared.2020.103447
8. Machado ÁS, Cañada-Soriano M, Jimenez-Perez I, Gil-Calvo M, Carpes FP, Perez-Soriano P, Priego-Quesada JI (2024) Distance and camera features measurements affect the detection of temperature asymmetries using infrared thermography. Quant InfraRed Thermogr J 21(2):69–81. https://doi.org/10.1080/17686733.2022.2143227
9. Machado ÁS, Priego-Quesada JI, Jimenez-Perez I, Gil-Calvo M, Carpes FP, Perez-Soriano P (2021) Influence of infrared camera model and evaluator reproducibility in the assessment of skin temperature responses to physical exercise. J Therm Biol 98:102913. https://doi.org/10.1016/j.jtherbio.2021.102913
10. Vardasca R (2019) Are the IR cameras FLIR ONE suitable for clinical applications? Thermol Int 23(3):95–102
11. Ring EFJ, Ammer K (2000) The technique of infrared imaging in medicine. Thermol Int 10:7–14
12. Vardasca R, Simoes R (2013) Current issues in medical thermography. In: Topics in medical image processing and computational vision. Springer, pp 223–237
13. Bach AJ, Stewart IB, Minett GM, Costello JT (2015) Does the technique employed for skin temperature assessment alter outcomes? A systematic review. Physiol Meas 36:R27
14. Howell KJ, Smith RE (2009) Guidelines for specifying and testing a thermal camera for medical applications. Thermol Int 19:3–14
15. ISO (2008) Particular requirements for the basic safety and essential performance of screening thermographs for human febrile temperature screening. TC121/SC3-IEC SC62D
16. Aggarwal N, Garg M, Dwarakanathan V, Gautam N, Kumar SS, Jadon RS, Gupta M, Ray A (2020) Diagnostic accuracy of non-contact infrared thermometers and thermal scanners: a systematic review and meta-analysis. J Travel Med 27(8):taaa193
17. Brazane S, Riou O, Delaleux F, Ibos L, Durastanti JF (2025) Management of thermal drift of bolometric infrared cameras: limits and recommendations. Quant InfraRed Thermogr J 22(1):54–69. https://doi.org/10.1080/17686733.2023.2290304
18. Ring EFJ, Jung A, Kalicki B et al (2015) New standards for fever screening with thermal imaging systems. Infrared Imaging

19. Sawka MN, Cheuvront SN, Kenefick RW (2012) High skin temperature and hypohydration impair aerobic performance. Exp Physiol 97:327–332. https://doi.org/10.1113/expphysiol.2011.061002
20. Cuddy JS, Hailes WS, Ruby BC (2014) A reduced core to skin temperature gradient, not a critical core temperature, affects aerobic capacity in the heat. J Therm Biol 43:7–12. https://doi.org/10.1016/j.jtherbio.2014.04.002
21. Racinais S, Havenith G, Aylwin P, Ihsan M, Taylor L, Adami PE, Adamuz M-C, Alhammoud M, Alonso JM, Bouscaren N, Buitrago S, Cardinale M, van Dyk N, Esh CJ, Gomez-Ezeiza J, Garrandes F, Holtzhausen L, Labidi M, Lange G et al (2022) Association between thermal responses, medical events, performance, heat acclimation and health status in male and female elite athletes during the 2019 Doha World Athletics Championships. Br J Sports Med 56(8):439–445. https://doi.org/10.1136/bjsports-2021-104569
22. Cheng T-Y, Deng D, Herman C (2012) Curvature effect quantification for in-vivo IR thermography. Int Mech Eng Congress Exposition 2:1–16. https://doi.org/10.1115/IMECE2012-88105
23. Grubišić I, Gjenero L, Lipić T et al (2011) Active 3D scanning based 3D thermography system and medical applications. In: 2011 proceedings of the 34th international convention MIPRO, pp 269–273
24. de Souza MA, Sanches IJ, Gamba HR, Nohama P (2013) Image fusion improvements applied at the generation of 3D thermal models. Conf Proc IEEE Eng Med Biol Soc 2013:3371–3374. https://doi.org/10.1109/EMBC.2013.6610264
25. Abreu de Souza M, Chagas Paz AA, Sanches IJ et al (2014) 3D thermal medical image visualization tool: integration between MRI and thermographic images. Conf Proc IEEE Eng Med Biol Soc 2014:5583–5586. https://doi.org/10.1109/EMBC.2014.6944892
26. Sutherland N, Marsh S, Priestnall G, Bryan P, Mills J (2023) InfraRed thermography and 3D-data fusion for architectural heritage: a scoping review. Remote Sens 15(9):Article 9. https://doi.org/10.3390/rs15092422
27. Vardasca R (2008) Template-based alignment and interpolation methods comparison of regions of interest in thermal images. In: Proceedings of the 3rd research student workshop. Faculty of Advanced Technology, University of Glamorgan, Glamorgan Business Center, UK, pp 21–24
28. Oniga M, Sultana A, Alexandrescu B, Orzan O (2025) Towards an integrated imaging for melanoma diagnosis: A review of multispectral, hyperspectral, and thermal technologies with preliminary system development. Comput Biol Med 185:109570. https://doi.org/10.1016/j.compbiomed.2024.109570
29. Di Gennaro SF, Toscano P, Gatti M, Poni S, Berton A, Matese A (2022) Spectral comparison of UAV-based hyper and multispectral cameras for precision viticulture. Remote Sens 14(3):Article 3. https://doi.org/10.3390/rs14030449
30. Ring EFJ, Ammer K, Wiecek B et al (2007) Quality assurance for thermal imaging systems in medicine. Thermol Int 17:103–106
31. Passos M d D, da Rocha AF (2022) Evaluation of infrared thermography with a portable camera as a diagnostic tool for peripheral arterial disease of the lower limbs compared with color Doppler ultrasonography. Arch Med Sci/Atherosclerotic Diseases 7:e66. https://doi.org/10.5114/amsad/150716
32. Qu Y, Meng Y, Fan H, Xu RX (2022) Low-cost thermal imaging with machine learning for non-invasive diagnosis and therapeutic monitoring of pneumonia. Infrared Phys Technol 123:104201. https://doi.org/10.1016/j.infrared.2022.104201
33. Villa E, Arteaga-Marrero N, Ruiz-Alzola J (2020) Performance assessment of low-cost thermal cameras for medical applications. Sensors 20(5). https://doi.org/10.3390/s20051321
34. Vardasca R, Gabriel J, Plassmann P, Ring EFJ (2014a) Towards a medical imaging standard capture and analysis software
35. Jones BF, Plassmann P (2002) Digital infrared thermal imaging of human skin. Eng Med Biol Mag IEEE 21:41–48

36. Lahiri BB, Bagavathiappan S, Jayakumar T, Philip J (2012) Medical applications of infrared thermography: a review. Infrared Phys Technol 55:221–235
37. Zaproudina N, Varmavuo V, Airaksinen O, Närhi M (2008) Reproducibility of infrared thermography measurements in healthy individuals. Physiol Meas 29:515. https://doi.org/10.1088/0967-3334/29/4/007
38. da Silva W, Machado ÁS, Kunzler MR, Jimenez-Perez I, Gil-Calvo M, Priego-Quesada JI, Carpes FP (2022) Reproducibility of skin temperature analyses by novice and experienced evaluators using infrared thermography. J Therm Biol 110:103345. https://doi.org/10.1016/j.jtherbio.2022.103345
39. Fernández-Cuevas I, Sillero-Quintana M, Garcia-Concepcion MA et al (2014) Monitoring skin thermal response to training with infrared thermography. New Stud Athl 29:57–71
40. Vardasca R, Gabriel J, Jones CD, Ring EFJ (2014b) A template based method for normalizing thermal images of the human body. QIRT2014
41. Barcelos EZ, Caminhas WM, Ribeiro E et al (2014) A combined method for segmentation and registration for an advanced and progressive evaluation of thermal images. Sensors 14:21950–21967. https://doi.org/10.3390/s141121950
42. Requena-Bueno L, Priego-Quesada JI, Jimenez-Perez I, Gil-Calvo M, Perez-Soriano P (2020) Validation of ThermoHuman automatic thermographic software for assessing foot temperature before and after running. J Therm Biol:102639. https://doi.org/10.1016/j.jtherbio.2020.102639
43. Hildebrandt C, Zeilberger K, Ring EFJ, Raschner C (2012) The application of medical infrared thermography in sports medicine. Ultrasound 10:2
44. Verderber L, da Silva W, Aparicio-Aparicio I, Germano AMC, Carpes FP, Priego-Quesada JI (2024) Assessment of alternative metrics in the application of infrared thermography to detect muscle damage in sports. Physiol Meas 45(9):095014. https://doi.org/10.1088/1361-6579/ad7ad3
45. Formenti D, Ludwig N, Rossi A, Trecroci A, Alberti G, Gargano M, Merla A, Ammer K, Caumo A (2018) Is the maximum value in the region of interest a reliable indicator of skin temperature? Infrared Phys Technol 94:299–304
46. Barnes RB (1963) Thermography of the human body. Science 140:870–877
47. Togawa T, Saito H (1994) Non-contact imaging of thermal properties of the skin. Physiol Meas 15:291. https://doi.org/10.1088/0967-3334/15/3/007
48. Fernández-Cuevas I, Bouzas Marins JC, Arnáiz Lastras J et al (2015) Classification of factors influencing the use of infrared thermography in humans: a review. Infrared Phys Technol 71:28–55. https://doi.org/10.1016/j.infrared.2015.02.007
49. Merla A, Mattei PA, Di Donato L, Romani GL (2010) Thermal imaging of cutaneous temperature modifications in runners during graded exercise. Ann Biomed Eng 38:158–163. https://doi.org/10.1007/s10439-009-9809-8
50. Formenti D, Ludwig N, Gargano M et al (2013) Thermal imaging of exercise-associated skin temperature changes in trained and untrained female subjects. Ann Biomed Eng 41:863–871. https://doi.org/10.1007/s10439-012-0718-x
51. Abate M, Di Carlo L, Di Donato L et al (2013) Comparison of cutaneous termic response to a standardised warm up in trained and untrained individuals. J Sports Med Phys Fitness 53:209–215
52. Ludwig N, Formenti D, Gargano M, Alberti G (2014) Skin temperature evaluation by infrared thermography: comparison of image analysis methods. Infrared Phys Technol 62:1–6
53. Ammer K (2011) Repeatability of identification of hot spots in thermal images is influenced by image processing. Thermol Int 21:40–46
54. Watmough DJ, Fowler PW, Oliver R (1970) The thermal scanning of a curved isothermal surface: implications for clinical thermography. Phys Med Biol 15:1. https://doi.org/10.1088/0031-9155/15/1/301

55. Swerdlow B, Dieter JNI (1992) An evaluation of the sensitivity and specificity of medical thermography for the documentation of myofascial trigger points. Pain 48:205–213. https://doi.org/10.1016/0304-3959(92)90060-O
56. Ohashi Y, Uchida I (2000) Applying dynamic thermography in the diagnosis of breast cancer. IEEE Eng Med Biol Mag 19:42–51. https://doi.org/10.1109/51.844379
57. Bogomilsky S, Hoffer O, Shalmon G, Scheinowitz M (2022) Preliminary study of thermal density distribution and entropy analysis during cycling exercise stress test using infrared thermography. Sci Rep 12(1):14018. https://doi.org/10.1038/s41598-022-18233-5
58. Khandakar A, Chowdhury MEH, Ibne Reaz MB, Md Ali SH, Hasan MA, Kiranyaz S, Rahman T, Alfkey R, Bakar AAA, Malik RA (2021) A machine learning model for early detection of diabetic foot using thermogram images. Comput Biol Med 137:104838. https://doi.org/10.1016/j.compbiomed.2021.104838
59. Morales-Ivorra I, Narváez J, Gómez-Vaquero C, Moragues C, Nolla JM, Narváez JA, Marín-López MA (2022) Assessment of inflammation in patients with rheumatoid arthritis using thermography and machine learning: A fast and automated technique. RMD Open 8(2):e002458. https://doi.org/10.1136/rmdopen-2022-002458
60. Torres-Galván JC, Guevara E, Kolosovas-Machuca ES, Oceguera-Villanueva A, Flores JL, González FJ (2022) Deep convolutional neural networks for classifying breast cancer using infrared thermography. Quant InfraRed Thermogr J 19(4):283–294. https://doi.org/10.1080/17686733.2021.1918514
61. Kovács P, Lehner B, Thummerer G, Mayr G, Burgholzer P, Huemer M (2020) Deep learning approaches for thermographic imaging. J Appl Phys 128(15) https://pubs.aip.org/aip/jap/article/128/15/155103/1077744
62. Wilson AN, Gupta KA, Koduru BH, Kumar A, Jha A, Cenkeramaddi LR (2023) Recent advances in thermal imaging and its applications using machine learning: a review. IEEE Sens J 23(4):3395–3407. https://doi.org/10.1109/JSEN.2023.3234335
63. McCoy M, Campbell I, Stone P et al (2011) Intra-examiner and inter-examiner reproducibility of paraspinal thermography. PLoS One 6:e16535. https://doi.org/10.1371/journal.pone.0016535
64. Priego Quesada JI, Carpes FP, Salvador Palmer R et al (2016a) Effect of saddle height on skin temperature measured in different days of cycling. Springerplus 5:205–214. https://doi.org/10.1186/s40064-016-1843-z
65. Calvo-Lobo C, San-Antolín M, García-García D, Becerro-de-Bengoa-Vallejo R, Losa-Iglesias ME, Cosín-Matamoros J, Casado-Hernández I, Martínez-Jiménez EM, Mazoteras-Pardo V, Rodríguez-Sanz D (2023) Intra- and inter-session reliability and repeatability of an infrared thermography device designed for materials to measure skin temperature of the triceps surae muscle tissue of athletes. PeerJ 11:e15011. https://doi.org/10.7717/peerj.15011
66. Priego-Quesada JI, Gandia-Soriano A, Pellicer-Chenoll MT, Catalá-Vilaplana I, Bermejo-Ruiz JL, Encarnación-Martínez A, Salvador-Palmer R, Cibrián Ortiz de Anda R (2021) Reproducibility of skin temperature response after cold stress test using the game ready system: preliminary study. Int J Environ Res Public Health 18(16):8295. https://doi.org/10.3390/ijerph18168295
67. Houdas Y, Ring EFJ (2013) Human body temperature: its measurement and regulation. Springer Science & Business Media
68. Tortora G, Grabowski (2003) Principles of anatomy and physiology, 10th edn. Wiley, New York
69. Ferreira JJA, Mendonça LCS, Nunes LAO et al (2008) Exercise-associated thermographic changes in young and elderly subjects. Ann Biomed Eng 36:1420–1427. https://doi.org/10.1007/s10439-008-9512-1
70. Priego Quesada JI, Carpes FP, Bini RR et al (2015a) Relationship between skin temperature and muscle activation during incremental cycle exercise. J Therm Biol 48:28–35. https://doi.org/10.1016/j.jtherbio.2014.12.005

71. Priego Quesada JI, Martínez N, Salvador Palmer R et al (2016b) Effects of the cycling workload on core and local skin temperatures. Exp Thermal Fluid Sci 77:91–99. https://doi.org/10.1016/j.expthermflusci.2016.04.008
72. Hillen B, Andrés López D, Marzano-Felisatti JM, Sanchez-Jimenez JL, Ortiz C, de Anda RM, Nägele M, Salvador-Palmer MR, Pérez-Soriano P, Schömer E, Simon P, Priego-Quesada JI (2023) Acute physiological responses to a pyramidal exercise protocol and the associations with skin temperature variation in different body areas. J Therm Biol 115:103605. https://doi.org/10.1016/j.jtherbio.2023.103605
73. Krustrup P, Ferguson RA, Kjær M, Bangsbo J (2003) ATP and heat production in human skeletal muscle during dynamic exercise: higher efficiency of anaerobic than aerobic ATP resynthesis. J Physiol 549:255–269. https://doi.org/10.1113/jphysiol.2002.035089
74. González-Alonso J (2012) Human thermoregulation and the cardiovascular system. Exp Physiol 97:340–346. https://doi.org/10.1113/expphysiol.2011.058701
75. Fujii N, Honda Y, Komura K et al (2014) Effect of voluntary hypocapnic hyperventilation on the relationship between core temperature and heat loss responses in exercising humans. J Appl Physiol 117:1317–1324. https://doi.org/10.1152/japplphysiol.00334.2014
76. Havenith G, Fogarty A, Bartlett R et al (2008) Male and female upper body sweat distribution during running measured with technical absorbents. Eur J Appl Physiol 104:245–255
77. Smith CJ, Havenith G (2011) Body mapping of sweating patterns in male athletes in mild exercise-induced hyperthermia. Eur J Appl Physiol 111:1391–1404. https://doi.org/10.1007/s00421-010-1744-8
78. Priego Quesada JI, Martínez Guillamón N, Cibrián O, de Anda RM et al (2015b) Effect of perspiration on skin temperature measurements by infrared thermography and contact thermometry during aerobic cycling. Infrared Phys Technol 72:68–76. https://doi.org/10.1016/j.infrared.2015.07.008
79. Priego-Quesada JI, Machado AS, Gil-Calvo M, Jimenez-Perez I, Ortiz C, de Anda RM, Salvador Palmer R, Perez-Soriano P (2020) A methodology to assess the effect of sweat on infrared thermography data after running: preliminary study. Infrared Phys Technol 109:103382. https://doi.org/10.1016/j.infrared.2020.103382
80. Ammer K (2009) Does neuromuscular thermography record nothing else but an infrared sympathetic skin response? Thermol Int 19:107–108
81. Zaidi H, Fohanno S, Polidori G, Taiar R (2007) The influence of swimming type on the skin-temperature maps of a competitive swimmer from infrared thermography. Acta Bioeng Biomech 9:47
82. Novotny J, Rybarova S, Zacha D et al (2015) The influence of breaststroke swimming on the muscle activity of young men in thermographic imaging. Acta Bioeng Biomech 17:121
83. Priego Quesada JI, Lucas-Cuevas AG, Gil-Calvo M et al (2015) Effects of graduated compression stockings on skin temperature after running. J Therm Biol 52:130–136. https://doi.org/10.1016/j.jtherbio.2015.06.005
84. Priego Quesada JI, Lucas-Cuevas AG, Salvador Palmer R et al (2016c) Definition of the thermographic regions of interest in cycling by using a factor analysis. Infrared Phys Technol 75:180–186. https://doi.org/10.1016/j.infrared.2016.01.014
85. de Andrade Fernandes A, dos Santos Amorim PR, Brito CJ et al (2014) Measuring skin temperature before, during and after exercise: a comparison of thermocouples and infrared thermography. Physiol Meas 35:189
86. James CA, Richardson AJ, Watt PW, Maxwell NS (2014) Reliability and validity of skin temperature measurement by telemetry thermistors and a thermal camera during exercise in the heat. J Therm Biol 45:141–149. https://doi.org/10.1016/j.jtherbio.2014.08.010
87. Buono MJ, Ulrich RL (1998) Comparison of mean skin temperature using "covered" versus "uncovered" contact thermistors. Physiol Meas 19:297–300
88. Psikuta A, Niedermann R, Rossi RM (2013) Effect of ambient temperature and attachment method on surface temperature measurements. Int J Biometeorol:1–9

89. Charkoudian N (2016) Human thermoregulation from the autonomic perspective. Auton Neurosci Basic Clin 196:1–2. https://doi.org/10.1016/j.autneu.2016.02.007
90. Keramidas ME, Kölegård R, Mekjavic IB, Eiken O (2015b) Hand temperature responses to local cooling after a 10-day confinement to normobaric hypoxia with and without exercise. Scand J Med Sci Sports 25:650–660. https://doi.org/10.1111/sms.12291
91. Eglin CM, Golden FS, Tipton MJ (2013) Cold sensitivity test for individuals with non-freezing cold injury: the effect of prior exercise. Extreme Physiol Med 2:16. https://doi.org/10.1186/2046-7648-2-16
92. Keramidas ME, Kounalakis SN, Eiken O, Mekjavic IB (2015a) Effects of two short-term, intermittent hypoxic training protocols on the finger temperature response to local cold stress. High Alt Med Biol 16:251–260. https://doi.org/10.1089/ham.2015.0013
93. Cholewka A, Stanek A, Sieroń A, Drzazga Z (2012) Thermography study of skin responsedue to whole-body cryotherapy. Skin Res Technol 18:180–187. https://doi.org/10.1111/j.1600-0846.2011.00550.x
94. Dębiec-Bąk A, Pawik Ł, Skrzek A (2016) Thermoregulation of football players after cryotherapy in thermography. J Therm Anal Calorim 123:1–12
95. Davey M, Eglin C, House J, Tipton M (2013) The contribution of blood flow to the skin temperature responses during a cold sensitivity test. Eur J Appl Physiol 113:2411–2417. https://doi.org/10.1007/s00421-013-2678-8
96. Coughlin PA, Chetter IC, Kent PJ, Kester RC (2001) The analysis of sensitivity, specificity, positive predictive value and negative predictive value of cold provocation thermography in the objective diagnosis of the hand–arm vibration syndrome. Occup Med 51:75–80. https://doi.org/10.1093/occmed/51.2.075
97. Antonio-Rubio I, Madrid-Navarro CJ, Salazar-López E et al (2015) Abnormal thermography in Parkinson's disease. Parkinsonism Relat Disord 21:852–857. https://doi.org/10.1016/j.parkreldis.2015.05.006
98. Nybo L (2010) Cycling in the heat: performance perspectives and cerebral challenges. Scand J Med Sci Sports 20(Suppl 3):71–79. https://doi.org/10.1111/j.1600-0838.2010.01211.x
99. Castellani JW, Young AJ (2016) Human physiological responses to cold exposure: acute responses and acclimatization to prolonged exposure. Auton Neurosci 196:63–74. https://doi.org/10.1016/j.autneu.2016.02.009
100. Périard JD, Travers GJS, Racinais S, Sawka MN (2016) Cardiovascular adaptations supporting human exercise-heat acclimation. Auton Neurosci 196:52–62. https://doi.org/10.1016/j.autneu.2016.02.002
101. Marins JCB, Moreira DG, Cano SP et al (2014a) Time required to stabilize thermographic images at rest. Infrared Phys Technol 65:30–35. https://doi.org/10.1016/j.infrared.2014.02.008
102. Johnson JM, Kellogg DL (2010) Local thermal control of the human cutaneous circulation. J Appl Physiol (1985) 109:1229–1238. https://doi.org/10.1152/japplphysiol.00407.2010
103. Akimov EB, Son'kin VD (2011) Skin temperature and lactate threshold during muscle work in athletes. Hum Physiol 37:621–628
104. Whay HR, Bell MJ, Main DCJ (2004) Validation of lame limb identification through thermal imaging. In: Proceedings of the 13th international symposium and 5th conference on lameness in ruminants, Maribor, Slovenia, pp 11–15
105. Stokes JE, Leach KA, Main DCJ, Whay HR (2012) An investigation into the use of infrared thermography (IRT) as a rapid diagnostic tool for foot lesions in dairy cattle. Vet J 193:674–678
106. Bouzas Marins JC, de Andrade FA, Gomes Moreira D et al (2014) Thermographic profile of soccer players' lower limbs. Rev Andal Med Deporte 7:1–6. https://doi.org/10.1016/S1888-7546(14)70053-X
107. Bertmaring I, Babski-Reeves K, Nussbaum MA (2008) Infrared imaging of the anterior deltoid during overhead static exertions. Ergonomics 51:1606–1619. https://doi.org/10.1080/00140130802216933

108. Vardasca R, Ring F, Plassmann P, Jones C (2012) Thermal symmetry of the upper and lower extremities in healthy subjects. Thermol Int 22:53–60
109. Sillero-Quintana M, Fernández-Jaén T, Fernández-Cuevas I et al (2015) Infrared thermography as a support tool for screening and early diagnosis in emergencies. J Med Imaging Health Inf 5:1223–1228
110. Sun P-C, Jao S-HE, Cheng C-K (2005) Assessing foot temperature using infrared thermography. Foot Ankle Int 26:847–853. https://doi.org/10.1177/107110070502601010
111. Haddad DS, Brioschi ML, Vardasca R et al (2014) Thermographic characterization of masticatory muscle regions in volunteers with and without myogenous temporomandibular disorder: preliminary results. Dentomaxillofac Radiol 43:20130440. https://doi.org/10.1259/dmfr.20130440
112. de Andrade Fernandes A, Marins JCB, de Andrade AGP, Albuquerque MR, Brito CJ, da Silva CD, do Valle MAAN, de Assis MG, Garcia ES, Pimenta EM (2024) Thermopixelgraphy: a new method for analyzing the skin temperature of the lower limbs through infrared thermography. J Therm Anal Calorim 149(22):12737–12745. https://doi.org/10.1007/s10973-024-13799-w
113. Savastano DM, Gorbach AM, Eden HS et al (2009) Adiposity and human regional body temperature. Am J Clin Nutr 90:1124–1131. https://doi.org/10.3945/ajcn.2009.27567
114. Chudecka M, Lubkowska A, Kempińska-Podhorodecka A (2014) Body surface temperature distribution in relation to body composition in obese women. J Therm Biol 43:1–6. https://doi.org/10.1016/j.jtherbio.2014.03.001
115. Korman P, Kusy K, Straburzyńska-Lupa A, Sillero-Quintana M, Zarębska E, Zieliński J (2024b) Exploring the correlation of skin temperature and body composition in athletes undergoing exhaustive physical exercise. J Therm Biol 123:103918
116. Johnson W, de Ruiter I, Kyvik KO et al (2014) Genetic and environmental transactions underlying the association between physical fitness/physical exercise and body composition. Behav Genet. https://doi.org/10.1007/s10519-014-9690-6
117. Fujimasa I, Saito I, Chinzei T (1997) Far infrared medical image database on World Wide Web. In: Proceedings of the 19th annual international conference IEEE on engineering in medicine biology society, vol 2, pp 652–653
118. Marins JCB, Fernandes AA, Cano SP et al (2014b) Thermal body patterns for healthy Brazilian adults (male and female). J Therm Biol 42:1–8
119. Ammer K (2015) Do we need reference data of local skin temperatures? Thermol Int 25:45–47
120. Chudecka M, Lubkowska A (2015) Thermal maps of young women and men. Infrared Phys Technol 69:81–87. https://doi.org/10.1016/j.infrared.2015.01.012

Editor Conclusion

Jose Ignacio Priego-Quesada

After reviewing all the chapters in this book and considering the experience gained in conducting research on infrared thermography in sports science, the editor would like to share a final conclusion regarding its content.

In general, there has been a growing interest among researchers in applying infrared thermography in sports science. This trend is well illustrated in Figure 1.12, which shows the number of publications on this topic from its inception in 1975 to 2025. The most extensively studied application has been the use of infrared thermography in assessing human thermoregulation during exercise and in sports medicine. However, while the number of studies is lower in other areas, such as clothing assessment or applications in animal sports (e.g., equestrian sports), these fields should not be underestimated, as infrared thermography has significant potential in these areas.

Although infrared thermography may seem easy to use, it is essential to have a fundamental understanding before employing it. Knowledge of the physical principles of infrared thermography, heat transfer, and thermoregulation is crucial for properly operating the infrared camera, identifying and addressing methodological issues, formulating logical hypotheses, and correctly interpreting thermal results. Furthermore, obtaining valid thermographic measurements requires following strict methodological protocols to ensure accurate data. In this regard, since the first edition of this book, the Thermographic Imaging in Sports and Exercise Medicine (TISEM) checklist, introduced in 2017, has become an important reference for verifying that key methodological aspects are properly considered.

However, further research is needed to determine optimal procedures for data acquisition and analysis in infrared thermography. Topics such as the use of low-cost cameras, alternative metrics for analysis, and the application of machine

J. I. Priego-Quesada
Department of Physical Education and Sports, University of Valencia, Valencia, Spain
e-mail: j.ignacio.priego@uv.es

J. I. Priego-Quesada (ed.), *Application of Infrared Thermography in Sports Science*, https://doi.org/10.1007/978-3-031-93311-0

learning techniques have been updated in this second edition, as they represent current challenges in the field. Although infrared thermography has been used in sports science for approximately 50 years, its widespread adoption in research laboratories has accelerated only in recent years, making it a relatively new technique compared to other instruments such as electromyography, photogrammetry, or indirect calorimetry. Therefore, continued research is necessary to improve all aspects of the research process, including methodology, image acquisition, data analysis, and result interpretation.

Another important aspect to consider is the significant variability in skin temperature dynamics across different studies. This is because skin temperature depends on multiple factors, including exercise characteristics, environmental conditions, thermoregulatory responses (such as core temperature, sweat rate, and blood flow), and individual attributes (such as sex, age, anthropometry, and body composition). However, this second edition introduces new findings that were previously unknown. For example, it has been observed that athletes in cyclic sports without underlying pathologies tend to exhibit thermal symmetry both before and after exercise, whereas athletes in acyclic sports and those with asymmetric techniques tend to have symmetrical temperature distributions before exercise but asymmetries afterward. Additionally, an increasing number of studies are monitoring skin temperature during exercise rather than just before and after, providing a more detailed understanding of thermoregulatory responses.

The chapters of this book provide extensive evidence of the applicability of infrared thermography across various fields in sports science. Infrared thermography has proven valuable for monitoring and preventing injuries in sports medicine. It also plays a crucial role in designing and assessing clothing and sports equipment. Additionally, it has a broad range of applications in equestrian sports, serving as a complementary diagnostic tool in veterinary practice and performance assessment. However, in other areas, such as psychophysiology or methodologies like dynamic thermography, research is still scarce, and their application in sports science remains minimal. Nevertheless, the chapters in this book highlight the great potential of infrared thermography in these emerging fields. As the application of infrared thermography in sports science continues to evolve, ongoing research is expected to expand its applications, making it necessary to update these chapters in future editions with new findings and advancements.

Finally, there is a pressing need for a stronger connection between laboratory research and real-world application in the field. Infrared thermography is beginning to be implemented in high-level sports clubs for daily athlete monitoring to reduce injury risk. However, despite its potential benefits, there is limited published data on this subject, making it difficult to establish solid scientific evidence regarding its effectiveness in injury prevention. Moreover, while some recent studies have successfully used infrared thermography to monitor athletes during competition, providing valuable insights, these studies remain relatively few. Additionally, as mentioned earlier, certain applications (e.g., psychophysiology) are still underutilized in sports settings. Therefore, it is crucial to establish an effective bridge between research and sports institutions to ensure that the advancements made in research laboratories are effectively translated into practical applications. This will ultimately help enhance athletic performance and reduce the risk of injury.

March, 2025.

Index

J. I. Priego-Quesada (ed.), *Application of Infrared Thermography in Sports Science*, https://doi.org/10.1007/978-3-031-93311-0

P

Q

R

S

MIX
Papier aus verantwortungsvollen Quellen
Paper from responsible sources
FSC® C105338

If you have any concerns about our products,
you can contact us on
ProductSafety@springernature.com

In case Publisher is established outside the EU,
the EU authorized representative is:
Springer Nature Customer Service Center GmbH
Europaplatz 3, 69115 Heidelberg, Germany

Printed by Libri Plureos GmbH
in Hamburg, Germany